Lecture Notes in Computer Science 5839

Commenced Publication in 1973
Founding and Former Series Editors:
Gerhard Goos, Juris Hartmanis, and Jan van Leeuwen

T0223380

Gary Geunbae Lee Dawei Song
Chin-Yew Lin Akiko Aizawa
Kazuko Kuriyama Masaharu Yoshioka
Tetsuya Sakai (Eds.)

Information Retrieval Technology

5th Asia Information Retrieval Symposium, AIRS 2009
Sapporo, Japan, October 21-23, 2009
Proceedings

 Springer

Volume Editors

Gary Geunbae Lee
Pohang University of Science and Technology, E-mail: gblee@postech.ac.kr

Dawei Song
The Robert Gordon University, E-mail: d.song@rgu.ac.uk

Chin-Yew Lin
Microsoft Research Asia, E-mail: cyl@microsoft.com

Akiko Aizawa
National Institute of Informatics, E-mail: aizawa@nii.ac.jp

Kazuko Kuriyama
Shirayuri College, E-mail: kuriyama@shirayuri.ac.jp

Masaharu Yoshioka
Hokkaido University, E-mail: yoshioka@ist.hokudai.ac.jp

Tetsuya Sakai
Microsoft Research Asia, E-mail: tetsuyasakai@acm.org

Library of Congress Control Number: 2009935701

CR Subject Classification (1998): H.3, H.4, F.2.2, E.1, E.2

LNCS Sublibrary: SL 3 – Information Systems and Application, incl. Internet/Web
and HCI

ISSN   0302-9743
ISBN-10   3-642-04768-8 Springer Berlin Heidelberg New York
ISBN-13   978-3-642-04768-8 Springer Berlin Heidelberg New York

Typesetting: Camera-ready by author, data conversion by Scientific Publishing Services, Chennai, India
Printed on acid-free paper   SPIN: 12769179   06/3180   5 4 3 2 1 0

Preface

Asia Information Retrieval Symposium (AIRS) 2009 was the fifth AIRS conference in the series established in 2004. The first AIRS was held in Beijing, China, the second in Jeju, Korea, the third in Singapore and the fourth in Harbin, China. The AIRS conferences trace their roots to the successful Information Retrieval with Asian Languages (IRAL) workshops, which started in 1996.

The AIRS series aims to bring together international researchers and developers to exchange new ideas and the latest results in information retrieval. The scope of the conference encompassed the theory and practice of all aspects of information retrieval in text, audio, image, video, and multimedia data.

AIRS 2009 received 82 submissions, from which we carefully selected 18 regular papers (22%) and 20 (24%) posters through a double-blind reviewing process. We are pleased to report that the conference proceedings include contributions from not only Asian countries, but also from Finland, Italy, Australia, UK and USA.

We are grateful to Elizabeth Liddy, chair of ACM SIGIR, for accepting to be the honorary conference chair, and to Hokkaido University for hosting the conference. We thank the Information Retrieval Facility, Microsoft Research Asia, Ricoh, Ltd., Global COE Program "Center for Next-Generation Information Technology Based on Knowledge Discovery and Knowledge Federation" and Sapporo International Communication Plaza Foundation for sponsoring the conference, and Springer for publishing the conference proceedings as part of their Lecture Notes in Computer Science (LNCS) series. We also thank ACM SIGIR and IPSJ SIGFI for giving the conference an "in cooperation with" status.

We thank the publication co-chairs, Akiko Aizawa and Kazuko Kuriyama, for compiling the camera-ready papers and liasing with Springer, and the finance chair, Takuya Kida, for successfully managing all money matters. We also thank the publicity co-chairs, Atsushi Fujii, Ian Soboroff, Dawei Song and William Webber, for advertising AIRS, and the area chairs and other program committee members for their high-quality and punctual reviews.

Finally, we acknowledge, and are inspired by, the many authors who submitted papers and who continue to contribute to this Asian community of IR research and development.

August 2009

Gary Geunbae Lee
Dawei Song
Chin-Yew Lin
Masaharu Yoshioka
Tetsuya Sakai

Organization

Steering Committee

Hsin-Hsi Chen	National Taiwan University, Taiwan
Wai Lam	The Chinese University of Hong Kong, China
Gary Geunbae Lee	Pohang University of Science and Technology, Korea
Alistair Moffat	University of Melbourne, Australia
Hwee Tou Ng	National University of Singapore, Singapore
Tetsuya Sakai	Microsoft Research Asia, China
Dawei Song	The Robert Gordon University, UK
Masaharu Yoshioka	Hokkaido University, Japan

Advisory Board

Mun Kew Leong	National Library Board, Singapore
Sung Hyon Myaeng	Information and Communication University, Korea
Kam-Fai Wong	The Chinese University of Hong Kong, China

Honorary Conference Chair

Elizabeth Liddy	Syracuse University, USA

General Co-chairs

Tetsuya Sakai	Microsoft Research Asia, China
Masaharu Yoshioka	Hokkaido University, Japan

Program Committee Co-chairs

Gary Geunbae Lee	Pohang University of Science and Technology, Korea
Chin-Yew Lin	Microsoft Research Asia, China
Dawei Song	The Robert Gordon University, UK

Organization Committee

Finance Chair   Takuya Kida, Hokkaido University, Japan

Publication Co-chair   Akiko Aizawa, National Institute of
                              Informatics, Japan
                            Kazuko Kuriyama, Shirayuri College, Japan

Publicity Co-chair   Atsushi Fujii, University of Tsukuba, Japan
                            Ian Soboroff, National Institute of Standards
                              and Technology, USA
                            Dawei Song, The Robert Gordon University,
                              UK
                            William Webber, University of Melbourne,
                              Australia

Area Chairs

Gareth Jones   Dublin City University, Ireland
Tie-Yan Liu   Microsoft Research Asia, China
David Losada   Universidad de Santiago de Compostela,
                              Spain
Teruko Mitamura   Carnegie Mellon University, USA
Ian Ruthven   University of Strathclyde, UK
Andrew Turpin   RMIT, Australia
Min Zhang   Tsinghua University, China

Hosted by

Hokkaido University, Japan

Program Committee

Gianni Amati   Fondazione Ugo Bordoni, Italy
Leif Azzopardi   University of Glasgow, UK
Peter Bailey   Microsoft Research, USA
Mark Baillie   University of Strathclyde, UK
Alvaro Barreiro   University of A Coruña, Spain
Bodo Billerbeck   RMIT University, Australia
Roi Blanco   University of A Coruña, Spain
Rui Cai   Microsoft Research Asia, China
Enhong Chen   University of Science and Technology, China
Hsin-Hsi Chen   National Taiwan University, Taiwan
Xuqi Cheng   Chinese Academy of Sciences, China
Seungjin Choi   Pohang University of Science and Technology,
                              Korea

Nick Craswell   Microsoft Research Cambridge, UK
Shane Culpepper   RMIT University, Australia
Efthimis N. Efthimiadis University of Washington, USA
David Elsweiler   Friedrich-Alexander-University of
                             Erlangen-Nürnberg, Germany
Juan M. Fernández Luna University of Granada, Spain
Colum Foley   Dublin City University, Ireland
Bin Gao   Microsoft Research Asia, China
Fred Gey   University of California, Berkeley, USA
Daqing He   University of Pittsburgh, USA
Ichiro Ide   Nagoya University, Japan
Hideo Joho   University of Glasgow, UK
Hiroshi Kanayama   Tokyo Research Laboratory, IBM, Japan
Jaana Kekäläinen   University of Tampere, Finland
Jeongwoo Ko   Google, USA
Wei Lai   Microsoft Research Asia, China
Mounia Lalmas   University of Glasgow, UK
Wai Lam   Chinese University of Hong Kong, Hong Kong
Ni Lao   Carnegie Mellon University, USA
Martha Larson   Technical University of Delft,
                             The Netherlands
Ting Liu   Harbin Institute of Technology, China
Yiqun Liu   Tsinghua University, China
Andrew MacFarlane   City University of London, UK
Stephane
   Marchand-Maillet   University of Geneva, Switzerland
Massimo Melucci   University of Padova, Italy
Helen Meng   The Chinese University of Hong Kong,
                             Hong Kong
Tatsunori Mori   Yokohama National University, Japan
Alessandro Moschitti   University of Trento, Italy
Jian-Yun Nie   University of Montreal, Canada
Eric Nyberg   Carnegie Mellon University, USA
Neil O'Hare   Dublin City University, Ireland
Nils Pharo   Oslo University College, Norway
Benjamin Piwowarski   University of Glasgow, UK
Dulce Ponceleón   IBM Almaden Research, USA
Thomas Roelleke   Queen Mary University of London, UK
Shinichi Satoh   National Institute of Informatics, Japan
Falk Scholer   RMIT University, Australia
Milad Shokouhi   Microsoft Research Cambridge, UK
Hideki Shima   Carnegie Mellon University, USA
Luo Si   Purdue University, USA
Mark Smucker   University of Waterloo, Canada

Table of Contents

Posters

Fully Automatic Text Categorization by Exploiting WordNet

Jianqiang Li, Yu Zhao, and Bo Liu

NEC Laboratories China
11F, Bldg.A, Innovation Plaza, Tsinghua Science Park
Haidian District, Beijing 100084, China
{lijianqiang,zhaoyu,liubo}@research.nec.com.cn

Abstract. This paper proposes a Fully Automatic Categorization approach for Text (FACT) by exploiting the semantic features from WordNet and document clustering. In FACT, the training data is constructed automatically by using the knowledge of the category name. With the support of WordNet, it first uses the category name to generate a set of features for the corresponding category. Then, a set of documents is labeled according to such features. To reduce the possible bias originating from the category name and generated features, document clustering is used to refine the quality of initial labeling. The training data are subsequently constructed to train the discriminative classifier. The empirical experiments show that the best performance of FACT can achieve more than 90% of the baseline SVM classifiers in F1 measure, which demonstrates the effectiveness of the proposed approach.

Keywords: WordNet, Text Categorization, Semantics.

1 Introduction

Supervised learning is the dominant approach for Text Categorization (TC) [15] [21]. Its performance depends heavily on the quantity and quality of hand-labeled documents. To reduce the burden of manual labeling, semi-supervised approaches [19] use the knowledge from both labeled and unlabeled data for classifier training.

This paper proposes FACT approach. Its underlying assumption is that the category name is specified in human-understandable words, which is generally true for many real applications with good human-computer interfaces. FACT employs the semantic of the category name and the hidden knowledge of the document set for automatic training data construction. First, the category name is extended as a set of representative keywords for each category, which serves as the Representative Profile (RP) of the category to initially label a set of documents. Second, the document clustering is used to refine the initial document labeling, which acts as a regulator to reduce the possible bias derived from the category name. Finally, the training data is constructed from the labeled documents. They are used to supervise the classifier learning.

The key of FACT approach is automatic document labeling. Its basic idea derived from following observation: Given the category name, the prerequisite for the human experts to manually label the documents is that they know the meanings of the

G.G. Lee et al. (Eds.): AIRS 2009, LNCS 5839, pp. 1–12, 2009.
© Springer-Verlag Berlin Heidelberg 2009

concepts implied by the category name. With the knowledge in mind to read the documents, the expert can assign the category label on them. The knowledge stored in their memories serves as the intermediate to link the categories and documents together. Similarly, with the availability of lexical databases such as WordNet, EuroWordNet, CoreNet, and HowNet, an intuitive way to simulate human's document labeling is to use these lexical resources to provide the same functionality as that of human's knowledge in their brains. In FACT, the description on the concepts implied in the category name and their definitions in WordNet are used as a bridge to provide the linkage between the category and the unlabeled documents.

2   Related Work

Both supervised [15] and semi-supervised [3] [8] [11] TC methods treat category name only as symbolic labels that assume no additional knowledge about them available to help building the classifier. So more or less, certain amount of manual data labeling is required. Our FACT is different. The semantics of the words appeared in the category name is used to supervise the classifier learning. Then, no manual labeling efforts is required in FACT.

FACT is closely related to unsupervised TC approaches. Without labeled documents, they utilize category names [1] or user-specified keywords [20] as the RP for training data building. Since it was hard for users to provide such keywords, [2] uses document clustering together with feature selection to find a set of important words to assist users for keyword selection. FACT is different since it utilizes WordNet to automatically generate a set of representative words as the extended features of the category.

Our work also relates to applying WordNet for automatic TC. [4] proposed to utilize the synonyms in WordNet to improve TC. Several similar techniques are reported to incorporate the synonyms [9], hypernyms [12], hyponyms [13], meronyms and holonyms [16] of words found in the training documents for classifier training. These researches mainly focus on incorporating WordNet to improve the TC model, where the labeled documents are still used. They are different from FACT since we need no labeled data.

3   Our FACT Approach

Given a set of categories C and a set of unlabeled documents D, our FACT consists of four steps: (1) Initial document labeling; (2) Refinement of the initial document labeling; (3) Training data construction; (4) Classifier building. Figure 1 is the flow chart of FACT.

3.1   Initial Document Labeling

Category names are used to initially label some documents in D. It includes three sub-steps: 1) Category name understanding; 2) RP generation; 3) Initial document labeling.

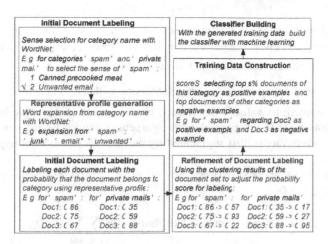

Fig. 1. The flow chart of FACT

1) Category Name Understanding

The goal of this sub-step is to find the relevant word senses of the word/phrase appeared in the category name. For each category name, a preprocessing is conducted. The words inside are tagged for their POS. Here, the word also refers to the simple phrase that can be found in WordNet. After the stop-words (conjunctions, prepositions, and pronouns) are eliminated, the remaining words are used to represent the concepts of the category. They will serve as the seed features to be extended as the RP of corresponding category.

WordNet organizes English nouns, verbs, adjectives, and adverbs into synonym sets, called synsets. Different relationships are defined to link these synsets. One simple method to extend the seed features is just to find all the synsets containing the seed features, then utilize the defined semantic relations to collect the semantically relevant words as the RP of the category. However, there are generally multiple senses (synsets) for each seed word. The homonyms or polysemes could introduce potential noise. So a sense ranking algorithm is given to determine which senses should be considered.

Similar to existing WSD methods [7] using the surrounding words in the sentence to select the correct sense, we propose a sense ranking approach for a word w in the name of category $c_j \in C$ by using the contextual words appeared in the names of category c_j (if the category name contains multiple words) and c_j's adjacent categories (i.e., its sibling, children, and parent categories defined in C, if there is a hierarchical structure inside). For the example given in Figure 1, *spam* and *private emails* are two categories. The words "private" and "email" is the contextual words of "spam".

Each word sense in WordNet is represented as a synset. It has a gloss that provides a linguistic micro-context for that sense. The senses of word w are ranked according to the relatedness of their linguistic micro-contexts with the contextual words of w found in the category names in C.

For a word w, assuming it has m senses $\{w^1, w^2, \ldots, w^m\}$, there are n contextual words $\{cw_1, cw_2, \ldots, cw_n\}$. Each word cw_i has m_i senses $\{cw_i^1, cw_i^2, \ldots, cw_i^{m_i}\}$, $1 \leq i \leq n$, the word sense ranks are calculated as follows:

1. For each word sense w^r, $1 \leq r \leq m$, its relatedness with cw_i^k, i.e., $R(w^r, cw_i^k)$, $1 \leq k \leq m_i$, is computed. Its value is determined by the cosine of the angle between their gloss vectors. Due to the possible data sparseness caused by the extremely short glosses, this measure is also augmented by the glosses of adjacent hypernyms and hyponyms.

2. Determine the relatedness between a word sense w^r and word cw_i by the sum of $R(w^r, cw_i^k)$, $1 \leq k \leq m_i$.

3. For each sense w^r, its rank is calculated: $Rank(w^r) = \sum_{1 \leq i \leq n} \sum_{k=1}^{m_j} f_r \times R(w^r, cw_i^k)$

where each sense of the contextual word is weighted by its frequency count f_r to indicate how important the word sense is.

The rank value reflects the extent to which this word sense shares information in common with its contextual words. The real meaning of the words in a category name might cover several senses, i.e., several senses defined in WordNet might be relevant to the concepts implied by the category name. So, different from traditional WSD that selects only one correct sense, we here need to choose several of them. Intuitively, we can select the top $t\%$ senses as the target senses for feature extension. However, the threshold value is difficult to tune. To handle such a problem, we use a compromised policy, i.e., the higher value of m, the lower value of t is set.

2) RP Generation

Based on the selected word senses, WordNet is used to construct a set of extended keywords for each category, which will serve as the RP rp_j of corresponding category $c_j \in C$. The basic idea is to use the multiple relations defined in WordNet to extract the semantically related words as extended features for each category.

For the category name with only one word: The rp_j is constituted by all the synonyms, hypernyms, hyponyms, meronyms, and holonyms that are identified using the selected senses and the semantic relations defined in WordNet. Also, the words from the derivationally related forms defined in WordNet for corresponding synsets are also used as the representative keywords.

For the category name with multiple words: Besides the keywords found by the multiple semantic relations in WordNet, new phrases are also constructed as a part of the rp_j by automatic synonym substitution. It means that the synonyms are utilized to substitute corresponding word to construct a new phrase. For example, a category name is "spam detection", the synonym of "spam" is "junk e-mail", then "junk e-mail detection" is also selected as the extended features of corresponding category.

Considering that the more relevant sense should contribute more to the TC model, we use the ranking value of each sense to assign the weight to the extended features from this sense, i.e., the word sense with higher rank value will play a more important role in identifying the relevant documents as training data. Intuitively, a keyword might be more discriminative for one class than the others. However, the experiments show that the TC performance of the final classifier is not sensitive to the weighting policies.

3) Initial Document Labeling

Once the rp for each category is constructed, we then apply it for probabilistic document labeling. The probabilistic labeling is based on the similarity of each document $d_i \in D$ with each rp_j. We use the popular cosine similarity metric to measure the similarity. It is based on the Vector Space Model (VSM) of the representations of the rp_j and d_i. $d_i \in D$ is compared with rp_j of category $c_j \in C$ using the cosine metric, and a score $s(d_i, c_j)$ is obtained to indicate similarity between d_i and rp_j.

Using the similarity scores between categories and documents, we can automatically generate a set of probabilistically labeled documents from D for corresponding category.

The initial probability value reflects to what extend a document belongs to the category. It is generated only based on the similarity sores between the document and the RP of corresponding category. There are many cases that one document has multiple probability values regarding to different categories. So, for each $c_j \in C$, assuming mx_j is the maximum $s(d_k, c_j)$, $p(c_j|s(d_i, c_j)) = s(d_i, c_j)/mx_j$, we use following formula to normalize the possibility value across multi-categories,

$$P(c_j \mid d_i) = \frac{P(c_j \mid s(d_i, c_j))}{\sum_{c \in C} P(c_j \mid s(d_i, c))}$$

3.2   Refinement of the Initial Document Labeling

Since the initial document labeling might be biased by the RPs, this step uses the document clustering to adjust the initial probability value of the label.

The adjustment is conducted by an alignment model based on the Bayesian inference. Assuming C' be the resultant cluster set and document d_i is clustered into cluster $c'_k \in C'$, based on the Bayesian probability, we have

$$P(c_j \mid d_i, c'_k) = \frac{P(c_j \mid d_i) P(c'_k \mid c_j)}{P(c'_k)},$$

where $P(c_j|d_i)$ can be obtained from the probabilistic document labeling, and the other two components could be obtained with following two equations:

$$P(c'_k \mid c_j) = \frac{\sum_{d \in c'_k} P(c_j \mid d)}{\sum_{d \in D} P(c_j \mid d)}, \quad P(c'_k) = \sum_{c \in C} P(c'_k \mid c),$$

3.3   Training Data Construction and Classifier Building

For each category, there is a list of documents with $P(c_j|d_i, c'_k) \geq 0$. A document could be labeled by multiple categories with certain probabilities. With these multi-labeled documents, the training data are generated automatically.

For each category, the policy for training data construction for category c_j can be described as: 1) The top $p^+\%$ documents of the list are selected as positive samples for c_j; 2) The documents at the top $p^-\%$ of the list from other categories are identified as reliable negative samples of c_j. Basically, higher $p^+(p^-)$ means more resultant training data but with lower quality; and lower $p^+(p^-)$ indicates less resultant training data but with higher quality. They need to be tuned carefully for the final classifier building.

There might be that multiple categories share the same documents as positive samples. To maximize the discriminative power of the training data, such shared documents only serve as the positive samples for the category with highest probability value.

With the constructed training data, classifiers are built by two discriminative methods, i.e., SVM [17] and TSVM [8].

4   Experiment and Evaluation

1)   Experiments Setup

Three English datasets, i.e, 20-NewsGroup (20NP), Reuters-21578, and WebKB, were used. WebKB data set consists of a collection of web pages gathered from university computer science departments, which are divided into seven categories. Our experiment uses four most populous categories: student, faculty, course and project—all together containing 4199 pages. Reuters-21578 is a collection of documents from the Reuters. As many researches [8][15] were conducted, only the most populous 10 classes from the whole dataset are used for our experiment, i.e., Earn, Acquisition, Grain, Crude, Trade, Interest, Ship, Wheat and Corn. There are totally 9296 documents covered by these classes. For each category, the ModApte split is adopted to obtain the training and test data. 20NP is a collection of approximately 20000 articles from 20 different UseNet discussion groups. There are 1000 news documents for each NP. For simplicity, we select two of the 4 main categories, i.e., *Science* (SCI), *Computing* (COMP), as the representation of this dataset. There are totally 9 subcategories, 4 in *Science* and 5 in *Computing*. Since we assume the category name should be specified as normal words in natural language, the abbreviation of the category name is transformed into its dictionary form, e.g., science for SCI, computing for COMP. For baseline approaches, the given training and test sets are used.

For each document, the title and body are extracted as a single feature vector. After the stop-words elimination and stemming, we select the traditional *tf-idf* term weighting scheme as our document representation model. In our experiment, the SVM-light package [8] is employed for the implementation of SVM and TSVM, where a linear kernel is used, and the weight C of the slack variables is set to default.

2)   Evaluation Criteria

We use the three standard criteria for binary TC classifier, precision, recall and F1 measure. For the multi-classification problem, we adopt both document-centric microaveraging F1 measure and category-centric macroaveraging F1. The accuracy, i.e., the proportion of documents with correct labels, is adopted to measure the quality of document labeling and the used training data.

3) Experiment Results

We conduct several experiments to tune the parameters of our algorithm.

Category name understanding: The compromised policy for selecting the top $t\%$ from the ranking result of m senses is given in Figure 2.

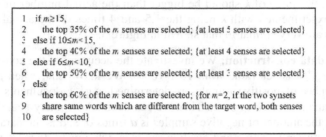

```
1   if m≥15,
2       the top 35% of the m senses are selected; {at least 5 senses are selected}
3   else if 10≤m<15,
4       the top 40% of the m senses are selected; {at least 4 senses are selected}
5   else if 6≤m<10,
6       the top 50% of the m senses are selected; {at least 3 senses are selected}
7   else
8       the top 60% of the m senses are selected; {for m=2, if the two synsets
9       share same words which are different from the target word, both senses
10      are selected}
```

Fig. 2. The policy for word sense selection

Actually, several experiments on word sense selection have been conducted to tune the settings of m and t. The results show that slight differences on the parameters defined in Fig. 2 have almost no effect on the final classification results.

Initial document labeling: We change the percentage $pc\%$ ranging from 10-100% to draw the curves of the quality of document labeling. The solid lines in Figure 3 are the result. In Figure 3, we find that the initial document labeling for the Reuters-21578 has the best quality, and WebKB is the poorest dataset. It indicates that, for Reuters-21578, its category names have the best quality to represent the content of its documents. However, the category names used in WebKB don't reflect appropriately its document contents. Also, we observe that, for Reuters-21578 and 20NP, the qualities of the document labeling decrease monotonously with pc. It is consistent with the initial assumption that smaller pc means more reliable labeled data. But for WebKB, after $pc>0.2$, the accuracy of the labeled document increases with pc. It is due to that there are biased keywords in the RPs.

Refinement of the probabilistic labeling: The k-means algorithm is adopted for document clustering. We use Weka [6] for its implementation. The double of the actual

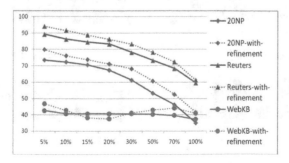

Fig. 3. Document labeling and refinement

number of classes of each dataset is used to set the value of k. The dotted lines in Figure 3 are the accuracies of the labeled documents after the refinement step. Averagely, there is 6-8% improvement. It shows the usefulness of our method for labeled data refinement.

The value setting of k does not matter much for FACT as long as it is not too small (generally, the value of k should be bigger than the actual number of the categories). We also experimented with k being the 1.5 and 4 times of the actual class numbers, the results are very similar to that shown in Figure 3.

Training data construction: We investigate the accuracy of the constructed training data when $p^+\%$ is set value from 10%-100%. We know that, when p^+ is given, the actual number of positive samples is determined, e.g., s documents are selected as positive samples for category c_j. Then we can decide how many negative samples are used, e.g., the amount of negative samples is a times of s. a is the parameter to reflect the ratio between negative and positive samples. Obviously, there is a bijective relation between a and p^-. For simplicity, in our implementation, we use a to represent the selected p^-.

Figure 4 shows the accuracies of the constructed training data based on different values of p^+ and a. *all* denotes all the positive samples of other categories are selected as negative samples of this category. We can observe that, with a fixed p^+(or a), the accuracies of the training data decrease with the increase of value a(or p^+). It means that, the more negative (or positive) samples is involved in the training data, the lower the quality of the training data is. Since the precision of negative samples is much higher than that of positive ones, the accuracy of training data is much higher than that of labeled data in Figure 3.

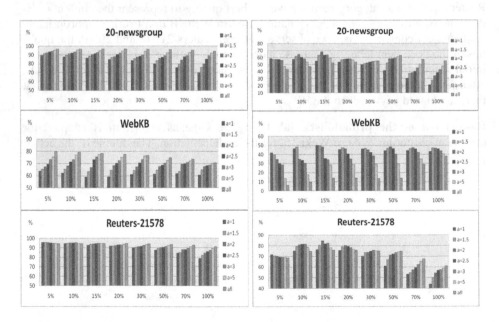

Fig. 4. Accuracy of the used training data   **Fig. 5.** Selection of parameters $p+$ and a

Classifier building: In our algorithm, p^+ is set to 15% and a is set to 1.5. The settings are based on empirical experiments which will be explained below.

We select **SVM** and **TSVM** as baseline approaches. For the baseline, the given training and test data from each of the three datasets are used for building and evaluating the classifiers. We implement our FACT approach into two versions, i.e., **SVM+WordNet** and **TSVM+WordNet**. SVM(TSVM)+WordNet means that all the documents in the training dataset are treated as unlabeled data to obtain the training data automatically; then SVM (TSVM) classifier is learned; and the test data are used for its evaluation.

Table 1. Comparison of FACT with baselines

Measure (%) Dataset and method		Macro-averaging			Micro-averaging		
		Prec	Rec	F1	Prec	Rec	F1
20NP	SVM(baseline)	91.1	65.5	76.3	91.2	63.8	75.1
	TSVM(baseline)	80.7	80.8	80.8	80.7	80.8	80.8
	SVM+WordNet	68.9	66.9	67.9	69.9	67.0	68.4
	TSVM+WordNet	14.6	95.4	25.4	14.6	85.8	25.0
	Without WordNet	26.6	81.0	40.1	26.6	48.6	34.4
Web KB	SVM (baseline)	92.3	74.5	82.5	92.3	72.4	81.2
	TSVM(baseline)	85.2	85.4	85.3	85.2	85.4	85.3
	SVM+WordNet	37.0	72.9	49.1	37.0	70.0	48.4
	TSVM+WordNet	36.8	87.4	51.8	36.8	85.8	51.5
	Without WordNet	27.3	97.9	42.7	27.3	54.5	36.3
Reuters-21778	SVM(baseline)	86.1	97.0	91.2	86.1	96.9	91.2
	TSVM(baseline)	92.7	92.5	92.6	92.7	91.7	92.2
	SVM+WordNet	85.7	84.6	85.1	85.7	83.4	84.5
	TSVM+WordNet	20.7	88.3	33.5	20.7	97.4	34.1
	Without WordNet	53.1	82.2	64.5	53.1	66.3	59.0

Table 1 illustrates the TC results. We observe that, when **SVM+WordNet** is used to categorize the Reuters-21578 and 20NP, FACT can achieve more than 90% of F1 performance of the baseline SVM methods. It proves the effectiveness of FACT.

The unsupervised method given in [1], which uses only the category name to bootstrap a TC classifier, achieves 65% F1 performance of the baseline SVM for 20NP (Although the document set between the two studies for 20NP were not totally the same, the overall ratios of training set to test set were almost exactly the same). The F1 measures are 0.74 for Reuters21578 and 0.65 for 20NP. Since we utilize the semantic knowledge of the category name defined in WordNet (as an external knowledge source) to bridge the categories and documents, as shown in Table 1, the **SVM+WordNet** classifier outperforms it. The TC classifier obtained by labeling words [2] has similar performance comparing to supervised approach. As mentioned in [1], its reason may be the easier TC task and the weaker NB classifier. Actually, the human intervention in selecting important words for each category might result in the high quality of the category's representative words. This fact also makes it possible that their TC results have better quality comparing with our FACT using WordNet to generate the representative words automatically.

For the phenomenon that SVM with inaccurate and insufficient training data could get satisfactory categorization results, we conjecture that the main reason is that the training data is more compact than the training data used for baseline method. Since we select top ranked documents from the initially labeled documents as training data, the similarity computing between the documents and RPs of the category realizes clustering for the documents in some sense. This can cause that the margin between positive and negative examples much wider than the baseline method. Thus, it might generate a more robust classifier for the document collection. Figure 6 shows this phenomenon.

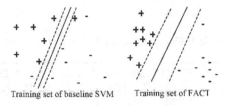

Fig. 6. Baseline SVM and FACT

However, for WebKB, FACT is worse than the baselines. It is because WebKB consists of web pages where many words have no semantic relation with the category name, e.g., Email, date, phone, etc. Then, the limitation of FACT is that it is sensitive not only to the expressiveness of the category name but also to whether the documents are semantically consistent to the category name. But considering the fact that for a real application with good human-computer interface, the category are generally specified clearly in a human-understandable way, it can make the generated features from WordNet and then the resulting training data (i.e., the correctness of the category labels) with good quality.

For the baselines, since the knowledge implied in the unlabeled documents is exploited, TSVM classifier outperforms SVM classifier. However, **SVM+WordNet** outperforms **TSVM+WordNet** a lot on Reuters-21578 and 20NP. This can be explained by that, since TSVM uses both the unlabeled and labeled data for classifier training, then for the resulting training data, the influence of the characteristic that negative and positive documents are separated with a wide margin is decreased a lot comparing that for SVM.

Regarding training data selection from the initial labeled documents, we conduct several experiments to select the value of p^+ and a using SVM. Figure 5 show the results. For 20NP and Reuters-21578, we can see that, when $p^+<0.5$, with the increase of a, the F1 measures increase at the beginning and reach the peak value when a is within 1~3, and then decrease slowly. However, for $p^+\geq0.5$, the F1 measure increase monotonously with a. We know that, (1) generally, when more training data are utilized, the classifier will get higher accuracy. However, for FACT, (2) when more labeled documents are utilized, more incorrectly labeled documents might be introduced into the training data, which will degrade the categorization results. The above phenomena can be explained by the tradeoff between these two trends of (1) and (2). As $p^+<0.5$, the positive samples have high quality. When a increases at the beginning, since the involved negative samples also with high quality, the trend (1) plays the major role. It causes the enhanced categorization performance with the increase of a.

When a reaches certain value, more noise data from the negative sample is introduced. It causes that trend (2) becomes the principle factor. And then, the performance began to decrease. When $p^+ \geq 0.5$, the low quality of positive samples always make trend (1) play the principle role. The performance of the classifier grows slowly. However, for WebKB, F1 doesn't have the similar behavior. The reason might be that its category name is not semantically consistent with document contents.

Figure 5 shows that, when p^+ is within 0.1~0.2, and a within 1~3, the learned classifier has almost the best performance. It is generally true for the three datasets. Then, in our implementation, we set p^+ and a to the medium values of their best performance ranges.

The contribution of WordNet: To evaluate the impact of WordNet on the results, additional experiments with SVM were done: **Without WordNet**, where the RP only contains the words from category names. The results are also shown in Table 1. Since WordNet bring more knowledge into the classifiers, in all the three datasets, **SVM+WordNet** outperforms **without WordNet** significantly. It demonstrates that using the training data from our automatic labeling approach can provide significant advantage than applying a simple query only consisting in the name of the category. Also, the contribution of WordNet for WebKB is not as notable as for 20NP and Reuters. It might be due to that WebKB's category names are not consistent with its documents, which make the contribution of generated features from WordNet is not stable (although it affect the results obtained without WordNet as well).

5 Conclusion

This paper proposes FACT for fully automatic TC. With the support of WordNet, the semantics of the category name are utilized for automatic document labeling. The document clustering is employed to reduce the possible biases derived from the category name and WordNet. The experiments show that, when the given category name has a clear representation of the topics described in the content of the documents, its performance is very close to the supervised method. Our future work will analyze its statistical significance and extend and evaluate the proposed approach for multilingual TC tasks.

References

1. Gliozzo, A.M., Strapparava, C., Dagan, I.: Investigating Unsupervised Learning for Text Categorization Bootstrapping. In: Proc. of EMNLP (2005)
2. Liu, B., Li, X., Lee, W.S., Yu, P.S.: Text Classification by Labeling Words. In: Proc. 19th Nat'l Conf. Artificial Intelligence (2004)
3. Blum, A., Mitchell, T.: Combining labeled and unlabeled data with co-training. In: Proc. of the Workshop on Computational Learning Theory (1998)
4. de Buenaga Rodriguez, M., Gomez-Hidalgo, J., Diaz- Agudo, B.: Using WordNet to complement training information in text categorization. In: Proc. of RANLP (1997)
5. Hotho, A., Staab, S., Stumme, G.: Wordnet Improves Text Document Clustering. In: Proc. of the Semantic Web Workshop at SIGIR (2003)

6. Witten, I.H., Frank, E.: Data Mining: Practical machine learning tools and techniques, 2nd edn. Morgan Kaufmann, San Francisco (2005)
7. Ide, N., Véronis, J.: Word sense disambiguation: The state of the art. Computational Linguistics 24(1), 1–40 (1998)
8. Joachims, T.: Transductive inference for text classification using support vector machines. In: Proc. 16th International Conf. on Machine Learning, pp. 200–209 (1999)
9. Kehagias, A., Petridis, V., Kaburlasos, V., Fragkou, P.: A comparison of word- and sense-based text classification using several classification algorithms. Journal of Intelligent Information Systems 21(3), 227–247 (2003)
10. Moldovan, D.I., Mihalcea, R.: Using WordNet and Lexical Operators to Improve Internet Searches. IEEE Internet Computing 4(1), 34–43 (2000)
11. Nigam, K., McCallum, A., Thrun, S., Mitchell, T.: Text classification from labeled and unlabeled documents using EM. Machine Learning, 103–134 (2000)
12. Scott, S., Matwin, S.: Text classification using WordNet hypernyms. In: Proc. Coling-ACL 1998, pp. 45–52 (1998)
13. Peng, X., Choi, B.: Document classifications based on word semantic hierarchies. In: Proc. of the International Conf. on Artificial Intelligence and Application (AIA 2005), pp. 362–367 (2005)
14. Banerjee, S., Pedersen, T.: An Adapted Lesk Algorithm for Word Sense Disambiguation Using WordNet. In: Gelbukh, A. (ed.) CICLing 2002. LNCS, vol. 2276, pp. 136–145. Springer, Heidelberg (2002)
15. Sebastiani, F.: Machine learning in automated text categorization. ACM Computing Surveys 34(1), 1–47 (2002)
16. Mansuy, T.N., Hilderman, R.J.: A Characterization of Wordnet Features in Boolean Models For Text Classification. In: AusDM 2006, pp. 103–109 (2006)
17. Vapnik, V.: The nature of statistical learning theory. Springer, Heidelberg (1995)
18. Chen, W., Zhu, J., Wu, H., Yao, T.: Automatic learning features using bootstrapping for text categorization. In: Gelbukh, A. (ed.) CICLing 2004. LNCS, vol. 2945, pp. 571–579. Springer, Heidelberg (2004)
19. Zhu, X.-J.: Semi-Supervised Learning Literature Survey (2007), http://pages.cs.wisc.edu/~jerryzhu/research/ssl/ semireview.html
20. Ko, Y., Seo, J.: Automatic text categorization by unsupervised learning. In: Proc. of COLING 2000 (2000)
21. Yang, Y., Liu, X.: A re-examination of text categorization methods. In: Proc. of SIGIR 1999 (1999)

A Latent Dirichlet Framework for Relevance Modeling

Viet Ha-Thuc and Padmini Srinivasan

Computer Science Department, The University of Iowa, Iowa City, IA52246, USA
hviet@cs.uiowa.edu,
padmini-srinivasan@uiowa.edu

Abstract. Relevance-based language models operate by estimating the probabilities of observing words in documents relevant (or pseudo relevant) to a topic. However, these models assume that if a document is relevant to a topic, then all tokens in the document are relevant to that topic. This could limit model robustness and effectiveness. In this study, we propose a Latent Dirichlet relevance model, which relaxes this assumption. Our approach derives from current research on Latent Dirichlet Allocation (LDA) topic models. LDA has been extensively explored, especially for discovering a set of topics from a corpus. LDA itself, however, has a limitation that is also addressed in our work. Topics generated by LDA from a corpus are synthetic, i.e., they do not necessarily correspond to topics identified by humans for the same corpus. In contrast, our model explicitly considers the relevance relationships between documents and given topics (queries). Thus unlike standard LDA, our model is directly applicable to goals such as relevance feedback for query modification and text classification, where topics (classes and queries) are provided upfront. Thus although the focus of our paper is on improving relevance-based language models, in effect our approach bridges relevance-based language models and LDA addressing limitations of both.

Keywords: LDA, topic models, relevance-based language models.

1 Introduction

Relevance is a key concept in retrieval theory [7][14]. Among the formal relevance models that have been proposed, relevance-based language models is perhaps the most popular one [9][10][11]. Given a topic of interest t, relevance-based language models estimate the probability distribution $p(w|R_t)$. The distribution is estimated by using a set of training documents. Nonetheless, these relevance-based language models have a limitation; they make an overly-strict assumption that all tokens in each training document are generated by a single topic to which the document belongs. This assumption is obviously not true in many practical cases. The example below is a part of a Wall Street Journal article judged relevant to the topic "machine translation" (TREC topic 63). As we see, many portions of it are non-relevant to the topic.

G.G. Lee et al. (Eds.): AIRS 2009, LNCS 5839, pp. 13–25, 2009.

Buried among the many trade issues that bedevil the U.S. and Japan is the 1 billion dollars of translation work done every year in Japan that could be done better and more efficiently in the U.S. And in the next two years, the dollar value of Japanese-to-English translations is expected to double. Think about it. Every car, videocassette recorder, boom box or stereo imported into the U.S. from Japan has operating and assembly instructions . . .

Alongside the development of relevance-based models, we observe a strong strand of research on Latent Dirichlet Allocation (LDA), that has been shown to be effective in many text-related applications [2][5][6][15]. LDA offers a strong theoretical framework within which we may consider each document as generated by a mixture of multiple topics. However, the topics discovered by LDA from a corpus are synthetic. In other words, if experts identified topics manually for a corpus, then these may have little or no correspondence with the synthetic topics identified by LDA. From a different perspective, we may say that probabilistic topic models are unable to model the concept of relevance to given topics of interest. Thus, not surprisingly LDA has not found use in applications such as relevance feedback based query modification. Our work shows how this can be done.

In this paper, we propose an approach that bridges relevance-based language models and LDA. Our approach allows us to address the limitation of relevance-based language models, specifically their assumption that all tokens of a relevant document are equally relevant to a topic. We do this by estimating the relevance model using the multiple-topic framework of LDA. In essence, we consider that although a document d may be relevant to a given topic t, it could still have non-relevant portions. Some portions could pertain to background information shared by many documents. Other non relevant portions while specific to d may be on themes other than t. Specifically, each document d is hypothesized to be generated by a combination of three topics: the topic t to which it is relevant, a background topic b representing the general language in the document set, and a third topic $t\_o(d)$ responsible for generating themes that though specific to d are neither b nor t. Because we consider this mixture of three topics, our model is able to identify just those portions of the document that are truly relevant to the topic t. In our work, these selected portions are the ones that contribute to the estimation of the relevance model $p(w|R_t)$. In this way, we utilize the Latent Dirichlet framework to solve for a limitation in relevance-based language models.

As in previous work in standard LDA [5][6][15], we also implement the inference process using Gibbs sampling [1][3]. A secondary contribution of this paper is that we exploit the "bag-of-words" assumption in order to reduce the computational complexity of the inference algorithm. Since token order in a document is not considered, we can re-arrange the tokens in any order that is convenient for the inference algorithm. In our case, we group tokens with the same stem into continuous segments because the topics of the tokens are sampled from the same distribution. That helps to reduce the running time of the sampling process. The proposed idea is also applicable for standard probabilistic topic models.

2   A Latent Dirichlet Relevance Model

2.1   Notation

A Vocabulary set (dictionary) V is a set of W possible words (terms) $V = \{word_1, word_2 \ldots word_W\}$. A token is a specific occurrence of a word in a document. Document d is a sequence of N_d tokens. A training set D_t of a topic of interest t is a set of $|D_t|$ relevant (or pseudo relevant) documents: $D_t = \{(w_1, d_1), (w_2, d_2) \ldots (w_{N_t}, d_{N_t})\}$, where $N_t = \sum_{d \in D_t} N_d$, w_i and d_i are word index and document index of the ith token. As we mentioned above, each document d in the training set of a topic of interest t is generated by a mixture of three topics $x_d = \{b, t, t\_o(d)\}$, where b denotes the background topic and $t\_o(d)$ denotes a document-leveled topic covering other themes rather t also mentioned in d. The topic mixing proportion of the three topics in d is represented by $\Theta_{d,z} = p(z|d)$ where $z \in x_d$. Each topic z is represented by a distribution over the vocabulary set denoted by: $\Phi_{z,w} = p(w|z)$ where $1 \le w \le W$.

2.2   Model Description

The proposed Latent Dirichlet relevance model is a generative model describing the process of generating relevant documents for K_0 given topics of interest. In this model, the language used to generate a document relevant to a topic of interest t is a combination of (1) the language reflecting the meaning of t itself, (2) the language of a general background topic, (3) the language reflecting themes other than t that are also mentioned in the document. For example, in the domain of computer science research papers, suppose that the training set for the topic *machine learning* (ML) includes d_1 a document about applying ML to *information retrieval* (IR) and d_2 a document about ML tools for the *banking industry*. The general background topic would be responsible for common words in English and common words in the domain such as "paper", "propose", "approach"... The distribution for topic ML, representing the meaning of ML, would likely give high probabilities to words like "learning", "training", "test"... Topic $t\_o(d_1)$ responsible for other themes in document d_1 would likely generate words relating to the IR aspects mentioned in d_1, while for d_2, $t\_o(d_2)$ would likely generate words such as "bank", "sales", "marketing" that are related to the banking industry emphasis in d_2. The process of generating relevant documents for K_0 topics of interest is formally described as follows:

1. Pick a multinomial distribution Φ_b for the background topic (b) from a W-dimensional Dirichlet distribution $Dir(\beta)$.

2. For each topic t in K_0 topics of interest:

2.1 Pick a multinomial distribution Φ_t for t from the W-dimensional $Dir(\beta)$.

2.2 For each document d relevant to t:

2.2.1 Pick a multinomial distribution $\Phi_{t\_o(d)}$ for the topic covering themes other than t that are also mentioned in d from the W-dimensional $Dir(\beta)$.

2.2.2 Pick a multinomial distribution Θ_d from a 3-dimensional $Dir(\alpha)$, each element of Θ_d corresponds to a topic in $x_d = \{b, t, t\_o(d)\}$

2.2.3 For each token in document d:

2.2.3.1 Pick a topic z among the three topics in x_d from multinomial Θ_d.

2.2.3.1 Pick a word from the corresponding multinomial distribution Φ_z.

The graphical model using plate notation in Fig. 1 describes this process. In the Figure, w_i (word index of a token i^{th}) and $x_d = \{b, t, t\_o(d)\}$ (topics generate document d) are observable variables and denoted by shaded circles; z_i (latent topic of a token i^{th}), Θ and Φ are hidden variables and denoted by un-shaded circles; α, β are hyper-parameters of Dirichlet distributions. In our model, values of α, β are pre-defined as in [15][16].

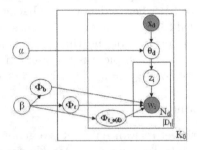

Fig. 1. Latent Dirichlet relevance model

Observe that unlike standard LDA describing how all documents in a corpus are generated, our model describes how *relevant* documents for a set of given topics are composed. Consequentially, each given topic of interest is explicitly associated with a multinomial distribution over the vocabulary. Therefore, we are able to explicitly model relevance.

The advantage of our model compared to relevance-based language models is that our model considers two more components b and $t\_o(d)$. The purpose of the background topic b is to explain words commonly appearing in training documents of all topics. That allows the distribution of the topic of interest t to be more discriminative. The purpose of $t\_o(d)$ is to explain words frequently appearing in the particular document d, but not in other training documents of topic t. That prevents the distribution of topic t from wasting its probability mass on these extra document-specific features. Thus, the consideration of document-specific $t\_o(d)$ topics minimizes the risk of t over-fitting the given set of training documents. We model the topic mixing proportion Θ_d and topic-word distribution Φ_z by latent variables which are assumed to be sampled from prior Dirichlet distributions. The explicit assumption about the prior sources of these variables provides complete generative semantics for the model [2][6][16]. Moreover, the mathematical property that the Dirichlet priors of $p(\Theta_d|\alpha)$ and $p(\Phi_z|\beta)$ are conjugate to their likelihoods (multinomial distributions) $p(z|\Theta_d)$ and $p(w|\Phi_z)$ results in the fact that their posteriors $p(\Theta_d|\alpha, \{z_i|$ for all tokens in doc $d\})$ and $p(\Phi_z|\beta, \{w_i|$ for all tokens generated by $z\})$ are also Dirichlet distributions. Mathematically that makes the inference feasible.

2.3   Inference

As in previous work on LDA, we also apply Gibbs sampling to infer latent variables. Formally, Gibbs sampling estimates $p(\{\Phi_z : \forall z\}, \{\Theta_d : \forall d\}, \{z_i : \forall i\} | \{x_d : \forall d\}, \{w_i : \forall i\})$, the conditional distribution of latent variables given observable ones by generating sequence of $(S+1)$ samples, where each sample contains values for all latent variables. The sampling algorithm is presented in Fig. 2. For the first sample, $\Phi_b^{(0)}$, $\Phi_t^{(0)}$, and $\Phi_{t\_o(d)}^{(0)}$ are initialized by smoothed term frequencies on all relevant sets, on the training set of t, and on document d, respectively. $\Theta_d^{(0)}$ for each document d is the uniform distribution i.e. $\Theta_d^{(0)} = \{1/3, 1/3, 1/3\}$. For each of the following S samples (Step 2, Fig. 2), each latent variable is randomly sampled from its posterior distribution given current values of all other variables. Specifically, latent topic of token i^{th} is sampled from its posterior distribution that is estimated by using values of Φ and Θ in the previous sample (Step 2.1), where w_i and d_i are word index and document index of token i^{th}. After sampling z_i for every token, we update the values for Φ and Θ by maximum likelihood principle (Steps 2.2 and 2.3). Given the (S+1) samples, we ignore the first S' samples (samples in the burn-in period), then select every P^{th} samples (i.e. samples S', (S'+P), (S'+2P)...) to estimate $\Phi_{z,w*} = p(word = w|topic = z)$ for all topics. Those distributions are estimated by averaging over $\Phi_{z,w}^{(s)}$ in these selected samples.

---

1. Initialize variables: $\Phi^{(0)}$ and $\theta^{(0)}$

2. For s = 0 to (S-1):

   2.1 For i = 1 to N: (N is the number of tokens)

   Sample $z_i^{(s+1)}$ from:

   $$p(z_i = z|\ w_i,\ d_i,\ others) = p(z_i = z|\ w_i,\ d_i,\ \theta^{(s)},\ \Phi^{(s)})$$

   $$= p(w_i|\ z_i = z,\ \Phi^{(s)})\ p(z_i = z|\ d_i,\ \theta^{(s)}) = \phi_{z,w_i}^{(s)}\ \theta_{d_i,z}^{(s)}$$

   2.2 For z = 1 to K: (K is the total number of topics)

   Estimate $\Phi_z^{(s+1)}$:

   For w = 1 to W: (W is the number of words)

   $$\Phi_{z,w}^{(s+1)} = p(w|z) = \frac{m_{z,w}^{(s+1)} + \beta}{\sum\limits_{w'=1}^{W}(m_{z,w'}^{(s+1)} + \beta)}$$

   2.3 For d = 1 to |D|: (|D| is the number of documents)

   Estimate $\theta_d^{(s+1)}$:

   For each z in $x_d = \{b, t, t\_o(d)\}$:

   $$\theta_{d,z}^{(s+1)} = p(z|d) = \frac{n_{d,z}^{(s+1)} + \alpha}{\sum\limits_{z' \in X_d}(n_{d,z'}^{(s+1)} + \alpha)}$$

Fig. 2. Gibbs sampling-based inference algorithm

Reducing Complexity: In the inference algorithm (Fig. 2), $z_i^{(s+1)}$ is conditionally independent of any other $z_k^{(s+1)}(k \neq i)$ (Step 2.1). So, we can re-arrange the sampling order in Step 2.1 in any order without affecting the final results. Exploiting this observation, in the preprocessing step, we re-arrange tokens in each document such that tokens from the same stem are consecutive. Since the latent topics for these tokens are sampled from the same posterior distribution, the re-arranging could reduce the complexity by a factor of $r = W_d/N_d$, where W_d is the average number of distinct stems, and N_d is the average number of distinct tokens in a document.

3 Experiments

3.1 Pseudo-relevance Feedback

In this section, we evaluate the effectiveness of our Latent Dirichlet relevance model (Dir Rel) on the task of pseudo-relevance feedback in comparison against a standard relevance-based language model (Rel LM). Our implementation of the Rel LM follows the description given in [7]. Our experiments are done using four corpora (Table 1). AP and WSJ contain newswire articles. For these corpora, we use 100 topics (title only) and partial judgments for these topics provided by TREC. 20 Newsgroup contains discussion posts. Each of the posts is labeled by one of 20 topics. Cora contains computer science abstract research papers. These papers are also manually assigned to topics. We use 20 topics for this corpus. For 20 Newsgroup and Cora, we have complete relevance judgments.

Table 1. Corpora

Corpus	# of documents	# of topics (queries)
TREC AP	242 918	100 (051-100, 151-200)
TREC WSJ	173 252	100 (051-100, 151-200)
20 Newsgroup	19 956	20
Cora	25 705	20

All documents are stemmed using the Porter stemmer [12] and indexed using Lucene. We do not remove stop words in this experiment. For a simple retrieval baseline, we use all Lucene default parameter settings. From the results returned by Lucene, the top 50 documents for each query are used to train Rel LM and Dir Rel. In each case the top ranked 50 words, with the highest probabilities estimated by each model, are used to expand the original query. These parameter values (50x50) have been tuned for Rel LM in previous work. We also use these values for our model. Tuning these values specifically for our model could result in a better performance. We leave this for future work. The expanded query is rerun using Lucene. The performances of the baseline retrieval and pseudo-relevance feedback by the two models are shown for each dataset in Tables 2-5. We measure averages across topics of precision at top 10, top 100 and top

1000 ranked documents. We also measure average precision (averaged across topics to yield MAP) and the total number of relevant documents retrieved ($\sharp\_rel\_ret$). As expected with only a single exception both feedback models are consistently better than the no feedback Lucene baseline (row 1 of the tables) for all measures. The only exception is for 20 Newsgroup for P@10 w.r.t. our model. Focusing just on MAP (the fifth column), notations α and β indicate statistically significant over the baseline and Rel LM ($p - value < 0.05$ by the paired t-test). The improvements against baseline for the Rel LM are generally in the range of 10% to 43%, while for Dir Rel are in 23% to 100%. In 3 of the 4 cases, the MAP improvements for Dir Rel against Rel LM are around 10%. The best improvement is observed in the 20 Newsgroup dataset (40%). In terms of precision, for example the P@100 score, we find that Dir Rel is consistently better than Baseline and Rel LM in all cases. Thus on the whole, we find that Dir Rel is successful at achieving improvements over the Rel LM, and both feedback models are, as expected, better than the no feedback baseline. These results support our contention that a) relevant documents may contain portions that are not relevant to the topic of interest and b) it is possible to build more robust relevance models using the Latent Dirichlet framework.

Table 2. Cora

	P@10	P@100	P@1000	MAP	MAP-Impr	#_rel_ret
Baseline	0.625	0.485	0.1795	0.2307	---	5010
Rel LM	0.65	0.5015	0.1967	0.2549$^\alpha$	10%	6667
Dir Rel	0.665	0.532	0.2124	0.2844$^{\alpha\beta}$	23%	7130

Table 3. 20 Newsgroup

	P@10	P@100	P@1000	MAP	MAP-Impr	#_rel_ret
Baseline	0.715	0.5905	0.273	0.1783	---	7875
Rel LM	0.73	0.612	0.3223	0.2548$^\alpha$	43%	15170
Dir Rel	0.67	0.625	0.3933	0.3568$^{\alpha\beta}$	100%	17621

Table 4. AP

	P@10	P@100	P@1000	MAP	MAP-Impr	#_rel_ret
Baseline	0.326	0.232	0.0778	0.1948	---	7783
Rel LM	0.372	0.2701	0.0887	0.2409$^\alpha$	23.7%	8864
Dir Rel	0.385	0.2895	0.0945	0.2650$^{\alpha\beta}$	36.0%	9444

Table 5. WSJ

	P@10	P@100	P@1000	MAP	MAP-Impr	#_rel_ret
Baseline	0.371	0.2311	0.0618	0.2340	---	6179
Rel LM	0.451	0.2689	0.0678	0.2817$^\alpha$	20.4%	6780
Dir Rel	0.482	0.2904	0.0713	0.3118$^{\alpha\beta}$	33.2%	7124

3.2   Perplexity

The goal of both relevance-based language models and our Latent Dirichlet relevance model is to estimate the unknown true relevance distribution $p(w|t)$ of some topic of interest t. A traditional measure for comparing the two estimations is perplexity. Perplexity indicates how well estimated distributions predict a new sequence of tokens drawn from the true distribution. Better estimations of the true distribution tend to give higher probabilities to test tokens. As a result, they have lower perplexity, which means they are less surprised by these tokens.

In our experiment such ideal test data is not available. Instead, for each topic (query) t, we approximate the new sequence of relevant tokens by using a held out set of 50 actual relevant documents that do not appear in the training set. We remove stop words from a standard list and also rare words in these relevant documents. Then, we use the remaining tokens as test data. Given estimated distributions $p_{RelLM}(w|t)$ and $p_{DirRel}(w|t)$ obtained from the previous experiment, we compute Perplexity (PPX) for each topic as follows:

$$PPX(TestData|t) = exp\{\frac{-1}{N} \sum_{w_i \in TestData} log(p(w_i|t))\} \tag{1}$$

where N is the number of tokens in the test data. Table 6 shows the average perplexity over 20 topics of Cora and 20 Newsgroup. We experiment on Cora and 20 Newsgroup since each topic of these corpora has hundreds of relevant documents. As we see, the perplexity of relevance distributions estimated by the proposed model is significantly lower than distributions estimated by relevance-based language models. The asterisk symbol ($*$) means that the difference between the two results is statistically significant (i.e. $p - value < 0.05$ by the paired t-test). This indicates that our Rel Dir is better able to predict unseen test data from the true distribution as compared to Rel LM. Again, the key difference here is that our model considers each document to be generated by a mixture of topics and not just the relevant topic alone.

Table 6. Average Perplexity

	Cora	20 Newsgroup
Rel LM	1364	4976
Dir Rel	942*	3134*

4   Deeper Analyses

In this section, we further analyze the key feature of our proposed model, i.e., the important fact that a document relevant to a given topic could also talk about other non-relevant themes and also have uninformative background terms. Our model's strength is that it automatically extracts relevant terms and rules out non-relevant background terms and terms belonging to other themes in each document. We illustrate this ability with the example below. The following is a relevant document in the training set for the topic of information retrieval (IR). The document seems to be about image retrieval in the medical domain. (Note: to make it more readable, we restore the stemmed words to the original forms.) After running the inference algorithm described in Fig.2, our model determines the latent topic of each token as shown in the example. Bold tokens are inferred to be generated by IR topic (i.e. are relevant terms), italicized tokens are inferred to be generated by the background topic (i.e. are non-relevant terms), underlined tokens are inferred tò be generated by $t\_o(d)$ (and so also non-relevant to t).

> *We present a* **principled** *method of obtaining a weighted* **similarity metric** *for* <u>3D</u> **image retrieval**, <u>firmly rooted</u> *in* **Bayes** *decision theory. The basic idea is to determine a set of most* <u>discriminative</u> *features by evaluating how well they perform on the task of classifying* **images** *according to* **predefined semantic categories**. *We propose this indirect method as a* <u>rigorous</u> *way to solve the difficult feature selection problem that comes up in most* **content** *based* **image retrieval** *tasks. The method is applied to* **normal** *and* <u>pathological neuroradiological CT</u> **images**, *where we take advantage of the fact that* **normal** <u>human brains</u> *present an approximate* <u>bilateral symmetry</u> *which is often* <u>absent</u> *in* <u>pathological brains</u>. *The* <u>quantitative</u> *evaluation of the* **retrieval** *system shows* **promising** *results.*

As we see all stop words as well as words popular in the domain such as "present", "method" "obtain" are inferred as background terms (recall that Cora contains computer science research papers). Most of the bold are really relevant to IR such as "similarity" "retrieval" "semantics". The $t\_o(d)$ terms identified by the model reflect the specific context of the document and contain almost nothing about the topic of IR.

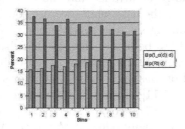

Fig. 3. Topic Contribution Proportion

A secondary hypothesis that we now explore is that the proportion of relevant (on topic) tokens in top retrieved documents is likely to be higher than in lower ranked ones. Analogously, the contributions of $t\_o(d)$ topics in lower ranked documents are likely to be more serious than in top ranked ones. To test this, we explore the contributions, in percentages, of the relevant topical component and the non-relevant component generated by $t\_o(d)$ over top 100 retrieved documents. We group the results by bins. Each bin contains 10 documents (i.e. the first bin in Fig. 3 includes the top 10 documents, the last bin includes documents from ranks $91 - 100$). Fig. 3 shows the result averaged over 20 topics on Cora. We see that proportion of relevant tokens in the first bin is 19% higher than in the 10th bin. Similarly, the contribution of $t\_o(d)$ topics in the last bin is 27% higher than in the first bin. The results on other datasets also have the same trend (not shown due to lack of space). Recall that the contribution proportions of topics in documents are modeled as a latent variable in our model, and are determined automatically by the inference algorithm.

5   Related Work

Our work proposed in this paper is related to two separate existing directions: relevance models, and probabilistic topic models.

Relevance-based language models [9], a popular approach for relevance modeling, expand a given topic (query) t to a multinomial distribution $p(w|R_t)$ of observing a word w in documents relevant to t. The probabilities are estimated by using a set of training documents: $p(w|R_t) = \sum_{D \in R_t} p(w|D)p(D|R_t)$, where $p(w|D)$ is a language model, and $p(D|R_t) = 1/|R_t|$ as assumed in previous work of Hiemstra et al. [7]. In our experiment, we use the same assumption for implementing relevance-based language models. A limitation of the relevance-based language models is that they are based on a strict assumption that if a document D is relevant to a topic, all tokens in the document are equally relevant to that topic. In [7][17], three-component mixture relevance models are proposed. Besides the relevance component (R_t), the authors introduce two additional components to capture the background (b) and local features (d) in documents. However, the model assumption that the mixing proportions of the three components $(\lambda_b, \lambda_{R_t}, \lambda_d)$ are known in advance and the same for all documents is not reasonable. For instance, in the case where we use top 50 retrieved documents for the query t as the training set, the contribution of relevance component in the first document is likely to be higher than in the 50th document.

Another approach to alleviate the problem of noises in training documents is to build relevance model on passages (usually windows of text) instead of the whole documents (Liu et al. [11]). However, the way that documents are broken into passages is rather ad-hoc and corpus specific. Moreover, all tokens in each passage are still considered equally relevant. As in the WSJ and Cora example documents we show above, relevant and non-relevant terms appear together even within a sentence.

In topic model literature, Hofmann [8] proposes probabilistic Latent Semantic Indexing (pLSI) modeling each document as a mixture of topics, where a topic is a multinomial distribution. Each word in a document is generated by a topic, and different words in the same document may be generated by different topics. Topics are automatically discovered from the corpus. One limitation of pLSI is that it is not clear how the mixing proportions for topics in a document are generated [2]. To overcome the limitation, Blei et al. [2] propose Latent Dirichlet Allocation (LDA). In LDA, topic proportion of every document is a K-dimensional hidden variable randomly drawn from the same Dirichlet distribution, where K is the number of topics. Thus, generative semantics of LDA are complete [16]. LDA and its variants have been applied in many applications such as finding scientific topics [6], E-community discovery [18], mixed-membership analysis [5] and ad-hoc retrieval for representing document language model [4][6].

However, a common problem of both pLSI and LDA is their inability to model the concept of relevance, which is key in information retrieval [7][13][14]. Consequently, there is no explicit mapping between the resulting topics generated by pLSI or LDA and the topics in the prior knowledge of human beings. Therefore, the approach could not be applied directly for problems, such as relevance feedback for query modification and text classification, where topics (classes and queries) are provided upfront.

Compared to these two sets of approaches, our Latent Dirichlet relevance model has the following advantages. First, our model explicitly takes the key concept of relevance in account, as in the relevance models [9]. Second, our model could be able to identify relevant and non-relevant terms in training documents. Only relevant terms contribute to the estimation of relevance models. Third, our model possesses complete generative semantics by treating document-topic mixing proportion (Θ_d) and topic-word distribution (Φ_z) as hidden random variables sampled from Dirichlet distributions as in the original LDA [2]. As a result, we could exploit the Latent Dirichlet theoretical framework to automatically infer both these variables by taking semantics of topics and content of each relevant document into account.

6   Conclusions

This paper presents a Latent Dirichlet relevance model that combines the advantages of both relevance-based language models [9] and probabilistic topic models [2][15]. Crucially, our model relaxes the strict assumption of relevance-based language models that if a document is relevant to a topic, the entire document is relevant to that topic. This is done by automatically identifying the non-relevant parts in the document. Second, in the context of research on probabilistic topic models, our model explicitly considers the notion of relevance by starting with given topics and estimating their distributions over the corpus vocabulary. We also propose the idea of exploiting the assumption of exchangeability for the tokens in a document ("bag-of-words" assumption) to reduce the computational complexity of the learning algorithm. This idea is not only applicable to our Latent Dirichlet relevance models, but also to conventional LDA.

Our preliminary experiments on pseudo-relevance feedback show the effectiveness the proposed model. The results obtained by the model are consistently better across all of the four corpora than the results of the baseline retrieval (23%-100% improvement in terms MAP) and relevance-based language models

(10%-40%). Our work on perplexity re-affirms the advantages of our model over relevance-based language models for the task of estimating the true unknown relevance model.

For future directions, we plan to apply the model for some other applications such as text classification without any human-labeled training data. Instead, we will use as training sets documents returned from a global search engine (e.g. Google) or an intranet search engine, retrieved by the topics themselves. The challenge of this approach is that there is a lot of noise (non-relevant portions) in the returned sets. The ability to automatically detect non-relevant parts in documents of our model is the key to tackling this challenge. Moreover, the background topic in our model could cover common word features of all given classes (topics of interest), so that each of these classes could spend its probability mass on its discriminative features that distinguish itself from the rest of the classes. The background topic could, therefore, increase the margins among the distributions of the classes. This idea is similar to SVM classification technique, but in our model it is not only applicable to case of two classes but also naturally applicable any set of classes.

References

1. Adrieu, C., Freitas, N., Doucet, A., Jordan, M.: An Introduction to Markov Chain Monte Carlo for Machine Learning. Machine Learning 50 (2003)
2. Blei, M., Ng, A., Jordan, M.: Latent Dirichlet Allocation. Journal of Machine Learning Research 3 (2003)
3. Casella, G., George, E.: Explaining the Gibbs Sampler. The American Statistician 46(3) (1992)
4. Chemudugunta, C., Smyth, P., Steyvers, M.: Modeling General and Specific Aspects of Documents with a Probabilistic Topic Model. In: Proceedings of the 20th NIPS (2006)
5. Erosheva, E., Fienberg, S., Lafferty, J.: Mixed-membership Models of Scientific Publication. In: Proceedings of National Academy of Science, PNAS (2004)
6. Griffiths, T., Steyvers, M.: Finding Scientific Topics. In: Proceedings of National Academy of Science, PNAS (2004)
7. Hiemstra, D., Robertson, S., Zaragoza, H.: Parsimonious Language Models for Information Retrieval. In: Proceedings of the 27th ACM SIGIR (2004)
8. Hofmann, T.: Probabilistic Latent Semantic Indexing. In: Proceedings of the 15th UAI (1999)
9. Lavrenko, V., Croft, W.B.: Relevance-based Language Models. In: Proceedings of the 24th ACM SIGIR (2001)
10. Lavrenko, V., Croft, W.B.: Relevance Models in Information Retrieval. In: Croft, B., Lafferty, J. (eds.) Language Modeling for Information Retrieval. Kluwer Academic Publishers, Dordrecht (2003)
11. Liu, X., Croft, B.: Passage Retrieval Based on Language Models. In: Proceedings of the 11th ACM CIKM (2002)
12. Rijsbergen, C., Robertson, S., Porter, M.: New Models in Probabilistic Information Retrieval, British Library Research and Development Report, 5587 (1980)
13. Robertson, S., Sparck-Jones, K.: Relevance Weighting of Search Terms. Journal of American Society for Information Science 27 (1988)

14. Sparck-Jones, A., Robertson, S., Hiemstra, D., Zaragoza, H.: Language Modelling and Relevance. In: Croft, B., Lafferty, J. (eds.) Language Modeling for Information Retrieval. Kluwer Academic Publishers, Dordrecht (2003)
15. Steyvers, M., Griffiths, T.: Probabilistic Topic Models. In: Landauer, T., et al. (eds.) Latent Semantic Analysis: A Road to Meaning. Lawrence Erlbaum, Mahwah (2006)
16. Wei, X., Croft, B.: LDA-based Document Models for Ad-hoc Retrieval. In: Proceedings of the 29th ACM SIGIR (2006)
17. Zhang, Y., Callan, J., Minka, T.: Novelty and Redundancy Detection in Adaptive Filtering. In: Proceedings of the 25th ACM SIGIR (2002)
18. Zhou, D., Manavoglu, E., Li, J., Giles, L., Zha, H.: Probabilistic Models for Discovering E-Communities. In: Proceedings of the 15th ACM WWW (2006)
19. Lucene, http://lucene.apache.org/

Assigning Location Information to Display Individuals on a Map for Web People Search Results

Harumi Murakami[1], Yuya Takamori[1,*], Hiroshi Ueda[2], and Shoji Tatsumi[2]

[1] Graduate School for Creative Cities, Osaka City University,
3-3-138, Sugimoto, Sumiyoshi, Osaka 558-8585 Japan
harumi@media.osaka-cu.ac.jp
http://murakami.media.osaka-cu.ac.jp/
[2] Graduate School of Engineering, Osaka City University,
3-3-138, Sugimoto, Sumiyoshi, Osaka 558-8585 Japan

Abstract. Distinguishing people with identical names is becoming more and more important in Web search. This research aims to display person icons on a map to help users select person clusters that are separated into different people from the result of person searches on the Web. We propose a method to assign person clusters with one piece of location information. Our method is comprised of two processes: (a) extracting location candidates from Web pages and (b) assigning location information using a local search engine. Our main idea exploits search engine rankings and character distance to obtain good location information among location candidates. Experimental results revealed the usefulness of our proposed method. We also show a developed prototype system.

Keywords: location information, Web people search, map interface, character distance, information extraction.

1 Introduction

Finding information about people on the Web is one of the most popular search activities. According to [1], 30% of all queries in Web searches include person names. Person name disambiguation, or distinguishing people with identical names, is becoming more and more important in Web searches. Most research into person name disambiguation concentrates on automatically separating Web pages for different people using clustering algorithms. However, if the list of search results is merely *person 1, person 2, . . . and so on*, users have difficulty determining which person clusters they should select.

This research locates person icons on a map to provide a user-interface to help users select person clusters that are separated into different people from the result of person searches on the Web. We assign one piece of representative

---

* Currently, Jupiter Telecommunications, Co., Ltd.

G.G. Lee et al. (Eds.): AIRS 2009, LNCS 5839, pp. 26–37, 2009.
© Springer-Verlag Berlin Heidelberg 2009

location information to an individual. Fig. 1 shows the aim of this research and a prototype interface. When a person name is input as a query, Web pages are classified into person clusters, and individual icons that express appropriate locations are displayed on a map. Users can select icons to display their location information to access searched Web pages.

In this paper, we assign one piece of location information to person clusters.

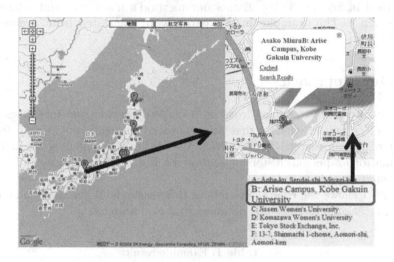

Fig. 1. Research aim

Many services extract location information from Web pages [2]. Most extract *addresses* using pattern-matching algorithms. For example, the following sequence, *number, avenue, city*, is treated as an address. These systems extract all such information from a Web page. However, existing systems have problems satisfying our requirements. First, they don't judge which address is the most suitable for particular people. Second, they don't treat multiple Web pages. Third, they cannot extract location information when addresses are not included in Web pages.

We propose a method to assign one piece of location information that is suitable for an individual using Yahoo! local search API [3] (hereafter Local search). Local search is one geocoding service that returns location coordinates based on an address or landmark query. For example, when a user inputs the following as an address query, *6-10-1, Roppongi, Minato-ku, Tokyo-to*, Local search returns *10-1, Roppongi-6-chome, Minato-ku, Tokyo-to* as an address, *35.65716694* as a latitude, and *139.73245194* as a longitude.

Our main ideas are as follows. First, we utilize landmarks as well as addresses to extract location information even from Web pages without addresses. Second, we use search engine rankings and character distance to assign suitable location information. Third, we use Local search to convert texts included in Web pages to location information. To obtain one piece of location information from Local search, we design several mechanisms. For the zero hit problem, we introduce two heuristics: a one-character deleting heuristic when the text is an address type, and a formal name inference heuristic when the text is a landmark type. A candidate list is another mechanism to avoid zero hits. For the multiple hits problem, we introduce a calculating context heuristic to select one result.

Below, in Section 2 we explain our method. The experimental results are described in Section 3. We discuss our method's usefulness and related work in Section 4. The examples in this paper were translated from Japanese into English for publication.

2 Our Approach

This research locates person icons on a map to provide user-interfaces to help users select person clusters that are separated into different people from the result of person searches on the Web. We obtain the following set of location information for person clusters: *a location label, a location address, and a latitude and longitude pair.*

Table 1 shows the example results of this research. When a person cluster is input, location information is obtained by Local search.

Table 1. Example results

Person cluster	Location Information		
	Location label	Location address	Lat/Lon
Asako Miura 3 (B)	Arise Campus, Kobe Gakuin University (landmark type)	Nishi-ku, Kobe-shi, Hyogo-ken	34.6.../135.0...
Asako Miura 11 (F)	13-17, Shinmachi-1-chome, Aomori-shi, Aomori-ken (address type)	13-17, Shinmachi-1-chome, Aomori-shi, Aomori-ken	40.8.../140.7...

For example, for the first cluster, Asako Miura 3 (associate professor at Kobe Gakuin University), a location label, *Arise Campus, Kobe Gakuin University*, a location address, *Nishi-ku, Kobe-shi, Hyogo-ken*, and a latitude and longitude set, *34.6.../135.0...* are obtained. This location information is a landmark type. For the second cluster, Asako Miura 11, the location label and the location address are *13-17, Shinmachi-1-chome, Aomori-shi, Aomori-ken*, and the location address is *40.8.../140....* This is an address type.

A method overview is shown in Fig. 2. Our method is comprised of two steps: (a) extracting location candidates from Web pages and (b) assigning location information using Local search.

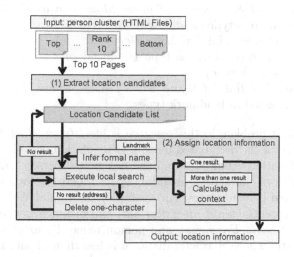

Fig. 2. Overview

2.1   Extracting Location Candidates

When a person cluster (HTML files obtained from person searches) is given, location candidates are extracted and a location candidate list is generated.

Determining web pages using search engine ranking. One difficult problem in this research is identifying which information is the most suitable for the designated person. If we only use one Web page, one very good piece of information may be obtained, but much good information may be overlooked. If we use all Web pages, all information might be obtained, but selecting the best information may become unwieldy. In this research we need to examine which Web pages should be treated. We propose using search engine rankings to determine Web pages to extract location information. We decided to use the top 10 Web pages to extract location information based on the experimental results in Section 3.

Extracting address and landmark candidates. When Web pages are given, after removing HTML tags, new line codes, and spaces from HTML files, we extract address and landmark candidates based on heuristics and morphological analysis. We used MeCab [4] for morphological analysis. Two heuristics are described below:

1. Extracting address candidate heuristic
   When a morphone that meets conditions (a) or (b) appears continuously more than once, we combine these morphones into a location candidate (address type): (a) whose type is judged *location*, *prefix*, or *number* by MeCab (b) which is included in the predefined term list such as: *hyphen, street, avenue, north, south, east*, or *west*.

For example, *3-3-138, Sugimoto, Sumiyoshi-ku, Osaka-shi, Osaka-fu,* is extracted as an address type candidate.

2. Extracting landmark candidate heuristic

   We extract a morphone whose type is judged *organization* as a location candidate (landmark type).

   For example, *City Hall of Kyoto City, Osaka Station, The University of Tokyo* are extracted as landmark types.

Again, one difficult problem in this research is identifying which information is the most suitable for the designated person. Our solution uses the character distance between location candidates and the designated person name. Character distance is number of characters (including spaces) between location candidates and the designated person names. For example, an original Web page contains *Asako Miura (Kobe Gakuin University)* and the distance between a location candidate *Kobe Gakuin University*, and person name *Asako Miura* becomes 2. Address candidates whose character distance is less than 71 and landmark candidates whose character distance is less than 31 are extracted from the top 10 Web pages.

Sorting location candidates. To avoid zero hits from a Local search, we order extracted location candidates by character distance to generate a location candidate list. When the character distance between two candidates is identical, those extracted from higher ranked pages become higher, and those extracted from the upper part of pages become higher.

2.2   Assigning Location Information

When a location candidate list is given, our method obtains one piece of location information using Local search. First, our method gives a location candidate from the top of the candidate list to Local search. When the candidate is a landmark type, a formal name inference heuristic is processed to change the abbreviation into a formal name to hit a Local search beforehand.

   When only one result is obtained from Local search, it becomes the answer. When more than one result is obtained, our method calculates the similarity between the candidate and obtained results, and the most similar result becomes the answer. We call this a calculating context heuristic. When the candidate is an address type and no result is obtained, the one-character deleting heuristic is applied to modify the candidate to hit a Local search. When no result is obtained after these processes, the next candidate in the list will be processed. When there is no answer after all candidates in the list are processed, *none* (*no location information*) is the output.

Formal name inference heuristic. When a location candidate is a landmark type, a formal name inference heuristic is applied once to change the abbreviations into formal names to get the Local search results. The formal name inference heuristic gives the candidate to the Yahoo! Web search API [5] (hereafter *Web search*) and gets the title of the first result. The title is modified to a formal name using a heuristic based on a stop list.

For example, when the candidate is *Todai* (abbreviation of *The University of Tokyo*), it is given to a Web search, and the title of the first result is *The University of Tokyo Homepage*. After *Homepage* is deleted using a stop list based heuristic, the candidate becomes *The University of Tokyo*.

One-character deleting heuristic. When no result is obtained for the address type candidate, the one-character deleting heuristic is repeatedly executed to delete the last character of the candidate and to repeatedly give it to Local search until the following stop condition is reached. The heuristic stops when the string no longer contains *ku* (town), *shi* (city), *to, do, fu,* or *ken* (these four terms denote prefectures).

For example, if address query *6-10-1, Roppongi, Minato-ku, Tokyo-to* obtains no result from a Local search, the next query becomes *6-10, Roppongi, Minato-ku, Tokyo-to*. Then the next query becomes *6, Roppongi, Minato-ku, Tokyo-to*.

Calculating context heuristic. When more than one result is obtained from a Local search, our method calculates context with a calculating context heuristic that calculates the similarity between the location candidate and the obtained results using vector space models. The most similar result becomes the answer.

For example, when the location candidate is *The University of Tokyo*, it is given to a Local search, and such multiple results are obtained as 1) The University of Tokyo, 2) The University of Tokyo Komaba Campus, 3) The University of Tokyo Cultural Development, and so on. We need to select a result. Our method can select an answer based on the Web page context. For example, if a person teaches at the main campus of The University of Tokyo, 1) is selected; if s/he works at the Komaba campus, 2) is selected.

The similarity between the candidate and the results is calculated using a cosine measure, based on a vector space model. Given candidate vector c and results vector r, similarity $sim(c, r_i)$ is defined as follows:

$$sim(c, r_i) = \frac{\sum_{j=1}^{t} w_{cj} w_{ij}}{\sqrt{\sum_{j=1}^{t} (w_{cj})^2} \sqrt{\sum_{j=1}^{t} (w_{ij})^2}} \tag{1}$$

Here t is the number of terms, w_{cj} is the weight of t_j in candidate vector c, and w_{ij} is the weight of t_j in result vector r. The terms are defined as the following morphones: (a) those included in the candidate list (when the candidate is an address type), (b) those included in the original Web page with a location candidate (when the candidate is a landmark type), and (c) those included in the Local search results. The weights of the terms are calculated by tf-idf. The result that is most similar and larger than 0.7 becomes the answer. When no result is larger than 0.7, the top result becomes the answer.

2.3   Example

There are 14 people for query *Asako Miura*. For the five people (Person clusters 2, 3, 6, 8, 11) with location information in Web pages, 100% (5/5) was correct. For nine people without location information in Web pages, no location information was assigned for eight people. In other words, 88.9% (8/9) was correct.

Asako Miura 3 is currently an associate professor at *Kobe Gakuin University*. There are several Kobe Gakuin University campuses, and our method outputs the correct answer, *Arise Campus, Kobe Gakuin University*, where her office is located. This good example shows the usefulness of our method. If we use all (i.e. 100) Web pages, *the Faculty of Human Science, Osaka University* will be displayed. If we use the frequency method (see Section 3.1), *the Faculty of Human Science, Osaka University* will also become the answer. Osaka University is her previous affiliation where she worked for many years. *Kyushu University*, which she occasionally visits for conferences, was the output from 15 pages. In this case, Kyushu University had the shortest character distance. No answer was provided when using the address method (see Section 3.1) because the Web pages had no address information.

Fig. 1 is an example interface. Icons indicating each person on a map and a list of location labels are displayed. Asako Miura 3 becomes Asako Miura B because person clusters with no location information assigned are not displayed on a map. When a user selects an icon or a list, information about the person is displayed with the location label: *Arise Campus, Kobe Gakuin University*. The user can display the original Web page that includes Kobe Gakuin University (cached) and the search results.

3   Evaluation

3.1   Method

Dataset. The twenty person names used in related work [6] were selected as queries. 100 HTML files were obtained for all 20 queries from Web searches [5]. Two subjects manually classified these Web pages into different people. 151 people were found in all 100 Web pages. For 13 person names, different people existed, and for seven person names, only one person existed.

The subjects extracted the location information from the person clusters by checking all Web pages. For 79 out of 151 people (52.3%), location information existed, and there was no appropriate location information for 72 people. The average number of Web pages for the former 79 people was 21.3 (SD=35.8) and 4.1 for the latter 72 (SD=14.6).

The evaluation measurement was as follows. First, current places (offices or homes) were judged as appropriate location information. When there was no current place, past places (offices or homes) were treated as appropriate location information. Places the person temporarily visited do not become location information. An address type and a landmark type are equally satisfactory.

Comparative methods. To evaluate the usefulness of search engine rankings and to determine the number of Web pages, we examined the number of Web pages (the top, top five, 10, 15, and 100). We call these methods: (a) top page, (b) top five pages, (c) top 10 pages (our method), (d) top 15 pages, and (e) top 100 pages.

To evaluate the usefulness of a combination of search engine rankings, landmarks, and character distance, we examined a method that only extracts addresses and a method that uses frequency to calculate location information. We used the top 10 pages because the former evaluation revealed that 10 pages was the best. We call the former (f) the address method (top 10 pages and only extracting addressess) and the latter (g) the frequency method (top 10 pages and using frequency).

Evaluation. Two subjects evaluated the assigned location information by the above methods. Precision, recall, F-measure, and total were calculated as follows:

$$precision = \frac{\text{number of people whose assigned LI is correct}}{\text{number of people whose LI is assigned}}$$

$$recall = \frac{\text{number of people whose assigned LI is correct}}{\text{number of people whose LI is included in Web pages}}$$

$$F - measure = \frac{2 \times precision \times recall}{precision + recall}$$

$$total = \frac{\text{number of correct answers}}{\text{number of people}}$$

With Local search, an assigned address and an address included in an original Web page sometimes differ. For example, when *3-3-138, Sugimoto-cho, Sumiyoshi-ku, Osaka-shi* is included in a Web page and Local search returns *Sugimoto 3-Chome, Sumiyoshi-ku, Osaka-shi*, we judged this correct. Another such example is *Fukuoka City*, which is included in a Web page; Local search returns an address of *city hall of Fukuoka City*.

3.2   Results and Discussion

Effect of search engine ranking. Fig. 3 shows precision, recall, F-measure, and total when the number of Web pages varies. Top 10 pages (our method) performed best in recall, F-measure, and total. The top page result was particularly inferior in recall due to the lack of information from only one page. The result of top 100 pages was also inferior on average for the following reasons: (a) location candidates included in lower ranked Web pages tend to be unrelated to designated people, and (b) the possibility of incorrect answers increases as the number of pages increases. The above results show the usefulness of our method to use top 10 Web pages as sources to extract location information.

Effect of landmarks and character distance. Fig. 4 compares other methods when the Web pages are in the top 10. Our method greatly outperforms the other methods in all evaluations. Comparison with the address method shows our method's usefulness using landmarks to extract useful information to assign location information. Comparison with the frequency method shows our method's usefulness to use character distance for person names rather than frequency.

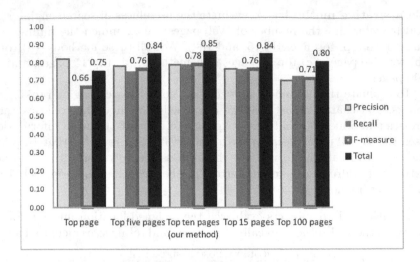

Fig. 3. Effect of search engine rankings

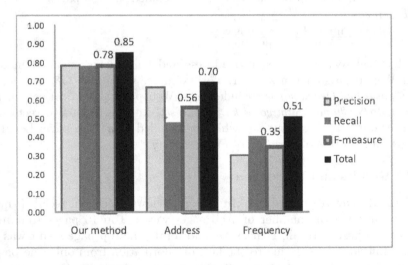

Fig. 4. Comparison with other methods

Answer analysis. We analyze the answers in detail. Out of 62 correct answers for assigned location information, 40 (64.5%) were landmark types and 22 (35.5%) were address types. This result suggests the usefulness of using landmarks. For 40 landmarks, 19 (47.5%) were universities, 6 (15.0%) were prefectural governments, 6 (15.0%) were train stations, 3 (7.5%) were high schools, and 6 (15.0%) were other. Most were public institutions. For people with answers, 12 were incorrect for assigned location information. 91.2% (11/12) chose incorrect candidates by character distance, 8.3% (1/12) chose incorrect results from Local search by calculating similarity. Inside the former 91.2% error, target candidates were extracted, but other candidates were chosen by character distance (91.0%, 10/11) and target candidates could not be extracted (9.0%, 1/11). A typical extraction error was that publishers were assigned because people with many pages often wrote books or were discussed in books. Overall, the main reason for errors included problems of (a) character distance and (b) extracting landmarks.

4   Related Work and Discussion

Wan et al. [7] also separated Web people search results and assigned titles to person clusters. We assign location information to display person icons on a map. Even though much work (e.g., [8], [9], [10]) separates Web pages into person clusters, it seldom assigns labels to person clusters.

The WWW9 WePS-2 workshop [11] evaluated a technique to extract attribute information. This task extracts 18 kinds of attribute values for target individuals whose names appear on each of the provided Web pages. However location is not included in the evaluation due to its ambiguity.

Our work is related to such clustering search engines as Clusty [12] which usually assigns keywords or phrases to a Web page cluster to help users select a cluster based on such information as term frequency and URLs.

Google's alternative to search results [13] displays multiple location information included in Web search results as an experimental service. It does not display one piece of representative location information. It mainly extracts city names and sometimes college names. We extract addresses (including city names) and landmarks (including college names).

Most existing systems and research extract multiple addresses contained in Web pages. Few systems extract landmarks. [14] extracted addresses, postal codes, and phone numbers contained in Web pages and converts them into coordinates. It does not extract landmarks, and we do not extract phone numbers. The discrepancy reflects different system aims. [14] gathered as much location information as possible, but we aim to output one good piece of location information for many people.

Morimoto et al. [15] extracted address pairs and descriptive text from Web pages and displays them on a map using a HTML structure. Much research utilizes HTML structure to limit the range of documents to judge the relationship between an object and an address. Instead of using HTML structure, we exploit character distance to judge the strength of the relationships. Our heuristics for extracting addresses are related to [16] because they use a Japanese morphological analysis and utilize location attached to nouns. [16] does not extract landmarks.

Many Web sites provide a facility that converts addresses or landmarks into coordinates with geocoding services including a Local search using simple pattern-matching. They return the first result (coordinates) based on an exact match. For example, when a user inputs *Hanshidai* (abbreviation of Osaka City University) it may fail because no Hanshidai exists in the geocoding databases. Our method copes with pattern-matching problems in three steps. First, a formal name inference heuristic (landmark type) and a one-character deleting heuristic (address type) are applied. Second, we use a candidate list to try another location candidate to get a result. Third, we apply a calculating context heuristic to identify one piece of location information from multiple results.

One advantage of our method is that no special location dictionary is needed because such existing tools and services as morphological analysis (MeCab), Yahoo! local search, and web search APIs are combined. A formal name inference heuristic, a one-character deleting heuristic, and a calculating context heuristic are powerful enough to utilize Local search to get one result. Our algorithm is simple and easy to implement. Although our algorithms are heuristic based, we do not need to modify or add new heuristics; therefore systems are easy to maintain.

The experimental results revealed that our method has the best recall, F-measure, and total performance among comparative methods. The above results suggest our method's usefulness. In addition, we successfully built a prototype interface to select a person cluster on a map. Our method is only inferior in precision to the top page method. If we need to build an interface with fewer incorrect icons (regardless how few), the top page method can be selected.

The limitations of our evaluation include the following. Since our evaluation is limited to Japanese, the proposed heuristics should be adjusted to other languages. We extracted addresses and landmarks, except for those addresses and landmarks that are not contained in the MeCab dictionary and the Local search database.

Future work includes the following. First, we need to improve the extraction of location candidates. For example, we should remove publishers by using a stop list. Second, new algorithms should be investigated to cope with errors caused by character distances. One possible solution may be to combine character distances with term-frequency, HTML structure, and/or syntactic analysis.

5   Conclusions

We proposed a method to assign person clusters with one piece of *representative* location information to display person icons on a map. The following are the main ideas of our method: (1) using landmarks as well as addresses to extract location information, (2) using search engine rankings and character distance to assign suitable location information, and (3) using Local search to convert texts included in Web pages to location information.

The experimental results revealed that our method has the best recall, F-measure, and total performance among comparative methods. The above results suggest our method's usefulness. In addition, we successfully built a prototype interface to select a person cluster on a map.

References

1. Guha, R., Garg, A.: Disambiguating people in search. Stanford University (2004)
2. Okilab.jp projects, http://okilab.jp/project/location/

3. Yahoo! Local search API,
   http://developer.yahoo.co.jp/webapi/map/localsearch/
   v1/localsearch.html
4. MeCab, http://mecab.sourceforge.net/
5. Yahoo! Web search API,
   http://developer.yahoo.co.jp/webapi/search/websearch/v1/websearch.html
6. Sato, S., Kazama, K., Fukuda, K.: Distinguishing between People on the Web
   with the Same First and Last Name by Real-world Oriented Web Mining. IPSJ
   Transactions on Databases 46(8), 26–36 (2005)
7. Wan, X., Gao, J., Li, M., Ding, G.: Person Resolution in Person Search Results:
   WebHawk. In: CIKM 2005. Proceedings of the Fourteenth ACM Conference on
   Information and Knowledge Management, pp. 163–170 (2005)
8. Bekkerman, R., McCallum, A.: Disambiguating Web Appearances of People in a
   Social Network. In: WWW 2005, Proceedings of the Fourteenth World Wide Web
   Conference, pp. 463–470 (2005)
9. Kozareva, Z., Moraliyski, R., Dias, G.: Web People Search with Domain Ranking,
   Text, Speech, and Dialogue. In: LNCS, pp. 133–140. Springer, Heidelberg (2008)
10. Artiles, J., Gonzalo, J., Sekine, S.: The SemEval-2007 WePS Evaluation: Establish-
   ing a Benchmark for the Web People Search Task. In: Proceedings of the Fourth
   International Workshop on Semantic Evaluations, pp. 64–69 (2007)
11. Task Definition of Attribute Extraction Subtask for WePS-2,
   http://nlp.uned.es/weps/weps2/WePS2_Attribute_Extraction.pdf
12. Clusty the clustering search engine, http://clusty.com/
13. Google Experimental Search, http://www.google.com/experimental/
14. McCurley, K.S.: Geospatial Mapping and Navigation on the Web. In: WWW 2001,
   pp. 221–229 (2001)
15. Morimoto, H., Fujimoto, N., Nagaya, T., Idehara, H., Hagihara, K.: A System for
   Web Retrieval of Address-Related Information. IEICE Trans. D J90-D(2), 245–256
   (2007)
16. Arai, I., Kawaguchi, Y., Fujikawa, K., Sunahara, H.: Geocrawler; Web Indexer for
   Store Search based on Geographical Information and Evaluation Information on
   Personal Web Sites. IPSJ Journal 48(7), 2319–2327 (2007)

Web Spam Identification with User Browsing Graph*

Huijia Yu, Yiqun Liu, Min Zhang, Liyun Ru, and Shaoping Ma

State Key Lab of Intelligent Technology and Systems
Tsinghua National Laboratory for Information Science and Technology
Department of Computer Science and Technology
Tsinghua University, Beijing, 100084, China P.R.
huijiayu@gmail.com

Abstract. Combating Web spam has become one of the top challenges for Web search engines. Most previous researches in link-based Web spam identification focus on exploiting hyperlink graphs and corresponding user-behavior models. However, the fact that hyperlinks can be easily added and removed by Web spammers makes hyperlink graph unreliable. We construct a user browsing graph based on users' Web access log and adopt link analysis algorithms on this graph to identify Web spam pages. The constructed graph is much smaller than the original Web Graph, and link analysis algorithms can perform efficiently on them. Comparative experimental results also show that algorithms performed on the constructed graph outperforms those on the original graph.

Keywords: Spam identification, TrustRank, User browsing graph.

1 Introduction

According to a study in 2009 [1], 68% of Web users use search engines frequently and 84.5% regard search engines as a dominant way to discover built-up Web sites. Although a search engine usually returns thousands of results for a certain query, most search engine users only view the first few pages in result lists according to [2]. As a consequence, ranking position has become a major concern of internet service providers. Hence Web spam pages use various techniques to achieve higher-than-deserved rankings in search engines' results, from which search engines and users suffers a lot.

State-of-the-art anti-spam techniques usually make use of Web page features, either content-based or hyper-link structure based, to identify spam pages. Currently, hyper-link-based anti-spam methods are based on hyperlink graph analysis. Well-known link-based detection algorithms include TrustRank [3], SpamRank[4], and so on. Generally, hyperlink analysis algorithms are based on two basic assumptions proposed by [5]: recommendation assumption and topic locality assumption. It assumes that if two pages are connected via a hyperlink, the linked page is recommended by the linking page (i.e., recommendation) and the two pages share a similar topic (i.e., topic locality).

* Supported by the Chinese National Key Foundation Research & Development Plan (2004CB318108), Natural Science Foundation (60621062, 60503064, 60736044) and National 863 High Technology Project (2006AA01Z141).

G.G. Lee et al. (Eds.): AIRS 2009, LNCS 5839, pp. 38–49, 2009.

However, in practical Web environment, hyperlinks can be easily added or removed by Web page authors or even by users (for Web2.0 sites). As a consequence, Web is filled with spam and advertising links so the assumptions as well as the hyperlink analysis algorithms meet lots of troubles in current Web environment. These entire situations make the link graph unreliable data source for hyperlink analysis algorithms. In order to solve this problem, Liu et al. [6] constructed a 'user browsing graph' with user's Web access log data. It is believed that link structure in user browsing graph is more reliable than hyperlink graph because users actually follow links in the browsing graph.

In this paper, we study the effectiveness of user browsing graph on spam filtering. The results are compared with whole hyperlink graph, and an improved hyperlink graph is proposed, which combines user browsing information and hyperlink information. Comparative experimental results show that the improved hyperlink graph with user behavior data improves spam classification performance by 5% compared to the same classifier which uses the whole hyperlink graph. What's more, the improved hyperlink graph contains much less vertices than the whole hyperlink graph, which will increase the efficiency of spam classifier.

The rest of the paper is organized as follows. In Section 2, we review related work in spam detection. Section 3 introduces how to construct user browsing graph and combined graph. Section 4 describes the calculation of TrustRank on user browsing graph and the distribution of the TrustRank values. Section 5 compares the performance of the TrustRank algorithm on several different graphs in spam detection. Conclusions and future work are given in Section 6.

2 Related Work

Most spam detection approaches utilize link and/or content information to help detect specific spam pages, just like evaluating page qualities with link and/or content information [12]. Gyöngyi et al. [3] propose a algorithm called TrustRank for link spamming detection. Wu et al. [9] use trust and distrust propagation through web links to help demote spam pages in search results. Ntoulas et al. [10] detect spam pages by building up a classification model which combines multiple heuristics based on page content analysis.

Recently, the wisdom of the crowd is paid much attention in Web search researches, e.g. [6, 7, 8]. For instance, Liu et al. [6] constructed 'user browsing graph' with Web access log data. They proposed a page importance estimation algorithm called BrowseRank which performs on the user browsing graph. It is believed that the link structure in user browsing graph is more reliable than hyperlink graph because users actually follow links in the browsing graph. With user behavior analyses into Web access logs, Liu et al. [8] propose a spam page detection algorithm based on Bayesian Learning, which can detect newly-appeared and various kinds of spam.

Most linking based spam detection algorithms are based on hyperlink graph. However, with the explosive growth of Web pages, the whole hyperlink graph is too large to be analyzed and contains lots of noise. In this work, we compare performance of linking analysis algorithm on different graph using the approach in [8].

3 Construction of the User Browsing Graph

3.1 Web Access Log Data Set

As search engines have developed, Web-browser tool bars have become more and more popular. Most toolbar servers collect anonymous click-through data from users' browsing behaviors, which is used in extensive studies of Web search by more and more researchers, e.g., [7] adopts click-through data to improve ranking performance, and [8] utilizes it to propose a Web spam identification algorithm. In this paper, we also make use of the data collected by search toolbars to construct user browsing graph, because of its low cost and no-interruption to users. Information shown in Table 1 can be recorded using browser toolbars by commercial search engine systems.

Table 1. Information recorded in Web access logs

Name	Description
Session ID	A randomly assigned ID for each user session
Source URL	URL of the page which the user is visiting
Destination URL	URL of the page which the user navigates to
Residence time	Residence time of the source page (in seconds)

3.2 User Browsing Graph Construction

$UG(V,E)$ is used to denote the browsing graph, in which V is the vertex set and E is the edge set. The construction process is described as Figure 1:

1. $V = \{\}, E = \{\}$
2. For each record in the Web access log, if the source URL is A and the destination URL is B, then
   if $A \notin V, V = V \cup \{A\}$;
   if $B \notin V, V = V \cup \{B\}$;
   if $(A,B) \notin E$
      $E = E \cup \{(A,B)\}, Weight\,(A,B) = 1$;
   else
      $Weight\,(A,B) + +$;

Fig. 1. Construction process of user browsing graph $UG(V,E)$

After the construction process, V includes all Web pages visited by users during the period when access logs were collected; and E records the users' browsing behaviors. Each edge in E is also assigned a weight that represents how many times Web users visited site B from site A.

With the help of a widely-used commercial Chinese search engine, Web access logs were collected from Aug.3rd, 2008 to Sep. 2nd, 2008 (30 days). Over 1.4 billion click-through events on 4.2 million Web sites were recorded in these logs. In our experiments, a site-level *UG (V, E)* was constructed with *V* containing 4,252,495 Web sites and *E* containing 10,564,205 edges. We constructed a site-level *UG (V, E)* because the site-level graph is more stable and it can avoid the problem of sparse data.

In order to show the advantages of user browsing graphs compared with over hyperlink graphs, we also construct a hyperlink graph for comparison. With the help of the same search engine which collected Web access log for us, we obtained hyperlink graph constructed by the log data which contains hyperlink relations of over 3 billion Web pages. Then we extracted a sub-graph for all the Web sites in *V* of *UG (V, E)*. The sub-graph is the hyperlink graph for these Web sites and we call this graph *extracted-HG(V,E)*. This graph was also constructed as a site-level graph to compare with the user browsing graph and contains 139,125,250 edges.

4 Trustrank on User Browsing Graph

4.1 Trustrank Algorithm

In 2004, Gyöngyi *et al.* [3] proposed TrustRank algorithm to semi-automatically separate reputable from spam pages. According to [3], TrustRank relies on an important empirical observation called approximate isolation of the good set: good pages seldom point to bad ones. In other words, TrustRank assumes that if page A pointing to page B then there exists recommending site B by A, so it's necessary to ensure sufficient reliabilities of the link relationship, which TrustRank computes on.

4.2 Seeds Set Selection

At the very beginning of TrustRank algorithm, we need to select a seed set. The purpose of seed selection is to identify desirable pages for the seed set. Pages in the seed set should be useful in identifying additional good pages. Two strategies for seed set selection were described in [3]. According to these two strategies, we defined the following seed set selection criteria for TrustRank on user browsing graph.

- Out-degree of a seed site should be neither too high nor too low, i.e., higher than P1 or lower than P2 (P1 < P2). Because it's difficult to ensure the quality of the out-linking if there are too many links coming out of site A. Seed site should also link to a relatively large number of other sites so that trust propagation can be finished as soon as possible.
- The PageRank value of the seed site should be higher than P3 to ensure the authority of the site.
- Search Engine cannot be seeds because links out from SE don't necessarily denote so-called recommendation.
- The seed set should not contain blog, forum sites and Web 2.0 sites. Because hyperlinks on these sites can be easily added or removed by Web users.

With these criterions, seed sites selected are all high-qualified, credible and popular Web sites in the link graph. These Web sites also contain appropriate amount of reliable out-links. With this method, we select altogether 1,153 Web sites to form the seed set.

4.3 Trust Score Calculation

After seed set construction, we can perform TrustRank algorithm on user browsing graph constructed in Section 3.2. We set parameter MB = 20 (iteration times), α = 0.85 (decay factor) in the algorithm. Then TrustRank values are calculated for sites both in user browsing graph $UG(V,E)$ and hyperlink graph $extracted\text{-}HG(V,E)$, respectively. In $UG(V,E)$, we can get 3,951,485 sites' TrustRank values, but only 2,658,345 sites' value can be obtained by algorithm on $extracted\text{-}HG(V,E)$. There are fewer sites which cannot be computed in $UG(V,E)$ than in $HG(V,E)$.

Some sites are not assigned a TrustRank value because these sites are isolated in the graph and hence aren't either directly or indirectly linked by the seed sites. We sampled 15,941 out of 1,594,150 (1%) non-valued sites in the $extracted\text{-}HG(V,E)$ and checked them manually finding that most of them were in low-quality, spam or non-GBK encoded sites.

4.4 Distribution of TrustRank Scores

Distributions of TrustRank scores obtained from $UG(V,E)$ and $extracted\text{-}HG(V,E)$ are shown in Figure 2.(a) and 1.(b). Because TrustRank values are little decimal fraction, we computed the value $LT = -Log_{10} (TrustRank(S))$ (S is a Web site) for all sites in the two graph.

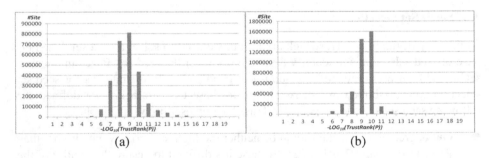

(a)   (b)

Fig. 2. Distributions of TrustRank scores on different graphs. (a): user browsing graph $UG(V,E)$. (b): hyperlink graph $extracted\text{-}HG(V,E)$.

From Figure 2, we find that most sites' TrustRank scores are between 10^{-6} and 10^{-12} in both graphs, but there are also some differences between these two distributions. For instance, LT values for most Web sites in $UG(V,E)$ are between 8 to 10 while a number of sites in $extracted\text{-}HG(V,E)$ get LT scores less than 8 or larger than 10. The average TrustRank score of $UG(V,E)$ is also smaller than that of $extracted\text{-}HG(V,E)$. These can be explained by the fact that links on user browsing graph are sparser than hyperlink graph.

5   Spam Identification Performance

5.1   Experiment Settings

From [6] we can see that specially-designed link analysis algorithm (called 'BrowseRank') could get better performance on user browsing graphs in page quality estimation than state-of-the-art link analysis algorithms which perform on hyperlink graphs. We want to find out whether this improvement comes from different algorithm design or the different graph structures. To answer this question, we build four graphs and compare how TrustRank algorithm performs on them. The details of the four graphs are shown in Table 2.

Table 2. User browsing graphs and hyperlink graphs constructed in our experiments

Graph	TrustRank
User Browsing Graph *UG(V,E)*	Constructed with Web access data from Aug.3rd, 2008 to Sept.2nd, 2008.
Hyperlink Graph *whole-HG(V,E)*	Constructed with over 3 billion pages (all pages in a certain search engine's index) and all hyperlinks among them
Hyperlink Graph *extracted-HG(V,E)*	Vertexes are from *UG(V,E)*. Edges among them are extracted from hyperlink relations in *whole-HG(V,E)*.
Combined Graph *CG(V,E)*	Vertexes are from *UG(V,E)*. Edges among them are from *UG(V,E)* combined with those from *extracted-HG(V,E)*.

After performing TrustRank algorithm on these graphs, we adopted two methods to evaluate the performance of Web spam identification. The primary one is based on ROC/AUC metric which is often used to evaluate performance of machine learning algorithms and classifiers, *e.g.* [11]. The other one is based on analysis into the distribution of TrustRank scores for Web spam sites on different graphs.

Table 3. Tags annotated in the page quality test sets

Tag	Description
High quality (HQ)	High quality pages
Low quality (LQ)	Low quality pages which contain few meaningful contents.
Web spam (SP)	Web spam pages which try to cheat search engine to get higher than deserved ranking in search result lists.
Non-GBK encoded (NG)	Pages which are written with encodes other than GBK (the most popular Chinese encode format)
Illegal (IL)	Pages which contain porn-related, gambling related issues or other illegal contents.

In order to evaluate the performance of Web spam detection, we construct three different test sets. Web sites in these sets were randomly selected from users' Web access logs at different time points and annotated with the tags shown in Table 3 by 3 product managers from a search engine company. The amount of Web sites in these three test sets are 646, 1000, and 1000 respectively. In the annotated results, about

39% are "high quality"; 19% are "Web spam" or "Illegal" and the others are annotated as "low quality", "Non-GBK encoded". As for the spam detection evaluation process, illegal sites are also regarded as spam sites because illegal sites usually adopt SEO or spamming techniques to improve their rankings in search result lists.

5.2 ROC/AUC Test

ROC/AUC metric is a method which is commonly used to evaluate performance of classifiers and it is adopted by many previous spam detection researches such as Web spam challenge workshop (http://webspam.lip6.fr/). With the Web spam annotation test sets proposed in Section 5.1, we chose ROC (Receiver Operating Characteristic) curves and corresponding AUC (Area under the ROC Curve) values to evaluate the performance of the TrustRank algorithm. The AUC has an important statistical property: the AUC value of a classifier is equivalent to the probability that the classifier will rank a randomly chosen positive instance higher than a randomly chosen negative instance.

Table 4 shows the spam page identification performances of BrowseRank algorithm [6] on *UG(V,E)* and TrustRank algorithms on 4 different graphs which we constructed in Table 2 as well as the BrowseRank algorithm on *UG(V,E)*.

Table 4. AUC/ROC values for Web spam page identification of TrustRank/BrowseRank algorithms

Graph	Test 1	Test 2	Test 3
BrowseRank on UG(V,E)	0.5806	0.7011	0.7131
TrustRank on UG(V,E)	0.59632	0.7406	0.7440
TrustRank on whole-HG(V,E)	0.66905	0.7934	0.7712
TrustRank on extracted-HG(V,E)	**0.69769**	**0.8209**	**0.8086**
TrustRank on CG(V,E)	0.68072	0.7376	0.7435

From Table 4 we can find several interesting results. Firstly, we can see that in all test sets TrustRank algorithms perform the best on *extracted-HG(V,E)* in the task of spam page identification. A possible reason is that, although *Extracted-HG (V,E)* shares a same vertex set of *UG(V, E)*, the edges among the vertexes inherit from the whole hyperlink graph, which are denser and more complete than *UG(V,E)*. So *extracted-HG(V,E)* can get better performance than *UG(V,E)* for TrustRank algorithm to identify spam sites. What's more, there is another result to support this explanation. *CG(V,E)*, a combination of *UG(V,E)* and *extracted-HG(V,E)*, which contains all edges in them, results better than *UG(V,E)*. It means both user browsing graphs and hyperlink graph can provide useful information that the other graph does not contain.

Secondly, in each of the three test sets, we found that BrowseRank doesn't perform as well as TrustRank on user browsing graph for the task of Web spam identification. Liu et al proposed in [6] that BrowseRank performs better than TrustRank and PageRank in spam fighting. However, according to our experimental results, TrustRank out-performs BrowseRank when both algorithms are performed on user browsing graph. It means traditional link analysis algorithms can perform as well as or even better than specially-designed algorithms on user browsing graphs.

Moreover, *Extracted-HG(V,E)*, as a sub-graph of *whole-HG(V,E)* (*extracted-HG(V,E)*) is better at identifying spam pages than the whole graph. This may be explained by the fact that *extracted-HG(V,E)* can be regarded as the user-accessed part of *Whole-HG(V,E)*. It shares the same vertex set of *UG(V,E)* and therefore reduces possible noises (i.e., unreliable Web sites) in the whole hyperlink graph.

5.3 Distribution of TrustRank Scores for Web Spam Sites

In order to compare spam detection performance of TrustRank algorithm on different graphs, we examined the distribution of spam sites' TrustRank scores on both graphs. As shown in Section 4.4, we computed the value $LT = -Log_{10} (TrustRank(S))$ (S is a Web site.) for Web sites in both graphs.

We can see that a low *LT* score (namely a high TrustRank score) means good site and a high *LT* score means spam. When we look into the Web sites with lowest and highest *LT* scores for the user browsing graph, we find that page quality accords with the TrustRank scores. Experimental results are shown in Table 5. Quality score of each Web site is annotated and examined by 2 different assessors, respectively. If 2 assessors give different annotation, one other assessor will join in to vote for net result.

Table 5. Quality estimation results for 100 Web sites with the highest/lowest TrustRank scores (calculated on *UG(V,E)*)

	Spam/Illegal	Low-quality	Non-GBK	High-quality
100 Web Sites with the highest TrustRank score	0	3	2	95
100 Web Sites with the lowest TrustRank score	10	79	9	2

From Table 5 we can see that most sites with highest TrustRank scores are high-quality ones. Meanwhile, most sites with lowest TrustRank scores are low-quality or spam ones. It means that TrustRank scores can represent the quality of Web sites and this validates the conclusion of [3] that TrustRank algorithm is effective in the detection of Web spam.

We further looked into the distribution of TrustRank scores while the algorithm was performed on different graphs. Figure 4 provides a side-by-side comparison of TrustRank on different graphs w.r.t ratio of bad sites in each bucket. In order to get a suitable sample set with enough data points, we adopted the sampling method proposed by Gyöngyi *et al.* in [3].

Firstly, We generated the list of sites in decreasing order of their TrustRank scores on user browsing graph *UG (V, E)*, and we segmented it into 20 buckets. Each of the buckets contained a different number of sites, with scores summing up to 5 percent of the total TrustRank score. Therefore, the first bucket contained 53 sites with the highest TrustRank scores, bucket 2 contains the next 126 sites, while the 20th bucket contains 3.9 million sites that are assigned the lowest TrustRank scores.

After that, we constructed a sample set of 1000 sites by selecting 50 sites randomly from each bucket. We manually examined the sampled Web sites and annotated whether they are spam or not. As for TrustRank scores obtained from *Whole-HG (V, E)* and *Extracted-HG (V, E)*, we defined their TrustRank buckets as containing the same number of sites as buckets on *UG (V, E)*. The vertical axis of the graph corresponds to the percentage of spam, low quality and illegal sites within a specific bucket. For instance, we can derive from Figure 3(a) that on the graph *Whole-HG (V, E)*, 30% of the Web sites in TrustRank bucket 9 are not high quality ones.

From Figure 3 (a)-(c) we see that TrustRank algorithm can perform better on *UG(V,E)* and *Extracted-HG(V,E)* than *Whole-HG (V, E)* graph. In particular, we note that there are all high quality Web sites in the top 6 buckets of *UG(V,E)* and *Extracted-HG(V,E)*, while there is a marked increase in spam concentration in the lower buckets. At the same time, it is surprising that almost 18% of the second *Whole-HG (V, E)*'s bucket is Web spam or low quality ones. It can be explained by the fact that hyperlinks in *Whole-HG(V,E)* are unreliable and the trust propagation process is therefore not reliable, either.

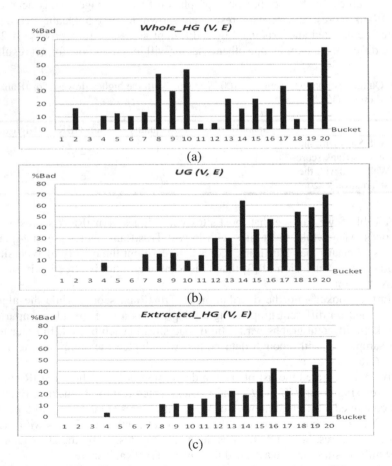

Fig. 3. Distribution of TrustRank scores for low-quality and spam sites on different graphs. (a): *whole-HG(V,E)*. (b): *UG(V,E)*. (c): *extracted-HG(V,E)*.

5.4   Performance of Spam Identification on Filtered User Browsing Graphs

According to the construction process of the user browsing graph given in section 3.2, there is an edge between sites A and B if there exists a certain user who visited B from A. The weight of this edge represents how many times Web users visit B from A. It is usually believed that more visit times mean more reliable recommendation.

For a given threshold N, we reduce the edges in the user browsing graph whose weights are no more than N. By this means, we get a graph $UG\_FN(V,E)$.

We calculated TrustRank scores for sites in $UG\_F1(V, E)$ and $UG\_F3(V, E)$ respectively, with the same parameter MB = 20 (iteration times), $\alpha = 0.85$ (decay factor) in algorithm. In $UG\_F1(V,E)$, we got 1,586,142 sites' TrustRank values, nevertheless only 774,029 sites' value can be obtained by algorithm on $HG\_F3 (V,E)$. Compared to 3,951,485 sites' TrustRank values in $UG (V, E)$, there are more than half sites which cannot be computed in the $UG\_F1 (V, E)$ than in unfiltered user browsing graph.

In order to find out whether filtration can improve reliability of user browsing graphs, we compared how TrustRank algorithm performed in the task of spam identification on the three graphs: $UG (V,E)$, $UG\_F1 (V,E)$ and $UG\_F3 (V,E)$.

Firstly, we adopted the approach proposed by Liu Y. et al. [8] as the baseline, which identified Web spam with user behavior features extracted from Web access logs. Second, we calculated TrustRank scores on three Graphs and syncretize them to the results of baseline approach. Test set includes 400 sites and we adopted the third set which contained 1000 sites described in section 5.1 as training set. Table 6 shows the accuracy of being spam for the top-ranked Web sites in the P(Spam) list on filtered graphs.

Table 6. Accuracy rate for Web spam page identification performance of TrustRank algorithms on filtered graphs

Graph	Spam Accuracy of top-ranked Web sites
$UG(V,E)$	0.8100
$UG\_F1(V,E)$	0.9133
$UG\_F3(V,E)$	0.9233

The results in Table 6 show that the accuracy rate increases with the rising threshold N. This means TrustRank algorithm performs better on filtered user browsing graphs, which proves "more visit times, more reliable recommendation". What's more, filtered graph is much smaller than source graph, which means lower cost for time and storage. With the swift growth of Web, we believe that filtered user browsing graphs can be helpful in improving the efficiency of spam identification.

We also want to find out whether the increase of threshold N results in failing to calculate some Web sites' TrustRank scores. By merging TrustRank and user behavior features with learning algorithm proposed in [8], we calculated sites' P(Spam) scores, representing probabilities of being Web spam. Then we sorted the Web sites in decreasing order of their P(Spam) scores. Table 7 shows how many top-ranked sites' TrustRank scores can be calculated in $UG\text{-}F1(V,E)$ but not in $UG\text{-}F3(V, E)$.

Table 7. Influence of threshold value N's increasing to Web spam page identification performance of TrustRank algorithms on filtered graphs

Amount of top-ranked Web sites according to P(Spam)	Number of sites with value in *UG-F1(V,E)* but not in *UG-F3(V, E)*
100	1
1000	1
2000	4
5000	8
10000	123

From Table 7 we can see that in the 5,000 sites at the top of the possible spam list, there are only 8 sites whose TrustRank scores can not be calculated on *UG-F3(V,E)* graph. It means that although we filtered some low weight edges from the user browsing graph, it does not influence the performance of spam detection a lot.

6 Conclusion

Current search engines are seriously threatened by malicious Web spam that attempts to obtain unbiased ranking in search result lists. State-of-the-art hyperlink-based anti-spam techniques do not work well on whole hyperlink graph because practical Web is filled with low quality and even spam hyperlinks due to the lack of editorial process in Web contents.

In this paper, we focused on spam detection techniques with user browsing graphs. Firstly, we constructed a user browsing graph with Web access log data which contained huge amount of user click-through information. Secondly, we adopted the approach proposed by Liu Y. et al. [8] as the baseline, which identified Web spam with user behavior features extracted from Web access logs, and calculated the TrustRank scores on different Graphs and syncretize them to the results of baseline approach. Then we compared the spam identification performances of above approach with TrustRank algorithm on different graphs: *UG(V,E)*, *CG(V,E)*, *extracted-HG(V,E)* and *whole-HG(V,E)*.

Experimental results show that, although vertices of the user browsing graph is much less than whole hyperlink graph, which means higher efficiency, the improved hyperlink graph with user access information (*extracted-HG(V,E)*) performs better than other graphs. From this result, we see that both of user browsing graph and hyperlink graph can provide useful information. And combining the vertex set of user browsing graph with edge set of hyperlink graph can improve performance in task of spam detection. What's more, TrustRank also outperforms BrowseRank algorithm while both algorithms were performed on user browsing graph in our results. We believe that the user browsing graph is useful for link-based spam detection algorithms such as TrustRank.

There are still several technical issues which need to be addressed as future work:

(1) There is noise in Web access log data, such as the clicks proposed by automated software programs (bots). Identification of this kind of access logs will improve the

reliability of the user browsing graph and corresponding spam detection performance. In the future, we will try to perform this kind of data cleansing before the user browsing graph construction process.

(2) According to our experimental results, $UG(V,E)$ and $Extracted\text{-}HG(V,E)$ achieve better detection performance than $whole\text{-}HG(V,E)$. What's more, sometimes $Extracted\text{-}HG(V,E)$, with $UG(V,E)$'s vertexes and $whole\text{-}HG(V,E)$'s edges, has the best effectiveness out of all graphs. It will be interesting to look into how to yield the greatest returns on investment of user browsing graph.

References

1. CNNIC (China Internet Network Information Center), the 23th report in development of Internet in China,
   http://www.cnnic.net.cn/uploadfiles/pdf/2009/1/13/92458.pdf
2. Silverstein, C., Marais, H., Henzinger, M., Moricz, M.: Analysis of a very large Web search engine query log. In: Proceedings of the 22nd Annual International ACM SIGIR Conference on Research and Development in Information Retrieval, pp. 6–12. ACM Press, California (1999)
3. Gyöngyi, Z., Garcia-Molina, H., Pedersen, J.: Combating Web spam with TrustRank. In: Proceedings of the 30th VLDB Conference, pp. 576–587. ACM Press, Toronto (2004)
4. Benczúr, A.A., Csalogány, K., Sarlós, T., et al.: SpamRank-Fully Automatic Link Spam Detection Work in progress. In: 1st international Workshop on Adversarial information Retrieval on the Web, Chiba (2005),
   http://airweb.cse.lehigh.edu/2005/benczur.pdf
5. Craswell, N., Hawking, D., Robertson, S.: Effective site finding using link anchor information. In: Proceedings of the 24th SIGIR Conference, pp. 250–257. ACM Press, New Orleans (2001)
6. Liu, Y., Gao, B., Liu, T., Zhang, Y., Ma, Z., He, S., Li, H.: BrowseRank: letting Web users vote for page importance. In: Proceedings of the 31st SIGIR Conference, pp. 451–458. ACM Press, Singapore (2008)
7. Bilenko, M., White, R.W.: Mining the search trails of surfing crowds: identifying relevant Websites from user activity. In: Proceeding of the 17th WWW Conference, pp. 51–60. ACM Press, Beijing (2008)
8. Liu, Y., Cen, R., Zhang, M., Ma, S., Ru, L.: Identifying Web spam with user behavior analysis. In: 4th international Workshop on Adversarial information Retrieval on the Web, pp. 9–16. ACM Press, Beijing (2008)
9. Wu, B., Goel, V., Davison, B.D.: Topical TrustRank: Using topicality to combat web spam. In: Proceedings of the 15th WWW Conference, pp. 63–72. ACM Press, Scotland (2006)
10. Ntoulas, A., Najork, M., Manasse, M., Fetterly, D.: Detecting spam web pages through content analysis. In: Proceedings of the 15th WWW Conference, pp. 83–92. ACM Press, Scotland (2006)
11. Svore, K., Wu, Q., Burges, C., Raman, A.: Improving Web Spam Classification using Rank-time Features. In: Proceedings of AIRWeb 2007, pp. 9–16. ACM Press, New York (2007)
12. Liu, Y., Zhang, M., Ma, S.: Web key resource page selection based on non content information. J. Transactions on Intelligent System 2(1), 45–52 (2007)

Metric and Relevance Mismatch in Retrieval Evaluation

Falk Scholer and Andrew Turpin

School of Computer Science and IT, RMIT University
GPO Box 2476v, Melbourne, Australia
{falk.scholer,andrew.turpin}@rmit.edu.au

Abstract. Recent investigations of search performance have shown that, even when presented with two systems that are superior and inferior based on a Cranfield-style batch experiment, real users may perform equally well with either system. In this paper, we explore how these evaluation paradigms may be reconciled. First, we investigate the DCG@1 and P@1 metrics, and their relationship with user performance on a common web search task. Our results show that batch experiment predictions based on P@1 or DCG@1 translate directly to user search effectiveness. However, marginally relevant documents are not strongly differentiable from non-relevant documents. Therefore, when folding multiple relevance levels into a binary scale, marginally relevant documents should be grouped with non-relevant documents, rather than with highly relevant documents, as is currently done in standard IR evaluations.

We then investigate relevance mismatch, classifying users based on relevance profiles, the likelihood with which they will judge documents of different relevance levels to be useful. When relevance profiles can be estimated well, this classification scheme can offer further insight into the transferability of batch results to real user search tasks.

1 Introduction

Information retrieval (IR) experiments based on the Cranfield methodology measure system performance using a set of queries and a test collection. The queries are run over the collection using a search system, and for each document that is returned, a human judge decides whether the document is relevant to the query, or not. The overall utility of the search system is then computed using a metric that aggregates the relevance judgements for documents in ranked lists returned by the system. In this batch evaluation approach, different search systems are compared based on how well they score on such metrics. For example, many papers report IR system comparisons using the TREC document collections, topics and judgements, using Mean Average Precision (MAP) or Precision at 10 documents retrieved (P@10) as the metric [23].

An alternate way to evaluate systems is to take a group of human users and ask them to perform search tasks with different systems, comparing outcome measures such as time to complete a task, success or failure on a task, or subjective measures like user satisfaction. Previous studies [1,2,8,9,12,18,19] have

G.G. Lee et al. (Eds.): AIRS 2009, LNCS 5839, pp. 50–62, 2009.

shown that attempting to transfer results from batch experiments to real users is difficult. That is, the systems rated as superior in the batch experiments may in fact not assist users in performing their tasks more quickly or more accurately than the systems that are rated more poorly in the batch experiments.

In this paper, we explore ways in which these two experimental paradigms may be reconciled. There are many possible causes for this seeming mismatch between batch and user-based experimental outcomes. We investigate two reasons using controlled batch and user experiments.

Mismatching metrics. It is possible that the metric used in a batch experiment to show that System A is superior to System B does not reflect the user task for which these systems will be employed. For example, if a batch experiment uses the MAP metric, which contains a recall component, but the user task is solely precision based, such as finding a single answer to a simple question, then differences between systems in the batch experiment may be meaningless in the user domain. On the other hand, if the batch experiment used a metric such as Precision at one document returned (P@1) or at three documents returned (P@3), then it is perhaps more likely that the batch results would carry over into the user domain. For example, Turpin and Scholer [19] used the MAP metric to choose superior systems, but then employed those systems on a precision-based user task and found that they did not outperform the inferior systems. When they re-analysed their data to choose systems based on the P@1 metric, it suggested that users performed better with the superior system. However, the analysis of P@1 was inconclusive because of the small number of systems for the non-relevant category of this metric. Motivated by this finding, we explicitly examine possible metric mismatch by using P@1 in our batch experiments, and a precision-based user outcome measure. We also extend this analysis to incorporate multiple levels of relevance, factoring in differences between non-relevant, relevant, and highly relevant documents.

Mismatching relevance profiles. Batch system results are based on relevance judgements assigned to documents by human assessors. However, it is possible that relevance judgements used in the batch experiments are made using different criteria, or on a different scale (whether perceptual or actual), than judgements that are made in a user study. For example, in this study we use TREC documents that are judged on a three-point scale: non-relevant (0), relevant (1), and highly relevant (2). The TREC judging criteria define level zero as being applicable where no "part of the document contains information which the assessor would include in a report on the topic"; while the distinction between level one and two was "left to the individual assessors to determine" [7]. If subjects in a user study receive identical instructions to the judges in the batch experiment, and carry out their evaluation in as similar an environment as possible, there is still scope for individuals to decide their own threshold on what information they would "include in a report", and to distinguish between the two categories of relevance. Even in the highly controlled TREC judging environment, the overlap between the relevance judgements of assessors is on average only about 45% [22], indicating that thresholds between relevance categories can differ even within relatively homogeneous populations where identical judging instructions are given. Therefore relevance mismatch, where users and batch judges have different expectations and preferences for documents of different relevance levels, may lead to conflicting results between batch and user results. We investigate the impact of relevance mismatch based on the *split agreement* approach [14], where users are classified into groups based on their responses to documents of different relevance levels.

These two possible explanations for differences between batch and user experiments are investigated through a user study. In Section 2 we survey related background work on experimental evaluation in IR. Our experimental methodology, including details of the user-based searching task, is explained in Section 3. Results are presented and discussed in Section 4, with conclusions and further work being considered in Section 5.

2   Background

The Cranfield paradigm of information retrieval evaluation involves using a search system to run a set of queries on a fixed collection of documents. For each potential answer that the search system returns, a human is required to judge the relevance of the particular document for the current query. This is the dominant framework for experimental IR, and is used, for example, in the ongoing series of Text REtrieval Conferences (TREC). TREC provides standard collections, queries, and relevance judgements so that the performance of different IR systems can be compared using common testbeds [23]. In TREC, queries are derived from topics that represent user information needs: topics consists of a *title* field (a small number of keywords, representative of what a user might type into a web search engine), a *description* (a longer statement of the topic, usually a single sentence), and a *narrative* (a short paragraph specifying further requirements) [6].

Based on the system search result lists and relevance judgements, different system performance metrics can be calculated. Many metrics that have been proposed in the literature focus on precision, which is the number of relevant documents that the search system has found as a proportion of the total number of documents that the system has returned. *Average precision* (AP) is calculated as the mean of the precision at each relevant item that occurs in a result list for a single query. Relevant documents that are not returned by the system contribute a precision of zero; this metric thus has a recall component, since the system is penalised for missing answers. Across a set of queries, the mean average precision (MAP) provides a single number that summarizes search performance, reflecting both the precision and the recall of the system [5].

Another widely-used class of performance metrics is the precision of a system at a particular cutoff point N in the search results list. For example, P@1 evaluates a system based on the relevance of the first item in the result list, while P@10 calculates the precision over the first 10 results. These metrics are popular for evaluating web search tasks, since users typically focus on results that occur early in the ranked list [17]. Analysis by Buckley and Voorhees has indicated that these P@N metrics require a relatively larger number of test queries, compared to other metrics such as MAP, in order to give stable results for the evaluation of batch experiments [4].

The most commonly used IR system performance metrics, such as those presented previously, treat relevance as a binary criterion: a document is either relevant, or it is not. Even where documents may have been judged on a multiple-level relevance scale, these levels are typically folded together into a binary classification before the metrics are calculated. However, studies of multiple levels of relevance have indicated that the traditional binary relevance assumption may not be appropriate where actual users of search systems are concerned [16,21]. In the TREC evaluation framework, the criterion for relevance states that if the document includes any reference to the topic, it should be counted as being relevant. This includes documents that are only *marginally* relevant, where the document does not contain information other than that contained in the topic description; in other words, these documents are largely useless from a user's perspective. Investigating the ability of users to judge documents of different relevance levels, Vakkari and Sormunen concluded that the likelihood of identifying highly relevant documents is much higher than for marginally relevant ones [21]. Further, analysis of 38 topics from TREC-7 and 8 by Sormunen showed that around 50% of documents that were judged as relevant under the TREC binary criterion were of this marginal category [16]. We investigate the effect of accounting for different levels of relevance has on the results of user-based and batch retrieval experiments.

The *cumulative gain* (CG) family of retrieval metrics are based on the idea that the relevance of documents is not equal: the usefulness to a user will depend on the level of relevance of an item [11]. This allows multiple levels of relevance to be incorporated in system evaluation, unlike the previously discussed metrics which assume a binary relevance scale. The CG values, where more highly relevant documents are rewarded by adding more to the overall performance score, can then be *discounted* (DCG) so that the further a document is from the top of a ranked list, the more heavily its relevance score is adjusted. In this paper, we investigate DCG@1 as a multiple-relevance level alternative to P@1. Since discounting is usually not applied at the first rank of the answer list, CG@1 and DCG@1 are equivalent.

The Cranfield paradigm of IR evaluation makes a number of simplifying assumptions about users: essentially, users and real search tasks are removed from the evaluation process, with both information needs and relevance being reduced to static components of the analysis. While this allows for repeatability of experiments, and the controlled evaluation of retrieval algorithms, it is widely acknowledged that these assumptions are significant simplifications of the actual retrieval process [10]. A number of studies have therefore investigated the relationship between system-centric retrieval performance metrics and the performance of users engaged in a range of different search tasks, which we briefly survey here. A relationship between the ability of users to find answer facets and high changes in the level of the bpref evaluation metric was found by Allan et al. [2]. Investigations by Hersh and Turpin found no relationship between MAP and user performance on an instance recall task [8], or a question answering task [18]. The relationship between simple web search tasks and MAP was investigated by Turpin and Scholer [19]; no relationship was found with a precision-oriented task, but a weak relationship was observed with a recall-oriented task.

Other recent studies have considered the relationship between result relevance and user satisfaction. Experiments by Huffman and Hochster showed that system performance measured by DCG@3 was related to user satisfaction for informational searches [9]; user satisfaction was measured by asking subjects to rate their overall search experience on a seven-point scale. Al-Maskari et al. [1] compared the precision and various cumulative-gain metrics of search results with user satisfaction. Here, users rated their satisfaction based on the accuracy, coverage and ranking of results. A high correlation was found between satisfaction and both the precision and CG metrics, while the correlation with nDCG was low. In a series of carefully controlled experiments, Kelly et al. [12] demonstrate a strong correlation between precision and user satisfaction; ranking also influenced user ratings, but to a lesser extent. In this paper, instead of using self-reported measures of satisfaction, we investigate user performance based on success in completing a simple search task, measuring the time taken to find a relevant document.

To construct user relevance profiles, Scholer, Turpin and Wu proposed the *split agreement* approach [14]. Here, users are analysed based on their rate of agreement when presented with documents at different TREC relevance levels. Users can deviate from TREC-like relevance behaviour in two ways: *generous* users have lower criteria for relevance than TREC judges, and are often satisfied even with non-relevant (level 0) documents. Conversely, *parsimonious* users have stricter relevance criteria than TREC judges, and are usually satisfied only with a highly relevant (level 2) document. Users who are *TREC-like* follow the assumed batch relevance profile, generally discarding level 0 documents, but liking level 1 and 2 documents.

Relevance profiles are established through repeated presentation of documents with different TREC relevance levels (unknown to the user). For each presented document, the user is asked to indicate whether they find the document to be relevant for a specified information need, or not. Across many presentations of documents, a response proportion can thus be calculated for each TREC relevance level. For example, a particular user may judge level 0, 1 and 2 documents to be relevant 6%, 63% and 94% of the time, respectively.

User classes are based on these proportions. Specifically, a generous user is defined as someone who judges level 0 documents to be relevant *more* than 50% of the time. A parsimonious user, on the other hand, judges level 1 documents to be relevant *less* than 50% of the time [14]. To investigate relevance mismatch, we attempt to classify users based on their relevance preferences, aiming to establish for each user whether their relevance profile is similar to that of TREC judges.

3   Experimental Methodology

This study investigates the relationship between the P@1 and DCG@1 system performance metrics and user performance on a web search task, and how this is affected by user perceptions of relevance. We use TREC data for the basis of our batch experiments, and a user study to collect data for a searching task.

Users and document collection. 40 experimental subjects were recruited from RMIT University by advertising on newsgroups and notice-boards. All subjects were required to complete entry questionnaires. Participants were university students undertaking undergraduate or postgraduate studies in computer science and information technology, and most were very familiar with online searching (the median response for searching frequency was that searches are conducted "once or more a day"). Subjects were from a variety of cultural backgrounds, but all had a reasonable grasp of the English language (a requirement for studying at RMIT University). Experiments were carried out in accordance with the guidelines of RMIT University Human Research Ethics Committee. Three of the 40 user study participants were unable to carry out all required aspects of the experiments, and are excluded from the analysis below.

The documents used for the searching task are from the TREC GOV2 collection, a 426 Gb crawl of the US .gov domain carried out in 2004 [6]. This collection was used for the TREC Terabyte tracks in 2004–2006, and has 150 associated search topics and corresponding relevance judgements, made by NIST assessors. The relevance judgements are on a three-level ordinal scale: not relevant (0); relevant (1); and highly relevant (2). According to the standard TREC judging approach, if any part of the document contains information that the assessor would include in a report on the topic, it should be judged relevant [7]. That is, relevant documents will include those that are only of marginal value, containing little or no information beyond what is already included in the topic statement.

Search systems. To investigate the relationship between user search performance and system performance as measured by a batch metric, we mimic batch experimental results by constructing ranked lists using the known TREC relevance levels of documents to achieve a given level of the performance metric under investigation. A set of ranked lists at a given level can be thought of as being generated by a search *system* that is engineered to always produce ranked lists that achieve a particular level of the metric, for any topic.

Given that the TREC relevance judgements have three levels, there are thus three possible systems for the DCG@1 metric, namely lists starting with a document of relevance level 0, 1, or 2. For P@1, a binary metric, these relevance levels are folded together: either level 0 compared to combined levels 1 and 2;

or combined levels 0 and 1 compared to level 2. To reduce variation, all system lists had identical TREC relevance scores assigned after the first position. The document relevance level allocations for complete system lists were

$$X, 1, 1, 1, 0, 2, 0, 0, 1, 0$$

where $X \in \{0, 1, 2\}$. Lists were constructed to a depth of 10 documents.

For the search task, 24 topics were chosen from TREC topics 700-850; the constraint for topic selection was that each topic must have the required number of documents at each relevance level to allow the construction of the appropriate lists. Documents were assigned to lists by relevance level, with candidate documents being drawn from the top 50 documents from the two runs with highest MAP scores submitted to the Terabyte track for 2004, 2005 and 2006; that is, they are documents that would feasibly be returned in response to the topic by a modern search system. Only documents of type "text/html" were retained, with other content types being discarded. Similarly, documents smaller than 750 bytes or larger than 100,000 bytes were discarded.

Search task and user interface. Users were asked to carry out a *precision-based* search task: to quickly find useful information about a topic. This type of search is common on the web, and can be considered to be a simple instance of the informational search categories identified by Rose and Levinson [13]. As a performance outcome, we measure the amount of time that a user needs to complete the task.

Specifically, the search scenario is that of a user being asked to find useful information about a topic:

> "Imagine that your boss has come running into the room and urgently needs information. He gives you a very quick topic description, and you have only a few minutes to find a document that is useful (that is, contains some information about the requested topic)."

The information needs were framed in a task-based scenario so as to ground them in a practical context; Borlund has demonstrated that searcher behaviour that is elicited through simulated search tasks may be similar to behaviour that is exhibited when engaged with real information needs [3].

A search session proceeded as follows. First, a subject was presented with an information need, comprised of the *narrative* and *description* fields of a TREC topic, at the top of the screen. Under the information need, a search interface was available. This was closely modelled on the search screens of popular Web search engines, and consisted of a text-box for the entry of search terms, together with a "search" button. After a user entered a query, they were presented with a results list of the required system level (that is, corresponding to one of the precision variants, as described previously). Users were not able to reformulate their queries.

Entries in the search results lists consisted of the document title, together with a short query-biased summary. The document summaries were generated following the approach of Turpin et al. [20], using the *title* field of the TREC topic as the query words. The document title was a hyperlink which, when clicked, opened the underlying document in a new window.

From the document window, in addition to being able to read the document, subjects were presented with two option buttons: "save", to mark the document as relevant to the information need; and "cancel", to close the document window and return to the search results list. Choosing to save a document brought up a confirmation dialogue box, which asked the subject to enter a brief description of why the document was considered to be relevant. After saving one document, the user is deemed to have completed that particular search task.

All interactions between users and the search system were written to a system log, including timestamps of when actions took place. Timings for the precision-based search task were calculated from when the user clicked the search button, until they chose to save their document.

Users were asked to carry out searches on the 24 topics three times, so that each topic would be completed with every system level. The experimental design ensured that users were presented with topics and systems in different orders, to account for possible biases and learning effects. Due to fatigue that was apparent in the last half hour of the user study, we only analyse the first 48 (out of 72) total searches for each user below. However, due to rotation in the experimental design, the results are balanced so that, across all searches, topic and system combinations were used an equal number of times.

4   Results

Based on the user study, we investigate whether system differences as shown by batch metrics that focus on the relevance of the top-ranked position in a search result list transfer successfully to an actual search task. We then examine relevance profiles, and whether these can help to explain the relationship between the two evaluation paradigms.

4.1   Comparing Batch and User Performance

To investigate our first hypothesis, that the P@1 and DCG@1 system performance metrics are closely matched to a precision-based user search task, we analyse the relationship between search system level and time taken to find a useful document. The mean and median times that a user needed to find a useful document with different systems are shown in Table 1. On average, the task time falls as the level of the system performance metric rises. A multifactorial analysis of variance (ANOVA) indicates that the effect of the different system levels is statistically significant ($p < 0.0001$). However, the time data from the user search task is truncated at zero, and so violates the normality assumption. Although ANOVA is generally robust, we therefore also analyse system effects using the Kruskal-Wallis test, a non-parametric alternative to ANOVA [15]; this supports the previous results, also showing a statistically significant effect for

Table 1. Average time (in seconds) for a user to save a relevant document using different systems

System	Mean	Median
0	117.86	89.55
1	112.62	80.89
2	98.00	70.45

system ($p < 0.0001$). Follow-up tests are required to distinguish which specific system levels lead to significant differences in performance.

There are three search systems, corresponding to documents at the first rank position with relevance level 0, 1 or 2. However, for batch system performance to be expressed using the P@1 metric, relevance needs to be folded into a binary scale, giving two ways of grouping relevance levels: folding level 1 and 2 documents together, as is commonly done in TREC; or, folding level 0 and 1 documents together. Differences between these system levels are examined using the Wilcoxon signed-rank test, a non-parametric test of the null hypothesis that the median values of two samples are the same. For both relevance groupings, the differences in search times are statistically significant ($p = 0.0002$ for 0 versus 1 and 2; $p < 0.0001$ for 0 and 1 versus 2) indicating that P@1 batch results transfer to the user task.

For DCG@1, multiple levels of relevance can be accounted for explicitly in the batch metric, so all three systems can be compared directly. User performance differs significantly between systems 0 and 2 ($p < 0.0001$), and between systems 1 and 2 ($p = 0.0046$). However, the difference between systems 0 and 1 is only weakly significant ($p = 0.0989$). These results indicate that there is a noticeable difference for the average time that users need to find a useful document using search systems with different DCG@1 levels. However, this effect is most noticeable when comparing non-relevant documents (system 0) and highly relevant documents (system 2). Marginally relevant documents (system 1) are also clearly differentiated from highly relevant documents, but are similar to non-relevant documents.

These differences between the three relevance levels strongly suggest that, for P@1, it is preferable to fold level 1 (marginally relevant) documents with level 0 (non-relevant) documents, since the difference between level 1 and 2 is much stronger than the difference between level 0 and 1.

4.2   Relevance Profiling Based on Split Agreement

To investigate the effect of relevance mismatch on the relative outcomes of batch and user experiments, we classify users based on their relevance preferences expressed during searching. While working through the search topics, users were able to view documents, and then either choose to save them (a relevance vote), or close them and continue searching (a non-relevance vote). We use these relevance decisions to classify users, using the split agreement approach outlined in Section 2. Recall that users are classified into three groups: TREC-like (their relevance profile matches the TREC judging scheme); generous (their threshold

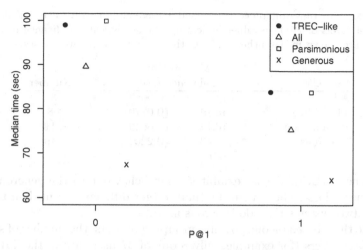

Fig. 1. Users categorised using the *split agreement* approach

for relevance is lower than that for TREC, so they are likely to accept level 0 documents as useful); and parsimonious (their threshold for relevance is higher than that for TREC, so they are unlikely to accept level 1 documents as useful).

If users can be successfully classified according to their relevance behaviour, then we would expect TREC-like users to show a larger difference in the time taken to find a useful document. That is, for users with relevance profiles that more closely match the criteria used in the batch experiment, the difference between retrieval systems as observed in the user task should be the most pronounced. Conversely, for users whose relevance profiles differ from the batch relevance judgements, the difference in retrieval systems should be less pronounced (generous users would be expected to be somewhat faster, no matter which system they are using; the opposite expectation holds for parsimonious users, who would be expected to be slower no matter which system they are using).

Note that here we are analysing relevance mismatch compared to the underlying batch experiment assumptions, based on the TREC relevance judgements: relevance is binary, with level 1 and level 2 documents grouped together into a single "relevant" category. That is, for the P@1 metric there are two possible outcomes, score of 0 (from level 0 documents), and a score of 1 (from level 1 or 2 documents).

Search times for different systems are shown in Figure 1. Across all users (represented by a triangle), task completion time falls when using a system with P@1 of 1, compared to 0. This difference is statistically significant, as shown in Table 2. Generous users are fast, whether they are using a system with a metric level of 0 or 1. As expected, the time taken to find a useful document is similar at both levels for this group, and the difference in batch metric does not lead to significantly different outcomes in the user task. Generous users are slowest when using the system where P@1 equals 0, and speed up when P@1 equals 1.

Table 2. Median time difference (in seconds) for a user to save a useful document for different levels of P@1 (*p*-values indicating the statistical significance of the time differences are shown in parentheses). Note that one user is in two classes.

User class	Median time difference (sec)		Number
All	14.26	(0.0002)	37
TREC-like	15.19	(0.0770)	8
Parsimonious	16.14	(0.0029)	11
Generous	3.43	(0.2209)	19

Users in the TREC-like class exhibit similar behaviour to the generous class. Both of these classes show a substantially larger difference in median time between the two systems than do generous users.

We note that, based on our post hoc grouping of users, the number of subjects in each class differs (for example, only 8 out of 37 users are in the TREC-like category, contributing to a weaker *p*-value despite the noticeable difference in median time). Nevertheless, it appears that relevance profiles can help to determine whether conclusions about batch P@1 values can be transferred to a user population: for generous users (who are satisfied with low-relevance documents as measured on the TREC scale, and don't differentiate strongly between any document levels), the differences are unlikely to hold. However, when the population consists of TREC-like or parsimonious users, the batch results are likely to be transferable.

5   Discussion and Conclusions

Batch evaluation is the dominant paradigm used to compare the performance of information retrieval systems. While a growing body of literature has suggested that there are mismatches between batch experiments based on widely used performance metrics such as MAP and actual search tasks are carried out by users, our results indicate the a simple performance metric such as P@1 can lead to search scenarios where the expected outcomes from batch experiments transfer directly to a precision-based user search task. This effect is statistically significant when relevance is treated as a binary criterion, as in the TREC framework. When multiple-level relevance judgements are available, DCG@1 is similarly effective at transferring expected batch experiment outcomes to a precision-oriented user search task. The difference in user performance is significant between the level 0 and 2, and level 1 and 2, relevance levels. However, it is only weakly significant between relevance levels 0 and 1. This suggests that the when multiple levels of relevance are folded into a binary scale, marginally relevant documents (level 1) should be grouped with non-relevant documents. This is in contrast with current standards used in IR evaluation, where marginally relevant documents are generally bundled with highly relevant documents.

The three system levels described above are intended to reflect possible scenarios of the DCG@1 and P@1 metrics. We note that, given the fixed distribution of relevance levels after rank position 1, our three defined *system levels* also correspond to particular values of other batch metrics. This holds for any metric that is only dependent on the relevance values of items within the top 10 positions of the ranked list (for example P@N or DCG@N for $N \leq 10$). However, for these metrics, the system levels defined for this study represent only a small range of the possible values that the metrics can take on. Therefore, the above conclusions from focusing on $N=1$ metrics should not be extended to the $N > 1$ alternatives directly. Moreover, metrics such as MAP, which include a recall component, will differ from topic to topic, since each topic considered will have a varying number of total relevant documents available. The conclusions from our experiments therefore do not transfer directly to such metrics.

We also investigated relevance mismatch, using split agreement to classify users into different relevance groups. Our analysis demonstrated that the transferability of batch experiment conclusions can differ between user classes; in particular, generous users, who have low thresholds for considering a document to be relevant, do not reflect the batch conclusions obtained form the P@1 metric.

The relevance profile analysis used the entire data obtained from the searching task; however, to be useful from a practical point of view, relevance matching should allow us to infer whether batch results are likely to successfully transfer to users, *without* requiring a full-scale user-study. In future work, we intend to investigate suitable approaches for estimating user relevance profiles with a minimum of effort. Naturally, there are many other possible causes of mismatch between batch experiments and user-based evaluations, including different levels of knowledge about the topics being searched on, age differences, cultural differences, and gender differences. We plan to incorporate these into the user classification approaches in future work.

References

1. Al-Maskari, A., Sanderson, M., Clough, P.: The relationship between IR effectiveness measures and user satisfaction. In: SIGIR, Amsterdam, Netherlands, pp. 773–774 (2007)
2. Allan, J., Carterette, B., Lewis, J.: When will information retrieval be "good enough"? In: SIGIR, Salvador, Brazil, pp. 433–440 (2005)
3. Borlund, P.: Experimental components for the evaluation of interactive information retrieval systems. Journal of Documentation 56(1), 71–90 (2000)
4. Buckley, C., Voorhees, E.M.: Evaluating Evaluation Measure Stability. In: SIGIR, Athens, Greece, pp. 33–40 (2000)
5. Buckley, C., Voorhees, E.M.: Retrieval system evaluation. In: Voorhees, E.M., Harman, D.K. (eds.) TREC: experiment and evaluation in information retrieval. MIT Press, Cambridge (2005)
6. Clarke, C., Craswell, N., Soboroff, I.: Overview of the TREC 2004 terabyte track. In: TREC 2004, Gaithersburg, MD (2005)
7. Clarke, C., Scholer, F., Soboroff, I.: The TREC 2005 terabyte track. In: TREC 2005. National Institute of Standards and Technology, Gaithersburg (2006)
8. Hersh, W., Turpin, A., Price, S., Chan, B., Kraemer, D., Sacherek, L., Olson, D.: Do batch and user evaluations give the same results? In: SIGIR, Athens, Greece, pp. 17–24 (2000)

9. Huffman, S.B., Hochster, M.: How well does result relevance predict session satisfaction? In: SIGIR, Amsterdam, Netherlands, pp. 567–574 (2007)
10. Ingwersen, P., Järvelin, K.: The Turn: Integration of Information Seeking and Retrieval in Context. Kluwer Academic Publishers, Dordrecht (2005)
11. Järvelin, K., Kekäläinen, J.: Cumulated gain-based evaluation of IR techniques. ACM Trans. Information Systems 20(4), 422–446 (2002)
12. Kelly, D., Fu, X., Shah, C.: Effects of rank and precision of search results on users' evaluations of system performance. Technical Report TR-2007-02, University of North Carolina (2007)
13. Rose, D.E., Levinson, D.: Understanding user goals in web search. In: WWW 2004, pp. 13–19. New York (2004)
14. Scholer, F., Turpin, A., Wu, M.: Measuring user relevance criteria. In: The Second International Workshop on Evaluating Information Access (EVIA 2008), Tokyo, Japan, pp. 47–56 (2008)
15. Sheskin, D.: Handbook of parametric and nonparametric statistical proceedures. CRC Press, Boca Raton (1997)
16. Sormunen, E.: Liberal relevance criteria of TREC – counting on negligible documents? In: SIGIR, Tampere, Finland, pp. 324–330 (2002)
17. Spink, A., Jansen, B.J., Wolfram, D., Saracevic, T.: From e-sex to e-commerce: Web search changes. IEEE Computer 35(3), 107–109 (2002)
18. Turpin, A., Hersh, W.: Why batch and user evaluations do not give the same results. In: SIGIR, New Orleans, LA, pp. 225–231 (2001)
19. Turpin, A., Scholer, F.: User performance versus precision measures for simple web search tasks. In: SIGIR, Seattle, WA, pp. 11–18 (2006)
20. Turpin, A., Tsegay, Y., Hawking, D., Williams, H.E.: Fast generation of result snippets in web search. In: SIGIR, Amsterdam, Netherlands, pp. 127–134 (2007)
21. Vakkari, P., Sormunen, E.: The influence of relevance levels on the effectiveness of interactive information retrieval. Journal of the American Society for Information Science and Technology 55(11), 963–969 (2004)
22. Voorhees, E.M.: Variations in relevance judgements and the measurement of retrieval effectiveness. Information Processing and Management 36(5), 697–716 (2000)
23. Voorhees, E.M., Harman, D.K.: TREC: experiment and evaluation in information retrieval. MIT Press, Cambridge (2005)

Test Collection-Based IR Evaluation Needs Extension toward Sessions – A Case of Extremely Short Queries

Heikki Keskustalo[1], Kalervo Järvelin[1], Ari Pirkola[1], Tarun Sharma[1], and Marianne Lykke[2]

[1] University of Tampere, Finland
[2] Royal School of Library and Information Science, Denmark
{heikki.keskustalo,kalervo.jarvelin,ari.pirkola}@uta.fi,
tarunbhu@yahoo.co.in, MLN@db.dk

Abstract. There is overwhelming evidence suggesting that the real users of IR systems often prefer using extremely short queries (one or two individual words) but they try out several queries if needed. Such behavior is fundamentally different from the process modeled in the traditional test collection-based IR evaluation based on using more verbose queries and only one query per topic. In the present paper, we propose an extension to the test collection-based evaluation. We will utilize *sequences* of short queries based on empirically grounded but idealized session strategies. We employ TREC data and have test persons to suggest search words, while simulating sessions based on the idealized strategies for repeatability and control. The experimental results show that, surprisingly, web-like very short queries (including one-word query sequences) typically lead to good enough results even in a TREC type test collection. This finding motivates the observed real user behavior: as few very simple attempts normally lead to good enough results, there is no need to pay more effort. We conclude by discussing the consequences of our finding for IR evaluation.

1 Introduction

Recent studies show that real users of information retrieval (IR) systems search by very short queries but may try out several queries in a session [1-5]. Such queries may consist of only 1-2 search keys. Smith and Kantor [3], and Turpin and Hersh [5] both found that real users successfully compensated for the performance deficiencies of retrieval systems by issuing more queries and/or reading more documents. In real life a searcher typically issues an initial query and inspects some top-N result documents. If no or an insufficient number of relevant documents are recognized, the user may repeatedly launch further queries until the information need is satisfied or (s)he gives up. This setting is different from the Cranfield style IR experiments based on verbose queries and one query per topic. IR evaluation focuses on the quality of the ranked result, measured in terms of available single query metrics, such as mean average precision or cumulated gain or its variants [2, 6]. These metrics do not directly pay attention to query formulation costs, that is, they encourage finding quality at any cost, and short queries are not rewarded for their minor formulation costs.

G.G. Lee et al. (Eds.): AIRS 2009, LNCS 5839, pp. 63–74, 2009.
© Springer-Verlag Berlin Heidelberg 2009

In the present paper, we take another look at user behavior and IR evaluation. In real IR situations there is a cost associated with initial query formulation and subsequent reformulation. Searchers optimize the total cost-and-benefit of their entire sessions. This may render sessions of short queries reasonable. They allow minimal query formulation costs while taking chances with the quality of results. We call such queries as trivial queries: they employ very few search keys in various combinations. Typical real searchers interact with IR systems using such trivial queries.

We show in this paper that trivial queries surprisingly quickly yield reasonable results. We utilize the TREC 7-8 test collection with 41 topics for which graded relevance assessments are available. We will define idealized trivial query strategies and run systematically constructed sessions seeking to find one relevant document (using two distinct relevance thresholds) in Top-10 of each query, and reformulating the query in case of failure. To render our simulation empirically well-founded, we collected data for query candidates from test persons. Our findings theoretically and experimentally motivate the observed real-life user behavior, which real users must have learned through experience, when interacting with IR systems. As few very simple attempts often lead to good enough results, there is no incentive to pay more effort. In Section 2, we review findings on user behavior and consider the costs and benefits of IR sessions before presenting the research problems. Section 3 discusses the construction of simulated sessions. Experimental results are given in Section 4 and discussed in Section 5. Section 6 presents our conclusions.

2 Session Costs and Benefits

2.1 User Behavior

Real searchers behave individually during search sessions. Their information needs may initially be muddled and change during the search process; they may learn as the session progresses, or switch focus. The initial query formulation may not be optimal and the searchers may need to try out different wordings [2]. In fact, it may be impossible for the searcher to predict how well the query will perform [7] because even if the query describes the topic well, it may be ambiguous [8] and retrieve documents not serving the particular information need. Therefore in real IR it is very common that the users may have to revise their topical queries.

Real-life searchers often prefer very short queries [1, 4]. They may also avoid excessive browsing [1, 9]. Jansen and colleagues [1] analyzed transaction logs containing thousands of queries posed by Internet search service users. They discovered that one in three queries had only *one* term. The average query length was 2.21 terms. Less than 4 % of the queries in Jansen's study had more than 6 terms. The average number of terms used in a query was even smaller, 1.45, in a study by Stenmark [4] focusing on *intranet* users.

The stopping decisions regarding browsing the retrieved documents depend on the search task and the individual performing the task [2]. Jansen and colleagues [1] observed that most users did not access results beyond the first page, i.e., the top-10 results retrieved. Therefore real life sessions often consist of sequences of very short queries. The data in Table 1 reflect these findings.

The data for Table 1 come from an empirical, interactive study comparing two search systems. Thirty domain experts each completed the same four realistic search tasks A – D simulating a need for specific information required to make a decision in a short time frame of several minutes. Each task formed a separate query session. The data represent the sessions of one of the systems, showing great variability between the tasks along various variables. Essentially, there were 2.5 queries per session and 2.4 unique keys per session. On average, each query had two keys and 0.9 filters (a geographic, document type or other condition). Only 10 among the 60 sessions employed four or more unique search keys. These searchers were precision-oriented, i.e., they quit searching soon after finding one or a few relevant documents. The four bottom lines report the frequency of the query strategies (S1-S3) that we shall define in Section 3.3. The total number of identified strategies (72) exceeds the number of sessions (60) because more than one strategy was employed in some sessions.

Table 1. Real-life session statistics based on 15 sessions for Tasks A-D (N=60) sessions [10]

Variable	A	B	C	D	Tot
Tot # queries per task	25	59	28	40	152
Avg queries in session	1.7	3.9	1.9	2.7	2.5
Avg # keys per session	1.5	3.9	1.9	2.2	2.4
Avg # keys per query	1.4	2.4	1.8	2.0	2.0
Avg # filters per query	1.2	1.1	0.8	0.7	0.9
S1 frequency	11	3	4	3	21
S2 frequency	2	4	3	4	13
S3 frequency	4	13	11	10	38
S1-S3 frequency sum	17	20	18	17	72

Real life searching of the kind described in Table 1 is fundamentally different from the Cranfield type IR evaluation scenario. In the traditional test collection-based evaluation a single query per topic exists and the queries used are longer (typically 7 to 15 search keys, see [1]). Because of these facts we will focus on trivial query sessions in the present study, including one-word queries for each topic.

2.2 Identifying Costs and Benefits

What explains the great difference between user behavior and effective laboratory queries? We believe that costs and benefits of IR interaction are currently not sufficiently taken into account to explain user behavior.

Early papers on IR evaluation had a comprehensive approach toward evaluation: Cleverdon, and colleagues [11] identified, among others, presentation issues and intellectual and physical user effort as important factors in IR evaluation. Salton [12] identified user effort measures as important components of IR evaluation. More recently, Su [13] compared 20 evaluation measures for interactive IR, including actual cost of search, several utility measures, and worth of search results vs. time expended.

Due to time pressure, documents retrieved in the top ranks may be of interest for real users [9, 14]. Järvelin et al. [2] extended the Discounted Cumulated Gain metric [6] into a session-based evaluation metric which evaluates multiple query sessions and takes the searcher's effort indirectly into account. Also the literature on usability has a comprehensive approach to costs and benefits, see, e.g., the ISO standard [15].

One may conclude that various costs and benefits of interactive IR systems have been brought up in the literature. The same does not hold on current IR evaluation practices. In interactive settings both costs and benefits are present and affect searcher behavior (e.g., through expectations). Therefore, interactive IR evaluation should incorporate the existing cost factors: search key generation cost, query execution cost, result scan cost, next result page access cost, and relevant document gain. Contemporary IR evaluation effectively assumes all costs as zero, thus focusing on benefits (the gain) at any cost. This hardly models real-life situations.

In the present paper we acknowledge that the query formulation costs may be a significant factor explaining user behavior. We will show that trivial queries are a reasonable alternative for the user because their formulation costs are minimal and their effectiveness competitive if sessions are allowed.

2.3 Research Problem

Our background assumption is that the observed user behavior [1, 4, 10] does satisfy real needs. Thus, the obvious question is whether it makes sense for the user to combine the use of short queries and generally "take their chances" with trivial queries and reformulate in case of a failure. Our overall research question therefore is: What is the effect of utilizing a sequence of trivial queries as a session compared to the traditional approach of utilizing one verbose query?.

In studying this problem, we make simplifying assumptions. First, we assume that the topical requests remain unchanged during a session: the simulated searcher neither learns nor switches focus during the session. Secondly, the simulated searcher is able to recognize the relevance of documents (see [16]). Third, the simulated searcher is assumed to scan the ranked list of documents from the top to bottom (see [17]). However, reflecting the observed searcher behavior, we focus on the Top-10 results. We will focus on the search task of finding a single relevant document (see [18]).

3   Constructing Simulated Sessions

We made use of the TREC 7-8 test collection, and real test persons to generate candidate queries and alternative search keys. The collection of these data is explained next and their properties analyzed thereafter, followed by the definition of the simulated session strategies, the retrieval protocol, and our evaluation method.

3.1   The Test Collection and Search Engine

We used the reassessed TREC test collection including 41 topics from TREC 7 and 8 ad hoc tracks [19]. The document database contains 528 155 documents organized under the retrieval system *Lemur*. The database index is constructed by lemmatizing the document words. The relevance judgments were based on topicality using a four-point scale: (0) irrelevant document: the document does not contain any information about the topic; (1) marginally relevant document: the document only points to the topic but does not contain more or other information than the topic description; (2) fairly relevant document: the document contains more information than the topic description but the presentation is not exhaustive; and (3) highly relevant document: the document discusses the themes of the topic exhaustively. In the recall base there are on the average 29 marginally relevant, 20 fairly relevant and 10 highly relevant documents for each topic [19].

3.2 Collecting the Query Data

As session-based collections do not currently exist we decided to construct one by ourselves on top of the TREC 7-8 test collection. 41 topics were analyzed intellectually by test persons to form query candidate sets. During the analysis the test persons did not interact with a real system. They probably would have been able to make higher quality queries had they had a chance to utilize system feedback.

A group of seven undergraduate information science students (Group A) and seven staff members (Group B) performed the analysis. Staff members having an extensive background regarding the specific test collection were excluded. Regarding each topic a printed topic description and a task questionnaire were presented for the test persons. Each of the 41 topics was analyzed twice - once by a student and once by a staff member. The users were asked to directly select and to think up good search words from topical descriptions and to create various query candidates.

First a two-page protocol explaining the task was presented by one of the researchers. Information in the description and narrative fields of the test collection topics were presented for the users. Descriptions regarding non-relevance of the documents were also omitted to make the task more manageable within the time limitation of 5 minutes per topic. The test persons were asked to mark up all potential search words directly from the topic description and to express the topic freely by their own words. Third, they were asked to form various query candidates (using freely any kinds of words) as unstructured word lists: (i) the query they would use first ("1st query"); (ii) the one they would try next, assuming that the first attempt would not have given a satisfactory result ("2nd query"). Finally, the test persons were asked to form query versions of various lengths: (iii) one word (1w), (iv) two words (2w), and (v) three or more words (3w+). The very last task was to estimate how appropriate each query candidate was using a four-point scale.

3.3 Simulated Session Strategies

Using the query data collected from the test persons we created four simulated session strategies (S1-S4) for the experiments. The strategies S1-S3 model *five-query sessions* (short queries), while Strategy S4 acts as a comparison baseline and utilizes *only one*, long query. Sessions longer than five queries are also relatively rare in real life, see [6] and Table 1. Data collected from the test persons were used in session strategies S1 to S3. In each session, the simulated searcher inspected at most 50 documents: in S1-S3 at most five distinct Top-10 results, and in S4 at most the single Top-50.

Session Strategy S1: One-word Queries Only
In S1 strategy we experiment *solely with one-word queries*. Unique individual words are selected randomly from various query types in the following order: "1st query",

"2nd query"; 1w, 2w, and 3w+ queries until five distinct words are collected. Within a session, if any given one-word query does not retrieve a (highly) relevant document within its top-10 ranks, we immediately try out the next one-word query. S1 is justified because (1) extremely short queries dominate in real life, and (2) the strategy was employed 21 times in the 60 real-live sessions of Table 1. Random selection obviously creates some bad one-word queries. We purposefully experimented with such a strategy to explore the effects of allowing bad queries within session - as may very often happen in real life.

Session Strategy S2: Incremental Query Extension
In S2 strategy we experiment with using incrementally longer queries in sessions. As stated above, our test persons were requested to form one-word (1w), two-word (2w), and longer (3w+) query versions, and the queries they would try first (1st query) and second (2nd query). Here we selected words *left to right* (i.e., not randomly) from each query version (using query versions in the order explained above) until a sequence of five, if possible, unique words w1,..., w5 is formed. These words are used to construct queries of varying lengths (i.e., w1; w1 w2; ...; w1 w2 w3 w4 w5) (from 1 to 5 words) for each topic. Within each topical session, the searcher starts with the one-word query. If a query does not retrieve the required (highly) relevant document, the next incrementally longer query is launched. S2 is justified because this strategy was employed in 13 times of the 60 real-live sessions of Table 1. It simulates a lazy searcher who tries to cope with minimal effort and adds one word at a time.

Session Strategy S3: Variations on a Theme of Two Words
In S3 *two core search keys* are fixed to represent the information need and several different third words are tried as variations. S3 is justified as this strategy was employed in 38 of the 60 real-live sessions of Table 1. According to Jansen et al. [1] modifications to successive queries are done in small increments by modifying, adding or deleting keys. We used first three words of the 3w+ query as the starting point, and varied randomly the third word by replacing it with distinct words selected from the 3w+, or from 1st, 2nd, 1w or 2w queries (in that order) if 3w+ ran out of words.

Session Strategy S4: Single Verbose Query
Session strategy S4 consists of a single verbose query. It contains all the words of the *description* and the *title* field (on the average 16.9 words). Thus, it represents traditional laboratory testing and serves as a baseline.

3.4   Retrieval Protocol and Evaluation

The run procedure went as follows:

1.   Based on session strategies S1 to S4 query sequences were constructed.
2.   Top-10 documents were retrieved for each query in S1-S3 and top-50 for S4.
3.   Success of each session strategy S1 to S4 was determined.

Stopping decisions often depend on the task, context, personality, and the retrieval results [10]. In this study, the stopping condition was defined as finding one relevant document. Failure was defined as inability to find a relevant document in a session.

Table 2. Effectiveness of session strategies S1 to S4 (User Group A and B) for 41 topics. Legend: number-in-cell denotes the ordinal of the first successful query in finding a relevant document within its top-10 ranks. For session S4 see text below. Hyphen denotes a failure to find a relevant document. Table on the left: *liberal* relevance threshold is used. Table on the right: *stringent* relevance threshold is used.

	Liberal Relevance							Stringent Relevance						
	S1		S2		S3		S4	S1		S2		S3		S4
Topic#	A	B	A	B	A	B	–	A	B	A	B	A	B	–
351	1	2	5	1	3	1	1	1	2	5	1	3	1	1
353	1	1	1	1	1	1	1	-	-	2	2	-	1	1
355	1	2	1	1	1	1	1	1	2	1	1	1	1	1
358	1	1	1	1	2	1	1	1	1	1	1	2	1	1
360	1	2	1	1	1	1	1	2	2	3	3	2	1	1
362	1	1	1	1	1	1	1	1	1	1	1	-	3	1
364	1	1	1	1	1	1	1	1	1	1	1	1	1	1
365	3	1	1	2	1	1	1	3	1	1	2	1	1	1
372	5	2	1	1	1	1	1	-	-	2	2	1	2	1
373	1	1	1	1	1	1	1	1	1	1	1	1	1	1
377	2	-	1	-	1	-	1	2	-	1	-	1	-	1
378	-	-	3	3	1	1	1	-	-	-	-	-	-	-
384	2	1	1	1	1	1	1	-	-	4	3	2	1	2
385	-	-	2	2	1	1	1	-	-	2	2	2	1	1
387	1	1	1	1	1	2	1	2	1	2	2	1	2	1
388	2	3	4	3	1	1	1	-	-	-	-	-	-	4
392	2	1	1	1	1	1	1	2	1	1	1	3	1	1
393	2	1	1	1	1	1	1	2	1	1	1	1	1	3
396	1	3	2	1	1	1	1	1	3	2	1	1	1	1
399	4	-	2	1	1	-	1	-	-	4	2	1	-	2
400	4	2	2	1	1	1	1	4	2	2	1	2	1	1
402	1	1	2	1	1	1	1	-	1	2	1	1	1	1
403	1	1	1	1	1	1	1	1	1	1	1	1	1	1
405	1	2	3	1	1	2	1	-	-	-	-	3	-	2
407	1	1	1	1	1	1	1	2	1	1	1	1	1	1
408	2	1	1	1	1	1	1	-	-	2	3	2	1	1
410	3	2	1	1	1	1	1	3	2	1	1	1	1	1
414	-	-	3	2	1	1	1	-	-	-	-	-	-	-
415	3	1	2	1	1	1	1	-	-	2	5	1	1	1
416	3	1	1	1	1	1	1	3	1	1	1	1	1	1
418	2	2	1	1	1	1	1	2	2	1	1	1	1	1
420	1	1	1	2	1	1	1	1	1	2	1	1	1	1
421	1	2	1	3	1	1	1	-	-	2	3	1	1	1
427	2	2	1	1	2	1	1	2	4	-	-	-	-	4
428	2	1	1	1	1	1	1	3	1	1	1	1	1	1
431	2	3	1	1	1	1	1	2	1	1	1	1	1	1
437	-	-	2	3	2	-	3	-	-	-	-	-	-	-
440	-	-	2	2	3	2	1	-	-	2	2	5	-	5
442	1	-	1	1	1	1	1	-	-	-	-	-	-	-
445	3	-	1	1	1	3	1	-	-	2	4	1	4	1
448	-	-	2	2	1	1		-	-	-	-	-	-	

4   Experimental Results

Above, general result for session strategies S1-S4 is presented for 41 individual queries (Table 2). Number 1 denotes that the first query in the session was successful in finding a relevant document for the topic within its top-10 ranks. Number 2 denotes the second query being successful etc. The table shows the effectiveness of session strategies based on both liberal and stringent relevance threshold. The columns for the two searcher groups A and B indicate the variability under the same strategy. Color-coding is used in cells in addition to the ordinal numbers for visual evaluation: black indicates the first query being successful, white no success at all, and grey scale success by a non-first query.

Table 3. Overall effectiveness of session strategies S1 to S4 (User Groups A and B). Top to bottom: average number of queries attempted per session; the count of successful sessions; and the percent of successful sessions.

Table 4. Pairwise statistical significance (+) of differences by Friedman's test for Searcher Group A using S1 – S3 and S4, p=0.01

Liberal Relevance						
S1		S2		S3		S4
A	B	A	B	A	B	-
2.3	2.4	1.5	1.5	1.2	1.4	1.0
35	31	41	40	41	38	41
85.4	75.6	100	97.6	100	92.7	100

Liberal Relevance			
	Strategies		
	S2-A	S3-A	S4
S1-A	+	+	+
S2-A		-	+
S3-A			-

Stringent Relevance						
S1		S2		S3		S4
A	B	A	B	A	B	-
3.1	2.9	2.2	2.2	2.2	2.0	1.6
23	23	33	32	31	30	36
60.5	60.5	86.8	84.2	81.6	78.9	94.7

Stringent Relevance			
	Strategies		
	S2-A	S3-A	S4
S1-A	+	+	+
S2-A		-	+
S3-A			-

Based on liberal relevance criteria, S1 (*one-word queries only*) is 15-25 percent units weaker than strategies S2-S4 in its success rate (Table 3). Strategies S2-S4 are equally good. The average number of queries in S1-S3 varies between 1.2 and 2.4 queries. Strategy S3 fairs almost as well as S4. On the stringent level S1 is clearly weaker than the other strategies. Surprisingly, S2 and S3 are only 10 – 15 percentage units weaker that S4. The average number of queries in S1-S3 varies between 2.0 and 3.1; the average number of pages browsed for S4 is 1.6.

According to Friedman's test the differences between the strategies are highly significant ($p < 0.001$). Table 4 gives *pairwise* results for Friedman's test and Searcher Group A. We observe that S1 is significantly different from others; S3 is not significantly different from S4, the baseline. The significance results for the group B are similar. Note that even when the results for some trivial query strategies are significantly worse than S4, the queries require much less effort.

4.1 Liberal Relevance Threshold

At the liberal relevance threshold the most successful strategy was the baseline session strategy S4 (*single verbose query*). As only one query candidate was formed in S4 strategy, number 1 in column S4 denotes the fact that the relevant document was found within the first results page (ranks 1 to 10); number 2 denotes second page, etc. All words of the title and description fields were used, thus rendering very long queries - an average query length of 16.9 words per query.

Among the trivial strategies S3 (*variations on a theme of two words*) was the most successful one. For 36 and 34 topics out of 41 (user group A and B, respectively) the very first query was successful. The strategy only failed three times (and only for group B). The session strategy S2 (*incremental query extension*) was also effective. For 27 and 29 topics (for A and B) the very first query (at this point a single key) was successful. Adding the second key to the query helped to find a relevant document for 9 and 7 additional topics (A and B), and the third key for 3 and 4 topics (A and B).

Strategy S2 only failed once (in group B). Strategy S1 (*one-word queries only*) was the least successful. The fact that the single query keys were selected randomly obviously hurt the performance. Yet, it failed only in 6 and 10 topics (A and B) out of 41.

4.2 Stringent Relevance Threshold

If liberal relevance threshold is used, low quality documents are accepted as relevant. These documents may be only marginally relevant and escape the reader's attention [16]. Finding one such document can hardly be justified as success even if the user only had to pay minimum effort. Therefore, in the remainder of this paper we only accept highly relevant documents as relevant. They discuss the themes of the topic extensively [19] thus better justifying the user's stopping decision in simulations.

Table 5 summarizes the results based on stringent relevance threshold. The recall base of the test collection did not contain highly relevant documents for 3 topics (#378, #414, #437), leaving 38 topics in the stringent relevance threshold case. Cumulative percentages are shown in the table regarding the share of successful topics for each session strategy.

Also when highly relevant documents are demanded, the baseline session strategy S4 (*single verbose query*) performs best. In 29 topics out of 38 a highly relevant document is found within the first page. Table 5 presents these success figures as percentages (29/38 = 76.3 %). For three topics the second page needs to be inspected in strategy S4; for one topic the third page; for two topics the fourth, and for one topic the fifth page. This strategy fails only for two topics.

Session strategy S3 (*variations on a theme of two words*) was the most successful one among strategies S1-S3 also when highly relevant documents are requested. For 21 and 26 topics out of 38 (for user groups A and B, respectively) the very first query was successful. For 7 and 8 topics (groups A and B) the strategy failed.

Table 5. Success of the session strategies by the ordinal of the query candidate. Figures express the share of the topics (cumulative %) for which a highly relevant document was found. *For explanation regarding session strategy S4 see text.

Que ry #	S1		S2		S3		S4*
	A	B	A	B	A	B	-
1st	23.7	39.5	44.7	47.4	55.3	68.4	76.3
2nd	47.4	55.3	76.3	68.4	71.1	73.7	84.2
3rd	57.9	57.9	78.9	78.9	78.9	76.3	86.8
4th	60.5	60.5	84.2	81.6	78.9	78.9	92.1
5th	60.5	60.5	86.8	84.2	81.6	78.9	94.7

Table 6. Strategy cost features for group A: strategy (S1-S4), expected number of search keys to enter (T) and queries to launch (Q) for one relevant document

Strat- egy	T	Q
S1	8.6	3.5
S2	4.3	2.3
S3	7.3	2.4
S4	16.9	1.0

The session strategy S2 (*incremental query extension*) was also effective. For 17 and 18 topics (A and B) the very first query (i.e., a singe key) attempted was successful. Adding the second key to the query helped to find a highly relevant document for 12 and 8 of the so far unsuccessful topics (A and B). Only in 5 cases for group A and in 6 cases for group B the incremental extension strategy failed.

The session strategy S1 (*one-word queries only*) was the least successful, yet remarkably, in more than half of the topics (57.9 %, Table 5, groups A and B) the strategy was successful after only three single-word queries were attempted.

We experimented also by measuring the effectiveness of S1 using traditional metrics (P@10 and non-interpolated average precision (AP)). We evaluated the effectiveness of all query candidate sets, 1st to 5th queries for group A and B), averaged over 38 topics, based on the top-1000 documents, and stringent relevance threshold. The highest value observed for P@10 was 7.4 % and for AP 11.3 %. The corresponding values for S4 (verbose queries) were 25.2 % and 19.5 %. Thus, if one query per topic is assumed, S1 queries are inferior, but they make sense if multiple queries are used.

5 Discussion

Real users of IR systems typically search by very short queries but may try out several queries [1, 4]. Test collection-based evaluation, e.g., like the one performed in TREC [20], typically employs longer queries and one query per topic. We analyzed the effectiveness of sessions of very short queries. We performed a simulation because it is difficult to use real interactive sessions and have control over multiple query/session types, avoid learning effects, and support repeatability of the experiment. The strengths of our approach include session strategies and intellectual word selections for queries based on an empirical ground truth.

We therefore had two sets of test persons to create realistic content for trivial queries. They did not interact with the retrieval system or test collection and thus did not use their own relevance assessments. This is justified as we studied idealized strategies. However, if the queries had lower quality than those in realistic situations, it only makes our argument stronger.

The idealized session strategies were constructed based on empirical data (Table 1). We set the limit of maximum of 5 queries per session because longer sessions are rare in real life. We set the limit of maximum of 10 documents per query results for strategies S1-S3 because scanning length in real life is limited [1]. The simplest strategy S1 (*one word queries*) was popular in our sample data (Table 1) and attempts to minimize the query formulation costs. The strategy S2 (*incremental query extension*) was less popular but nevertheless present in the sample data. The strategy S3 (*variations of three word queries*) seeks to fix a focus by two keys and vary by trying out different third keys. This was the most popular strategy in our sample data and to some degree corresponds to Bates' Berry-picking strategy [21]. S4 represents a long TREC-type of single query strategy, and did not occur in our data.

Table 6 shows the expected number of search keys and queries, when each strategy is successful. Regarding S1-S3, we assume that for unsuccessful topics the searcher would in desperation launch one more query (#6), a successful one represented by S4 containing on average 16.9 search keys, to guarantee comparable performance. Strategies S1-S3 yield a low query formulation cost in the number of search terms. If the query launching costs and result scanning unit costs are minor, strategies S1-S3 make sense to users. They mean low formulation costs while taking chances with the result. We focused on the limited task of finding one relevant document but our simulation method fits well to tasks where more than one relevant document is required.

6 Conclusion

Log analyses reveal that real users often try out sessions of several short queries. Traditional laboratory evaluation is not well-suited to study this phenomenon.

We demonstrated session-based batch evaluations utilizing test collections and query data collected from test persons. We focused on studying the effectiveness of sessions of very short queries. We assumed searchers requiring one relevant document, browsing a limited length of results, and using a limited set of session strategies. Short query sessions turned out to be successful.

Future evaluation should model processes where the searcher may try out several queries for a topic, and use broader costs and benefits than the ones focusing on the quality of the retrieved result. This may help toward resolving the current disparity of the observed searcher behavior and the assumptions of laboratory experiments.

Acknowledgements

Lemur and ENGTWOL were used in the study. See http://www.lemurproject.org. ENGTWOL: Copyright © 1989-1992 Atro Voutilainen and Juha Heikkilä. TWOL-R: Copyright © Kimmo Koskenniemi and Lingsoft plc.1983-1992. This work was funded by Academy of Finland under grants #120996 and #124131.

References

1. Jansen, M.B.J., Spink, A., Saracevic, T.: Real Life, Real Users, and Real Needs: A Study and Analysis of User Queries on the Web. Inf. Proc. Man. 36(2), 207–227 (2000)
2. Järvelin, K., Price, S.L., Delcambre, L.M.L., Nielsen, M.L.: Discounted Cumulated Gain Based Evaluation of Multiple-Query IR Sessions. In: Macdonald, C., Ounis, I., Plachouras, V., Ruthven, I., White, R.W. (eds.) ECIR 2008. LNCS, vol. 4956, pp. 4–15. Springer, Heidelberg (2008)
3. Smith, C.L., Kantor, P.B.: User Adaptation: Good Results from Poor Systems. In: Proc. ACM SIGIR 2008, pp. 147–154 (2008)
4. Stenmark, D.: Identifying Clusters of User Behavior in Intranet Search Engine Log Files. JASIST 59(14), 2232–2243 (2008)
5. Turpin, A., Hersh, W.: Why Batch and User Evaluations Do Not Give the Same Results. In: Proc. ACM SIGIR 2001, pp. 225–231 (2001)
6. Järvelin, K., Kekäläinen, J.: Cumulated Gain-Based Evaluation of IR Techniques. ACM TOIS 20(4), 422–446 (2002)
7. Swanson, D.: Information Retrieval as a Trial-and-Error Process. Library Quarterly 47(2), 128–148 (1977)
8. Sanderson, M.: Ambiguous Queries: Test Collections Need More Sense. In: Proc. ACM SIGIR 2008, pp. 499–506 (2008)

9. Azzopardi, L.: Position Paper: Towards Evaluating the User Experience of Interactive Information Access Systems. In: SIGIR 2007 Web Information-Seeking and Interaction Workshop, p. 5 (2007)
10. Lykke, M., Price, S.L., Delcambre, L.M.L., Vedsted, P.: How doctors search: a study of family practitioners' query behaviour and the impact on search results (in press, 2009)
11. Cleverdon, C.W., Mills, L., Keen, M.: Factors determining the performance of indexing systems, vol. 1 - design. Aslib Cranfield Research Project, Cranfield (1966)
12. Salton, G.: Evaluation Problems in Interactive Information Retrieval. Inf. Stor. Retr. 6, 29–44 (1970)
13. Su, L.T.: Evaluation Measures for Interactive Information Retrieval. Inf. Proc. Man. 28(4), 503–516 (1992)
14. Hersh, W.: Relevance and Retrieval Evaluation: Perspectives from Medicine. JASIS, 201–206 (April 1994)
15. ISO: Ergonomic Requirements for Office Work with Visual Display Terminals (VDTs), Part 11: Guidance on Usability. ISO 9241-11:1998 (E) (1998)
16. Vakkari, P., Sormunen, E.: The Influence of Relevance Levels on the Effectiveness of Interactive Retrieval. JASIST 55(11), 963–969 (2004)
17. Joachims, T., Granka, L., Pan, B., Hembrooke, H., Gay, G.: Accurately Interpreting Click-through Data as Implicit Feedback. In: Proc. ACM SIGIR 2005, pp. 154–161 (2005)
18. Price, S.L., Nielsen, M.L., Delcambre, L.M.L., Vedsted, P.: Semantic Components Enhance Retrieval of Domain-specific Documents. In: Proc. ACM CIKM 2007, pp. 429–438 (2007)
19. Sormunen, E.: Liberal Relevance Criteria of TREC - Counting on Negligible Documents? In: Proc. ACM SIGIR 2002, pp. 324–330 (2002)
20. Voorhees, E., Harman, D.: TREC: Experiment and Evaluation in Information Retrieval. MIT Press, Cambridge (2005)
21. Bates, M.J.: The Design of Browsing and Berrypicking Techniques for the Online Search Interface (1989),
    http://www.gseis.ucla.edu/faculty/bates/berrypicking.html

Weighted Rank Correlation in Information Retrieval Evaluation

Massimo Melucci

University of Padua

Abstract. In Information Retrieval (IR), it is common practice to compare the rankings observed during an experiment – the statistical procedure to compare rankings is called rank correlation. Rank correlation helps decide the success of new systems, models and techniques. To measure rank correlation, the most used coefficient is Kendall's τ. However, in IR, when computing the correlations, the most relevant, useful or interesting items should often be considered more important than the least important items. Despite its simplicity and widespread use, Kendall's τ little helps discriminate the items by importance. To overcome this drawback, in this paper, a family τ_* of rank correlation coefficients for IR has been introduced for discriminating the rank correlation according to the rank of the items. The basis has been provided by the notion of gain previously utilized in retrieval effectiveness measurement. The probability distribution for τ_* has also been provided.

1 Introduction

In Information Retrieval (IR), it is common practice to compare the rankings observed during an experiment with the rankings produced by (i) a competitor system, (ii) the same system but with different parameters or (iii) the system which correctly ranks all the items (e.g. a human) and is then considered the best. Examples of (i) regard the comparison of the algorithms for various techniques used in IR, such as query expansion, stemming, or graph link-based webpage ranking. An example of (ii) is the correlation between the webpage rankings computed by PageRank at different damping factors or numbers of iterations calculated through the power method. An example of (iii) is the comparison of the ranking of IR systems produced by using the relevance assessments collected by the human assessors with the ranking produced by using the relevance assessments collected through the manual runs. The comparisons aim at helping the experimenter to decide whether the compared rankings are "approximately" the same or they are significantly different. The measurement of the degree to which two rankings are the same is called rank correlation in the literature of Statistics. Rank correlation is crucial in IR because it helps decide the success of new models and techniques.

As in IR the rankings are often compared with a reference based on relevance assessments or preferences provided by human judges, this paper is focussed on the comparison between an observed ranking and a reference ranking which puts

G.G. Lee et al. (Eds.): AIRS 2009, LNCS 5839, pp. 75–86, 2009.
© Springer-Verlag Berlin Heidelberg 2009

the item in the correct order. Thus, it is assumed in this paper that the reference ranking is the best ordering of the items. Examples are the documents ranked by relevance assessments, the items ranked by preferences, or the terms ranked by degree of synonymity – for the sake of simplicity, in the following, "relevance" refers to "preference", "importance", "usefulness" and to similar terms.

Suppose a ranking of $n \geq 2$ items, e.g. documents, is observed – this is called observed ranking. The items are conventionally numbered from 1 to n and thus item i is represented as the natural number i. For example, item 5 is ranked at the second position in the observed ranking $(4, 5, 2, 1, 3)$. The ranking to which the observed ranking is compared is called *reference* ranking and is conventionally represented by the increasing sequence $(1, \ldots, n)$ which puts i at rank i.

The most used rank correlation coefficient (RCC) is Kendall's τ. Kendall's τ can be viewed as a function of the number of exchanges of two items necessary to transform the observed ranking to the reference ranking – the higher the number of exchanges necessary to transform the observed ranking to the reference ranking, the less correlated the two rankings are. Intuitively, the idea is similar to the bubble-sort algorithm. Formally, Kendall's τ is defined as the ratio of the difference between the number of concordant pairs and the number of discordant pairs in the observed ranking with respect to the reference ranking to the total number of pairs [1]. From this definition, it follows that

$$\tau = 2p - 1$$

where

$$p = \frac{C}{n(n-1)/2}$$

is the proportion of concordant items to the total number of pairs, which is $n(n-1)/2$, and C is the number of concordant pairs in the n-item observed ranking.

An important property of τ is its limiting probability distribution. Under the null hypothesis of random, uniformly distributed rankings, and then of incorrelation, τ is approximated by a Normal random variable with mean zero and known sampling variance, thus permitting the experimenter to infer about the significance of the RCCs by using the Normal probability distribution tables. The main advantage of τ is the rapid convergence to normality even with small samples [1, 2] – however, this is only one of the reasons why this coefficient is largely used in IR, the others being the simple underlying idea, the sixty year-long history, and the wide availability of routines for calculating it in many statistical or mathematical software packages.

The problem is that the dis/concordances should be treated differently depending on the relevance of the items, the latter being an important issue in IR when some measure of relevance is often attributed to the items. In this regard, despite its simplicity, Kendall's τ is inappropriate in discriminating the correlation involving the items on the top of the reference ranking from that involving the bottom ranked items. As a consequence, when the items are ranked, say, by relevance in the reference ranking, τ is insensitive to the relevance of the items.

Actually, in IR, the concordances or discordances involving the most relevant (or useful, interesting, important) items should often be considered more important than the concordances involving the least relevant. The importance of the degree of relevance of the items is evident, for example, when non-binary relevance is recorded and document rankings are compared – the concordances of the most relevant items are the most crucial and then the weights given to these concordances should be higher than those given to the other concordances. Therefore, the RCCs should weight the concordances differently depending on the ranks where the items occur in the reference ranking. Since the reference ranking is often arranged by relevance, the treatment of dis/concordances depending on the item relevance means that the dis/coconcordances are treated differently depending on the ranks of the items in the reference ranking.

In this paper, a family of RCCs for IR, called τ_* has been introduced to give different weights according to the rank of the items in the reference ranking. Each instance of τ_* differs from the others for a series of $n - 1$ weights used to give different importance to the ranks where the concordant pairs occur. The *main contribution* of this paper is the use of the gains [3] in order to provide a conceptual basis to the assignment of the weights of τ_*. Thus, the definition of the weights of τ_* is very simple, easy to use and intuitively understandable. The importance of the use of the gains is due to their role in estimating the probability that a relevant item is picked at random among those ranked before a fixed item, thus generalizing the notion used with τ_{AP} in which all the relevant items are treated equally.

The probability distribution for τ_* has also been provided, thus permitting to decide if τ_* computed using a given series of weights is significantly different from zero, that is, if the observed ranking is different from the reference ranking and if the difference mostly affects the most relevant items – this is the second contribution of the paper. In this way, the experimenter can decide whether the use of a tested technique, model or method is significantly different from the technique, model or method used for another ranking, especially when the top ranked items are deemed the most crucial, thus grounding the comparison of experimental results on a sound inferential statistical basis. Although τ_{AP} is more sensitive to the concordances at the top-ranks than τ, it is not provided with a probability distribution. However, in this paper, it is shown that, as τ_{AP} is an instance of τ_*, and then it is approximately Normal with mean zero and known sampling variance.

As the paper is mainly theoretical, it provides the basis for the future experiments and applications. It is structured as follows. Section 2 briefly reviews the most relevant related work. Section 3 describes τ_*. Section 4 illustrates the problem of weighting in rank correlation. In Section 5 the criterion to define the weights of τ_* is presented.

2   Related Work

In [4] the use of weights giving different importance to the concordances or the discordances depending on the ranks of the items was considered. To this end, a RCC called τ_w was defined and the normality of the distribution probability was proved. Due to τ_w, it can be shown that τ_* is a special case of τ_w and τ is a special case of τ_*.

The problem of giving more importance to the concordances occurring on the top was also addressed in [5] where a modified τ, called τ_{AP}, was proposed. Those authors showed that τ_{AP} may be greater than τ when the concordance between the reference ranking and the observed ranking occurs at the top ranks, while τ_{AP} may be less than τ when the concordance between the reference ranking and the observed ranking occurs at the bottom ranks. In this paper, it has been shown that τ_{AP} is special case of τ_*, and then of τ_w.

In [6], various rank correlation measures are investigated and classified. Using a preference criterion, the authors suggested which RCC to use. Our work differs from that article since our aim is to generalize τ and τ_{AP}, thus leveraging the known properties and advantages of these two coefficients and proposing a general coefficient which can be tailored to the needs of the evaluation of specific retrieval tasks. This paper, moreover, anchors τ_* to the notion of gain as τ_{AP} has been anchored to the notion of Average Precision.

Many are the papers which have used rank correlation, and in particular Kendall's τ. The issue of giving more importance to the concordances at the top ranks raised in [7–11] as regards to ranking and advertising in the Web, in [12–21] as regards to the problem of evaluating IR systems, in [22, 23] as for distributed IR (the problem was to select some of the best sources from a networked system), in [24] as for social search, and in [25–27] as for feature selection, indexing and crawling (the problem was to select the best features, the best indexed documents from a posting list or to crawl the best websites). In [5, 28], some research results on rank correlation and IR were surveyed.

3   The Family of Rank Correlation Coefficients

From the computation of τ, one realizes that C is the sum of the number of concordant pairs computed over all the items where an item of the pair is fixed. That is, $C = C(2) + \cdots C(n)$ where $C(i)$ is the number of items less than i-th item in the observed ranking (see also [5]). For example, $C(6) = 3$ in $(4, 7, 2, 10, 3, 6, 8, 1, 5, 9)$.

It can be shown that

$$p = \frac{2C}{n(n-1)} = \sum_{i=2}^{n} \left(\frac{i-1}{n(n-1)/2} \right) p(i)$$

where

$$p(i) = \frac{C(i)}{i-1}$$

is the proportion of concordant items at rank i. The Average Precision (AP) correlation introduced in [5] was defined as

$$\tau_{AP} = 2p' - 1 \,,$$

where
$$p' = \sum_{i=2}^{n} \frac{1}{n-1} p(i) \,,$$

thus highlighting the fact that τ and τ_{AP} belong to the same family of RCCs which is defined as

$$\tau_* = 2p_* - 1 \qquad p_* = \sum_{i=2}^{n} w_i p(i) \,,$$

where

$$\sum_{i=2}^{n} w_i = 1 \qquad w_1 = 0 \qquad w_i > 0 \qquad i = 2, \ldots, n \,,$$

that is, a mixture of proportions of concordant items computed at every rank $i = 2, \ldots, n$ where the w_i's form a series of mixture weights.[1]

Suppose a reference ranking and an observed ranking are to be compared and consider the following experiment performed on the urn including all the items distributed according to the w_i's. First, pick at random an item i other than the top ranked from the reference ranking with probability w_i. Second, pick at random with probability $\frac{1}{i-1}$ an item ranked before i in the observed ranking. Finally, return 1 if the two picked items are ranked in accordance with the reference ranking. The expected value of this process is p_*.

Note that, when using τ_{AP}, the probability that an item i other than the top ranked is picked at random from the reference ranking is uniform and equal to $\frac{1}{n-1}$. This view is derived from Average Precision which underlies τ_{AP}. Indeed, the picking of an item at random corresponds to the selection of a relevant document – as the relevant documents have the same degree of relevance (i.e., relevance is binary), the probability that an item is picked is uniform. On the contrary, τ_* is based on a non-uniform probability (i.e., w_i) of picking an item, thus suggesting the idea that a non-binary or graded relevance underlies its definition – this is indeed the idea described in the following sections.

4 The Problem of Weighting Rank Correlation in Information Retrieval

Let us consider the following toy examples in order to intuitively illustrate the differences between the RCCs – of course, the example may seem a little contrived, but it is a sort of counter-example to show that the RCCs do not provide the same result. Suppose $(1, \ldots 10)$ be the reference ranking and that three observed rankings differ from each other by the rank of the items affected by exchanges:

$$x = (\underline{2}, \underline{1}, 3, 4 \ldots) \qquad y = (\ldots 3, \underline{5}, \underline{4}, 6, \ldots) \qquad z = (\ldots, 8, \underline{10}, \underline{9})$$

---

[1] Note that $w_1 = 0$ by definition since $n \geq 2$ and $p(1)$ is undefined.

At first sight x is very similar to the reference one, however, the two top ranked items are reversed. While the end user would appreciate the difference because the top ranked document in the observed ranking is not on the top, $\tau = 0.96$ and so the two rankings are considered highly correlated.

Although y is a little more similar to the reference ranking – two middle ranked items are reversed while the top three items are unchanged, $\tau = 0.96$ which again suggests concordance, thus indicating that the decision about concordance is insensitive to the rank of the reversed items – indeed, τ does not change even for z. Let us compute τ_{AP} instead: $\tau_{AP} = 0.78$ for x while $\tau_{AP} = 0.94, 0.98$ for y, z, respectively, thus confirming that this RCC is sensitive to the rank of the reversed items.

Let us now consider τ_* and let the weights be, for example,

$$(0.20, 0.18, 0.16, 0.13, 0.11, 0.09, 0.07, 0.04, 0.02) .$$

It follows that $\tau_* = 0.60, 0.93, 0.99$ for x, y, z, respectively, thus showing that, thanks to the decreasing series of weights, the role played in τ_* by the ranks of the reversed items is even more important than in τ and τ_{AP}. This happens because the probability that a top ranked item is picked is higher than the probability of a bottom ranked item. Therefore, the discordances at the top ranks will be highly weighted as in x, while the discordance in z will be little weighted because the probability that the bottom ranked item is picked is low.

This is of course relevant to IR since the criteria used for assigning the weights to the number of concordant items at every rank determine the value of the RCC. Therefore, if τ suggests that two rankings are equivalent, it might be that they agree at the bottom ranks rather than at the top ranks, the latter being a conclusion not always welcome when comparing rankings which are in contrast appreciated if they place the best items on the top.

τ_{AP} reduces this drawback because it gives equal weight to all the concordances. Indeed, a decreasing series of $p(i)$'s is sufficient to say that $\tau_{AP} > \tau$. However, this condition is not necessary and τ_{AP} may fail to discriminate when this series is not decreasing. Let the observed rankings be $u = (1, 3, 2, 4, 5)$ and $v = (1, 2, 4, 5, 3)$. In terms of the number of exchanges, u is more correlated to the reference ranking than v. Indeed, $\tau = 0.80$ for u and $\tau = 0.60$ for v. Nevertheless, τ_{AP} provides the same value (i.e. 0.75) for both the observed rankings, thus not discriminating as τ does. The reason lies in the proportion of concordant items. When these probabilities are computed, one obtains the series $p(2) = 1, p(3) = \frac{1}{2}, p(4) = 1, p(5) = 1$ for u and the series $p(2) = 1, p(3) = 1, p(4) = 1, p(5) = \frac{1}{2}$ for v.

However, while τ_{AP} is based on the notion of Average Precision, thus making it more sound, τ_* does in contrast still lack of such a basis. The next section provides this basis.

5   A Methodology for Weighted Rank Correlation in Information Retrieval

The mixture weights thus play an important role in weighting rank correlation. At this point, there are two main methodological issues to which a great deal of attention should be paid. From the one hand, an issue is about how the mixture weights w_i's should be defined in order to compare two rankings from a IR perspective. On the other hand, the probability distribution of the RCC family has to be defined in order to compare two rankings and to decide if they are significantly different from a statistical point of view. To address these two methodological issues, in this paper, two methodological results are introduced, that is, the the notion of Cumulative Gain presented in [3] and weighted Kendall's τ presented in [4].

5.1   The Weighted Kendall's Rank Correlation Coefficient

The weighted kendall's τ, that is, τ_w has been illustrated in [4] and is defined in this paper as

$$\tau_w = \frac{\sum_{i=1}^{n} \sum_{j=1, i\neq j}^{n} v_{ij} I_{ij}}{\sum_{i=1}^{n} \sum_{j=1, i\neq j}^{n} v_{ij}} \ ,$$

where I is an indicator function such that $I_{ij} = 1$ if items i, j are concordant (i.e. they are ranked in the observed ranking in the same order as the order in the reference ranking), $I_{ij} = -1$ if items i, j are discordant (i.e. they are ranked in the observed ranking in the opposite order as the order in the reference ranking) and v_{ij} is the weight assigned to the concordance or discordance occurring between items i and j. This RCC assigns different weights depending on the ranks i, j. After a few passages, it can be shown that

$$\tau_w = 2\frac{\sum_{i=1}^{n} \sum_{j=1, i\neq j, I_{ij}=1}^{n} v_{ij}}{\sum_{i=1}^{n} \sum_{j=1, i\neq j}^{n} v_{ij}} - 1$$

where the term multiplied by 2 resembles the p of τ. Supposes that $v_{ij} = v_{ji}$, that is, the con/discordance weights are symmetric. In this case,

$$\tau_w = 2\frac{\sum_{i=1}^{n} \sum_{j<1, I_{ij}=1}^{n} v_{ij}}{\sum_{i=1}^{n} \sum_{j<1}^{n} v_{ij}} - 1 \ .$$

Suppose, now, that the weights are constant for every i, that is, $v_{ij} = v_i$ and then are independent of j – this means that the concordance (or discordance) between, say, the fifth item and the first item is equally treated as the concordance between the fifth item and the fourth item. This is actually what is assumed by many RCCs such as τ and τ_{AP}. It follows that (when $v_{ij} = v_i$)

$$\tau_w = 2\frac{\sum_{i=2}^{n} C(i)v_i}{\sum_{i=2}^{n} (i-1)v_i} - 1 \ .$$

As $C(i) = (i - 1)p(i)$, it is found that

$$\tau_w = 2\frac{\sum_{i=2}^n (i - 1)v_i p(i)}{\sum_{i=2}^n (i - 1)v_i} - 1 .$$ (1)

If

$$w_i = \frac{(i - 1)v_i}{\sum_{i=2}^n (i - 1)v_i}$$

one can easily see that $\tau_w = \tau_*$.

It may be noted that, when $v_{ij} = 1$, the weighted Kendall's τ_w turns into τ, and when $v_{ij} = \frac{1}{i-1}$, it turns into τ_{AP}.

Hence, the weighted Kendall's τ_w can be utilized as a basis to decide the mixture weights of τ_*. Indeed, the definition of the v_i's is sufficient to achieve the mixture weights of τ_*. At this aim, one needs a basis to assign a value to the v_i's before calculating the mixture weights. As the v_i's need not to be normalized, that basis does only define the v_i's as non-negative real values. In the next section, the basis proposed in this paper is illustrated.

5.2   Gains in Information Retrieval Evaluation

The notion of gain and of cumulative gain for IR evaluation was proposed in [3]. In this section, this notion is briefly reviewed before illustrating how it has been exploited to define the basis for computing the mixture weights of τ_* (and, in general, of τ_w).

The value $G(r_i)$ of the item ranked at i is the degree of relevance, interest or usefulness (in general, the "gain") of the item to the user; for example, when graded relevance is used, $G(r_i)$ may range between, say, 0 and 3. There are search scenario where, for example, the users are asked to provide judgments from 0 to 7.

The notion of gain is useful in IR thanks to the cumulative gain (CG) of item ranked at i, that is, $CG(r_i)$, which is the sum of the gains computed over all the items ranked not after r_i. Let for example be $2, 3, 0, 1, 2, 1, 0, 0, 0, 0$ the relevance degrees of ten documents ranked by, say, similarity to a query. The degree of the fourth item is $G(r_4) = 1$, while its cumulative gain, that is, $CG(r_4)$ is $2 + 3 + 0 + 1 = 6$.

5.3   Weighted Rank Correlation Using Gain

The selection of the mixture weights of τ_* might somehow be arbitrary – an experimenter may choose a series of mixture weights different from that chosen by another experimenter and the evaluation results may hardly be comparable. This may turn out to be a danger whenever competitor observed rankings are compared with the reference ranking, especially if the details of the experiments are incomplete or imprecise. The problem is to define, in a principled way, the mixture weights so that the concordances, and the proportion of concordant items occurring at the top ranks are given higher mixture weight than those occurring at the bottom ranks. For making the choice of the mixture weights less arbitrary, the notions of gain and of cumulative gain for IR evaluation proposed in [3] and briefly reviewed in Section 5.2, have been utilized in this paper.

The basic idea underlying the utilization in this paper of the notions of gain and of cumulative gain for IR evaluation proposed in [3] has been that w_i should measure the *preference* expressed by the user in picking the item at rank i which is then used to count the number of concordant items ranked before i. The more relevant, useful or interesting the item is perceived, the higher the probability this item is considered for measuring the concordance. In this way, the concordances with the items perceived more relevant than others are weighted higher than the concordances with the least relevant items. This criterion is rendered as

$$v_i = \frac{G(r_i)}{i - 1} , \tag{2}$$

that is, the concordances with an item ranked at i will be weighted proportionally to the gain of the item. It follows that

$$w_i = \frac{G(r_i)}{\sum_{i=2}^{n} G(r_i)} , \tag{3}$$

thus showing that the mixture weight depends on both the degree of relevance of r_i and the rank of r_i in the observed ranking.

Suppose that r_i has been picked at random with probability w_i. If $G(r_i)$ is relatively high, then the dis/concordances will highly be weighted – the concordances will have a relatively high $p(i)$ which is multiplied by w_i. Suppose also that the reference ranking is based on G, that is, the items are ranked by G in the reference ranking so that the most relevant are on the top of the list. When the number $C(i)$ of concordances at rank i with the reference ranking is high, many other items that are at least as relevant as the picked item are ranked on top of the observed ranking, thus indicating that the observed ranking is highly correlated with the reference ranking. The number of concordances between pairs of highly relevant items will be weighted higher than number of concordances between pairs of less relevant items due to $G(i)$ in Equation 3.

Suppose, for example, that the items of the reference ranking $(1, 2, 3, 4, 5)$ are provided with the gains $3, 3, 2, 1, 1$. Thus, the w_i's become $\frac{3}{7}, \frac{2}{7}, \frac{1}{7}, \frac{1}{7}$. There are 7 concordant pairs both in the observed ranking $(1, 2, 5, 4, 3)$ and in the observed ranking $(3, 1, 2, 5, 4)$. The value of τ_* for the first observed ranking is then $2(\frac{3}{7} \times 1 + \frac{1}{7} \times \frac{2}{3} + \frac{1}{7} \times \frac{2}{3} + \frac{2}{7} \times \frac{2}{4}) - 1$, while the value of τ_* for the second ranking becomes $2(\frac{2}{7} \times 0 + \frac{3}{7} \times \frac{1}{2} + \frac{1}{7} \times \frac{3}{3} + \frac{1}{7} \times \frac{3}{4}) - 1$.

5.4 The Probability Distribution of the Family of Rank Correlation Coefficients

To judge the significance of an observed value of rank correlation with respect to a reference ranking, it is necessary to compare the observed ranking with that which would be observed if the ranking were formed randomly, i.e. it could be observed with uniform probability. Suppose that the null hypothesis of incorrelation holds. In [4] it was shown that τ_w is approximately Normal. In particular, when $v_{ij} = v_{ji} = v_i$, it can be shown that τ_* is approximately Normal with mean zero and variance $4\sigma_v^2/9n\bar{v}$ where $\bar{v} = \left(\sum_i v_i\right)/n$ and $\sigma_v^2 = \left(\sum_i v_i^2\right)/n$.

As the v_i's can easily defined using the $G(r_i)$'s, the sampling distribution is readily available and the observed value of the τ_* can be checked against the Normal distribution for testing the null hypothesis of non-correlation, that is,

$$\tau_* \sim N(0, 4\sigma_G^2/9n\bar{G}) \tag{4}$$

where $\bar{G} = \sum_i \frac{G(r_i)}{i-1}/n$ and $\sigma_G^2 = \sum_i \left(\frac{G(r_i)}{i-1}\right)^2/n$.

After substituting v_i with $\frac{1}{i-1}$, the limiting probability distribution of τ_{AP} can easily be obtained.

The relevance of the limiting convergence to the Normal distribution is that the experimenter can infer if an observed ranking, and then the algorithm which produced it, is significantly different from the reference ranking produced by an alternative algorithm.

6   Conclusions and Future Work

While rank correlation has extensively been studied in many different domains, it was rarely addressed by paying attention to the particular issues of IR. One of these issues is related to the importance of the items, the latter being an aspect often not considered in other domains because the rankings are relatively short or the items are not provided with a degree of relevance. In IR, on the contrary, the items are often provided with a degree of relevance, thus making the error of placing a highly relevant item on the bottom worse than the error of placing a little relevant item.

The proposed RCC family can have an impact on evaluation since provides the experimenter with a measure to compare rankings when the degree of relevance are non-binary and has to play a role in the comparison. However, τ_* cannot always be applied. Indeed, there may be situations in which the degree of relevance is not available or does not make sense; for example, when comparing retrieval systems, it is difficult to say that a system is more important than another.

References

1. Kendall, M.: A new measure of rank correlation. Biometrika 30(1/2), 81–93 (1938)
2. Sillitto, G.: The distribution of Kendall's τ coefficient of rank correlation in rankings containing ties. Biometrika 34(1/2), 36–40 (1947)
3. Jarvëlin, K., Kekäläinen, J.: Cumulated gain-based evaluation of IR techniques. ACM Transactions on Information Systems 20(4), 422–446 (2002)
4. Shieh, G.: A weighted Kendall's tau statistic. Statistics and Probability Letters 39(1), 17–24 (1998)

 5. Yilmaz, E., Aslam, J., Robertson, S.: A new rank correlation coefficient for information retrieval. In: Proceedings of the ACM International Conference on Research and Development in Information Retrieval (SIGIR), pp. 587–594 (2008)
 6. Fagin, R., Kumar, R., Sivakumar, D.: Comparing top k lists. SIAM Journal of Discrete Mathematics 17(1), 134–160 (2003)
 7. Amento, B., Terveen, L., Hill, W.: Does authority mean quality? Predicting expert quality ratings of web documents. In: Proceedings of the ACM International Conference on Research and Development in Information Retrieval (SIGIR), pp. 296–303 (2000)
 8. Baeza-Yates, R., Castillo, C., Marín, M., Rodríguez, A.: Crawling a country: better strategies than breadth-first for web page ordering. In: Proceedings of the World Wide Web Conference, pp. 864–872 (2005)
 9. Broder, A.Z., Lempel, R., Maghoul, F., Pedersen, J.: Efficient pagerank approximation via graph aggregation. Journal of Information Retrieval 9(2), 123–138 (2006)
10. Broder, A., Fontoura, M., Josifovski, V., Riedel, L.: A semantic approach to contextual advertising. In: Proceedings of the ACM International Conference on Research and Development in Information Retrieval (SIGIR), pp. 559–556 (2007)
11. Hauff, C., Murdock, V., Baeza-Yates, R.: Improved query difculty prediction for the web. In: Proceedings of the ACM Conference on Information and Knowledge Management (CIKM), pp. 439–448 (2008)
12. Sanderson, M., Joho, H.: Forming test collections with no system pooling. In: Proceedings of the ACM International Conference on Research and Development in Information Retrieval (SIGIR), pp. 25–29 (2004)
13. Aslam, J., Pavlu, V., Yilmaz, E.: A statistical method for system evaluation using incomplete judgments. In: Proceedings of the ACM International Conference on Research and Development in Information Retrieval (SIGIR), pp. 541–547 (2006)
14. Voorhees, E.: Evaluation by highly relevant documents. In: Proceedings of the ACM International Conference on Research and Development in Information Retrieval (SIGIR), pp. 74–82 (2001)
15. Bailey, P., Craswell, N., Soboroff, I., Thomas, P., de Vries, A.P., Yilmaz, E.: Relevance assessment: Are judges exchangeable and does it matter? In: Proceedings of the ACM International Conference on Research and Development in Information Retrieval (SIGIR), pp. 667–674 (2008)
16. Büttcher, S., Clarke, C.L.A., Yeung, P.: Reliable information retrieval evaluation with incomplete and biased judgements. In: Proceedings of the ACM International Conference on Research and Development in Information Retrieval (SIGIR), pp. 63–70 (2008)
17. Sakai, T.: Alternatives to bpref. In: Proceedings of the ACM International Conference on Research and Development in Information Retrieval (SIGIR), pp. 71–78 (2007)
18. Sakai, T.: Comparing metrics across trec and ntcir: The robustness to system bias. In: Proceedings of the ACM Conference on Information and Knowledge Management (CIKM), pp. 581–590 (2008)
19. Carterette, B., Pavlu, V., Kanoulas, E., Aslam, J., Allan, J.: Evaluation over thousands of queries. In: Proceedings of the ACM International Conference on Research and Development in Information Retrieval (SIGIR), pp. 651–658 (2008)
20. Carterette, B.: Robust test collections for retrieval evaluation. In: Proceedings of the ACM International Conference on Research and Development in Information Retrieval (SIGIR), pp. 55–62 (2007)

21. Webber, W., Moffat, A., Zobel, J.: Score standardization for inter-collection comparison of retrieval systems. In: Proceedings of the ACM International Conference on Research and Development in Information Retrieval (SIGIR), pp. 51–58 (2008)

22. Callan, J., Connell, M.: Query-based sampling of text databases. ACM Transactions on Information Systems 19(2), 97–130 (2001)

23. Caverlee, J., Liu, L., Bae, J.: Distributed query sampling: a quality-conscious approach. In: Proceedings of the ACM International Conference on Research and Development in Information Retrieval (SIGIR), pp. 340–347 (2006)

24. Song, Y., Zhuang, Z., Li, H., Li, Q., Lee, W.C., Giles, C.: Real-time automatic tag recommendation. In: Proceedings of the ACM International Conference on Research and Development in Information Retrieval (SIGIR), pp. 515–522 (2008)

25. Carmel, D., Yom-Tov, E., Darlow, A., Pelleg, D.: What makes a query difficult? In: Proceedings of the ACM International Conference on Research and Development in Information Retrieval (SIGIR), pp. 6–11 (2006)

26. de Moura, E., dos Santos, C., Fernandes, D., da Silva, A., Calado, P., Nascimento, M.: Improving web search efficiency via a locality based static pruning method. In: Proceedings of the World Wide Web Conference, pp. 235–244 (2005)

27. Geng, X., Liu, T.Y., Qin, T., Li, H.: Feature selection for ranking. In: Proceedings of the ACM International Conference on Research and Development in Information Retrieval (SIGIR), pp. 407–414 (2007)

28. Melucci, M.: On rank correlation in information retrieval evaluation. SIGIR Forum 41(1) (June 2007)

Extractive Summarization Based on Event Term Temporal Relation Graph and Critical Chain

Maofu Liu[1], Wenjie Li[2], and Huijun Hu[1]

[1] College of Computer Science and Technology,
Wuhan University of Science and Technology, Wuhan, P.R.China
{liumaofu,huhuijun}@wust.edu.cn
[2] Department of Computing, The Hong Kong Polytechnic University, Kowloon, Hong Kong
cswjli@comp.polyu.edu.hk

Abstract. In this paper, we investigate whether temporal relations among event terms can help improve event-based extractive summarization and text cohesion of machine-generated summaries. Using the verb semantic relation, namely *happens-before* provided by VerbOcean, we construct an event term temporal relation graph for source documents. We assume that the maximal weakly connected component on this graph represents the main topic of source documents. The event terms in the temporal critical chain identified from the maximal weakly connected component are then used to calculate the significance of the sentences in source documents. The most significant sentences are included in final summaries. Experiments conducted on the DUC 2001 corpus show that extractive summarization based on event term temporal relation graph and critical chain is able to organize final summaries in a more coherent way and accordingly achieves encouraging improvement over the well-known tf*idf-based and PageRank-based approaches.

1 Introduction

Extractive summarization selects the most representative sentences from source documents. Under the extractive summarization framework, event has been regarded as an effective concept representation in recently emerged event-based summarization, which extracts salient sentences from single document or multiple documents and re-organizes them in a machine-generated summary according to how important the events that sentences describe are. With regard to the definition of events, in conjunction with the common agreement that event contains a series of happenings, we formulate events as "[Who] did [What] to [Whom] [When] and [Where]" at sentence level in the context of event-based summarization. In this paper, we focus on "did [What]" and approximately define verbs and action nouns in source documents as *event terms* that characterize or partially characterize event occurrences.

Notice that in addition to the quality and the quantity of the informative contents conveyed in the extracted sentences, the relations among the extracted sentences, such as temporal relations in news articles, and structure of the final summary text should be a matter of concern. The sentence relations in source and summary text, if appropriately defined and identified, are a good means to reflect the text cohesion, i.e. the way of getting the source and extracted text to "hang together" as a whole and the

G.G. Lee et al. (Eds.): AIRS 2009, LNCS 5839, pp. 87–99, 2009.

indicator of text unity. In the literature, text cohesion has been modeled by lexical cohesion in terms of the semantic relations existing between not only pairs of words but also over a succession of a number of nearby related words spanning a topical unit of the text. These sequences of related words are called lexical chains and tend to delineate portions of the text that have a strong unity of meaning.

Lexical chains have been investigated for extractive summarization in the past. They are regarded as a direct result of units of text being "about the same thing" and having a correspondence to the structure of the text. Normally, nouns or noun compounds are used to denote the things and compute lexical chains (i.e. lexical chains are normally noun chains). In this paper, we assume that the source text describes a series of events via the set of sentences and take both informative content and structure of the source text into consideration. We look for the event term *temporal critical chain* in the event term temporal relation graph and use it to represent the source text and to generate the final summary. We concentrate on verb chains, other than noun chains, aiming at improving event-based summarization and lexical cohesion of the generated summaries. Here, event terms and event term chain characterize informative content and text cohesion, respectively.

To compute the event term temporal critical chain, event terms are connected to construct an *event term temporal relation graph* based on the *happens-before* relations provided in VerbOcean. This type of relations indicates that the two verbs refer to two temporally disjoint intervals or instances [16]. The DFS-based (Depth-First Search) algorithm is applied in searching the temporal critical chain. Then the event terms in the temporal critical chain are used to evaluate sentences. The sentences with the highest significance scores are extracted to form the final summary.

The remainder of this paper is organized as follows. Section 2 reviews related work. Section 3 introduces the proposed event-based summarization approach based on event term temporal relation graph and critical chain. Section 4 then presents experiments and discussions. Finally, Section 5 concludes the paper and suggests the future work.

2   Related Work

Daniel et al. [1] pilot the study of event-based summarization. They recognize a news topic as a series of sub-events according to human understanding of the topic and investigate whether identifying sub-events in a news topic can help capture essential information in order to produce better summaries. Filatova and Hatzivassiloglou [2] then define the concept of atomic events as a feature that can be automatically extracted. Atomic events are defined as the relations between the important named entities. The proposed approach is claimed to outperform conventional tf*idf approach and experimental results indicate that event is indeed an effective feature for producing better summaries. Allan et al. [3] present a list of events within the topic in the order those events are reported and produce a revised up-to-date summary at regular time intervals. Afantenos et al. [5] discuss the techniques to summarize events happened in predictable time synchronously. Relations between events are defined on the axes of time and information source. Lim et al. [4] group source documents on time slots by the time information given in newspaper articles or publication dates. They build the local or global term cluster of each time slot and use it to identify a topical sentence as the representative for the time slot. Jatowt and Ishizuka [6] introduce the approaches to monitor the trends of dynamic web documents, which are different versions of documents on the time line. Based on distributions, terms are scored in order to identify whether they are popular and active. They employ a simple regression analysis of word frequency and time. Wu et al. [7] investigate whether time features help improve event-based summarization. After anchoring events on the time line, two different statistical measures, $tf*idf$ and x^2, are employed to identify importance of events on each date.

The concept of lexical chain is originally proposed to represent the discourse structure of a document by Morris and Hirst [8]. They define lexical chain as a cluster of semantically related terms and construct lexical chains manually from Roget's Thesaurus according to the distance between the occurrences of related nouns and use lexical chain as an indicator of lexical cohesion of the text structure and the semantic context for interpreting words, concepts and sentences. Barzilay and Elhadad [9] first introduce lexical chain in single document summarization. They produce a summary of an original text relying on a model of the topic progression in the text derived from lexical chains. The lexical chains are computed using relatedness of nouns determined in terms of the distance between their occurrences and the shape of the path connecting them in the WordNet thesaurus. Following the same line of thought, Silber and McCoy [10, 11] employ lexical chains to extract important concepts from the source document and make lexical chains a computationally feasible candidate as an intermediate representation. Doran et al. [12] highlight the effect of lexical chain scoring metrics and sentence extraction techniques in summary generation. Zhou et al. [13] adapt lexical chains derived from WordNet based on noun compounds, noun entries and name entities to multi-document and query based summarization. Reeve et al. [14, 15] apply lexical chain based summarization approach to biomedical text. They build concept chains to link semantically-related concepts. The concepts are not derived from WordNet but domain-specific semantic resources, such as UMLS Metathesaurus and semantic network. The resulting concept chains are used to identify candidate sentences useful for summarization.

At present, the applications of temporal information in summarization are mostly based on time information in the source document or publication date. Meanwhile, the lexical chains mentioned above are all based on nouns, derived from WordNet or domain specific knowledge base. In this paper, we derive temporal information from the temporal relations among event terms and regard the event term temporal chains as the immediate representation of the source documents.

3 Summarization Based on Event Term Temporal Relation Graph and Critical Chain

In this section, we first illustrate how the event term temporal relation graph is constructed based on *happens-before* relation in VerbOcean. Then we explain how the event term temporal critical chain is determined from the temporal graph. Finally, sentence selection based on the significance of the event terms in the temporal critical chain is introduced.

3.1 Event Term Temporal Relation Graph Construction

In this paper, we introduce VerbOcean, a broad-coverage repository of semantic verb relations, into event-based extractive summarization. Different from other thesaurus like WordNet, VerbOcean provides five types of semantic relations among verbs at finer level. This just fits in with our idea to introduce event term semantic relations into summarization. In this paper, only *happens-before* temporal relation is explored. When two events happen, one may happen before the other. This is defined as *happens-before* temporal relation in VerbOcean. Examples of *happens-before* relations are illustrated below.

"wit" *happens-before* "record"
"move" *happens-before* "run"

The happens-before temporal relations on a set of event terms can be naturally represented by a graph, called event term temporal relation graph. We formally define the event term temporal relation graph connected by temporal relation as G=(V, E), where V is a set of event term vertices and E is a set of edges temporally connecting event terms. Fig.1 below shows a sample of event term temporal relation graph built from a DUC 2001 document set.

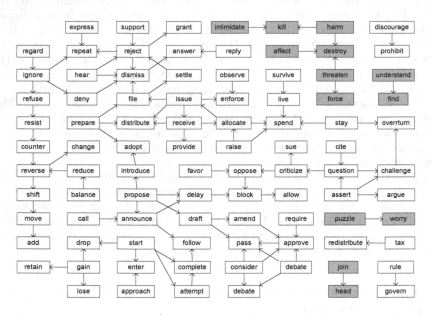

Fig. 1. Event term temporal graph based on happens-before relation

As we know, the graph is directed if the relation has the property of the anti-symmetric and undirected otherwise. Certainly, the event term temporal relation graph is a directed graph because *happens-before* relation in VerbOcean clearly exhibits the conspicuous anti-symmetric property. For example, one may "question" something and then decide to "criticize" it for some reason. The event represented by the term "question" *happens-before* the one represented by the term "criticize". So, a directed edge from "question" to "criticize" appears in Fig. 1.

The *happens-before* relation is also anti-reflexive because each event term cannot *happens-before* itself. This means that there is no self-loop at each term vertex. There are no parallel edges between any two adjacent term vertices either. Therefore, we can say that the event term temporal relation graph is a simple directed graph. In fact, *happens-before* relation is a transitive one, though there may not be an explicit edge from event term et_a to et_c when the edges both from et_a to et_b and from et_b to et_c exist. For example, there is no edge from "regard" to "repeat", but the edges exist from "regard" to "ignore" and from "ignore" to "repeat" in Fig. 1.

3.2 Event Term Temporal Critical Chain Identification

The event term temporal graph based on *happens-before* relation is not a fully connected graph. For example, there are eight sub-graphs or components in the graph illustrated in Fig. 1. Among them, a maximal weakly connected component, which contains the maximal number of the event terms, can be found. We assume that the event terms in the maximal weakly connected component reflect the main topic of original documents since such a component normally involves much more connected (i.e. relevant) event terms than any other components on the graph. Referring back to Fig. 1, the maximal weakly connected component contains 118 event terms, while the second largest weakly connected component contains only 8. Note that, for illustration purpose, only a partial maximal weakly connected component is shown in Fig. 1.

Some maximal weakly connected sub-graphs are cyclic. The graph in Fig. 2(a), a part of maximal weakly connected sub-graph in Fig. 1, is cyclic. We can see that there are cyclic relations among the three event terms "issue", "receive" and "allocate". In such a situation, the edge whose terminal term vertex has the maximal in-degree is removed in order to avoid the infinite loop in the identification of the event term temporal critical chain. Anyway, the terminal term can still be reached from other term vertices. The connection remains. For example, we remove the edge from "receive" to "allocate" in Fig. 2(a) and obtain a directed acyclic graph in Fig. 2(b).

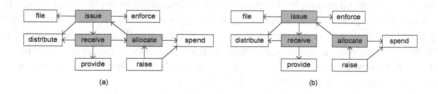

(a)   (b)

Fig. 2. The cyclic and acyclic graph part

From the directed acyclic graph, we extract all the source vertices and the sink vertices. The source vertex is defined as a vertex with successors but no predecessors, i.e. the edges being incident out of it but no edge being incident on it. On the contrary, the sink vertex is defined as a vertex with predecessors but no successors, i.e. the edges being incident on it but no edge being incident out of it. All source vertices and sink vertices in the directed acyclic graph of the maximal weakly connected component in Fig. 1 is shown in Fig. 3.

source vertices:
{regard, call, prepare, balance, hear, express, support, start, approach, require, debate, assert, cite, stay, survive, reply, observe, raise, gain, propose}

sink vertices:
{add, allow, answer, challenge, retain, lose, enter, change, repeat, adopt, provide, sue, spend, grant, overturn, argue, enforce, distribute }

Fig. 3. The source vertices and sink vertices of the directed acyclic graph

All directed paths from each source vertex to each sink vertex in the directed acyclic graph of the maximal weakly connected component are computed by the DFS-based algorithm. The longest path is defined as the event term temporal critical chain. The directed paths in Fig. 4 (a)-(d) are four event term chains found in Fig. 1 and the one in Fig. 4(a) is identified as the event term temporal critical chain.

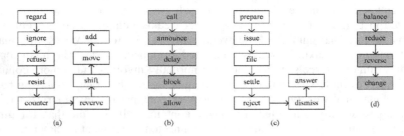

Fig. 4. The event term temporal critical chain for Fig. 1

Afterwards, we evaluate the sentences that contain the event terms in the chain and determine which ones should be extracted into the final summary based upon the event term temporal critical chain computation.

3.3 Sentence Selection

To apply the event term temporal critical chain in summarization, we need to identify the most significant sentences that best describe the event terms in the chain. Considering terms are the basic constitution of sentences, term significance is computed first. Since this paper studies event-based summarization, we only consider event terms and compute the significances of the event terms in the maximal weakly connected component of an event term temporal relation graph.

Two parameters are used in the calculation of event term significance. One is the occurrence of an event term in source documents. The other is the degree of an event term in an event term temporal relation graph. The degree of a term vertex in the directed graph is the sum of the in-degrees and out-degrees of that term.

For each event term in the temporal critical chain, it is likely to locate more than one sentence containing this term in source documents. We extract only one sentence for each event term to represent the event in order to avoid repeating the same or quite similar information in the summary. For sentence selection, the sentence significance is computed according to the event terms contained in it.

Based on event term occurrences, the significance of a sentence is calculated as

$$SC_i = \frac{TFS_i}{TFS_m} \qquad (1)$$

where TFS_i and TFS_m are the sum of the term occurrences of the ith sentence and the maximum of all the sentences that contain event terms in the temporal critical chain, respectively.

Alternatively, we can use degrees of event terms to calculate the significance of a sentence.

$$SC_i = \frac{DS_i}{DS_m} \qquad (2)$$

where DS_i and DS_m are the sum of the term degrees of the ith sentence and maximum of all the sentences which contain event terms in the temporal critical chain, respectively.

It should be emphasized here that the event terms under concern in Equations (1) and (2) must be the ones in the maximal weakly connected component of the event term temporal relation graph.

4 Experiment and Discussion

We evaluate the proposed summarization approach based on the event term temporal relation graph and critical chain on the DUC 2001 corpus. The corpus contains 30 English document sets. Among them, 10 sets are observed to contain descriptions of event sequences. They are the main concern of this paper. All final summaries are generated in 200 words length.

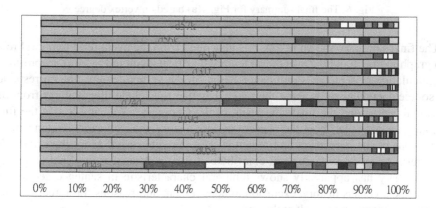

Fig. 5. The vertex number of the weakly connected components

The number of term vertices in the weakly connected component for each of 10 sets is illustrated in Fig. 5. Each bar denotes all weakly connected components of an event term temporal relation graph and the length of each series represents the vertex number in the weakly connected component. In Fig. 5, we can see that the numbers of the term vertices in maximal weakly connected components are actually much larger than the numbers of the term vertices in other components on the event term temporal relation graphs. For example, the maximal weakly connected component of the document set "d30e" almost includes all the vertices on its event term temporal relation graph. The other three document sets, i.e. "d05a", "d13c" and "d50i", occupy above 90% vertices. The minimum of the vertex number of the maximal weakly connected component in the "d04a" document set also includes almost 30% vertices. The assumption that the main topic of a document set can be represented by the maximal weakly connected component is not unreasonable. The maximal weakly connected component roughly describes all the events about main topic of the source documents in detail.

For 190 years, said Sen. Daniel Patrick Moynihan (D-N.Y.), the federal government has counted all inhabitants without regard to citizenship in accordance with the Constitution's provisions.
Groups which have filed suit to ignore the aliens contend large concentrations of them could result in some states gaining seats in the House of Representatives at the expense of other states.
Asking people about their status likely would result in people lying or refusing to participate in the count, officials say, resulting in a potential undercount of residents in many areas.
But he added that he is "optimistic, cautiously optimistic," that House conferees would resist the Senate-approved ban and not force Bush to veto the legislation.
Sen. Pete Wilson (R-Calif.) countered that excluding illegal residents from the decennial census is unfair to the states that have suffered from a huge influx of immigration beyond the legal limits.
The Senate's action was sharply criticized by Undersecretary of Commerce Michael Darby, but he voiced hope that it would be reversed by a Senate-House conference.
There could be enough of them to shift seats away from at least five states to Sun Belt states with large numbers of illegal residents.
The amendment was adopted after the Senate voted 58 to 41 against a move to reject it, and 56 to 43 against scuttling it as unconstitutional.

Fig. 6. The final summary for Fig. 4(a) based on vertex degree

The final summary shown in Fig. 6 is generated from the event term temporal relation graph in Fig. 1 and the event term temporal critical chain in Fig. 4(a) according to the calculation of sentence significance based on the vertex degree. The corresponding source document set is about the topic "whether to exclude illegal aliens from the decennial census and the final vote result of the Congress". We can see that the final summary indeed talks about the resident census history, the exclusion announcement, the reasons of the agreement and rejection, and the vote result. More important, the event term temporal critical chain contributes to the cohesion of the final summary. The summary has apparently shows temporal characteristic in sentence sequences, from resident census history, exclusion announcement and reasons for agreement and rejection to Senate vote result at the end.

To evaluate the quality of generated summaries, an automatic evaluation tool, called ROUGE [17] is used. The tool presents three ROUGE values including unigram-based ROUGE-1, bigram-based ROUGE-2 and ROUGE-W which is based on longest common subsequence weighted by the length.

In Fig. 7, the tf*idf approach is based on term frequency and inverse document frequency (a well-known statistical feature used in automatic summarization). We calculate tf*idf weights for all the words excluding stop-words and evaluate sentence significance using the SUM of tf*idf weights of all words occurring in the sentence. For the term occurrence approach, Equation (1) is adopted to calculate the significance of the sentence containing the event terms in the temporal critical chain using term occurrences. The comparative experiment results for the selected ten document sets are illustrated below.

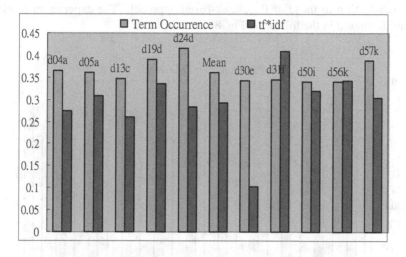

Fig. 7. ROUGE-1 scores on Term Occurrence and tf*idf

Except for the document sets "d31f" and "d56k", the experiment results based on event term occurrence in the critical temporal chains are all better than those based on tf*idf. In particular, event term occurrence achieves about 47.2% improvement on ROUGE-1 comparing to tf*idf for the document set "d24d".

Table 1. ROUGE scores on Term Occurrence and tf*idf

	tf*idf	Term Occurrence	Improvement
ROUGE-1	0.29222	0.36118	23.6%
ROUGE-2	0.04494	0.06254	39.2%
ROUGE-W	0.10099	0.12877	27.5%

Table 1 above shows the average ROUGE scores of event term occurrence and tf*idf approaches on the ten selected document sets. The ROUGE-2 score of the event term occurrence approach is comparatively about 39.2% better than the tf*idf approach. Our approach also shows more advantageous than the tf*idf approach on both ROUGE-1 and ROUGE-W. We highlight the ROUGE-2 scores here because we take semantic relevance between the events into account but tf*idf does not.

Google's PageRank [18] is one of the most popular ranking algorithms. It is a kind of graph-based ranking algorithm deciding on the importance of a node within a graph by taking into account the global information recursively computed from the entire graph. After constructing the event term temporal relation graph, we can also use the PageRank algorithm to calculate the significance of the terms in the graph. Because the calculations of sentence significance using PageRank and vertex degree are both based on the links among the vertices in the graph, we compare the ROUGE scores of the event term temporal critical chain based summarization using the vertex degree in sentences selection to those of PageRank-based approach. The experiment results of them are illustrated in the following Fig. 8.

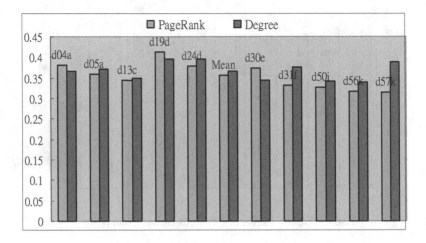

Fig. 8. ROUGE-1 scores on PageRank and Degree

Except for the document sets "d31f" and "d57k", the experiment results of event term degree approach are quite close to those based on the PageRank algorithm. For "d57k", the ROUGE-1 score of the event term degree approach is 0.38744 while the same of the PageRank algorithm is only 0.31385. It is about 23.4% of improvement. Table 2 below compares the average ROUGE scores of the two approaches. The ROUGE-1 score of the event term degree approach is about 2.5% above the ROUGE-1 score of the PageRank algorithm while the ROUGE-2 scores of them are quite similar. This is mainly because both the degree and the PageRank approaches take the semantic relevance between event terms into consideration in the calculation of significances of event terms and sentences.

Table 2. ROUGE scores on PageRank and Degree

	PageRank	Degree	Improvement
ROUGE-1	0.35645	0.36546	2.5%
ROUGE-2	0.06403	0.06490	1.6%
ROUGE-W	0.12504	0.13021	4.1%

Table 3. ROUGE scores on Term Occurrence and Degree

	Term Occurrence		Degree	
	Ten Sets	Twenty Sets	Ten Sets	Twenty Sets
ROUGE-1	0.36118	0.30453	0.36546	0.30099
ROUGE-2	0.06254	0.04390	0.06490	0.04449
ROUGE-W	0.12877	0.10861	0.13021	0.10803

Table 3 finally compares the average ROUGE scores of the ten document sets selected for the experiments above and the other twenty document sets in the DUC 2001 corpus with term occurrence and event term degree respectively. The ROUGE-1 average scores of the twenty document sets are much worse than the average scores of the ten document sets, i.e. about 15.7% lower using term occurrence and 17.6% lower using event term degree. The similar conclusions can be drawn on ROUGE-2 and ROUGE-W. This suggests that the proposed event-based approaches indeed can handle those documents describing events or event sequences much better. But they may not suit event irrelevant topics.

While the previously presented results are evaluated on 200 word summaries, now we move to check the results in the other three different sizes, i.e. 50, 100 and 400 words and the experiments results based on Degree and PageRank are in Table 4.

Table 4. ROUGE scores on PageRank and Degree with different summary lengths

Degree	50	100	400	PageRank	50	100	400
ROUGE-1	0.22956	0.28987	0.42475	ROUGE-1	0.20477	0.27250	0.42907
ROUGE-2	0.02550	0.04471	0.10186	ROUGE-2	0.02063	0.03933	0.10393
ROUGE-W	0.09655	0.11557	0.13763	ROUGE-W	0.08759	0.10849	0.13839

The experiment results in Table 4 show that our approach using event term degree makes much better results in 50 and 100 words summaries. We can also find that our approach prefers shorter summaries comparing with the PageRank approach in 400 words summaries. Of course, we need to test on more data in the future.

5   Conclusions and Future Work

In this paper, we investigate whether temporal relation between events helps improve performance of event-based summarization. By constructing the event term temporal graph based on the semantic relations derived from the knowledge base VerbOcean and computing weakly connected components, we find that the maximal weakly connected component can often denote the main topic of the source document.

Searching in the maximal weakly connected component with DFS-based algorithm, we can discover an event term temporal critical chain. The event terms in this chain are supposed to be critical for generating the final summary. In experiments, the significance of these terms is measured by either term occurrence in source documents or the degree in the constructed graph. The ROUGE results are promising.

Term occurrence significantly outperforms tf*idf and term degree is also comparative to well-known PageRank.

In the future, we will introduce the other types of VerbOcean verb semantic relations into event-based summarization. Besides the temporal critical chain in the event term relation graph, we will investigate the other possible event term temporal chains in the semantic computation. We also plan to combine the surface statistical features and the semantic features during the selection of representative sentences in order to generate better summaries.

Acknowledgements

The work described in this paper was supported partially by a grant from the Research Grants Council of the Hong Kong Special Administrative Region, China (Project No. CERG PolyU 5217/07E) and Research Grant from Hubei Provincial Department of Education (No. 500064).

References

1. Daniel, N., Radev, D., Allison, T.: Sub-event based Multi-document Summarization. In: Proceedings of the HLT-NAACL Workshop on Text Summarization, pp. 9–16 (2003)
2. Filatova, E., Hatzivassiloglou, V.: Event-based Extractive Summarization. In: Proceedings of ACL 2004 Workshop on Summarization, pp. 104–111 (2004)
3. Allan, J., Gupta, R., Khandelwal, V.: Temporal Summaries of News Topics. In: Proceedings of the 24th Annual International ACM SIGIR Conference on Research and Development in Information Retrieval, pp. 10–18 (2001)
4. Lim, J.M., Kang, I.S., Bae, J.H., Lee, J.H.: Sentence Extraction Using Time Features in Multi-document Summarization. In: Information Retrieval Technology: Asia Information Retrieval Symposium (2004)
5. Afantenos, S.D., Karkaletsis, V., Stamatopoulos, P.: Summarizing Reports on Evolving Events; Part I: Linear Evolution. In: Proceedings of Recent Advances in Natural Language Processing (2005)
6. Wu, M., Li, W., Lu, Q., Wong, K.F.: Event-Based Summarization Using Time Features. In: Gelbukh, A. (ed.) CICLing 2007. LNCS, vol. 4394, pp. 563–574. Springer, Heidelberg (2007)
7. Jatowt, A., Ishizuka, M.: Temporal Web Page Summarization. In: Proceedings of the 5th International Conference on Web Information Systems Engineering, pp. 303–312 (2004)
8. Morris, J., Hirst, G.: Lexical Cohesion Computed by Thesaurus Relations as an Indicator of the Structure of Text. Computational Linguistics 17(1), 21–48 (1991)
9. Barzilay, R., Elhadad, M.: Using Lexical Chains for Text Summarization. In: Proceedings of ACL 1997/EACL 1997 Workshop on Intelligent Scalable Text Summarization, pp. 10–17 (1997)
10. Silber, H.G., McCoy, K.F.: Efficient Text Summarization Using Lexical Chains. In: Proceedings of the 5th International Conference on Intelligent User Interfaces, pp. 252–255 (2000)
11. Silber, H.G., McCoy, K.F.: Efficiently Computed Lexical Chains as an Intermediate Representation for Automatic Text Summarization. Computational Linguistics 28(4), 487–496 (2002)

12. Doran, W., Stokes, N., Dunnion, J., Carthy, J.: Assessing the Impact of Lexical Chain Scoring Methods and Sentence Extraction Schemes on Summarization. In: Gelbukh, A. (ed.) CICLing 2004. LNCS, vol. 2945, pp. 627–635. Springer, Heidelberg (2004)
13. Zhou, Q., Sun, L., Lv, Y.: ISCAS at DUC 2006. In: Proceedings of Document Understanding Conference 2006 (2006)
14. Reeve, L.H., Han, H., Brooks, A.D.: BioChain-Lexical Chaining Methods for Biomedical Text Summarization. In: Proceedings of the 2006 ACM Symposium on Applied Computing, pp. 180–184 (2006)
15. Reeve, L.H., Han, H., Brooks, A.D.: The Use of Domain-Specific Concepts in Biomedical Text Summarization. Information Processing and Management 43(6), 1765–1776 (2007)
16. Timothy, C., Patrick, P.: VerbOcean: Mining the Web for Fine-Grained Semantic Verb Relations. In: Proceedings of Conference on Empirical Methods in Natural Language Processing (2004)
17. Lin, C.Y., Hovy, E.: Automatic Evaluation of Summaries using N-gram Cooccurrence Statistics. In: Proceedings of HLTNAACL, pp. 71–78 (2003)
18. Page, L., Brin, S., Motwani, R., Winograd, T.: The PageRank CitationRanking: Bring Order to the Web. Technical Report, Stanford University (1998)

Using an Information Quality Framework to Evaluate the Quality of Product Reviews

You-De Tseng and Chien Chin Chen

Department of Information Management, National Taiwan University
No. 1, Sec. 4, Roosevelt Road, Taipei, 10617 Taiwan(R.O.C)
r96725044@ntu.edu.tw, paton@im.ntu.edu.tw

Abstract. The prevalence of Web2.0 makes the Web an invaluable source of information. For instance, product reviews composed collaboratively by many independent Internet reviewers can help consumers make purchase decisions and enable manufactures to improve their business strategies. As the number of reviews is increasing exponentially, opinion mining is needed to identify important reviews and opinions for users. Most opinion mining approaches try to extract sentimental or bipolar expressions from a large volume of reviews. However, the mining process often ignores the quality of each review and may retrieve useless or even noisy reviews. In this paper, we propose a method for evaluating the quality of information in product reviews. We treat review quality evaluation as a classification problem and employ an effective information quality framework to extract representative review features. Experiments based on an expert-composed data corpus demonstrate that the proposed method outperforms state-of-the-art approaches significantly.

Keywords: Text Mining, Classification, Opinion Mining.

1 Introduction

"How do people feel?" "What opinions do people have?" These questions are frequently asked by people when making decisions. In the past, public opinion was often hard to gauge because most media, such as news services or advertisements, were simply one-way communication channels. However, with the advent of Web2.0, many online collaboration tools, e.g., weblogs and discussion forums, are being developed to allow Internet users to exchange opinions and share valuable knowledge. In [18], the author observes that Internet users are often willing to divulge personal information and are forthcoming in presenting their personal viewpoints honestly. This kind of behavior has an indirect *word of mouth*[1] effect on marketing because users' opinions posted on the Web have a huge impact on consumer decisions [1]. Many e-commerce websites, such as Amazon[2], are aware of the word of mouth effect and offer users a platform to post their product reviews. However, as the number of reviews is growing exponentially, users are finding it increasingly difficult to find

---

[1] http://en.wikipedia.org/wiki/Word_of_mouth
[2] http://www.amazon.com/

G.G. Lee et al. (Eds.): AIRS 2009, LNCS 5839, pp. 100–111, 2009.

desired information. To alleviate this information overload problem, *opinion mining* techniques have been devised to extract and summarize meaningful opinions from reviews.

A major task of opinion mining is to identify sentimental (or bipolar) text units in review documents. A text unit can be a word, a sentence, a paragraph, or even the whole document, depending on the granularity of opinion mining. Generally, supervised approaches like classification, which categorizes text units as sentimental or non-sentimental, can extract the core opinions expressed in reviews efficiently [11, 17]. [2, 7] also identify opinion targets (i.e., product features) and compile summaries of reviews to help users understand the advantages and disadvantages of a product. Obviously, the opinions of an expert or someone with in-depth knowledge will be more authoritative than those of ordinary people. [11] analyzes opinion discourses to identify opinion holders. Opinion units are then weighted according to the holders' authority.

While many opinion mining approaches try to identify and analyze opinions from reviews, few works consider the quality of reviews. As Web 2.0 encourages knowledge sharing, there are no constraints on review writing. Consequently, the quality of reviews varies enormously. We observe from Amazon that many reviews simply contain emotional expressions, such as, *"I love this camera and it is really nice."* Such reviews lack constructive expressions and should not be included in the opinion mining process. Some websites do consider the quality of reviews, and provide rating systems to rank reviews according votes submitted by users. Figure 1 shows a review rating on Amazon, where 121 out of 128 voters thought the review was helpful.

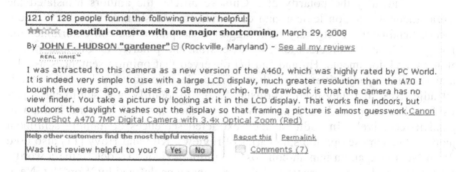

Fig. 1. A review rating on Amazon

Even with a rating system, ratings still suffer from imbalance vote bias, winner circle bias, and early bird bias, which make the system impracticable [14]. Thus, there is an urgent need for quality evaluation mechanisms to help users or opinion mining algorithms identify informative reviews. In this paper, we treat the evaluation of product reviews' quality as a classification problem and employ a multiclass support vector machine (multiclass SVM) [23] model to categorize reviews. In addition, we adopt a mature information quality framework [8], which has been widely used in many domains over the last twenty years, to define meaningful review features for classification. Experiments demonstrate that the proposed method outperforms state-of-the-art systems significantly. In addition, the learned model help identify the factors that are critical for compiling informative reviews.

The remainder of the paper is organized as follows. Section 2 contains a review of related works. In Section 3, we introduce the information quality framework and apply it to the problem of review quality classification. We evaluate the system performance in Section 4. Then, in Section 5, we summarize our conclusions and avenues for future research.

2 Related Work

2.1 Opinion Mining

Opinion extraction and polarity identification are two major tasks in opinion mining. Depending on the granularity of the opinion mining approach, an opinion can be a word, a sentence, a paragraph, or even a complete review document. Most approaches rely on a human-composed opinion lexicon. Turney [21] gathered seven positive and seven negative words as an opinion dictionary and proposed the use of pointwise mutual information to calculate the degree of co-occurrence of a word with the words in the lexicon. A word has a positive orientation if it tends to co-occur with the positive words; otherwise, it has a negative orientation. Dave et al. [2] used information retrieval techniques to extract sentiment n-gram features from a set of positive and negative product reviews. Then, based on the features, classification algorithms can be employed to extract and classify sentiments or opinions in new reviews. Ku et al. [13] dealt with the opinion mining problem in a bottom-up manner. To identify the polarity of a Chinese article, the authors translated the General Inquirer[3] opinion lexicon and combined the translations with the Chinese Network Sentimental Dictionary[4]. The aggregated opinion lexicon determines the polarity of Chinese characters and, by extension, the polarity of Chinese words, sentences, and documents. Hu and Liu [7] observed that opinion sentences usually contain sentiment adjectives; thus, their opinion lexicon contained a set of sentiment adjectives, and they used WordNet [5, 16] to identify new sentiment adjectives found in reviews. Those adjectives were then inserted into the lexicon to expand it recursively. In addition, they extracted sentences containing sentiment adjectives to compose opinion summaries of reviews. Kim and Hovy [11] also used of WordNet to expand a human-composed opinion lexicon. To determine a word's polarity, the word was represented as a set of synonyms defined by WordNet. Next, a Naive Bayes classifier was employed to assign the polarity of the word. Their method also identified opinion holders by selecting name entities close to topic phrases in opinion sentences. In practice, opinion words are context dependent and can belong to any part-of-speech [3]. For example, the word *"simple"* has a positive orientation in the sentence *"The user interface of this PDA is simple."* However, it conveys a negative sentiment in *"The story of this movie is too simple."* Ding et al. [3] expanded the opinion lexicon in [7] by adding opinion verbs and nouns. They also considered the context information in sentences for opinion mining. Their experiments demonstrated that the accuracy of opinion mining can be improved by using a holistic opinion lexicon.

---

[3] http://www.wjh.harvard.edu/~inquirer/
[4] http://134.208.10.186/WBB/EMOTION_KEYWORD/Atx_emtwordP.htm

2.2 Review Quality Evaluation

Zhang and Varadarajan [24] collected a set of product reviews on Amazon along with the corresponding helpfulness votes given by users. Based on the votes, the authors defined the utility of a review as the ratio of users (voters) who thought it was helpful and employed SVM regression to approximate the utility of the reviews. The resulting regression function was then used to estimate the utility of new reviews. The authors demonstrated through experiments that the shallow syntactic features of reviews are highly correlated with utility estimation. Kim et al. [12] also employed SVM regression to predict the helpfulness (i.e., utility) of a review; and used five categories of review features, namely, structural, lexical, syntactical, semantic, and meta-data features, to construct a regression function. They found that the unigrams, length, and rating stars of a review have a significant impact on users' assessments of the review's utility. Liu et al. [14] conducted a detailed survey of reviews at Amazon and found that users' votes were influenced by three types of bias: imbalance vote bias, winner circle bias, and early bird bias. Thus, methods that employ users' votes as training examples, e.g., [12] and [24], are affected by the above types of bias and are unreliable. Rather than make predictions based on biased votes, Liu et al. treated review quality evaluation as a classification problem. In addition, they used an expert-composed data set to train an unbiased SVM classifier, which categorized reviews as either high-quality or low-quality.

The above methods adopt a diverse set of features to evaluate the quality of reviews. However, many of the features are lexical and syntactically oriented, so they hardly reflect the intrinsic characteristics of reviews. In this paper, we apply an effective framework of information quality to derive information-oriented features for evaluating the quality of reviews. Using the information-oriented features improves the evaluation performance, and resolves important factors in review compositions.

3 Method

3.1 Definition of Review Quality

We regard review quality evaluation as a classification problem and employ an information quality framework to derive informative review features for classification. Five categories of review quality, namely "*high-quality*", "*medium-quality*", "*low-quality*", "*duplicate*", and "*spam*", are defined according to the specification of review quality in [14] and the definition of spam reviews in [9]. A high-quality review must provide complete and timely information about a product. It must also contain a large number of opinions to help readers make purchasing decisions. The content of a medium-quality review is relevant to a product, but it is not informative enough. Although such reviews are useful, they hardly persuade readers to make decisions. A low-quality review contains little information about a product, or the information is too objective to judge the value of the product. A review is considered a duplicate if its content is very similar to a review posted previously. It may be a fake review or a repeat review posted by mistake. Finally, a spam review only provides comments about product-irrelevant matters, such as brands and services. Otherwise, it is an advertisement or a question-answer type of review.

3.2 Classification Models

The support vector machine (SVM) is a state-of-the-art machine learning algorithm for classification problems [19]. In this study, we employ two multiclass SVM-based approaches: One-Versus-All SVM and Single-Machine Multiclass SVM.

One-Versus-All SVM (OVA SVM): This approach decomposes a multiclass classification problem into N independent binary classifiers. When training a binary classifier for a quality class, the training reviews in the class are regarded as positive examples, and remaining reviews are considered negative examples. To classify a new review, each classifier computes a score indicating the degree of association (or margin) between the review and the corresponding class. Then, the review is assigned to the class with the largest score. We implement the one-versus-all approach with the SVM^{light} binary SVM tool [10]. The RBF kernel is selected because of its superior classification performance.

Single-Machine Multiclass SVM (SMM SVM): Rather than combine the results of independent binary classifiers, the single-machine approach constructs a classification function by considering N classes simultaneously [23].

$$\min \quad \frac{1}{2}\sum_{n=1}^{N}(\underline{w}_n \cdot \underline{w}_n) + C\sum_{i=1}^{l}\sum_{n \neq y_i}\xi_i^n \tag{1}$$

s.t.

$$(\underline{w}_{y_i} \cdot \underline{x}_i) + b_{y_i} \geq (\underline{w}_n \cdot \underline{x}_i) + b_n + 2 - \xi_i^n,$$
$$\xi_i^n \geq 0, \quad i = 1,...,l, \quad n \in \{1,...,N\}, \tag{2}$$

where $\{(\underline{x}_1,y_1),...,(\underline{x}_l,y_l)\}$ is a set of l training examples. Each review is represented as a high-dimensional feature vector $\underline{x}_i$, and y_i is its class label; $\underline{w}_n$ and b_n are, respectively, the weight vector and intercept of class n learned from the training examples; ζ's are the slack variables for the training examples; and C is a regularization term to control overfitting. The acquired $\underline{w}$'s and b's then assign a class label to a test review $\underline{x}_{test}$.

$$f(\underline{x}_{test}) = \arg\max_n [(\underline{w}_n \cdot \underline{x}_{test}) + b_n], \quad n = 1,...,N. \tag{3}$$

We employ the $SVM^{multiclass}$ tool [20] in our experiments; and use linear kernels because non-linear kernels are time-consuming for multiclass problems. Moreover, the experiments reported in [6] demonstrate that linear kernels are comparable to non-linear kernels in many complex and large problems.

3.3 Information Quality-Based Learning Features

Information quality (IQ) methodology investigates the characteristics of information items and derives item features considered informative from the perspective of information consumers [22]. In the last twenty years, many IQ frameworks have been developed for various application domains [4]. For instance, Zhu [25] assessed the quality of web pages in terms of an IQ framework to enhance the retrieval perform-ance of information systems. Meanwhile, Wang and Strong [22] developed a two-stage survey to collect features considered important by information consumers, and proposed a hierarchical IQ framework that organized the features along different dimensions. Eppler and Wittig [4] commented that Wang and Strong's framework, shown in Table 1, attempts to strike a balance between theoretical consistency and practicability. In this study, we treat a product review as an information item and derive informative review features in terms of the hierarchical framework. Some dimensions are not considered because they are not applicable to product reviews. We use following nine dimensions and fifty features for SVM training and testing.

Table 1. Wang and Strong's hierarchical IQ framework [22]

IQ category	IQ dimensions
intrinsic IQ	believability, accuracy, objectivity, reputation
contextual IQ	value-added, relevancy, timeliness, completeness, appropriate amount of information
representational IQ	interpretability, ease of understanding, representational consistency, concise representation
accessibility IQ	accessibility, access security

Believability (D1): This dimension is the extent to which an information item (i.e., a review) is credible, or regarded as true. Jindal and Liu [9] observed that reviews whose product ratings are extremely high or low are likely to be radical reviews. We therefore measure the deviation of a review's product rating from the average to assess its believability.

- The rating deviation of a review (f_1).

Objectivity (D2): This dimension is the extent to which an information item is biased. Apparently, subjective opinions in reviews help readers to make decisions. We therefore apply Hu and Liu's algorithms [7] to extract opinion sentences and measure this dimension in terms of review opinions.

- The number of opinion sentences (f_2), positive sentences (f_3), negative sentences (f_4), and neutral sentences (f_5) in a review.
- The percentage of opinion sentences (f_6), positive sentences (f_7), negative sentences (f_8), and neutral sentences (f_9) in a review.
- The percentage of positive sentences (f_{10}) and negative sentences (f_{11}) in all opinion sentences of a review.
- The cosine similarity between the tf-idf vectors [15] of a review and the product description (f_{12}). The larger the similarity, the more objective the review will be.

Reputation (D3): This dimension is the extent to which the author of a review is trusted or highly regarded. Reviews written by authoritative reviewers are certainly influential. We measure this dimension based on the reviewer's publications and the ranking given by e-commerce websites.

- The number of reviews written by the reviewer (f_{13}).
- The ranking of the reviewer (f_{14}).

Relevancy (D4): This dimension is the extent to which the content in a review is useful for decision-making. Helpful product reviews should provide a large amount of product information. We consider the following statistics to assess the relevance of a review.

- The number of the product name (f_{15}), brand names (f_{16}), website names (f_{17}), and other product names (f_{18}) in a review.
- The percentage of the product name (f_{19}), brand names (f_{20}), website names (f_{21}), and other product names (f_{22}) in a review.
- The number of opinion sentences containing the product name (f_{23}), brand names (f_{24}), website names (f_{25}), and other product names (f_{26}) in a review.
- The percentage of opinion sentences containing the product name (f_{27}), brand names (f_{28}), website names (f_{29}), and other product names (f_{30}) in a review.

Timeliness (D5): This dimension is the extent to which the information in a review is timely and updated. Old or duplicate reviews cannot reflect the value of a product in time; thus, the quality of information is low.

- The degree of duplication of a review (f_{31}), defined as the maximum cosine similarity between the tf-idf vectors of the review to those of reviews published previously.
- The interval (in terms of the number of days) between the current review and the first review of the product (f_{32}).

Completeness (D6): This dimension is the extent to which the information in a review is complete and covers various aspects of a product. High quality reviews should cover all kinds of product features and specifications.

- The number of kinds of product features (f_{33}), brand names (f_{34}), websites (f_{35}), and product names (f_{36}) mentioned in a review.

Appropriate Amount of Information (D7): This dimension is the extent to which the volume of information in a review is sufficient for decision-making. The more product information included in a review, the higher the information quality of the review will be.

- The number of product features (f_{37}), opinion-bearing words (f_{38}), words (f_{39}), sentences (f_{40}), and paragraphs (f_{41}) in a review.
- The average frequency of product features in a review (f_{42})
- The number of sentences that mention product features in a review (f_{43}).

Ease of Understanding (D8): A comprehensible review should state opinions about a product directly and clearly; and it should not contain rarely used or misspelled words.

- The number of misspelled words in a review (f_{44}).
- The average document frequency [15] of review words (f_{45}). The average will be low if the review contains several rarely used words or misspellings.
- The position of the first opinion sentence in the review (f_{46}).

Concise Representation (D9): This dimension represents the conciseness of a review, and complements the dimension of the appropriate amount of information. Including a lot of information may result in a review that is too long.

- The average length of sentences (f_{47}) and paragraphs (f_{48}) in a review.
- The average number of sentences (f_{49}) and opinion sentences (f_{50}) in a paragraph of a review.

4 Performance Evaluations

4.1 Data Preprocessing and Annotation

We selected the reviews of ten popular digital cameras and ten mp3 players at Amazon for evaluation. For each product, the first 150 reviews (in order of publication date) were collected to construct an evaluation corpus. Two human experts annotated the reviews independently based on the quality classes defined in Sec. 3.1. Inconsistent annotations were resolved through discussions between the annotators and a third person (i.e., the first author of the paper) to establish a ground-truth, as shown in Table 2. The kappa statistics between the annotators for digital cameras and mp3 players are 0.7253 and 0.7928, respectively, and are good enough to conduct reliable evaluations.

Table 2. The statistics of the data corpus

ground-truth	high	medium	low	duplicate	spam	total
digital cameras	113	297	1053	13	24	1500
mp3 players	134	297	1007	37	25	1500
total	247	594	2060	50	49	3000

The evaluations are conducted as follows. First, we examine the performance of our IQ dimensions. Then, the most effective dimension combination is compared with the following three feature sets: 1) the shallow syntactic features of Zhang's method [24]; 2) the lexical features of Kim's method [12], along with the length, unigram, and rating stars of a review; and 3) the informativeness features of Liu's method [14]. We use these feature sets for comparison because they have proven effective in review mining tasks. In addition, we assess the performance of the bag-of-words model [15] in which each uniqe term is treated as a feature. The method is also regarded as a baseline system. To convert the selected reviews into IQ-based feature vectors, we first remove stopwords [15] in the reviews and check the remaining terms for misspellings by using the WordNet, Wiktionary[5], and Google's spell check function[6]. Next, we apply Hu and Liu's algorithm [7] to extract product features, opinion words, opinion sentences, and sentence polarities from the reviews. Then, the extracted entities are manually examined to filter out false alarms. For each compared method, 10-fold cross-validation [15] is adpoted to derive credible results. Macro/micro-average precision, recall, and the F1 score [15] are used as evaluation metrics. However, as each review can only belong to one quality category, the micro-average precision, recall, and F1 scores are equivalent; thus, we only consider the micro-average F1 scores.

4.2 IQ Dimension Evaluations

Tables 3 and 4 show the performances of the IQ dimensions using SMM SVM and OVA SVM, respectively. As shown from Table 2, the 'spam' and 'duplicate' categories are very small, so a simple mis-classification in these categories would cause a huge variation in the macro-averaging performance. In contrast, the micro-average F1 score is insensitive to category sizes and is thus appropriate for evaluating the overall performance of each dimension. We assess the effect of IQ dimensions iteration by iteration. In the first iteration, we individually examine the performance of each IQ dimension. Generally, all dimensions produce similar F1 scores, so we only list the performance of the best dimension in the first row of each table. In the i'th iteration ($2 \le i \le 9$), the set of dimensions selected in the (i-1)'th iteration serves as the basis. Next, we examine each remaining dimension combined with the basis and show the performance of the best combination in the i'th row. For instance, the third row in Table 3 shows the performance of the top-3 effective dimensions, {objectivity, reputation, information}, in SMM SVM. For each row, a one-tail paired t-test is applied to determine whether combining each dimension with the basis improves the system performance significantly. The symbol '*' indicates that combining a dimension improves the performance significantly, while the symbol '#' indicates the opposite.

Table 3. The effect of IQ dimensions using SMM SVM with the linear kernel

	digital cameras				mp3 players			
	mac-precision	mac-recall	mac-F1	mic-F1	mac-precision	mac-recall	mac-F1	mic-F1
objectivity (D2)	0.697	0.440	0.535	0.768	0.715	0.358	0.471	0.742
+ reputation (D3)	*0.745	****0.486	***0.585	****0.832	*0.742	****0.404	****0.514	****0.782
+ information (D7)	****0.868	****0.575	****0.691	****0.900	*0.800	****0.465	****0.584	****0.849
+understanding (D8)	0.866	**0.582	0.696	0.906	*0.842	0.480	*0.605	***0.862
+ timeliness (D5)	0.856	*0.593	0.699	0.906	**0.845**	0.483	**0.608**	**0.862**
+ believability (D1)	0.862	**0.597**	*0.704	**0.908**	0.842	0.480	0.605	0.862
+ relevancy (D4)	**0.874**	0.595	**0.707**	0.907	#0.792	0.478	0.595	0.858
+ completeness (D6)	##0.837	0.587	##0.688	0.905	0.791	0.486	0.601	0.862
+ concise (D9)	0.833	0.577	0.681	0.901	0.788	**0.488**	0.599	0.862

*, **, ***, and **** represent right-tail paired t-tests with α=0.1, 0.05, 0.025, and 0.01, respectively.
#, ##, ###, and #### represent left-tail paired t-tests with α=0.1, 0.05, 0.025, and 0.01, respectively.

---

[5] http://en.wiktionary.org/wiki/Wiktionary:Main_Page
[6] http://www.google.com/support/toolbar/bin/answer.py?answer=32703&hl=en

Table 4. The effect of IQ dimensions using OVA SVM with the RBF kernel

	digital cameras				mp3 players			
	mac-precision	mac-recall	mac- F1	mic-F1	mac-precision	mac-recall	mac-F1	mic-F1
information (D7)	0.838	0.586	0.686	0.908	0.821	**0.508**	0.622	0.870
+ concise (D9)	0.861	*0.592	0.700	0.910	*0.832	0.506	0.623	*0.875
+ objectivity (D2)	**0.866**	**0.597**	**0.706**	**0.914**	0.836	0.505	0.623	**0.876**
+ believability (D1)	0.866	0.597	0.706	0.914	0.836	0.505	0.623	0.876
+ completeness (D6)	0.865	0.597	0.705	0.914	0.831	0.508	**0.624**	0.876
+understanding (D8)	0.864	0.595	0.703	0.912	0.831	0.507	0.623	0.876
+ timeliness (D5)	0.819	0.583	0.680	0.906	0.842	0.502	0.624	0.872
+ relevancy (D4)	0.818	0.580	0.678	0.906	**0.844**	0.499	0.623	0.871
+ reputation (D3)	#0.783	#0.571	#0.657	#0.901	0.836	0.495	##0.617	#0.867

It is noteworthy that one IQ dimension (i.e., the appropriate amount of information) is sufficient for the RBF kernel to achieve superior performances. However, combining this dimension with the other IQ dimensions does not improve the system performance overall. This is because non-linear kernels have a powerful modeling ability so that a few discriminative features and dimensions are sufficient to construct accurate classifiers. Note that 'objectivity' and 'the appropriate amount of information' are in the top-3 effective dimensions of the both SVM approaches. This indicates that degree of sentiment and the amount of product information are critical criteria for judging the quality of a review. The most effective combinations for the linear kernel and the RBF kernel are {D2, D3, D5, D7, D8} and {D2, D7, D9} respectively, which are used in the following comparisons. We observe that the other dimensions (i.e., other than the most effective combinations) contain diverse features. For instance, the brand names and product names mentioned in the evaluated reviews vary a great deal, so the features of the 'completeness' and 'relevance' dimensions are too sparse to contribute the system performance. Consequently, the performances of employing all IQ dimensions have little difference to those generated by the most effective combinations.

4.3 Comparisons with Other Methods

Tables 5 and 6 show the performances of the compared methods. Generally, all the compared methods outperform the baseline method in terms of the micro-average F1 scores. The proposed method achieves the best performance and the improvement over each of the compared methods is statistically significant in terms of the one-tailed paired t-test. Liu's approach [14] is a state-of-the-art method for classifying the quality of reviews. The set of informativeness features is information-oriented; hence its performance is good and comparable. The sets of shallow syntactic features and lexical features are effective in determining a review's helpfulness. Helpfulness is the ratio of helpful votes given by readers, but it is not exactly equivalent to the review quality. Therefore, the performances of those features are inferior and are even worse than the performance of the baseline when using the linear kernel.

Table 5. The comparison results using SMM SVM with the linear kernel

	digital cameras				mp3 players			
	mac-precision	mac-recall	mac- F1	mic-F1	mac-precision	mac-recall	mac-F1	mic-F1
baseline	****0.424	****0.323	****0.365	****0.670	****0.361	****0.257	****0.297	****0.580
shallow syntactic [24]	****0.279	****0.283	****0.275	****0.608	****0.213	****0.239	****0.220	****0.298
lexical [12]	****0.595	****0.362	****0.448	****0.744	****0.472	****0.306	****0.369	****0.683
informativeness [14]	0.853	***0.562	*0.676	****0.886	***0.794	0.463	*0.581	**0.844
D2+D3+D5+D7+D8	**0.856**	**0.593**	**0.699**	**0.906**	**0.845**	**0.483**	**0.608**	**0.862**

Table 6. The comparison results using OVA SVM with the RBF kernel

	digital cameras				mp3 players			
	mac-precision	mac-recall	mac- F1	mic-F1	mac-precision	mac-recall	mac-F1	mic-F1
baseline	****0.696	****0.412	****0.516	****0.782	****0.711	****0.389	****0.502	****0.757
shallow syntactic [24]	****0.796	****0.577	****0.666	****0.888	0.817	0.491	0.610	****0.852
lexical [12]	**0.821	****0.576	***0.674	****0.895	****0.801	**0.515**	**0.626**	***0.856
informativeness [14]	0.836	0.590	0.688	****0.904	****0.800	**0.488	***0.601	****0.856
D2+D7+D9	**0.866**	**0.597**	**0.706**	**0.914**	**0.836**	0.505	0.623	**0.876**

The inferior performances of the baseline method highlight the difficulty of evaluating the quality of reviews. Since reviews are sentimental and information-oriented, evaluation systems must consider both textual and semantic characteristics of reviews to measure review quality. Our method examines various factors of reviews in detail. The superior evaluation performances indicate that the derived features and dimensions based on a theoretical IQ framework are highly representative of the reviews' characteristics.

5   Conclusion

In this paper, we have presented a method for evaluating the quality of product reviews. We regard a review as an information item and apply a theoretical IQ framework to derive representative review features and dimensions. Experiments show that our method can accurately classify reviews in terms of their quality, and that it outperforms existing methods significantly. In the future, we will examine the correlation between features and filter out redundant features to improve the system's efficiency. We will also apply the proposed method to various styles of opinion documents, such as blog entries and forum threads, to assess the quality of the information they provide.

Acknowledgments. The authors would like to thank the anonymous reviewers for their valuable comments and suggestions. This work was supported in part by NSC 97-2221-E-002-225-MY2 and JAID S09800251079.

References

1. Chevalier, J.A., Mayzlin, D.: The Effect of Word of Mouth on Sales: Online Book Reviews. Journal of Marketing Research 43(3), 345–354 (2006)
2. Dave, K., Lawrence, S., Pennock, D.M.: Mining the Peanut Gallery: Opinion Extraction and Semantic Classification of Product Reviews. In: WWW, pp. 519–528 (2003)

3. Ding, X., Liu, B., Yu, P.S.: A Holistic Lexicon-Based Approach to Opinion Mining. In: WSDM, pp. 231–240 (2008)
4. Eppler, M.J., Wittig, D.: Conceptualizing Information Quality: A Review of Information Quality Frameworks from the Last Ten Years. In: ICIQ, pp. 83–96 (2000)
5. Fellbaum, C.: WordNet: an Electronic Lexical Database. MIT Press, Cambridge (1998)
6. Hsu, C.W., Lin, C.J.: A Comparison of Methods for Multiclass Support Vector Machines. IEEE Transactions on Neural Networks 13(2), 415–425 (2002)
7. Hu, M., Liu, B.: Mining and Summarizing Customer Reviews. In: SIGKDD, pp. 168–177 (2004)
8. Huang, K.T., Lee, Y.W., Wang, R.Y.: Quality Information and Knowledge. Prentice Hall PTR, Upper Saddle River (1998)
9. Jindal, N., Liu, B.: Opinion Spam and Analysis. In: WSDM, pp. 219–230 (2008)
10. Joachims, T.: Making Large-scale SVM Learning Practical. In: Schökopf, B., Burges, C., Smola, A. (eds.) Advances in Kernel Methods - Support Vector Learning. MIT-Press, Cambridge (1999)
11. Kim, S.M., Hovy, E.: Determining the Sentiment of Opinions. In: ICCL, pp. 1367–1373 (2004)
12. Kim, S.M., Pantel, P., Chklovski, T., Pennacchiotti, M.: Automatically Assessing Review Helpfulness. In: EMNLP, pp. 423–430 (2006)
13. Ku, L.W., Liang, Y.T., Chen, H.H.: Opinion Extraction, Summarization and Tracking in News and Blog Corpora. In: AAAI-CAAW, Technical Report SS-06-03, pp. 100–107 (2006)
14. Liu, J., Cao, Y., Lin, C.Y., Huang, Y., Zhou, M.: Low-Quality Product Review Detection in Opinion Summarization. In: EMNLP-CoNLL, pp. 334–342 (2007)
15. Manning, C., Raghavan, P., Schütze, H.: An Introduction to Information Retrieval. Cambridge University Press, Cambridge (2008)
16. Miller, G., Beckwith, R., Fellbaum, C., Gross, D., Miller, K.: Introduction to WordNet: An On-line Lexical Database. International Journal of Lexicography 3(4), 235–244 (1990)
17. Pang, B., Lee, L., Vaithyanathan, S.: Thumbs up? Sentiment Classification using Machine Learning Techniques. In: EMNLP, pp. 79–86 (2002)
18. Roed, J.: Language Learner Behavior in a Virtual Environment. Computer Assisted Language Learning 16(2–3), 155–172 (2003)
19. Steinwart, I., Christmann, A.: Support Vector Machines. Springer, New York (2008)
20. Tsochantaridis, I., Joachims, T., Hofmann, T., Altun, Y.: Large Margin Methods for Structured and Interdependent Output Variables. Journal of Machine Learning Research 6, 1453–1484 (2005)
21. Turney, P.D.: Thumbs Up or Thumbs Down? Semantic Orientation Applied to Unsupervised Classification of Reviews. In: ACL, pp. 129–159 (2002)
22. Wang, R.Y., Strong, D.M.: Beyond Accuracy: What Data Quality Means to Data Consumers. Journal of Management Information Systems 12(4), 5–33 (1996)
23. Weston. J., Watkins. C.: Multi-class Support Vector Machines. Technical Report CSD-TR-98-04, Royal Holloway, University of London, Department of Computer Science (1998)
24. Zhang, Z., Varadarajan, B.: Utility Scoring of Product Reviews. In: CIKM, pp. 51–57 (2006)
25. Zhu, X., Gauch, S.: Incorporating Quality Metrics in Centralized/Distributed Information Retrieval on the World Wide Web. In: SIGIR, pp. 288–295 (2000)

Automatic Extraction for Product Feature Words from Comments on the Web[*]

Zhichao Li, Min Zhang, Shaoping Ma, Bo Zhou, and Yu Sun

State Key Laboratory of Intelligent Technology and Systems,
Tsinghua National Laboratory for Information Science and Technology,
Department of Computer Science and Technology, Tsinghua University,
Beijing, 100084, China P.R.
lizhichaoxyz@sohu.com, {z-m,msp}@tsinghua.edu.cn,
{zhoubo2000,sunorrain}@gmail.com

Abstract. Before deciding to buy a product, many people tend to consult others' opinions on it. Web provides a perfect platform which one can get information to find out the advantages and disadvantages of the product of his interest. How to automatically manage the numerous opinionated documents and then to give suggestions to the potential customers is becoming a research hotspot recently. Constructing a sentiment resource is one of the vital elements of opinion finding and polarity analysis tasks. For a specific domain, the sentiment resource can be regarded as a dictionary, which contains a list of product feature words and several opinion words with sentiment polarity for each feature word. This paper proposes an automatic algorithm to extraction feature words and opinion words for the sentiment resource. We mine the feature words and opinion words from the comments on the Web with both NLP technique and statistical method. Left context entropy is proposed to extract unknown feature words; Adjective rules and background corpus are taken into consideration in the algorithm. Experimental results show the effectiveness of the proposed automatic sentiment resource construction approach. The proposed method that combines NLP and statistical techniques is better than using only NLP-based technique. Although the experiment is built on mobile telephone comments in Chinese, the algorithm is domain independent.

Keywords: Resource constructing, product feature, opinion word.

1 Introduction

Before deciding to buy a product, many people tend to consult others' opinions on it. Web provides a perfect platform which one can get information. Many customers record their comments on products on the Websites, forums or blogs. Reading the comments, one concerned about a product can find out its main advantages and disadvantages. However, only few comments cannot give a convincing suggestion and

---

[*] Supported by the Chinese National Key Foundation Research & Development Plan (2004CB318108), Natural Science Foundation (60621062, 60503064, 60736044) and National 863 High Technology Project (2006AA01Z141).

G.G. Lee et al. (Eds.): AIRS 2009, LNCS 5839, pp. 112–123, 2009.
© Springer-Verlag Berlin Heidelberg 2009

persons do not have sufficient energy to browse more. Therefore, how to automatically manage the numerous comments and suggest the potential customers is becoming a research hotspot recently. The main target is mining the customers' opinions to the products. The opinions are classified into positive, negative and neutral. Naturally, customers prefer the products with more positive comments than those with more negative ones.

There are three levels for such opinion mining task: document level, product level and feature level. To the document level, a whole document only generates a single opinion, which is coarse and inaccurate. Because one document may contain not only one product, and have different opinion polarities to each, a general classification to the document is not appropriate. Then it comes to product level, which is generating an opinion polarity for a given product or a brand. Since nothing can be consummate, there are always some advantages and disadvantages for a product. So feature level opinion can be more exact to express the customers' attitude. It also helps potential customers understand the product more clearly and definitely. Feature stands for an *attribute* or a *component* of a product. For example: "待机时间(standby time)" is an attribute of a mobile telephone, while "显示屏(screen)" is a component, both of them can be named as "**feature words**". In the examples following, "The standby time of Nokia is long enough" expresses a positive opinion while "The screen is too small" expresses a negative one.

*The **standby time** of Nokia is long enough.*	*Ex. 1*
诺基亚的待机时间足够长	
*The **screen** is too small.*	*Ex. 2*
显示屏太小了.	

Therefore, to analyze a comment, we need to find the feature words the document contains and the opinion words that embellish the feature words. By confirming the polarity of the opinion words, the main viewpoint of the comment is discovered. With so many comments owning clear polarity of opinions for a given product, customers can easily get to know the product deeply.

We construct a sentiment resource for a given domain to offer efficient and effective utilities to analyze the comments. The sentiment resource can be considered as a dictionary that contains a list of feature words and several opinion words with polarity tag for each feature. This paper mainly introduces the algorithm to extraction the feature words and opinion words. The polarity of opinion words determination is taken as future work. Although the product description by the company shows a lot of feature words of the product, the words are not abundant. Web users sometimes take informal words to describe a feature. Therefore, mining the feature words is necessary.

There are three benefits to construct such a dictionary beforehand. First, after holding such a dictionary, a new comment will be easily dealt with. We can extract the feature words and most opinion words from the new comment through matching the items in the dictionary quickly. Computing the polarity of the opinion words for every new comments offline will save much time. Second, processing a large set of comments to find the feature words and opinion words can not only use the NLP (Natural Language Processing) techniques but also the statistical characteristics. It gets a better performance than processing comments one by one online. Supposing that we do not have the dictionary, when we get a new comment to process, we hardly take any statistical characteristics to make the performance better. Third, the dictionary is easy to maintain. When a new feature word or a new opinion word discovered, it can be easily added into the resource.

We use both NLP technique and statistical method to extract the feature words and the opinion words. Through our life experiences, the feature words are usually nouns or noun phrases, while the opinion words are usually adjectives. Tagging part-of-speech for all the words in the comments is helpful to the task. We can predefine some patterns to extract the target words with certain natural language characteristics. But this will bring a lot of noises. However, by using the different statistical characteristics of the words in the comments and in the whole Web background corpus, most noises can be removed. To get a higher recall, we also use unknown word finding techniques to extract more feature words. After filtering the noises and adding the new feature words, the performance is improved significantly in term of f-measure.

After describing about the related works in section 2, the data set and the main algorithm are introduced in section 3 and section 4 respectively, including feature words extraction, unknown word finding, and the usage of the background corpus, et al. Section 5 shows the evaluation of the performance. The conclusion and the future work are discussed in the last section.

2   Related Work

There are several researches to opinion mining for the product. The work in [2, 3] are based on document level. Feature level opinion mining also gains the interesting of researchers [4-11]. How to get feature words more exactly is the main problem in this task.

M. Hu and B. Liu [4, 5, 6] used association rule mining to find all frequent itemsets which are sets of words or phrases that occur together. CBA which was based on the Apriori algorithm is used. The words and the phrases extracted are considered as feature words. After two pruning phases (compactness pruning and redundancy pruning) to increase the precision and an infrequent feature identification phase to increase the recall, they got a good performance finally.

A. Popescu and O. Etzioni [7] built a system named OPINE. It is built on top of KnowItAll which was a Web-based domain-independent information extraction system. The system first extracts noun phrase and then filtered with point-wise mutual information value between the phrase and meronymy discriminators associated with the product class. They increased both precision and recall compared with M. Hu [6].

J. Yi and W. Niblack [8] extracted definite base noun phrases at the beginning of sentences followed by a verb phrase as the feature words. A definite base noun phrase is a noun phrase by specifically patterns preceded by the definite article "the". The method was called bBNP (Beginning definite Base Noun Phrases) heuristic. To filter the noises, they used a document set that did not focus on the product. The terms appeared more in the documents focused on the product than in the ones did not were kept, while others were removed. They only used precision to evaluate their method and got a very high precision at the top 20 feature words.

C. Scaffidi, K. Bierhoff and et al. [9] considered noun and two nouns that occur successively as the feature word candidates. Comparing with the random section of English text, the feature words often occurred far more frequently in the comment text. Using the distribution of the words in the random section of English text to compute the probability that it occurred n times in the comment text, the less the probability was, the more possible it was a feature word. The Red Opal system can return quite high precision when few feature words returned.

There are also some researchers on feature extraction in Chinese. B. Wang and H. Wang [10] created a bootstrapping method to extract feature words and opinion words in Chinese product comments. They used only a few manual tagged training data to generate a Naive Bayesian classifier. The features to train are only natural language characteristics, such as "is there an adverb in the right". The algorithm is iterative; the terms tagged by former round were added to training set to train the latter round classifier. The experiment indicated that few manual tagged data can also bring high performance on both feature words extraction and opinion words extracting.

Q. Su, X. Xu and et al. [11] also used noun and noun phrase (two or more adjacent nouns) as feature words candidates. Unlike English, there is no definite article "the" in Chinese comments to help to filter noises. But there are some Chinese own boundary indicators such as "的(of)". They clustered the feature words and the opinion words respectively to eliminate the problem that hard to mine the implicit feature words. The feature words embellished by the same opinion word may be clustered together to stand for a feature.

Our work is different at these points: first, we also employ verb and verb phrase to be feature word candidates, it can improve the recall; second, we use the statistical characteristic of candidates on background corpus to filter noises; third, we use unknown word finding techniques to extract more feature words.

3 Data Set

The experiment was built on mobile telephone domain in Chinese. We gathered comments from 2 web sites: http://www.bibifa.com/ and http://dp.cnmo.com/. There are totally 6405 comments (6.18 MB plain text data). The average length of the comments is 994 byte (nearly 500 Chinese characters). There are two types of the data in each comment. One is tagged by the customer with "advantage" or "disadvantage", while the sentences are short but have strong sentiment. The other is descriptive text.

Unlike English, Chinese does not make use of any white space characters between words. Therefore, we should segment the sentence into words by a tool ICTCLAS [12]. It can take the corpus and tag the words with part-of-speech. Following is an example. The tags "n" "d" "a" stand for noun, adverb and adjective respectively.

Original text:	外形太大，电池性能较差！	*Ex. 3*
Segmented text:	外形/n 太/d 大/a ，/wd 电池/n 性能/n 较/d 差/a ！/wt	*Ex. 4*
Translation text:	The profile is too big. The battery performance is poor.	

4 Algorithm

4.1 Feature Words Extraction

We use both NLP techniques and statistic methods to extract the feature words. First, using NLP techniques to acquire the feature words candidates, and then filter the noises with statistic characteristics.

In most of the researches, only nouns and noun phrases are extracted as the feature word candidates. But taking the particularity of Chinese into account, sometimes a verb can also be regarded as a feature word. It is not so much far to find an example (Ex. 5).

> *The **operating** is simple; the **reaction** is quick.*   *Ex. 5*
> 操作/v 简单/a , /wd 反应/vi 快/a

The words "操作(operating)" and "反应(reaction)" are both verbs, although they are nouns in English translation. However, postulating them as feature words is property. Therefore, for extracting feature words with more coverage, we use four patterns shown in Table 1.

Table 1. Feature words extraction patterns

Patterns	Examples
Noun	外形/n (figure)
Verb	反应/vi (reaction)
Noun + Noun	音乐/n 功能/n (music function)
Verb + Noun	拍摄/v 效果/n (screen effect)

We can establish such an assumption that the term has a higher probability to be a feature word when it occurs more times in the comments. It is fairly explicit, because the main features of the product must be the hotspots of the customer discussions. Obviously, not all the nouns, verbs and noun phrases are feature words. Extracting all such terms must bring lots of noises. We can simply remove the terms with low occurrence frequency to filter the noises, but it is not very effective. So we employ three other filter methods to remove the noises:

1) Import a rule that there should be an adjective on the right of the feature word.
2) Use the frequency of the terms in the background corpus.
3) Use the unknown word finding techniques.

4.1.1 Adjective Rule

As the purpose of extracting feature words and opinion words is to construct a dictionary, if no opinion words appear with the feature word, the feature word is not very valuable in our resource. And the most syntax that the customers use to express their opinion is "feature word + adjective". (Ex.1 Ex.3 Ex.4 and Ex. 5) This syntax is easy to express an opinion and accords to oral language. While the comments on the Web are often not very official text, it occurs quite often. Usually, customers add some adverb between the feature word and the adjective. Therefore, before extracting the feature words, stopwords and adverbs should be removed first. The adjective rule can be defined as: if in all the comments there is not any adjective occurs on right of the candidate term, the term is not a feature word. As the rule, the feature word candidates without an adjective on the right are removed to aspire after a higher precision. The main reason to follow this rule is to remove the verbs that cannot be feature words.

4.1.2 Background Corpus

Because the comments we use to mining feature words are from the Web, it is ineluctable that many common words that occur everywhere on the Web are mistakenly extracted as feature words. For example: "网址(Website)", "人们(people)". If the feature word candidate occurs too often on the whole Web text, we can doubt it is a true feature word for the domain we are concerned about justly. We regard the whole Web text as a background corpus. To get the occurrence frequency of the terms, Sogou Lab Internet Vocabulary [13] is used. The Internet Vocabulary is from the statistic analysis of the Chinese Web corpus indexed by Sogou search engine (http://www.sogou.com/) in October 2006. It is related to over 100 million Web pages containing more than 150,000 words with high frequency. The vocabulary gives the POS tag and occurrence frequency for each word. The common words that may be noises are usually nouns or verbs, while the phrases are seldom noises. We use the background corpus to filter the single nouns or verbs.

We define a feature named **TFP** (*term frequency proportion*) as follow:

$$TFP(t) = n / \ln(N) \tag{1}$$

Where n is the occurrence frequency of the term t in the comments that we use to mine the feature words, N is the occurrence frequency in the background corpus given by the Internet Vocabulary.

With the higher TFP, the candidate has more probability to be a true feature word. For example, the word "生活(life)" occurs in the comments 168 times and 251,581,894 times in the background corpus, while "杂音(murmur)" occurs 151 times in the comments and 856,288 times in the background corpus. "杂音(murmur)" has a higher TFP value (11.05) than "生活(life)" (8.69). Although "杂音(murmur)" appears less than "生活(life)" in the comments, it has more probability to be a true feature word. Therefore we select a threshold to confine the terms that have low TFP value. If TFP is less than the threshold, the term will be removed from the candidate set.

4.1.3 Unknown Word Finding Technique

As we use nouns, verbs and noun phrase to be the feature word candidates, the potential feature words that are not tagged as these POS will be missed. Especially, the mistake of Chinese word segmentation will cause the missing more familiarly. The feature word that is quite relative to the product or quite technological may not be segmented correctly by the common segmentation tool. It can influence the performance of feature word extraction. For example, the word "蓝牙(Bluetooth)" is segmented as "蓝(blue)/a 牙(tooth)/n", not "蓝牙(Bluetooth)/n". Although it is a feature word, it cannot be extracted from the comments. To solve this problem, we employ an unknown word finding technique.

Z. Luo and R. Song [14] use context-entropy to find the unknown words. According to their investigation, "significant terms in specific collection of texts can be used frequently and in different contexts. On the other hand sub-string of significant term almost locates in its corresponding upper string (that is, in fixed context) even through it occur frequently." We select the feature word "蓝牙(Bluetooth)" to explain this. Fig.1 shows the contexts of "蓝牙(Bluetooth)" and "牙(tooth)". We can see on the left of "蓝牙(Bluetooth)", the contexts are various, while on the left of "牙(tooth)" the contexts are almost only "蓝(blue)" which is dominating. It means that in the comments of mobile telephone the word "牙(tooth)" is just a substring of the word "蓝牙 (Bluetooth)". "蓝牙(Bluetooth)" can stand for a significant term.

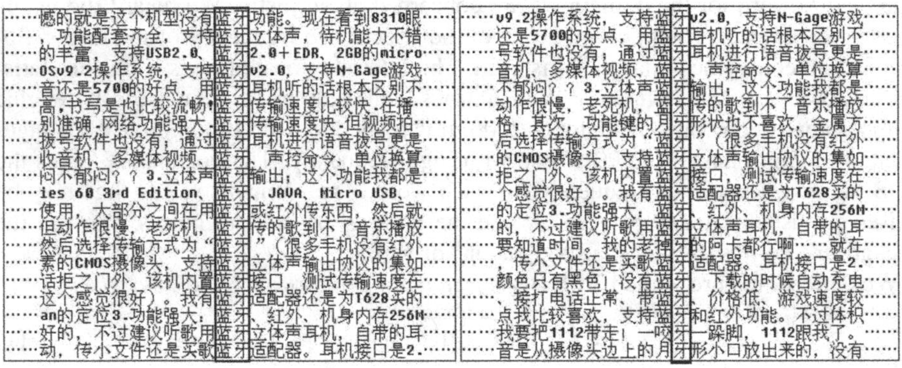

Fig. 1. The contexts of the word "蓝牙" and "牙"

To scale the chaos degree of the various contexts, left-context-entropy and right-context-entropy are defined. Assume ω as a term which appears n times in the corpus, $\alpha=\{a_1,a_2,...,a_s\}$ and $\beta=\{b_1,b_2,...,b_t\}$ as a set of left and right side contexts of ω in the corpus. Left-context-entropy and right-context-entropy of ω can be defined as:

$$LCE(\omega) = -\frac{1}{n}\sum_{i=1}^{s} C(a_i,\omega)\ln\frac{C(a_i,\omega)}{n} \tag{2}$$

$$RCE(\omega) = -\frac{1}{n}\sum_{i=1}^{t} C(\omega,b_i)\ln\frac{C(\omega,b_i)}{n} \tag{3}$$

Where $C(a_i, \omega)$ is count of co-occurrence of a_i and ω in the corpus and $C(\omega,b_i)$ is count of co-occurrence of ω and b_i.

We consider a feature word candidate joint with the word occurred on its left in the comments as a joint word. The difference of the left-context-entropy between original feature word candidate and the joint word can used to judge if the candidate is only a substring of the joint word which can be a true feature word. If the LCE of the original candidate is low, and the LCE of the joint word is higher, we can regard that the joint word is a feature word more possibly than the original candidate. We can select an LCE threshold; if the LCE of the joint word is over threshold greater than the LCE of the original candidate, the original candidate is replaced by the joint word in the candidate set. The reason why we do not use the right-context-entropy is we have already employed the adjective rule on the right of the candidate. The candidate cannot joint the right term.

4.2 Opinion Words Extraction

In this task, we regarded only adjectives as opinion words. This step is simple. When extracting the feature words, the opinion words are also generated. The opinion words are extracted when we use adjective rule to filter the feature words noises. The candidate with an adjective on its right is considered as a feature word, while the adjective is an opinion word to this feature word. As describing in section 1, every opinion word must be along with a feature word to express a polarity. Therefore, an adjective without a feature word on the left is not a significant opinion word.

5 Evaluation

We manually pick up the feature words from the comments to organize the test data set to evaluate the performance of the feature words extraction algorithm. There are totally 1328 feature words in the test set TEST. The result feature word set that are created by the algorithm is signed as RESULT. Precision, recall and f-measure defined as follow are used.

$$precision = \frac{|TEST \cap RESULT|}{|RESULT|}, recall = \frac{|TEST \cap RESULT|}{|TEST|},$$

$$f\_measure = \frac{2 * precision * recall}{precision + recall}$$

$$(4)$$

The performance of only using the patterns and adjective rule is shown in Fig. 2. Totally 2596 feature words have been extracted. Ranking the feature words by their occurrence frequency descending, the horizontal axis n stands for the set with top n results. From the illustration, we can find that at the top of the result list, there is a higher precision. It indicates that most of words with the high occurrence frequency are true feature words, which is in accordance with the assumption we take before in section 4.1. The recall curve in Fig. 2 is nearly linear, which indicates that the occurrence frequency of true feature words is nearly uniform distribution in the comments. Cutting the words whose occurrence frequency is below any threshold to maintain a high precision may depress the recall markedly. This can be proved by the f-measure curve; when all the results are kept, the f-measure can get the highest value.

Fig. 3 shows the performance that uses the background corpus to filter the noises. The result feature word set is ranking the feature words by their TFP value descending. The horizontal axis n also stands for the set with top n results. Comparing with the performance without filtering (ref. Fig. 2), the precision curve and the recall curve are both different. The precision curve is gently at the top, which indicates that most of the words with high TFP value are true feature words. They are most the phrases. The recall curve is gently at the bottom of the result list, which indicates most of the words with low TFP value are not true feature words, and cutting them will not influence the recall badly. There is a maximum point in the f-measure curve; keeping the top n results in the maximum point in the set can get the highest f-measure 0.6817. The words with low TFP value are removed.

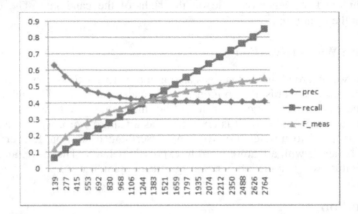

Fig. 2. The performance that only uses the patterns without any filtering

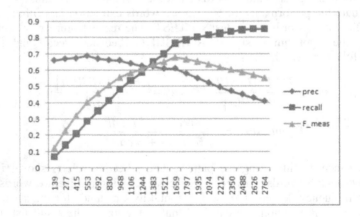

Fig. 3. The performance that uses the background corpus to filter the noises

To show the validity of using verbs and verb phrases as feature word candidates, we compare the performance between not using verbs and verb phrases, not using verbs and using all. The precision, recall and f-measure in Fig. 4 are the highest performance after filtering noises. The gray vertical bar stands for the performance of only using nouns and noun phrases (nouns + nouns) as candidates. It is the lowest among the three. When we added the verb phrases (verbs + nouns) pattern, the performance increases evidently. The white vertical bar stands for this performance. The black vertical bar for using all the

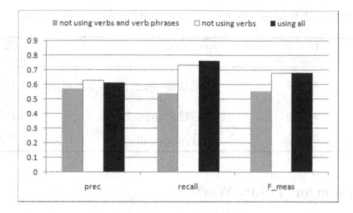

Fig. 4. Comparison of the performances between not using verbs and verb phrases, not using verbs and using all

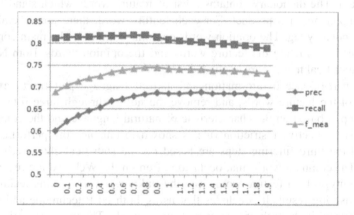

Fig. 5. The precision, recall and f-measure curves with the LCE threshold increasing

patterns is a little better than the white at f-measure and recall, which indicates that using verbs as feature words can extract more, but the noises are growing as well.

Next, we evaluate the validity of the unknown word finding technique. We control the threshold to observe the precision and the recall. If the LCE threshold is low, many terms will be replaced, and then the precision will become lower. Fig. 5 illustrates the curves with the LCE threshold increasing. The horizontal axis stands for the threshold. When the threshold is low, many joint words with a low LCE may replace the original candidate, which makes the precision low. However, when the threshold is high, very few joint words that are true feature words could replace the original candidates which are not. The recall is decreasing. When we select a proper LCE threshold 0.8, a higher f-measure 0.7407 is achieved.

Table 2 gives some examples for the unknown word finding technique using. The original candidates are not complete attribute or component of the product in the specific domain (Some of them do not have corresponding English translation in the domain context, which are marked as N/A in the table).

Table 2. Examples for the unknown word finding technique using

Original candidate	LCE of original candidate	Joint word	LCE of joint word
信功能 (N/A)	0.5069	短信功能(Short message function)	4.4261
性(N/A)	3.2736	兼容性(Compatibility)	4.1672
牙(tooth)	0.1392	蓝牙(Bluetooth)	6.5181

6 Conclusion and Future Work

We aim at automatically constructing the sentiment resource by mining customers' product comments. The sentiment resource can be considered as a dictionary for a given domain. The dictionary contains a list of feature words which stand for attributes or components of a product. For each feature word, there are several opinion words with polarity tag. The opinion words contain sentiment. The main contribution of this paper is extracting the feature words and the opinion words. Both NLP technique and statistical method are applied.

We use part-of-speech information and natural language patterns to extract the candidates of the feature words, and remove the noises through three filtering steps. The first step is based on the characteristic of natural language and the comments on the Web. The adjective restriction rule is shown to enhance the performance. The second and the third filtering steps are based on the statistical characteristic of the Web text. The common words that occur too often on the Web text are regarded with low probability to be a feature word. And the unknown word finding technique also brings a satisfying result. It concludes that using both NLP technique and statistical method is helpful in the task of extracting feature words. The use of statistical characteristic can be taken in the process of sentiment resource constructing. While dealing with a single comment, no statistical characteristic can be used. That is why we emphasize the importance and validity of resource construction.

Furthermore, although the experiment is built on the comments of mobile telephone domain in Chinese, the algorithm does not refer to any information of the domain. So it is a domain independent.

The future work can be expanded as the following:

1) For the aspect of feature word extraction, the more agile background corpus might be used to improve the timeliness of the vocabulary.

2) Polarity determination of opinion word will be studied in the future. For example, using the existing nature language tools such as HowNet [1] and considering the context of the opinion words, we can tag the polarity of the opinion words.

3) For the aspect of application of the sentiment resource, how to use the resource to analyze the comments is also an interesting task.

References

1. HowNet, http://www.keenage.com/
2. Ye, Q., Shi, W., Li, Y.: Sentiment Classification for Movie Reviews in Chinese by Improved Semantic Oriented Approach. In: Proceedings of the 39th Annual Hawaii international Conference on System Sciences, HICSS, January 04 - 07, vol. 03, p. 53.2. IEEE Computer Society, Washington (2006)
3. Li, J., Sun, M.: Experimental Study on Sentiment Classification of Chinese Review using Machine Learning Techniques. In: Proceedings of IEEE International Conference on Natural Language Processing and Knowledge Engineering 2007, pp. 393–400 (2007)
4. Hu, M., Liu, B.: Mining Opinion Features in Customer Reviews. In: Proceedings of Nineteenth National Conference on Artificial Intelligence, San Jose, California, USA, July 2-29, pp. 755–760. AAAI Press, Menlo Park (2004)
5. Hu, M., Liu, B.: Mining and summarizing customer reviews. In: Proceedings of the Tenth ACM SIGKDD international Conference on Knowledge Discovery and Data Mining, KDD 2004, Seattle, WA, USA, August 22 - 25, pp. 168–177. ACM, New York (2004)
6. Liu, B., Hu, M., Cheng, J.: Opinion observer: analyzing and comparing opinions on the Web. In: Proceedings of the 14th international Conference on World Wide Web, WWW 2005, Chiba, Japan, May 10-14, pp. 342–351. ACM, New York (2005)
7. Popescu, A., Etzioni, O.: Extracting product features and opinions from reviews. In: Proceedings of the Conference on Human Language Technology and Empirical Methods in Natural Language Processing, Human Language Technology Conference. Association for Computational Linguistics, Vancouver, British Columbia, Canada, October 06-08, pp. 339–346. Morristown, NJ (2005)
8. Yi, J., Niblack, W.: Sentiment Mining in WebFountain. In: Proceedings of the 21st international Conference on Data Engineering (Icde 2005), ICDE, April 05-08, vol. 00, pp. 1073–1083. IEEE Computer Society, Washington (2005)
9. Scaffidi, C., Bierhoff, K., Chang, E., Felker, M., Ng, H., Jin, C.: Red Opal: product-feature scoring from reviews. In: Proceedings of the 8th ACM Conference on Electronic Commerce, EC 2007, San Diego, California, USA, June 11-15, pp. 182–191. ACM, New York (2007)
10. Wang, B., Wang, H.: Bootstrapping both Product Properties and Opinion Words from Chinese Reviews with Cross-Training. In: Proceedings of the IEEE/WIC/ACM international Conference on Web intelligence, Web Intelligence, November 02 - 05, pp. 259–262. IEEE Computer Society, Washington (2007)
11. Su, Q., Xu, X., Guo, H., Guo, Z., Wu, X., Zhang, X., Swen, B., Su, Z.: Hidden sentiment association in chinese Web opinion mining. In: Proceeding of the 17th international Conference on World Wide Web, WWW 2008, Beijing, China, April 21 - 25, pp. 959–968. ACM, New York (2008)
12. ICTCLAS, http://www.nlp.org.cn/
13. Sogou Lab Internet Vocabulary, http://www.sogou.com/labs/dl/w.html
14. Luo, Z., Song, R.: An Integrated Method for Chinese Unknown Word Extraction. In: Proceedings of 3rd ACL SIGHAN Workshop on Chinese Language Processing, Barcelona, Spain, pp. 148–155 (2004)

Image Sense Classification in Text-Based Image Retrieval

Yih-Chen Chang and Hsin-Hsi Chen

Department of Computer Science and Information Engineering
National Taiwan University, Taipei, Taiwan
ycchang@nlg.csie.ntu.edu.tw, hhchen@csie.ntu.edu.tw

Abstract. An *image sense* is a graphic representation of a concept denoted by a (set of) term(s). This paper proposes algorithms to find image senses for a concept, collect the sense descriptions, and employ them to disambiguate the image senses in text-based image retrieval. In the experiments on 10 ambiguous terms, 97.12% of image senses returned by a search engine are covered. The average precision of sample images is 68.26%. We propose four kinds of classifiers using text, image, URL, and expanded text features, respectively, and a merge strategy to combine the results of these classifiers. The merge classifier achieves 0.3974 in F-measure (β=0.5), which is much better than the baseline and has 51.61% of human performance.

Keywords: Image sense disambiguation, Text-based image retrieval, Word sense disambiguation.

1 Introduction

In web image search, users submit text queries to express their image needs and search engines return ranked images. Two major issues have to be dealt with in an ambiguous query. If one sense of a query term dominates the other senses, most of the retrieved images will contain the dominated senses, and images of the other senses will be relatively few and even hard to be found. Meanwhile, images of different senses may be mixed together in the result list of search engines. Thus, users may have to browse the list to find the requested images. Mixing up the two issues makes text-based image retrieval more challenging. Figure 1 shows an example of web image search result of query term *jaguar*. *Animal* jaguar and *car* jaguar dominate the result list, and the top 10 images belong to these two senses. In contrast, *game system* jaguar and *operating system* jaguar rank behind the common image senses, e.g., the first images for these two senses appear at the ranks 74 and 172, respectively, and become hard to be found.

The basic issues to disambiguate the images in text-based image retrieval are how many image senses there may be, and what the image senses they are. Here, an *image sense* is defined to be a graphic representation of a concept denoted by a (set of) term(s). For similar word sense disambiguation (WSD) problem, we can look up a linguistic resource to determine the possible senses for a given word, and guess the most appropriate sense from the context. Due to the lack of word-image ontology, we cannot deal with these issues by simple lookup. Even if there is such transmedia

G.G. Lee et al. (Eds.): AIRS 2009, LNCS 5839, pp. 124–135, 2009.
© Springer-Verlag Berlin Heidelberg 2009

Fig. 1. Result List of Ambiguous Query Term *jaguar*

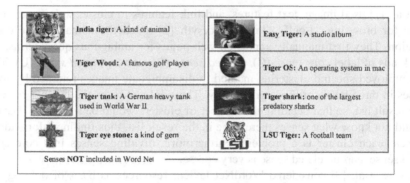

Fig. 2. Coverage Problem of Image Senses

ontology, the coverage problem is more serious than that in text-based ontology. In the image result list of a text query, there may be many unknown senses. For example, the word *tiger* has two senses in WordNet – say, *animal* and *person*. In contrast, the result list for query term *tiger* may contain images related to *tiger tank*, which is a German heavy tank used in World War II, *tiger shark*, which is one of the largest predatory sharks, and so on. Figure 2 shows six senses not included in WordNet, i.e., tiger tank, tiger eye stone, easy tiger, tiger OS, tiger shark, and LSU tiger, for query term *tiger*.

The web provides rich images for different senses to increase coverage. The problem is how to collect enough noiseless sample images for each sense without too much human intervention. In WSD, we all know the context of a word can restrict its sense to some degree. This idea can be employed to collect sample images. If the context related to an image sense can be found and added to a query term, then the ambiguity degree of text-based image retrieval will be decreased. For example, the sense of images will be narrowed down to *person*, when the words like *wood* and *golf* are added to query term *tiger*. Now the problem is transformed into how to collect the related words for each sense and how many words are needed to collect the sample images.

The set of sample images along with their text form a sense description. Several issues have to be considered, when a collection of sense descriptions are employed for image sense disambiguation. Firstly, the contextual information for each returned image is very few. Secondly, the words to be disambiguated in WSD problem often belong to one of the given senses. In contrast, the images returned by a search engine may not belong to any senses in the precompiled collection because irrelevant images may be retrieved. That makes image sense disambiguation harder.

This paper is organized as follows. Section 2 introduces the related works. Section 3 proposes a method to collect sample images and their text descriptions from the web. Section 4 applies the results to disambiguate the image senses. Section 5 concludes the remarks.

2   Related Works

Cai et al. [1] used image, text features, and link features to cluster web image search results for browsing. Loeff, Alm and Forsyth [2] touched on image sense disambiguation. They first manually annotated web pages for ambiguous query terms, with three kinds of label *core sense* (C), *related sense* (R), and *unrelated sense* (U). Then, they employed spectral clustering method to disambiguate images returned by search engines in three levels of sense granularity. They did not know the number of senses in the result list, so that deciding the number of clusters would be a problem. Besides, they did not know what senses there were in the result list, thus finding information to represent each cluster is a problem. Furthermore, dividing senses into core sense, related sense, and unrelated sense is very coarse.

Zinger, et al. [3] considered WordNet lexical resources as basis of ontology and created large-scale image ontology. The image collections were acquired through web-based image mining. Fluhr, et al. [4] constructed general purpose ontology, where each node is decorated with multilingual and multimodal data. Chang and Chen [5] extended WordNet to word-image ontology. The previous work submitted text queries to web image search engines to acquire images. Lexical ambiguity makes the alignment of lexical terms in WordNet and the retrieved images challenging.

3   Image Sense Representation

An image sense is described by a set of text terms and sample images. This section discusses how to collect text terms first and then sample images. An example *tiger* is considered as a *seed word* to demonstrate the algorithms.

3.1   Finding Text Terms to Label Senses

3.1.1   Finding Context Terms
A seed word of a specific sense appears in the similar contexts. If we regard a context as a set of terms, then the contexts that strongly relate to a sense of a seed word

co-occur with it very often. The terms that co-occur with a given seed word with high frequency are considered as *context terms* for this word. Algorithm 1 extracts context terms for a given seed word.

Algorithm 1. *FindingContextTerms(sw,CT)*
Input: A seed word *sw*
Output: A set *CT* of context terms for *sw*
Method:
1. Submit *sw* to a web page search engine, and retrieve the snippets.
2. Preprocess the snippets by changing all characters into the lowercase, removing symbols, stemming, and filtering out the stop words.
3. Count the frequency of the remaining terms and choose the terms passing a dynamic threshold as context terms. The threshold is set to the frequency of *sw* multiplying 0.01.

At step 1, we use a web page search engine instead of an image search engine. At step 2, we employ a special stop word list to exclude words which often appear in web pages.

3.1.2 Finding Related Terms of Context Terms

Because a seed word is ambiguous, its context terms co-occur with different senses of the seed word. They are partitioned into groups to disambiguate different uses. Two context terms are defined to have a relationship if they relate to the same sense of a seed word. For each context term, we can collect the highly co-occurring terms from the web in the similar way as that done on seed word in Section 3.1.1., use the information to group context terms and filter out context terms not related to any sense. A context term may relate to more than one sense of the seed word. That may confuse the interpretation of an image sense later. In the following, we will assign it to the most common sense only. For example, the context term *sport* of the seed word *tiger* relates to a famous golf player (i.e., Tiger Woods) and a football team (i.e., LSU Tiger). The most common assignment will be the former instead of the latter. Algorithm 2 shows how to find the related terms of context terms.

At step 7, we check the collocation strengths of *ct* and all context terms in RT_i instead of ct_i only. The context terms checked earlier, i.e., those terms of higher collocation strength with ct_i, are relatively easy to be added to RT_i because they need less number of checks. That ensures the terms added to RT_i are related to only one sense. In the previous example, ct_i is *sport*. If we already add *golf* to RT_i, then we cannot add *football* into RT_i later because the collocation strength between *golf* and *football* is not strong enough to include *football* in the same group as *golf*. Therefore, *tiger* and *sport* together tend to denote the sense of *golfer player*. Table 1 lists some context terms of the seed word *tiger*, and their related terms. For each row, the first (an underlined term) and the rest denote a context term and its related terms.

Algorithm 2. *FindingRelatedTerms(sw,CT,RT)*
Input: A seed word *sw* and a set *CT* of context terms $ct_1, ct_2, \ldots, ct_n$ for *sw*
Output: A set *RT* of relationships $RT_1, RT_2, \ldots, RT_n$, where each set RT_i contains a context term ct_i and its related terms selected from *CT*
Method:
1. Initialize RT_i to $\{ct_i\}$ for i=1 to *n*.
2. Construct query q_i with two query terms *sw* and ct_i for i=1 to *n*.
3. Submit q_i (for i=1 to *n*) to a web page search engine, and retrieve the snippets.
4. Preprocess the snippets in the similar way as Step 2 of Algorithm 1.
5. For each q_i, count the frequency f_{ij} of context term ct_j (for i=1 to *n*) in the snippets retrieved for q_i.
6. Construct a collocation matrix G=$[g_{ij}]_{n\times n}$ as follows.

$$g_{ij} = \frac{f_{ij}}{\sum_{k=1}^{n} f_{ik}} \text{ for } i \neq j.$$

Let g_{ij}=0, when i=j. The normalized value g_{ij} indicates the collocation strength between ct_i and ct_j.
7. For each context term ct_i, filter out context term ct_j whose $g_{ij}<2/n$, sort the remaining context terms by descending order of their collocation strengths, and check the context terms *ct* in the sorted sequence as follows. If the collocation strengths of *ct* and all context terms in current RT_i>2/n, then add *ct* to RT_i. That is, *ct* is considered as one of the related terms of ct_i.

Table 1. Some Context Terms of the Seed Word *tiger*, and Their Related Terms

<u>wood</u> golf picture photo baby	<u>wildlife</u> tour india national bengal
<u>os</u> mac apple update software system	<u>baby</u> wood
<u>mac</u> os apple feature	<u>tank</u> history

3.1.3 Grouping and Sense Labeling

Algorithm 3 deals with the grouping of context terms based on their related terms. At step 2, we check if a context term is a related term of its related terms. If the strict double checking satisfies, we merge the related groups into a larger one. At step 3, we remove the groups composed of only one context term and its related terms. The context terms (along with related terms) in the same group will denote a sense. At step 4, we select representative terms from each group and use them to label the corresponding group. Table 2 shows some of 13 groups generated using the relationships found by Algorithm 2. Each entry of a group contains a context term (underlined) and its related terms. The sense label of each group is also shown in bold after group id.

3.2 Finding Sample Pages and Sample Images for Image Senses

Given a word, we can find its senses labeled by text terms with the algorithms proposed in Section 3.1. The text sense labels are submitted to web page search engine and image search engine to collect sample pages and sample images. Table 3 demonstrates top 5 sample images for senses of *tiger*.

Algorithm 3. *GroupingAndSenseLabeling(sw,CT,RT,TS,SL)*

Input: A seed word *sw*, a set *CT* of context terms ct_1, ct_2, ..., ct_n for *sw*, and a set *RT* of relationships RT_1, RT_2, ..., RT_n

Output: A set *TS* of senses TS_1, TS_2, ..., TS_m, where TS_i is a subset of *RT* and denotes a sense of *sw*, and a set *SL* of sense labels SL_1, SL_2, ..., SL_m, where SL_i is a subset of *CT* and denotes a sense label of *sw*.

Method:
1. Initialize *TS* to be an empty set.
2. For each context term ct_i (*i*=1 to *n*), check if there are related terms in RT_i belonging to some sets in *TS*.
   a. Some related terms, say ct_{i1}, ct_{i2}, ..., ct_{ip}, satisfy this condition
      Double check if ct_i is a related term of ct_{i1}, ct_{i2}, ..., or ct_{ip}. If the relationships exist in sets TS_{i1}, TS_{i2}, ..., TS_{iq}, then merge TS_{i1}, TS_{i2}, ..., TS_{iq} into a new group. If there do not exit any such relationships, then regard RT_i itself as a new group and put it into *TS*.
   b. Not any related terms satisfy this condition
      Regard RT_i itself as a new group and put it into *TS*.
3. Remove those singleton sets from *TS*.
4. For each remaining TS_i, those terms appear in more than half of the relationships RT_{i1}, RT_{i2}, ..., RT_{ik} are selected as a sense label SL_i of this group.

Table 2. Groups and Sense Labels for the Seed Word *tiger*

Group 1 (label: **tiger os mac apple software system computer**)	Group 3 (label: **tiger wood golf**)
os mac apple update software system	wood golf picture photo baby
mac os apple feature	photo picture wood art cat
system mac os apple software computer	tour wood pga game offer
apple mac os computer system software	golf wood pga tour sport play
feature mac os	sport wood golf open
technology computer	pga wood tour game golf
computer apple software system technology	picture photo wood
software mac os apple system computer	game wood pga tour video golf play
update os apple release software system	play wood golf game video
	video wood game play golf
	offer tour india
	open wood golf sport
	baby wood
Group 5 (label: **tiger india wildlife bengal**)	Group 7 (label: **tiger java release**)
india tour wildlife national bengal project	release apple java
national wildlife india bengal	java release
project india save	
wildlife tour india national bengal	
Group 8 (label: **tiger great shark white**)	Group 9 (label: **tiger history tank**)
great shark white	history tank
shark great white specie	tank history

Table 3. Top 5 Sample Images for Senses of *Tiger*

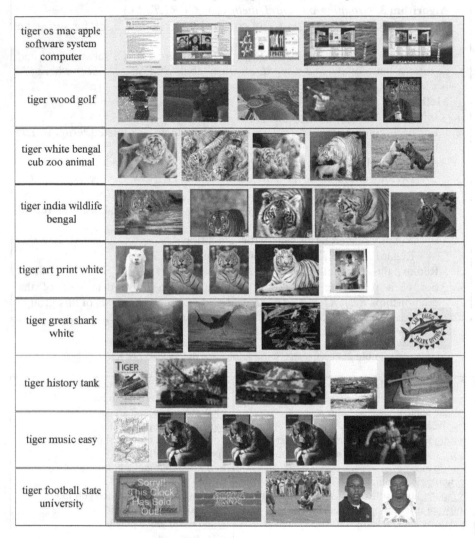

tiger os mac apple software system computer	
tiger wood golf	
tiger white bengal cub zoo animal	
tiger india wildlife bengal	
tiger art print white	
tiger great shark white	
tiger history tank	
tiger music easy	
tiger football state university	

3.3 Experiments and Discussion

In the experiments, we evaluate the results from the two measures: *coverage* and *precision*. The first metric concerns if all the image senses in the image result list are covered by the senses we found. The second metric concerns the quality of the sample images we collected. Total 10 ambiguous words including *tiger*, *bass*, *bike*, *mouse*, *jaguar*, *junk*, *star*, *plane*, *bat*, and *pick* are adopted.

We submit each word to Google image search engine, and retrieve the top 300 images. Total 3,000 images are annotated by human. There are four possible annotations shown as follows.

(1)   An image is relevant to the target word, and also relate to an image sense we found.
(2)   An image is relevant to the target word, but its sense is not included in the image sense list we found.
(3)   An image is irrelevant to the target word.
(4)   An image is fuzzy, and is not decided.

Table 4 shows the evaluation of sense coverage. The number enclosed in parentheses is total senses found by our method. On the average, 97.12% of the image senses are covered. Most of the missing image senses are not very common. When an image sense is not popular, the frequency of terms related to this sense may not be high.

Table 4. Evaluation of Sense Coverage

Query Term	Tiger (9)	Bass (5)	Bike (11)	Mouse (6)	Jaguar (8)
Coverage	98.00%	100.00%	99.00%	100.00%	99.33%
Query Term	Junk (10)	Star (15)	Plane (11)	Bat (12)	Pick (13)
Coverage	87.29%	91.95%	100.00%	100.00%	95.62%
Average: 97.12%					

For each word, we submit all the senses found by our method to Google image search engine, and retrieve the top 100 images for each sense. Assessors check if the retrieved images belong to the corresponding sense. If assessors think the query is too fuzzy to make decision, the group of images retrieved is considered as noise. Table 5 shows the precision of sample images. The number (n/m) enclosed in parentheses denotes n of m senses found are undecidable by assessors. In the experiments, 10% of senses are undecidable and the average precision is 68.26%.

Table 5. Precision of Sample Images

Query Term	Tiger (0/9)	Bass (0/5)	Bike (0/11)	Mouse (0/6)	Jaguar (0/8)
Precision	74.78%	83..40%	53.36%	81.33%	89.75%
Query Term	Junk (1/10)	Star (6/15)	Plane (0/11)	Bat (1/12)	Pick (2/13)
Precision	66.50%	45.00%	78.09%	58.41%	52.00%
Average: 68.26%					

4   Image Sense Classifiers

As discussed in Section 3.2, we collect sample pages and sample images for a target term. We will use them to classify the images returned by image search engine. Search engine provides three kinds of information for each returned image, including a thumbnail, a source web site, and a short snippet. Text features, URL features, and image features are extracted from sample pages, sample images, and the returned images for classification. Five types of classifiers are presented.

4.1 Feature Extraction and Classification Algorithms

Assume there are n image senses, $s_1, s_2, ..., s_n$, for a target term v. Text features and website features are extracted from sample pages (snippets) for each sense. The sample pages are preprocessed in the similar way as the step (2) of Algorithm 1. Let T_i be a set of m_i terms $t_{i1}, t_{i2}, ..., t_{im_i}$ in the sample pages of sense s_i. They are called text features.

The first classifier called *text classifier* employs text features only. We count the frequency f of term t_{ij} in the sample pages, and the frequency g of each term in the short snippet of a returned image G. The following procedure assigns G to the sense of the highest score if the score is also larger than a threshold. If all the scores are less than the threshold, we regard G as an irrelevant image.

$$\text{if } \max_{1\leq i\leq n} \sum_{t\in T_i \text{ and } t\in G} f(t)\times g(t) > \varepsilon \text{ then } \hat{s} = \arg\max_{1\leq i\leq n} \sum_{t\in T_i \text{ and } t\in G} f(t)\times g(t) \text{ else } \hat{s} = \text{irrelevant}$$

An alternative procedure considers inverse sense frequency (*ISF*) defined as follows. N is the number of senses for a given target word, and $sf(t)$ is the number of senses containing t in the sample pages.

$$ISF(t) = log(\frac{N}{sf(t)})$$

The revised procedure is shown as follows.

$$\text{if } \max_{1\leq i\leq n} \sum_{t\in T_i \text{ and } t\in G} f(t)\times g(t)\times ISF(t) > \varepsilon \text{ then } \hat{s} = \arg\max_{1\leq i\leq n} \sum_{t\in T_i \text{ and } t\in G} f(t)\times g(t) \text{ else } \hat{s} = \text{irrelevant}$$

The second classifier called *image classifier* employs image features. Image features are extracted from each sample image. We first segment a sample image into 32×32 blocks, and compute an average RGB value for each block. An image is represented as a vector of 3,072 (32×32×3) average values. We represent a thumbnail as a vector of size 3,072 similarly. Assume there are n_i sample images $G_{i1}, G_{i2}, ..., G_{in_i}$ for image sense s_i. The following procedure determines the image sense which G belongs to. We compute the Euclidian distance of sample images and G. The image sense containing a sample image with the smallest distance with G will be considered. If the distances of all the sample images and G are larger than a threshold, then G will be assigned to irrelevant category.

$$\text{if } \min_{1\leq i\leq n}(\min_{1\leq j\leq n_i}(dist(G,G_{ij}))) < \varepsilon \text{ then } \hat{s} = \arg\min_{1\leq i\leq n}(\arg\min_{1\leq j\leq n_i}((dist(G,G_{ij}))) \text{ then } \hat{s} = \text{irrelevant}$$

The third classifier called *URL classifier* employs website features. We postulate that images coming from the same web site share the same sense. This idea is similar to one sense per discourse in WSD [6]. The discourse here is the website in which images occur. We extract the top level URLs of all the sample images and the target image G, count how many sample images of each sense have the same URL as G, and select the sense of the highest count. This procedure is defined as follows, where U_i is a set of top-level URLs of the sample pages of sense s_i, and u_G denotes top-level URL of G.

$$\text{if } \max_{1 \le i \le n} (cardinality(\{u \mid u \in U_i, u = u_G\})) > 0$$

$$\text{then } \hat{s} = \underset{1 \le i \le n}{argmax} \, (cardinality(\{u \mid u \in U_i, u = u_G\})) \text{ else } \hat{s} = \text{irrelevant}$$

Besides the fundamental URL classifier, we also consider the rank of each sample image returned by search engine. The revised procedure is shown as follows. The function $rank(u)$ returns the rank of sample image u. The larger the reciprocal Rank of u is, the higher score it contributes.

$$\text{if } \max_{1 \le i \le n} \sum_{u \in U_i, u = u_G} \frac{1}{\text{rank}(u)} > 0 \text{ then } \hat{s} = \underset{1 \le i \le n}{argmax} \sum_{u \in U_i, u = u_G} \frac{1}{\text{rank}(u)} \text{ else } \hat{s} = \text{irrelevant}$$

The fourth classifier called *expanded text classifier* uses top-level URL u_G to expand the short snippet of a returned image G for a target term v. We submit "$v \, u_G$" to text search engine, collect the returned snippets, regard them as expanded text, and adopt the same procedure as the first classifier to determine the suitable sense.

The fifth classifier adopts voting strategy to merge the results of the above four classifiers. A sense s will get a vote 1 for an image G when a classifier labels G with s. If a classifier determines G to be irrelevant, then the irrelevant label gets vote 0.5. Finally, the label (a sense or an irrelevant mark) of the highest vote will be proposed.

4.2 Experiments and Discussion

In the experiments, we adopt the following 7 metrics to evaluate the performance of image sense classifiers.

(1) $$\text{Recall} = \frac{\text{\# of relevant images with correct annotation}}{\text{\# of relevant images}}$$

(2) $$\text{Precision} = \frac{\text{\# of relevant images with correct annotation}}{\text{\# of images annotated with a sense}}$$

(3) $$\text{Fmeasure}_1 = \frac{2 \times \text{Recall} \times \text{Precision}}{\text{Recall} + \text{Precision}}$$

(4) $$\text{Fmeasure}_{0.5} = \frac{1.25 \times \text{Recall} \times \text{Precision}}{\text{Recall} + 0.25 \times \text{Precision}}$$

(5) $\text{ClassNum} = $ Number of classes that a classifier proposes

(6) $$\text{Purity} = \frac{\displaystyle\sum_{c=1}^{\text{ClassNum}} \frac{\text{\# of correct images annotated as class c}}{\text{\# of images annotated as class c}}}{\text{ClassNum}}$$

(7) $$\text{AvgNum} = \frac{\displaystyle\sum_{c=1}^{\text{ClassNum}} \text{\# of images annotated as class c}}{\text{ClassNum}}$$

Recall, precision, and F-measure are common metrics used in information retrieval. ClassNum denotes the number of classes that a classifier proposes. A useful classifier will deal with as many senses as possible. Purity is similar to precision, but different classes have different weights. For example, the dominate senses have the higher weight in precision. AvgNum is the average size for each class that a classifier proposes. Here, a useful classifier should provide enough images for each class. Fmeasure$_{0.5}$ is adopted because precision is more important than recall in this task.

Table 6 shows the performance of different classifiers. Here, C# and Avg# abbreviate average *ClassNum* and average *AvgNum*, respectively. Two baselines are experimented for comparisons. Baseline 1 neglects the unpopular senses, and classifies images to the most common sense. That is, C#=1 in baseline 1. The recall of the baseline 1, i.e., 0.5665, shows that there always exists a dominating image sense for each experimental query term. Baseline 2 assigns one of the possible senses to an image at random. Baseline 1 is better than baseline 2 in near all the measures. However, only the major sense is considered in baseline 1. That is impractical. Comparatively, baseline 2 deals with large number of possible senses (C#=10).

Table 6. Performance of Different Classifiers

Classifier	Prec	Recall	F1	F0.5	C#	Purity	Avg#
Baseline 1	0.4033	0.5665	0.4680	0.4438	1	0.4033	298
Baseline 2	0.1407	0.1928	0.1618	0.1483	10	0.1407	33.09
Text w/o *ISF*	0.4012	0.2482	0.2727	0.2938	6.5	0.4048	24.48
Text w/ *ISF*	0.4273	0.2604	0.3119	0.3396	10	0.3481	15.34
Image	0.3021	0.2391	0.2648	0.2756	9.7	0.2738	19.47
URL w/o ranks	0.4758	0.1353	0.2053	0.2495	8.6	0.4051	6.41
URL w/ Ranks	0.4544	0.1277	0.1941	0.2362	8.9	0.4141	6.66
Expanded Text	0.3696	0.4062	0.3826	0.3773	10	0.2940	25.49
Merge	0.4283	0.3666	0.3860	0.3974	10	0.3729	20.41
Human	0.7892	0.7435	0.7622	0.7700	9.9	0.5942	20.72

The text classifier with *ISF* is better than that without *ISF* in most of the measures. C#=10 means that the text classifier with *ISF* can deal with all the senses of each query term. In contrast, the text classifier without *ISF* can only disambiguate 65% of senses (refer to C# column). Compared with baseline 1, the text classifier with *ISF* can deal with 10 times more senses with competitive performance. Image classifier is worse than text classifier, but is still better than baseline 2.

The recall and Avg# of URL classifier is the worst of all models since most of images come from the unseen web sites. In contrast, its precision and Purity is the best of all models. That confirms the postulation of one sense per discourse: images in the same web sites share the same sense. The performance of the URL classifier with and without ranks is quite similar. That shows the ranks of snippets may not be a powerful factor.

Comparing the pure text classifier and the expanded text classifier, we can find the latter is better than the former. In particular, the recall is improved 55.99% at the expense of 13.50% precision decrease. The expanded text helps reduce the short snippet problem. When all the classifiers are merged together by voting strategy, the performance is the best. The last row of Table 6 also lists human performance for comparison. Given a set of image senses and a set of images, a user is asked to assign each image to a sense cluster. In this experiment, F1, F0.5 and Purity of human are 0.7622, 0.7700, and 0.5942, respectively. The merge classifier achieves 54.27%, 49.31%, 50.64%, and 51.61% of human performance in precision, recall, Fmeasure1, and Fmeasure0.5, respectively.

We analyze the performance of various models under different query terms to find the influence factors. Baseline 1 depends on the dominating degree of most common senses. Baseline 2 is influenced by the number of image senses and the number of irrelevant images. The quality of training samples is affected by the number of irrelevant images. The performance of classifiers is sensitive to the quality of training samples, the number of image senses, number of irrelevant images and the classification strategies.

5 Concluding Remarks

This paper investigates the image sense disambiguation in web image retrieval. A method to find unknown image senses is proposed. Total 97.12% of image senses returned by a search engine are covered. We also collect sample pages and sample images of each sense without human annotation. The average precision of sample images is 68.26%. We propose four kinds of classifiers using text, image, URL, and expanded text features, respectively, and a merge strategy to combine the results of these classifiers. For classifying unseen images, the merge classifier achieves 0.3974 in Fmeasure$_{0.5}$, which is much better than the baseline (0.1483). It has 51.61% of human performance. There are still spaces for further improvement. The irrelevant images influence performance very much. How to filter out irrelevant images is important for image sense disambiguation. How to integrate the collection of sense descriptions to an existing ontology like WordNet will be investigated.

References

1. Cai, D., He, X., Li, Z., Ma, W.Y., Wen, J.R.: Hierarchical Clustering of WWW Image Search Results Using Visual, Textual and Link Information. In: The 12th Annual ACM International Conference on Multimedia, pp. 952–959 (2004)
2. Loeff, N., Alm, C.O., Forsyth, D.A.: Discriminating Image Senses by Clustering with Multimodal Features. In: The COLING/ACL on Main Conference Poster Sessions, pp. 547–554 (2006)
3. Zinger, S., Millet, C., Mathieu, B., Grefenstette, G., Hède, P., Moëllic, P.A.: Extracting an Ontology of Portrayable Objects from WordNet. In: The 1st MUSCLE/ImageCLEF Workshop on Image and Video Retrieval Evaluation (2005)
4. Fluhr, C., Grefenstette, G., Popescu, A.: Toward a Common Semantics between Media and Languages. In: The 2006 International Workshop on Research Issues in Digital Libraries (2006)
5. Chang, Y.C., Chen, H.H.: Approaches of Using Word-Image Ontology and an Annotated Image Corpus as Intermedia for Cross-Language Image Retrieval. In: Peters, C., Clough, P., Gey, F.C., Karlgren, J., Magnini, B., Oard, D.W., de Rijke, M., Stempfhuber, M. (eds.) CLEF 2006. LNCS, vol. 4730, pp. 625–632. Springer, Heidelberg (2007)
6. Yarowsky, D.: Unsupervised Word Sense Disambiguation Rivaling Supervised Methods. In: ACL, pp. 189–196 (1995)

A Subword Normalized Cut Approach to Automatic Story Segmentation of Chinese Broadcast News

Jin Zhang[1], Lei Xie[1], Wei Feng[2], and Yanning Zhang[1]

[1] Shaanxi Provincial Key Laboratory of Speech and Image Information Processing
School of Computer Science, Northwestern Polytechnical University, Xi'an, China
[2] Dept. of Computer Science and Engineering, The Chinese University of Hong Kong
{jzhang,lxie}@nwpu-aslp.org, wfeng@ieee.org, ynzhang@nwpu.edu.cn

Abstract. This paper presents a subword normalized cut (N-cut) approach to automatic story segmentation of Chinese broadcast news (BN). We represent a speech recognition transcript using a weighted undirected graph, where the nodes correspond to sentences and the weights of edges describe inter-sentence similarities. Story segmentation is formalized as a graph-partitioning problem under the N-cut criterion, which simultaneously minimizes the similarity across different partitions and maximizes the similarity within each partition. We measure inter-sentence similarities and perform N-cut segmentation on the character/syllable (i.e. subword units) overlapping n-gram sequences. Our method works at the subword levels because subword matching is robust to speech recognition errors and out-of-vocabulary words. Experiments on the TDT2 Mandarin BN corpus show that syllable-bigram-based N-cut achieves the best F1-measure of 0.6911 with relative improvement of 11.52% over previous word-based N-cut that has an F1-measure of 0.6197. N-cut at the subword levels is more effective than the word level for story segmentation of noisy Chinese BN transcripts.

1 Introduction

Story segmentation is an interesting task aiming at partitioning a text, audio or video stream into a sequence of topically coherent segments known as stories. The increasing availability of multimedia data is fostering a new wave of semantic access to the media content. Story segmentation is an important *prerequisite* since various tasks, e.g., topic tracking, summarization, information extraction, indexing and retrieval, usually assume the presence of individual topical 'documents'. Specifically, for a broadcast news (BN) retrieval system, users expect short clips of relevant news stories rather than an entire news stream in response to their specific queries. However, manual segmentation requires annotators to go through the whole stream, which costs tremendous labor. As the exponential proliferation of multimedia content on the Internet, automatic story segmentation techniques are highly in demand and this need will continue to rise.

G.G. Lee et al. (Eds.): AIRS 2009, LNCS 5839, pp. 136–148, 2009.
© Springer-Verlag Berlin Heidelberg 2009

Previous efforts on automatic story segmentation focus on three categories of cues: (1) video cues such as anchor face and frame similarity [1], (2) acoustic/prosodic cues such as significant pauses and pitch resets from audio [1,2], and (3) lexical cues from text transcripts. Compared to other two types of cues, lexical cues are more popular since it works on texts and speech recognition transcripts of multimedia sources. Main lexical approaches include word cohesiveness, e.g. TextTiling [3], the use of cue phrases [4] and statistical modeling. TextTiling is based on a straightforward observation that different topics usually employ different sets of words. As a result, pairwise similarity measure between consecutive sentences can be used across the text, and a local similarity minimum implies a possible topic shift.

These pairwise-similarity-based approaches achieve satisfying performance when the documents have sharp variations in lexical distribution, such as synthetic collections by concatenation of random texts [5]. However, in real-world collections, e.g. broadcast news and spoken lectures, transitions between topics are usually smooth and distributional variations are subtle. Recently, graph-based algorithms have drawn much attention for modeling real-world discourses[1]. In many natural language processing applications, entities can be naturally represented as nodes in a graph and relations between them can be represented as edges. Research has shown that graph-based representations of linguistic units, as diverse as words, sentences and documents, give rise to novel and efficient solutions in a variety of tasks. Malioutov *et al.* [6] proposed a minimum cut (Min-cut) approach for spoken lecture segmentation. They abstracted a text into a weighted undirected graph, where the nodes correspond to sentences and edges represent the pairwise sentence similarities. The segmentation task thus shifts to a graph partitioning problem that optimizes the normalized cuts (N-cut) criterion [7], where the similarity within each partition and the dissimilarity across different partitions are both considered. This approach takes into account long-range changes in lexical distribution and outperforms the state-of-the-art similarity-based segmentation, such as the one developed by Choi *et al.* [8].

One of the major challenges of story segmentation on multimedia documents (e.g. BN) is that story boundaries have to be detected on inaccurate texts transcribed from audio via a large vocabulary continuous speech recognizer (LVCSR). The inevitable speech recognition errors pose significant difficulties in lexical-based story segmentation since noisy texts break lexical cohesion. Speech recognition errors result from adverse acoustic conditions, diverse speaking styles, and absence of in-domain vocabulary. The existence of out-of-vocabulary (OOV) words (i.e., words outside the vocabulary of the speech recognizer) is more common for Chinese than other languages such as English. Chinese OOV words are largely named entities (e.g. Chinese person names and transliterated foreign names) that are keys to topic discrimination. Recently, the partial matching merit of subword lexical units [9], has been successfully applied to TextTiling-based automatic story segmentation of Chinese BN [10,11]. At the subword level,

---

[1] http://www.textgraphs.org/ws07

the incorrectly recognized words may include several subword units correctly recognized and thus recover lexical cohesion in noisy transcripts.

In this paper, we propose a subword N-cut approach to automatic story segmentation of Chinese broadcast news. Graph-based segmentation is conducted on character and syllable sequences of Chinese LVCSR transcripts. We find that employing N-cut framework at the subword levels is more effective than the word level due to the partial matching merit of subwords in Chinese BN. The proposed approach demonstrates the potential of graph cut approach in story segmentation. Moreover, the effectiveness of the graph cut approach deserves further investigation since broadcast news has a different genre with lectures. For instance, story segmentation in BN punctuates a news program into distinct topic units while spoken lecture segmentation probes sub-topic changes within a lecture that has a unique central topic.

2 Corpus

We perform a data-oriented study with the standard TDT2 Mandarin BN corpus that contains about 53 hours of VOA Mandarin Chinese BN audio[2]. The 177 audio recordings are accompanied with manually annotated meta-data (including story boundaries) and word-level speech recognition transcripts. The TDT2 audio was transcribed by the Dragon LVCSR with word, character and base syllable error rates of 37%, 20% and 15%, respectively. We adopt a home-grown Pinyin lexicon to get the syllable sequences of words. We separate the corpus into two non-overlapping parts: a development set of 90 recordings for parameter tuning and a set of 87 for story segmentation testing. According to TDT2, a detected story boundary is considered correct if it lies within a 15-second tolerant window on each side of a manually-annotated reference boundary.

3 Normalized Cut for Automatic Story Segmentation

Lexical-based story segmentation holds the view that the inter-sentence similarity tends to be high within the same story and a inter-story boundary tends to exhibit a low similarity. Fig. 1 illustrates the text *Dotplotting* [6] of a broadcast news transcript from the TDT2 corpus introduced in Section 2. This figure plots the cosine similarity scores between every pair of sentences in the text. A 'sentence' is defined as a fixed number of consecutive non-overlapping terms (e.g. words) in the transcript. This is because: 1) real sentence boundaries are not readily available in the speech recognition transcripts and sentence segmentation is another challenging task that is out of the focus of this paper; 2) the number of shared terms between two long sentences and between a long and a short sentence would probably yield incomparable similarity scores [10]. The optimal length of the sentence is tuned on the training set. The intensity of a pixel (i, j) reflects the degree to which the ith sentence in the text is similar to the jth

---

[2] http://projects.ldc.upenn.edu/TDT2/

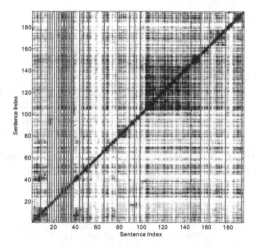

Fig. 1. Sentence similarity dotplot for a BN transcript in the TDT2 corpus

sentence. As shown in Fig. 1, lower intensity means higher similarity. The vertical red lines denotes the manually annotated story boundaries. From this figure, we can clearly observe that the difference in similarity between intra-story regions and inter-story boundaries. Lexical approaches, such as TextTiling [3] and lexical chaining [12] are all based on this observation.

In the next, we present a lexical approach that regards broadcast news segmentation as a graph-partitioning task aiming at minimizing the normalized-cut criterion. Normalized-cut has been successfully used in image segmentation [7]. Recently, Malioutov et al. [6] have introduced it to spoken lecture segmentation and have achieved superior performance over conventional lexical approaches. In this paper, we show that it can be used at the subword level to push forward the state-of-the-art of automatic story segmentation of Chinese BN.

3.1 Graph Representation of Text

We depict a text into a weighted undirected graph $\mathcal{G} = (\mathcal{V}, \mathcal{E})$, where the set of nodes $\mathcal{V}$ corresponds to sentences and $\mathcal{E}$ is the set of weighted edges between each pair of nodes, as shown in Fig. 2 (a). The weight of a edge, $w(i, j)$, define a similarity measure between sentences s_i and s_j, where higher scores indicate higher similarities. Fig. 2 (a) considers long-term similarities between nodes by constructing a fully connected graph. However, considering all pairwise relations may be problematic and a *cutoff* should be performed, i.e., edges exceeding a certain threshold distance is discarded [6]. Fig. 2 (b) shows a graph representation which discards edges between sentences whose distance exceed two. As shown later in Fig. 5, we empirically observe that an appropriate cutoff is especially important for story segmentation in BN. In a BN program, stories related to the same topic (i.e., a breaking news) may often re-occur for several times. Fig. 3 shows a sentence similarity dotplot of a BN transcript from the TDT2 corpus.

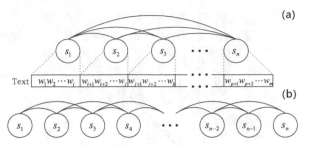

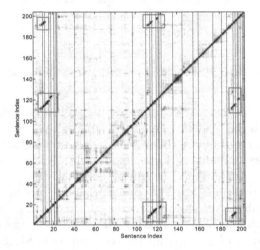

Fig. 2. Fully connected graph representation of text (a) and discarding edges between sentences whose distance exceeds two (b)

Fig. 3. Sentence similarity dotplot for a news program in TDT2, where dark points in the rectangles indicate high similarities between news stories with the same topic

The stories focusing on the same topic are reported at the beginning, middle and the end of the program, leading to high sentence similarities between these stories (i.e., dark points in the rectangles in Fig. 3). Using fully connected graph, these re-occurring topics may implicitly decrease the inter-class distance between the re-occurred topics and inbetweening topics. Therefore, a fully connected graph may be useful to a topic tracking task, while an appropriate cutoff is essential and more applicable to a news story segmentation task.

3.2   Similarity Measure

The weight of edges $w(i, j)$ in the graph denotes a similarity measure between sentences s_i and s_j. Cosine similarity is usually used in story segmentation, which is defined by the cosine of the angle between two sentences[3]:

---

[3] Exponentiated cosine similarity has been used in a real application to avoid losing numerical precision when summing a series of very small scores[6].

$$w(s_i, s_j) = \cos(\mathbf{v}_i, \mathbf{v}_j) = \frac{\mathbf{v}_i \cdot \mathbf{v}_j}{\|\mathbf{v}_i\| \times \|\mathbf{v}_j\|} = \frac{\sum_{t=1}^{T} v_{t,i} v_{t,j}}{\sqrt{\sum_{t=1}^{T} v_{t,i}^2}\sqrt{\sum_{t=1}^{T} v_{t,j}^2}} \tag{1}$$

where $\mathbf{v}_i$ and $\mathbf{v}_j$ are the term frequency vectors for sentences s_i and s_j, respectively. $v_{t,i}$ is the tth element of $\mathbf{v}_i$, i.e., the term frequency of word t registered in the vocabulary (with size of T).

We further use similarity score smoothing to avoid possible story boundary false alarms caused by temporal low similarity points within the same story. Although sentence similarities incline to remain high within the same news story and low at inter-story borders, the individual similarity scores can be highly variable due to rigid matching on term repetitions in similarity computation. For example, a sudden very low score in the neighborhood of a sentence (may caused by very infrequent lexical terms, use of synonyms instead of repetitions, etc.) within a story may induce a story transition false alarm. Therefore, we use smoothed term frequency vectors achieved by exponentially weighted moving average. We add counts of words that occur in m adjacent sentences to the current sentence term frequency vector, i.e.,

$$\tilde{\mathbf{v}}_i = \sum_{j=i}^{i+m} e^{-\alpha(j-i)} \mathbf{v}_j \tag{2}$$

where α is used to control the degree of smoothing.

3.3 Normalized Cut Criterion and Dynamic Programming Solution

The graph partitioning task can be simply defined as dividing the graph $\mathcal{G} = (\mathcal{V}, \mathcal{E})$ into two disjoint partitions, $\mathcal{P}_1$ and $\mathcal{P}_2$, where $\mathcal{P}_1 \cup \mathcal{P}_2 = \mathcal{V}$ and $\mathcal{P}_1 \cap \mathcal{P}_2 = \emptyset$, by removing edges connecting the two partitions. The degree of dissimilarity between the two partitions is defined as the *cut*, which is the sum of the weights of the crossing (removed) edges between the two partitions:

$$cut(\mathcal{P}_1, \mathcal{P}_2) = \sum_{u \in \mathcal{P}_1, v \in \mathcal{P}_2} w(u, v) \tag{3}$$

The optimal partitioning of a graph is the one that minimizes the *cut* value, which is called *Min-cut*. Regarding our story segmentation work, Min-cut is to split the sentences of a text into two maximally dissimilar classes, i.e., by choosing $\hat{\mathcal{P}}_1$ and $\hat{\mathcal{P}}_2$ to minimize:

$$(\hat{\mathcal{P}}_1, \hat{\mathcal{P}}_2) = \arg \min_{(\mathcal{P}_1, \mathcal{P}_2)} \sum_{u \in \mathcal{P}_1, v \in \mathcal{P}_2} w(u, v) \tag{4}$$

The Min-cut criterion only can make sure that similarity between the two partitions is minimized. It thus favors cutting small sets of isolated nodes in the graph [7]. To avoid partitioning out small sets of nodes, Shi *et al.* [7] have proposed *normalized cut* (N-cut). N-cut ensures the two partitions are themselves homogeneous by accounting for intra-partition similarity. It computes the normalized cut cost as a fraction of the total edge connections to all the nodes in the graph:

$$Ncut(\mathcal{P}_1, \mathcal{P}_2) = \frac{cut(\mathcal{P}_1, \mathcal{P}_2)}{assoc(\mathcal{P}_1, \mathcal{V})} + \frac{cut(\mathcal{P}_1, \mathcal{P}_2)}{assoc(\mathcal{P}_2, \mathcal{V})} \tag{5}$$

where $assoc(\mathcal{P}_1, \mathcal{V}) = \sum_{u \in \mathcal{P}_1, v \in \mathcal{V}} w(u, v)$ is the normalization term. It has been shown that, by minimizing Eq. (5), we can simultaneously minimize the similarity across different partitions and maximize the similarity within each partition [7].

N-cut based story segmentation is to partition 'the bag of sentences' into K disjoint sets $A_1, A_2, \ldots, A_K$, so that the similarity among the vertices in a set A_i is maximized and across different sets A_i, A_j the similarity is minimized. The K-way N-cut criterion [7] is adopted:

$$Ncut_K(\mathcal{V}) = \frac{cut(A_1, \mathcal{V} - A_1)}{assoc(A_1, \mathcal{V})} + \frac{cut(A_2, \mathcal{V} - A_2)}{assoc(A_2, \mathcal{V})} + \ldots + \frac{cut(A_K, \mathcal{V} - A_K)}{assoc(A_K, \mathcal{V})}, \tag{6}$$

where $\{A_1 \ldots A_K\}$ forms a partition of the graph, and $\mathcal{V} - A_K$ is the rest set of the entire graph except partition K.

The problem of minimizing normalized cuts on a graph is a NP-complete task [7]. However, finding story boundaries is a linear problem, i.e., all of the nodes (sentences) between the leftmost and the rightmost nodes of a particular partition should belong to the that partition (story). With this linearity constraint, a dynamic programming (DP) solution [6] can be used to exactly minimize the N-cut cost in polynomial time:

$$C[i, n] = \min_{i \le j \le n} \left[C[i - 1, j - 1] + \frac{cut[A_{j,n}, \mathcal{V} - A_{j,n}]}{vol[A_{j,n}]} \right], i \le j \le n, i > 1, \tag{7}$$

$$B[i, n] = \operatorname*{argmin}_{i \le j \le n} \left[C[i - 1, j - 1] + \frac{cut[A_{j,n}, \mathcal{V} - A_{j,n}]}{vol[A_{j,n}]} \right], i \le j \le n, i > 1, \tag{8}$$

$$\text{s.t.} \quad C[1, n] = \frac{cut[A_{1,n}, \mathcal{V} - A_{1,n}]}{vol[A_{1,n}]}, 1 \le n \le N, \tag{9}$$

$$B[1, n] = 1, 1 \le n \le N, \tag{10}$$

where $C[i, n]$ is the N-cut segmentation of the first n sentences into i segments. The ith segment, $A_{j,n}$, begins at node s_j and ends at node s_k. $B[i, n]$ is the back-pointer table that is used to recover the optimal sequence of the segment boundaries. Eq. (9) and (10) indicate the initial condition, i.e. the N-cut value of the trivial segmentation of a text into one segment and the first segment starts with the first node. The time complexity of the DP solution is $O(KN^2)$, where K is the number of partitions and N is the number of nodes in the graph.

4   The Subword N-Cut Approach

4.1   Merits of Use of Subwords in Chinese

Lexical-based story segmentation approaches usually involve word matching, e.g., word frequency count in similarity measure [3], [6] and linking word repetitions in lexical chaining [12]. However, the inevitable speech recognition errors may induce severe word matching failures, resulting in incorrect lexical similarity measures. However, the recognition error rates at subword levels are much lower than the word level, as can be seen from Section 2. At subword levels, we can conduct partial matching and this will partially recover the relations among words. This partial matching merit is especially important for Chinese LVCSR transcripts.

Chinese is highly different from western languages such as English both in written and spoken forms. A Chinese word is formed by one to several component characters, and there is no space between words serving as word delimiters in a Chinese text. In fact, 'word' is not defined explicitly in Chinese and word segmentation is definitely not unique. As a result, the same string of characters may be segmented into different word sequences in different places in a LVCSR transcript. In some cases, even more than one word sequences are both syntactically valid and semantically meaningful. For example, the word 北韩 (North Korea) is segmented to 北 (North) and 韩 (Korea) and they both occur in a news story in the TDT2 Mandarin corpus. In this case, it is impossible to relate them by rigid word matching. The high flexibility of Chinese word segmentation is easy to cause word matching failures. However, the above problem can be solved by character matching because different segmentations still share the same component characters.

A Chinese character is pronounced as a tonal syllable. In Mandarin, about 1200 phonologically allowed tonal syllables correspond to over 6500 commonly used simplified Chinese characters. When tones are disregarded, the number is reduced to only about 400, known as base syllables. This indicates that there are a large number of *homophones* sharing the same base syllable. Tones are often mis-recognized by the speech recognizer, which contributes a lot to the recognition errors. In Chinese LVCSR transcripts, it is common that a word is substituted by another character sequence with the same or similar pronunciations, in which homophone characters are the probable substitutions. Table 1 shows some word matching failures due to speech recognition errors. Rigid word matching cannot link the original word and their substitutions together. However, matching at subword levels can recover their connections.

Table 1. Recognition error samples from TDT2. Subword units for partial matching are underlined. English translations are in brackets.

Original word	ASR error	Base syllable sequence
阿尔及利亚	鲍尔 激励 要	a _er_ _ji_ _li_ ya
(*Algeria*)	(*Bauer drive want*)	bao _er_ _ji_ _li_ yao
奥尔布莱特	二 步 莱特	ao _er_ _bu_ _lai_ _te_
(*Albright*)	(*two step Wright*)	_er_ _bu_ _lai_ _te_
赈济	震级	_zhen_ _ji_
(*relieve*)	(*quake magnitude*)	_zhen_ _ji_
股市	故事	_gu_ _shi_
(*stock exchange*)	(*story*)	_gu_ _shi_

Table 2. OOV word samples from TDT2. Subword units for partial matching are underlined. English translations are in brackets.

Character sequence	Base syllable sequence
OOV word: 王有才(*a Chinese name*)	*wang you <u>cai</u>*
ASR output 当 有 财(*when have money*)	*dang you <u>cai</u>*
王 油菜(*king rape*)	*wang you <u>cai</u>*
邦友 才(*national friendship talent*)	*bang you <u>cai</u>*
OOV word: 莱温斯基(*Lewinsky*)	*lai wen <u>si</u> ji*
ASR output 来 文 斯 基(*come article this base*)	*lai wen <u>si</u> ji*
来 问 司机 (*come ask driver*)	*lai wen <u>si</u> ji*
来 的 司机(*show-up driver*)	*lai de <u>si</u> ji*

Flexible word-building in Chinese makes the limited Chinese characters to produce unlimited words. Hence, there does not exist a commonly accepted Chinese lexicon. Consequently, the OOV problem is more pronounced in Chinese LVCSR transcripts, especially in the BN program that focuses on timely events. Many OOV words in BN are named entities (NE). An OOV word appeared in different places of a spoken document may share part of the characters or be substituted by several totally different character strings with the same (or partially same) syllable sequence. For example, foreign proper names are common OOV words in Chinese spoken documents as they are transliterated to Chinese character sequences based on the pronunciations (i.e. phonetic transliteration). As a result, speech recognizer may return different character sequences with the same or similar pronunciations, probably their homophones. Matching at syllable level can recover these highly-topic-related OOV words due to partial matching. Some examples are shown in Table 2.

4.2   Subword N-Cut

Motivated by the merits of subwords in lexical matching in Chinese BN transcripts, we propose a N-cut story segmentation approach on different Chinese subword representations, i.e., character and syllable n-gram units. Given a sentence composed of a sequence of words $\{w_1 w_2 w_3 \ldots w_Q\}$ and the sequence of their component characters or syllables $\{c_1 c_2 c_3 \ldots c_L\}$, the overlapping subword n-gram is defined in Table 3. Higher order subword overlapping n-grams ($n \geq 4$) can be formed accordingly. Overlap between subwords is used to reduce the possibility of missing any useful information embedded in the subword sequence.

For the same LVCSR transcript of a broadcast news program, we observe that the syllable-bigram-based dotplot is clearer than the word-based dotplot, as illustrated in Fig. 4. It means syllable-bigram-based inter-sentence similarity decreases across different partitions and increase within partitions, leading to clearer story boundaries. Based on this observation, we propose to perform subword N-cut on the subword n-gram representation of sentences (Table 1)

Table 3. Forming character/syllable overlapping n-grams from a word sequence

n-gram	Forming rule	Example (character & syllable)
word	$w_1\ w_2\ w_3 \cdots w_Q$	俄罗斯 第一 副总理 名 模 作 福 e-luo-si di-yi fu-zong-li ming mo zuo fu
unigram	$c_1\ c_2\ c_3 \cdots c_L$	俄 罗 斯 第 一 副 总 理 名 模 作 福 e luo si di yi fu zong li ming mo zuo fu
bigram	$c_1c_2\ c_2c_3 \cdots c_{L-1}c_L$	俄罗 罗斯 斯第 第一 一副 副总 总理 理名 名模 模作 作福 e-luo luo-si si-di di-yi yi-fu fu-zong zong-li $\cdots$
trigram	$c_1c_2c_3\ c_2c_3c_4 \cdots$ $c_{L-2}c_{L-1}c_L$	俄罗斯 罗斯第 斯第一 第一副 一副总 副总理 总理名 $\cdots$ e-luo-si luo-si-di si-di-yi di-yi-fu yi-fu-zong $\cdots$

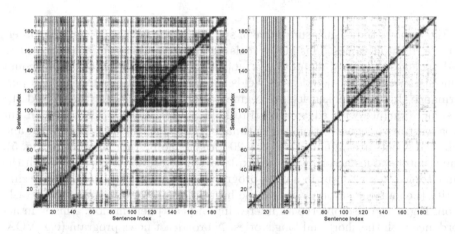

Fig. 4. Sentence similarity dotplots for an LVCSR transcript in the TDT2 corpus. Left: word-based dotplot (same in Fig. 1), right: syllable-bigram-based dotplot.

according to the N-cut procedure described in Section 3. In the similarity computation defined in Eq. (1), $v_{t,i}$ denotes the term frequency of the corresponding overlapping n-gram unit t in sentence s_i.

5 Experiments and Analysis

We carried out story segmentation experiments to evaluate the proposed subword N-cut approach with several state-of-the-art lexical-based approaches, i.e., (1) word-based TextTiling [3], (2) subword-based TextTiling [10], (3) word-LSA-based TextTiling [8], (4) subword-LSA-based TextTiling [11] and (5) word-based N-cut [6]. Recall, precision and their harmonic mean, i.e. F1-measure, are used as evaluation criteria. Empirical parameter tuning was first performed on the development set of TDT2, which selects parameters achieving the best F1-measure of story segmentation. Empirical parameters include the sentence length (i.e. word block length), the cutoff value, α and m for similarity smoothing in Eq. (2).

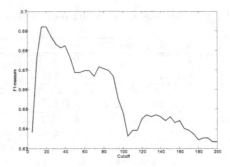

Fig. 5. The cutoff value vs F1-measure for syllable-bigram-based N-cut

Fig. 5 shows the F1-measure curve of syllable-bigram-based N-cut with different graph cutoff values. This plot reveals the difference between broadcast news and spoken lectures. Malioutov *et al.* [6] pointed out that taking into account long-distance lexical dependencies yields substantial gains in segmentation performance. In their spoken lecture segmentation work, they have achieved the best performance with edge cutoff threshold at 100 and 200 sentences. In our broadcast news segmentation, the optimal cutoff values for all testing word and subword N-cuts are around 15–35 (15–20 for syllable-bigram, as show in Fig. 5). Small cutoff values accord with our explanation in Section 3.1: stories with the same topic often re-occur in a news program and considering long-distance similarity (i.e., a large cutoff value) can be harmful to the news segmentation task. From Fig. 5, we also observe an interesting bi-peak phenomenon. This is in accordance with the short and long stories. A broadcast news program (e.g. VOA, the source of TDT2) is usually composed of brief news reporting headlines and detailed news that focus on intensive reports on news events. The first peak with small cutoff (around 15–20) fits the short brief news and the second peak with big cutoff (around 120–140) fits the long detailed news. The highest F1-measure is achieved by a small cutoff is due to the large number of brief news in TDT2 (over 3/5 of the corpus).

The experimental results on the test set are summarized in Fig. 6 and Fig. 7. Results indicate that the syllable-bigram-based N-cut achieves the best F1-measure of 0.6911 with relative improvement of 11.5% over Malioutov's word-based N-cut(F1=0.6197) [6] and 4.74% over the character-bigram-LSA-based TextTiling (F1=0.6598) [11]. In general, syllable/character unigram and bigram N-cuts show superior performance, while trigram and 4-gram N-cuts present inferior performance. This is mainly due to the fact that (1) the probability of long sequences with correctly recognized characters/syllables is smaller than short character/syllable units; (2) the most frequently used words in Chinese are bi-character. Trigram and 4-gram LSA based TextTiling achieves better F1-measure as compared with other trigram and 4-gram approaches because of LSA's noise-removal feature [11].

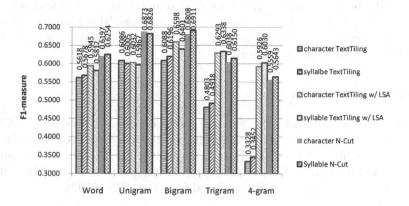

Fig. 6. Experimental results on the TDT2 test set

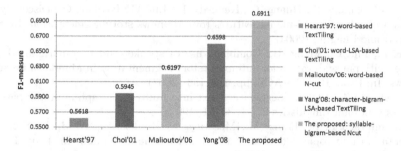

Fig. 7. Best performance comparison on TDT2 corpus: Hearst'97 [3], Choi'01 [8], Mallioutov'06 [6], Yang'08 [11] and the proposed approach

6   Summary and Future Work

This paper has modeled Chinese BN segmentation as a graph partitioning task under N-cut criterion that aims to simultaneously optimize the sentence similarity within each story and dissimilarity across different stories. Motivated by the robustness of subword units in partial matching of Chinese words, we have proposed to perform N-cut on character/syllable overlapping n-grams of noisy broadcast news transcripts with speech recognition errors. We conclude from the experiments that the proposed method can effectively improve the automatic story segmentation performance on Chinese BN. Syllable-bigram-based N-cut achieves the best F1-measure of 0.6911 with relative improvement of 11.5% over Malioutov's word-based N-cut that has an F1-measure of 0.6197.

Current K-way N-cut approach cannot automatically determine the granularity of a resultant segmentation. The number of stories (K) has to be set a priori. We plan to introduce the graduated graph cuts (GGC) approach [13] to BN segmentation, which can automatically determine the optimal number of partitions and keep the best granularity of the whole segmentation. Since latent semantic analysis (LSA) has the merit of concept matching, we also plan to introduce LSA to subword-based N-cut to achieve more robustness in story segmentation of noisy speech recognition transcripts.

Acknowledgements

This work was supported by the National Natural Science Foundation of China (60802085 and 60872145), the Research Fund for the Doctoral Program of Higher Education in China (20070699015), the Cultivation Fund of the Key Scientific and Technical Innovation Project, Ministry of Education of China (No.708085) and the NPU Foundation for Fundamental Research (W018103).

References

1. Hsu, W., Chang, S., Huang, C., Kennedy, L., Lin, C., Iyengar, G.: Discovery and fusion of salient multi-modal features towards news story segmentation. In: SPIE Electronic Imaging (2004)
2. Xie, L., Liu, C., Meng, H.: Combined use of speaker-and tone-normalized pitch reset with pause duration for automatic story segmentation in Mandarin broadcast news. In: Proc. HLT-NAACL, pp. 193–196 (2007)
3. Hearst, M.: TextTiling: Segmenting text into multi-paragraph subtopic passages. Computational Linguistics 23(1), 33–64 (1997)
4. Dharanipragada, S., Franz, M., Mccarley, J., Roukos, S., Ward, T.: Story segmentation and topic detection in the broadcast news domain. In: Proc. DARPA Broadcast News Workshop (1999)
5. Choi, F.Y.Y.: Advances in domain independent linear text segmentation. In: Proc. NAACL, pp. 26–33 (2000)
6. Malioutov, I., Barzilay, R.: Minimum cut model for spoken lecture segmentation. In: Proc. ACL, pp. 25–32 (2006)
7. Shi, J., Malik, J.: Normalized cuts and image segmentation. IEEE Transactions on Pattern Analysis and Machine Intelligence 22(8), 888–905 (2000)
8. Choi, F., Wiemer-Hastings, P., Moore, J.: Latent semantic analysis for text segmentation. In: Proc. EMNLP (2001)
9. Ng, K., Zue, V.W.: Subword-based approaches for spoken document retrieval. Speech Communication 32(3), 157–186 (2000)
10. Xie, L., Zeng, J., Feng, W.: Multi-scale TextTiling for Automatic Story Segmentation in Chinese Broadcast News. In: Li, H., Liu, T., Ma, W.-Y., Sakai, T., Wong, K.-F., Zhou, G. (eds.) AIRS 2008. LNCS, vol. 4993, pp. 345–355. Springer, Heidelberg (2008)
11. Yang, Y., Xie, L.: Subword latent semantic analysis for texttiling-based automatic story segmentation of chinese broadcast news. In: Proc. ISCSLP, pp. 358–361 (2008)
12. Stokes, N., Carthy, J., Smeaton, A.: Select: A lexical cohesion based news story segmentation system. Journal of AI Communication 17(1), 3–12 (2004)
13. Feng, W., Liu, Z.Q.: Self-validated and spatially coherent clustering with net-structured MRF and graph cuts. In: Proc. ICPR, vol. 4, pp. 37–40 (2006)

Japanese Spontaneous Spoken Document Retrieval Using NMF-Based Topic Models

Xinhui Hu, Hideki Kashioka, Ryosuke Isotani, and Satoshi Nakamura

National Institute of Information and Communications Technology, Japan
{xinhui.hu,hideki.kashioka,ryosuke.isotani,
satoshi.nakamura}@nict.go.jp

Abstract. In this paper, we propose a document topic model (DTM) which is based on the non-negative matrix factorization (NMF) approach, to explore Japanese spontaneous spoken document retrieval. Each document is interpreted as a generative topic model, belonging to many topics. The relevance of a document to a query is expressed by the probability of a query word being generated by the model. Different from the conventional vector space model where the matching between query and document is at the word level, the topic model complete its matching in the concept or semantic level. So, the problem of term mismatch in the information retrieval can be improved, that is, the relevant documents have possibilities to be retrieved even if the query words do not appear in them. The method also benefit the retrieval of spoken document containing "term misrecognitions", which is peculiar to the speech transcripts. By using this approach, experiments are conducted on a test collection of corpora of spontaneous Japanese (CSJ), where some of the evaluating queries and answer references are suited to retrieval in semantic level. The retrieval performance is improved by increasing the number of topics. When the topic number exceeds a threshold, the NMF's retrieval performance surpasses the tf-idf-based vector space model (VSM). Furthermore, compared to the VSM-based method, the NMF-based topic model also shows its strongpoint in dealing with term mismatch and term misrecognition.

Keywords: spoken document retrieval, non-negative matrix factorization, document topic model.

1 Introduction

The search and retrieval of a document is generally conducted by matching keywords in the query to those in the target documents. When the keywords are found in a document, the document is regarded to be relevant to the input query. A fundamental problem of information retrieval (IR) is *term mismatch*. A query is usually a short and incomplete description of the user's information need. Users and authors of documents often use different terms to refer to the same concepts and this produces an incorrect relevance ranking of documents with regard to the information need expressed in the query.

For spoken document retrieval (SDR), it faces a new problem besides the term mismatch. The SDR is generally carried out by using textual approaches to speech

G.G. Lee et al. (Eds.): AIRS 2009, LNCS 5839, pp. 149–156, 2009.

transcripts. The transcripts are generally obtained by utilizing automatic speech recognition systems. However, because of the limitation of current speech recognition technology, the transcript produced by the speech recognition process always contains errors. In SDR, terms misrecognized will not match the query and the document representations. Naturally, this hinders the effectiveness of the SDR system in a way similar to the term mismatch. Here, we call this problem as the term misrecognition.

Advanced users need tools that can find underlying concepts and not just search for keywords appearing in the query. It is widely acknowledged that the ability to work with text on a semantic basis is essential to modern information retrieval systems. Topic models are very popular for presenting the content of documents. Recently, researches on these aspects are becoming booming. The probabilistic latent topic modeling approaches, such as probabilistic latent semantic analysis (PLSA) [1] have been demonstrated effective in the tasks of spoken document retrieval. Chen [2] proposed a word topic model (WTM) to explore the co-occurrence relationship between words, as well as the long-span latent topical information, for language modeling in spoken document retrieval and transcription, and verified that the WTM is a feasible alternative to the existing models, i.e. PLSA.

Non-negative matrix factorization (NMF) [3] is also an approach in latent semantic space. It is a type of dimension reduction technique, and has distinct features of preserving the original data as well as the non-negative of the original data. Different from the other similar decomposition approaches such as singular value decomposition (SVD) [4], the NMF uses non-negativity constraints; the decomposition is purely additive; no cancellations between components are allowed, so they lead to a parts-based representation. Also, the NMF computation is based on a simple iterative process, it is therefore advantageous for applications involving data sparseness, like large vocabulary speech recognition. It is regarded to be suitable for finding the latent semantic structure from the document corpus and to identify document clusters in the derived latent semantic space. We adopt the NMF-based document topic model (DTM) approach for spontaneous spoken document retrieval (SDR) in this study. Since the approaches of latent semantic indexing are based on the semantic relations, a relevant document can be retrieved even if a query word does not appear in that document. So this feature can be used to compensate for the speech recognition errors. In this study, the focuses are mainly on dealing with the term misrecognitions, investigating the effectiveness of this DTM for SDR. The comparisons are conducted between this model and the conventional vector space model (VSM), since we presently limit on investigating the difference between the semantic matching and the keyword matching.

The rest of this paper is organized as follows: In Section 2, based on our previous work, we briefly introduce how to build the term-document matrix stochastically using N-best sequence. In Section 3, we describe the document topic model for information retrieval, and explain the method to construct the topic model by using the factorized matrices of the NMF, and show how to compute relevance of target document to the retrieving query using this topic model. In Section 4, the experimental setups and results are reported, highlighting the comparison between the proposed method and the conventional tf-idf-based VSM. Finally, in Section 5, we present our conclusions, discuss the characteristics of NMF in retrieval, especially when dealing with the term misrecognitions.

2 Term-Document Matrix Built on N-Best

The system presented here operates in two phases combining speech-based processing and text-based processing.

In the speech-based processing phase, the spoken documents are transcribed by an automatic speech recognizer (ASR). The transcription of the ASR is in the form of an N-best list, in which the top N hypotheses of the ASR results are stored in the recognition result lattice. The reason to select the N-best is that it needs less computation and less memory than the original lattice in search a recognition hypothesis. The usage of N hypotheses is to utilize those correct term candidates hidden in other hypotheses, and to compensate the effectiveness of term misrecognitions.

In the text-based processing phase, the term-document matrix used for NMF is built on an updated tf-idf-based vector space model (VSM). In tf-idf-based VSM, term frequency tf, which is defined as the number of a term occurs in a document and the inverse document frequency idf, are the two fundamental parameters. For the N-best, we introduce a stochastic method to compute these two parameters. This method is described as follows:

Let D be a document modeled by a segment of the N-Best. P(w|o,D) is defined as the posterior probability or confidence of a term w at position o in D in order to refer to the occurrence of w in the N-Best.

The tf is evaluated by summing the posterior probabilities of all occurrences of the term in the N-Best. Furthermore, we update it with Robertson's 2-Poisson model as follows.

$$tf(D,w) = \frac{tf'(D,w)}{tf'(D,w) + \dfrac{length(D)}{\Delta}} \tag{1}$$

Where the tf' is the conventional term frequency, and is defined as follows:

$$tf'(w,D) = \sum_{D}\sum_{i=1}^{N} K(i) * P(w|o_i,D) \tag{2}$$

$$K(i) = (N+1-i) \Big/ \sum_{t=1}^{N} t \tag{3}$$

The length(D) is the length of document D, Δ is the average length of the whole document set. Similarly, the idf is calculated on the basis of the posterior probability of w, as shown in the following equation:

$$idf(w) = \log(N_D / \sum_{D \in C} O(w,D)) \tag{4}$$

Here,

$$O(w,D) = \begin{cases} 1, & \text{if } tf'(w,D) > 0.5 \\ 0, & \text{otherwise} \end{cases} \tag{5}$$

N_D is the total number of documents contained in the corpus. C is the entire document set of the corpus.

The term-document matrix A for NMF is finally built by using *tf idf*. By using the processing of NMF, a topic model is constructed, and is used for computing the relevance of target documents to the input query.

3 NMF-Based Document Topic Model for Spoken Document Retrieval

3.1 Document Topic Model and Information Retrieval

In information retrieval (IR), the relevance measure between a query Q and a document D can be expressed as $P(D|Q)$. By applying the Bayes theorem, it can be transformed into:

$$P(D \mid Q) = \frac{P(Q \mid D)P(D)}{P(Q)} \tag{6}$$

With the invariability of $P(Q)$ over all documents, and assuming that document probability $P(D)$ has a uniform distribution, ranking the documents by the $P(D|Q)$ can be realized using $P(Q|D)$, the probability of query Q being generated by the document D. If the query Q is composed of a sequence of terms (or words) $Q = w_1 w_2 ... w_{Nq}$, the $P(Q|D)$ can be further decomposed as a product of the probabilities of the query words generated by the document :

$$P(Q \mid D) = \prod_{w_i} P(w_i \mid D) \tag{7}$$

Each individual document D can be interpreted as a generative document topic model (DTM), denoted as M_D, and is embodied with K latent topics. Each latent topic is expressed by the word distribution of the language. So two probabilities are associated with this topic model: the probability of a latent topic given a document and the probability of word in a latent topic. So the probability of a query word w_i generated by D is expressed by

$$P_{DTM}(w_i \mid M_D) = \sum_{k=1}^{K} P(w_i \mid k)P(k \mid M_D) \tag{8}$$

Where $P(w_i \mid k)$ denotes the probability of a query word w_i occurring in a specific latent topic k, and $P(k \mid M_D)$ is the posterior probability of the topic k generated by the document model M_D.

Therefore, considering on the equation (7) and (8), the likelihood of a query Q generated by D is thus represented by

$$P_{DTM}(Q|M_D) = \prod_{w_i} \left[\sum_{k=1}^{K} P(w_i|k)P(k|M_D) \right]$$

(9)

In this study, we compare the retrieval performance of the NMF with the conventional vector space vector (VSM) where the similarity between the query Q and document D is computed by following equation:

$$sim(D,Q) = \frac{D \bullet Q}{|D| \parallel Q|}$$

(10)

3.2 Link NMF to Topic Model

Let A be the matrix produced in section 2 to stand for relationships among the terms and documents, with dimension $m \times n$. Let S be the sum of all elements in A. Then $\overline{A} = A/S$ forms a normalized table to approximate the joint probability $p(w,d)$ of term w, and document d.

NMF is a matrix factorization algorithm [3] that finds the positive factorization of a given positive matrix. Assume that the given document corpus consists of K topics. The general form of NMF is defined as:

$$\overline{A} \approx GH$$

(11)

The matrix factorization of $\overline{A}$ will result in an approximation by a product of two non-negative matrices, G and H with dimension $m \times k$ and dimension $k \times n$ respectively. So from the equation (11), the joint probability $p(w,d)$ can be expressed by

$$p(w,d) = \overline{A} = \sum_{k=1}^{K} G_{w,k} \cdot H_{k,d}$$

(12)

To normalize G by $\alpha_k = \sum_w G_{w,k}$, H by $\beta_k = \sum_d H_{k,d}$, and define $p(k) = \alpha_k \beta_k$, the $p(w,d)$ can be rewritten as:

$$p(w,d) = \sum_{k=1}^{K} \frac{G_{w,k}}{\alpha_k} \frac{H_{k,d}}{\beta_k} p(k) = \sum_{k=1}^{K} \hat{G}_{w,k} \hat{H}_{k,d} \, p(k)$$

(13)

Each entity $g_{w_i,k} (1 \leq k \leq K)$ of the matrix $\hat{G}$ accounts for the probability of a word w_i that would be generated by a latent topic k., that is $P(w_i|k)$ of the equation (9).

Each entity $h_{k,n}$ of the matrix $\hat{H}$ accounts for the probability of document D by a latent topic k, that is $P(M_D|k)$.

The $P(k|M_D)$ of equation (8) can be obtained by using Bayes theorem

$$p(w, d) = \sum_{k=1}^{K} \frac{G_{w,k}}{\alpha_k} \frac{H_{k,d}}{\beta_k} p(k) = \sum_{k=1}^{K} \hat{G}_{w,k} \hat{H}_{k,d} p(k) \tag{14}$$

Based on the above equations, the equation (9) for relevance can be computed by the matrices G and H, the factorized matrices of the NMF.

4 Experiments

4.1 Experimental setups

A test collection of CSJ is used for evaluation. The CSJ (Corpus of Spontaneous Japanese) is the result of a Japanese national project on 'Spontaneous Speech Corpus and Processing Technology ' [5]. It contains 658 hours of speech consisting of approximately 7.5 million words. The speech materials were provided by more than 1,400 speakers of various ages. About 95% of the CSJ corpus is devoted to spontaneous monologues, such as academic presentations and public speaking, including manual transcriptions. This test collection is developed by the Japanese Spoken Document Processing Working Group [6], with the aim of evaluating the retrieval of spoken document retrieval systems. This collection consists of a set of 39 textual queries, the corresponding relevant segment lists, and transcriptions by an automatic speech recognition (ASR) system, allowing retrieval of 2702 spoken documents of the CSJ. The large vocabulary continuous ASR system use an engine in which the acoustical model is trained by a corpus in the domain of travel [8], but the language model is trained by the manually-built transcript of the CSJ corpus. The word accuracy of the recognition system is evaluated as 60.5%. Because the criteria in determining relevant is not merely dependent on query's keywords, the semantic content needs to be taken into consideration. For examples, for keyword sequence of "ペット 効用 目的" which corresponds to query text "[HN101801] ペットを飼うことの効用または目的について述べている箇所を探したい(*search the utterances about the purposes or effects of raising pets*", keyword "ペット(pet)" appears in all of its answer files, but "効用(effect), "目的(purpose)" are misrecognized or only appear in just 2 files with low *tf*. That means that when using the VSM, these answers can be hit mainly dependent on the first keyword "ペット(pet)".

For the data structure of test transcript, three types of transcripts of the same spoken documents are used for evaluations.

(1) N-best (here 10-best is used, denoted as *nbst*).
(2) 1-best (denoted as *1bst*).
(3) Manual transcript (denoted as *tran*).

The mean average precision (**MAP**) is used as the performance measure in this study.

4.2 Experimental Results

4.2.1 Retrieval Performance with Topic Number

Figure 1 shows the MAPs of retrieving *nbst* using the proposed NMF-based model, and conventional tf-idf-based VSM in different topics number. As the topic number increases, the *nbst's* MAP increases nearly monotonously. After the topic number is over a threshold (here it is 700), the MAP of the NMF-based model surpasses the VSM. Therefore it can be concluded that the NMF mainly functions in the high dimensional semantic space.

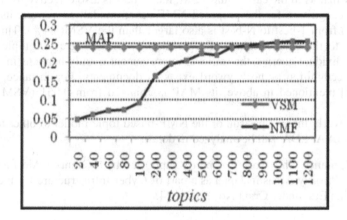

Fig. 1. MAPs in different topic numbers

4.2.2 Effectiveness on different Data Type

Table 1 shows the retrieval performance for different data types using the NMF and VSM methods. In this experiment, the number of topics was selected to be 1000.

For all of 3 data types, the NMF method proved to be superior to the VSM. For example, the improvement of NMF to the VSM is 5.5% for the N-best transcripts. Although the improvement is not so large, the significance of the NMF is that its retrieval is on the semantic level, so it has a potential ability to deal with the problem of misrecognition. Meanwhile, the performance of systems that use the N-best are better than the those that use the *1-bst*, they show the same characteristics as in other research on lattice-based spoken document retrieval that the N-best can search or retrieve correct speech segment by utilizing multiple recognition hypotheses even if the 1-best one is incorrect [7].

Table1. Retrieval performance of different data type (for NMF, dimension=1000)

	1bst	nbst	tran
NMF	0.240	0.255	0.285
VSM	0.233	0.241	0.253

5 Conclusions

In this paper, we proposed a NMF-based document topic model to explore the Japanese spoken document retrieval. By experiments on the CSJ spoken corpus, the retrieval performance of the NMF-based topic model is found to be steadily improved with the increases in the number of topics. When the topic number becomes sufficiently large, the NMF-based model outperforms the conventional tf-idf-based VSM. However, this fact also reveals that the merit of NMF-based topic model for retrieval is conditional on the number of topics.

We show that as in the case of the VSM, the N-best is also effective to compensate for the misrecognition for the proposed NMF-based model. Moreover, its improvement (6.2%) from 1-best to N-best is also larger than the VSM(3.4%). This achievement is due to the characteristics of topic model – matching at the topic level. By analyzing individual queries, the retrieval improvement mainly happens in those containing misrecognition or no keyword exists in documents. For instance, for query [HN101801] mentioned in above, its MAP is changed from 0.051 (VSM) to 0.128 (NMF).

In future work, the comparison of the NMF-based topic model to other topic models such as PLSA, LDA will be analyzed in detail.

Acknowledgments. This work was partly supported by a Grant-in-Aid for Scientific Research on Priority Areas in Japan as a part of Cyber Infrastructure for the Information Explosion Era, under Grant No. 19024074.

References

1. Hoffmann, T.: Probabilistic latent semantic indexing. In: Proceedings of the SIGIR 1999, pp. 50–57 (1999)
2. Chen, B.: Word topic models for spoken document retrieval and transcription. ACM Transactions on Asian Language Information Processing (TALIP) 8(1), 1–27 (2009)
3. Lee, D., Seung, H.S.: Algorithms for non-negative matrix factorization. In: NIPS, vol. 13 (2001)
4. Kita, K., Tuda, K.H., Sisibori, M.: Information Retrieval Algorithms. Kyoritu Press (2003)
5. Maekawa, K.: Corpus of spontaneous Japanese: its design and evaluation. In: SSPR 2003 (2003)
6. Akiba, T., Aikawa, K., Itoh, Y., Kawahara, T., Nanjo, H., Nishizaki, H., Yasuda, N., Yamashita, Y., Itou, K.: Test collection for spoken document retrieval from lecture audio data. In: Proceedings of the LREC 2008 (2008)
7. Saraclar, M., Sproat, R.: Lattice-based Search for Spoken Utterance Retrieval. In: Proc. of HLT-NAACL, pp. 129–136 (2004)
8. Nakamura, S., Markov, K., Nakaiwa, H., Kikui, G., Kawai, H., Jitsuhiro, T., Zhang, J., Yamamoto, H., Sumita, E., Yamamoto, S.: The ATR multilingual speech-to-speech translation system. IEEE Trans. on Audio, Speech, and Language Processing 14(2), 365–376 (2006)

Finding 'Lucy in Disguise': The Misheard Lyric Matching Problem

Nicholas Ring and Alexandra L. Uitdenbogerd

School of Computer Science and Information Technology, RMIT University
GPO Box 2476V, Melbourne 3001, Australia
alu@cs.rmit.edu.au

Abstract. We investigated methods for music information retrieval systems where the search term is a portion of a misheard lyric. Lyric data presents its own unique challenges that are different to related problems such as name search. We compared three techniques, each configured for local rather than global matching: edit distance, Editex, and SAPS-L — a technique derived from Syllable Alignment Pattern Searching. Each technique was selected based on effectiveness at approximate pattern matching in related fields. Local edit distance and Editex performed comparably as evaluated with mean average precision and mean reciprocal rank. SAPS-L's effectiveness varied between measures.

1 Introduction

The internet provides great potential for customers to locate and purchase songs that they remember from years earlier. It is a frequent occurrence that the title of a song is not known, but a portion of lyrics is remembered. To locate the songs a web search may be carried out using portions of the lyrics as the query. We aim here to investigate methods for improving retrieval effectiveness where the portions of the lyrics used are misremembered or misheard. Misheard lyrics can stick surprisingly persistently in people's minds as will be very apparent to those who have witnessed somebody unashamedly singing "Lucy in disguise with lions" along to the well-known Beatles song.

An appealing approach to solving this problem would be to conduct a Google search using either an exact quoted query made up of a portion of the lyric, or a handful of unique words remembered from the lyric. This approach has a good chance of success if the lyric has been remembered correctly. In the case where the lyric has been misheard, success is less likely. As a successful quoted query requires an exact portion of the lyric, this approach is inappropriate for solving the misheard lyric problem. A query made up of a handful of words is susceptible to the same problem, words from the misheard lyric that do not appear in the actual lyric will produce results not related to the desired lyric.

The approaches used here were selected based on their effectiveness at approximate pattern matching in related fields where they are often used when exact matching fails to produce useful results. We compared three techniques configured for local rather than global matching: edit distance, Editex, and Syllable

G.G. Lee et al. (Eds.): AIRS 2009, LNCS 5839, pp. 157–167, 2009.

Alignment Pattern Searching - Local (SAPS-L). The aim was to determine which technique would provide the most relevant results, where relevance is judged by a small panel of assessors and relevance candidates are selected using the pooling method favoured by the Text Retrieval Conference, TREC[4]. Due to the nature of the data, the query string (the *pattern*) is significantly shorter than the entire lyric (the *text*). An approximate alignment between the text and pattern is likely to occur at some location within the text, but not necessarily commencing at the beginning. For this reason it is appropriate to search for a local alignment rather than a global alignment.

All techniques were successful in retrieving relevant answers to queries. Local edit distance and Editex performed comparably in both measures of retrieval effectiveness. The t-test did not confirm that there was a statistically significant difference between the effectiveness of these techniques. SAPS-L performed better on the binary measure, mean reciprocal rank, than on mean average precision (MAP).

2   Previous Work

To our knowledge there is no previous work published on the specific problem of lyric matching for misheard lyrics, so we discuss related textual pattern matching problems, and other work involving lyric retrieval.

Zobel and Dart[18] assessed the retrieval effectiveness of a series of edit distance techniques, phonetic techniques, and combinations of the two on personal name matching. Editex was found to be very effective for phonetic matching in this area. Edit distance was also among the leading techniques in this study.

Kondrak and Dorr[8] investigated the problem of ensuring that new drugs are not marketed with names that sound like existing drugs for the purpose of avoiding mistakes in a hospital setting. A variety of matching techniques, including some mentioned above, were used to search for global alignments of single word patterns in single word texts. They found Editex to be one of two superior techniques.

Knees et al. [7] sought an answer to the problem of retrieving lyrics of arbitrary songs from online sources. In the process of doing this they detailed issues related to ensuring the reliability and accuracy of lyrics retrieval. On comparing versions of lyrics from multiple sources it was found that they differed in: the spellings of words; semantic content due to misunderstandings; multiple versions and covers containing altered content; annotations detailing background vocals, spoken text and sounds; annotations detailing song meta-data; and abbreviations. Each of these differences could result in a reduction of retrieval effectiveness when present and require careful consideration when building a collection data set for lyric information retrieval. While it might seem that the same techniques that are used for multiple lyric alignment could be applied to misheard lyric matching, the problems differ in that a lyric query is likely to be short compared to the song length, leading to a larger number of mismatching characters between the strings.

Kleedorfer et al. [6] applied various clustering techniques to identify clusters of songs with the same topic, whereas Mayer et al. [11] extracted features from lyrics to apply to musical genre classification. Logan et al. [9] also explored the use of lyrics for genre clustering, by applying Probabilistic Latent Semantic Analysis (PLSA). They found that using lyric data for genre prediction was better than random but not as good as using audio features. Building on their work, Neumayer and Rauber [14] achieved higher classification accuracy using lyric features than audio ones, and the combination of lyric and audio features gave the best results. In addition there is some work on aligning lyrics with audio [13] as well as extracting lyrics from singing in order to improve query by singing systems [5,12,16,17].

Similar problems to misheard lyric matching need to be solved for speech recognition, in that word disambiguation is required. The main difference between the two domains is that for the misheard lyric matching problem addressed here, a text-based representation is assumed to exist for both the query and answer text. In the speech recognition case speech signals are captured and compared in some form with representations of words. Word disambiguation is partly achieved with statistical models that predict how likely a particular word is given previous ones [10].

3 Data Collections

In terms of pattern matching research, the problem of matching misheard lyrics to actual song lyrics is new, and thus required the creation of a new collection. Suitable data is available on the web, which became a source of both the collection and an abundant source of seemingly genuine queries.

3.1 Query Set

The strings that make up the query set were obtained from a user-generated misheard lyrics website[1]. Contributors submit a pair of strings where the first is a portion of the correct lyric of a song and the second is what they misheard the first string as, in sometimes comical fashion. The site is maintained and junk submissions are periodically removed. The site yielded 4124 portions of misheard lyrics of which every 82nd was used, for a total of 50. The average query length was 8.27 words in the full set and 7.42 in the set of 50. Each query represents a potential search query where exact matching would not produce relevant results.

3.2 Collection

The collection was made up of unabridged lyrics obtained from a web lyrics database[2]. The base collection is a subset of the lyrics available from that source. Where not already present, the lyrics corresponding to each of the songs referred to by the queries were added to the collection.

---

[1] http://www.kissthisguy.com
[2] http://lyrics.astraweb.com/

The lyrics required some filtering before use in the experiment. To avoid excessive and error-prone hand-editing, lyrics containing an excess of noise (in the form of annotations, metadata, etc.) were removed from the set. The lyrics that remained were converted to lower case. Hyphenated compound words were broken into two words separated by a single space. All punctuation characters were stripped and whitespace was substituted for a single space. Digits were substituted for their written form. For example '3' was replaced with 'three', likewise '911' was replaced with 'nine one one'. As a result of these changes, each lyric occupied a single line of space-delimited lower-case words. This process was for the most part successful although it was later noted that there remained items in the lyrics which did not correspond to their pronunciation. Abbreviations and acronyms proved problematic as a reliable method of removing them was not found in time. Many abbreviations and acronyms differ in spelling from their pronunciation. The phonetic techniques may have been disadvantaged by their presence in the collection. It is also possible that pronunciation may be different in sung versions compared to rendered versions where numbers exist. Consider the number 4500, which could be pronounced "forty-five hundred" (as in the Status Quo song), "four five oh oh", "four five zero zero", or "four thousand five hundred".

The process described above resulted in a data set containing the lyrics for 2359 songs, being approximately 2.3MB and made up of almost half a million words.

3.3   Relevance Judgements

The relevance set was created using the methods introduced by the TREC conference[4] although on a much smaller scale. The set is made up of items selected by a panel of four assessors from a pool of candidates and ranged in number between 1 per query up to 6 per query. The relevance candidates for each of the 50 queries were created by pooling together the most relevant 15 results from each of the three techniques. Each query-candidate pair was judged by four assessors. All possible results that were not included in the pool of candidates were considered not relevant. The candidates and the corresponding query were read aloud by the first author to the assessors, who were instructed to select all of the candidates they believed were relevant. A candidate could be considered relevant if the assessor believed that it could be misheard as the query. For example, the candidate 'dude looks like a lady' could be considered relevant if the assessor believed that it could be misheard as 'doodoos like a lady'. Only those items that were selected by two or more of the assessors were used in the relevance set for measuring the average precision of the results. This resulted in an average of 2.5 relevant answers per query.

4   Matching Techniques

Three techniques for determining approximate pattern matches on text were compared, each being configured for local matching: edit distance, Editex, and SAPS-L.

$$edit(0,0) = 0$$
$$\mathbf{edit(i,0) = edit(i-1,0) + d(s_{i-1}, s_i)}$$
$$edit(0,j) = edit(0,j-1) + d(t_{j-1}, t_j)$$
$$edit(i,j) = min[edit(i-1,j) + d(s_{i-1}, s_i),$$
$$edit(i,j-1) + d(t_{j-1}, t_j),$$
$$edit(i-1,j-1) + r(s_i, t_j)]$$

Fig. 1. The original Editex recurrence relation

$$edit(0,0) = 0$$
$$\mathbf{edit(i,0) = 0}$$
$$edit(0,j) = edit(0,j-1) + d(t_{j-1}, t_j)$$
$$edit(i,j) = min[edit(i-1,j) + d(s_{i-1}, s_i),$$
$$edit(i,j-1) + d(t_{j-1}, t_j),$$
$$edit(i-1,j-1) + r(s_i, t_j)]$$

Fig. 2. The local alignment variation of Editex

Techniques not present for the construction of the candidates were not considered for comparison due to the effect discussed by Büttcher, et al. [2], that an evaluation scheme based on the pooling method will discriminate against any technique that did not contribute to the pool.

The edit distance measure[15] is the simplest of the three techniques and uses dynamic programming to process its input. It quantifies the similarity between two strings with the number of insertions, deletions, and substitutions required to transform the first string into the second. As it is the most general of the three it provides good results on a wide variety of data and does not rely on the characteristics of any one language. For our experiments we used equal penalties for substitutions, insertions and deletions, with matches not contributing to the score.

Editex[18] combines edit distance with character groups similar to those found in phonetic matching techniques like Phonix and Soundex (both discussed in Zobel and Dart [18]). The recurrence relation that lies at the heart of the algorithm combines some of the characteristics of the phonetic searching techniques with the dynamic programming edit distance technique. Editex takes into consideration that repeated letters often have a single sound. For example, 'll' results in a single 'l' sound when pronounced. The Editex algorithm was demonstrated by Zobel and Dart[18] on data sets made up of personal names. For data of this type it was preferable to seek to match the pattern over the entire text because both were of similar size. For the data used here, where the text is many times longer than the patterns, it is more appropriate to seek a local match, that is, a region occurring anywhere within the text that matches the pattern. The original Editex is shown in Figure 1 and the local alignment modification is shown in Figure 2. The local alignment variation alters one of the initial conditions of the table. Alignments are not penalised for beginning any number of characters into the text. The weights used for our experiments are identical to those used by Zobel and Dart[18], that is: $r(a, b)$ has a score of 0 if a and b are identical, 1 if they are in the same letter group, and 2 otherwise; and $d(a, b)$ is almost the same as $r(a, b)$ except when $a \neq b$ and a is equal to the letter h or w, then $d(a, b) = 1$. The Editex letter groups are shown in Figure 3.

$$0 \quad 1 \quad 2 \quad 3 \quad 4 \quad 5 \quad 6 \quad 7 \quad 8 \quad 9$$
aeiouy bp ckq dt lr mn gj fpv sxz csz

Fig. 3. Editex letter groups

SAPS-L was derived from Syllable Alignment Pattern Matching[3] (SAPS) which combines edit distance with a unique three phase process for both text and pattern which breaks them down to a series of phonetic syllables. The first phase uses Phonix transformation rules to produce the phonetic form of the data. The second segments the data into syllables. The third performs a dynamic programming algorithm to determine the level of similarity between pattern and text, favouring letter matches and corresponding syllables. In including this technique we had hoped to retrieve results that while not letter-by-letter matches, had similar sound and rhythm to the query.

Phonetic transformations, like Phonix, work by performing a series of substring substitutions on a source text with the aim of producing phonetically spelled output text. This process is of great utility to phonetic matching techniques as the English language features many heterographic homonyms, word pairings with the same pronunciation but different spelling. With an edit distance algorithm the words "rough" and "ruff" are unlikely to be considered similar despite identical pronunciation. Phonetic transformations correct this issue by attempting to give homophones a common spelling.

SAPS-L was derived and used rather than SAPS due to the singular focus on personal names present in the original SAPS technique. SAPS-L differs from SAPS in the phonetic transformation rules used in phase 1, in the syllable segmentation rules used in phase 2, and an alteration to the initial conditions of the recurrence relation used in phase 3 much like the one made to Editex. Rubin and Gong trialled SAPS on personal names[3] and was able to use the Phonix transformation rules in the first phase of processing. The Phonix rules are tailored to suit the characteristics of personal names and were inappropriate for the data used here as many of the transformations do not occur in English words. An example is the substitution of 'vskie' for all instances of 'wsk'. This is appropriate for personal names of some nationalities, but not for the English language, and particularly not the subset of English that can be found in lyrics from popular music. The altered rules can be found in Table 1. The changes made to the

Table 1. The SAPS-L rule set, showing the substitution value, the characters that will be substituted, depending on whether they occur at the start, middle or end of a word. V can be any vowel, C can be any consonant, [chars] represents the possible values for a single character, and [!chars] represents the only disallowed values for a single character.

SUB	START	MIDDLE	END	SUB	START	MIDDLE	END
g	dg	dg	dg	s	ps		
ko	co	co	co	sV		zV	zV
ka	ca	ca	ca	ts			tz
ku	cu	cu	cu	hrew	hrough	hrough	hrough
si	cy	cy	cy	[!dh]uf	[!dh]ough	[!dh]ough	[!dh]ough
s[ei]	sc[ei]			eek			ique
shal	cial	cial	cial	ker			quer
si	ci	ci	ci	esk			esque
se	sce	sce	sce	rk		rq	rq
se	ce	ce	ce	kwV	quV	quV	quV
klV	clV	clV	clV	g	gh		
k	ck	ck	ck	e			Vgh
krV	chrV			si	cy		
krV	crV			nks		nx	nx
r	wr			r	rh		
kt	ct	ct	ct	ith		yth	yth
f	ph	ph	ph	nV	[mp]nV		
sk	sch			Vsl	Vstl	Vstl	Vstl
tl		btl	btl	oh			eaux
ay	eigh	eigh	eigh	eksi	exci	exci	exci
ite	ighte	ighte	ighte	Vks	Vx	Vx	Vx
ite	ight	ight	ight	eks	x	x	x
ort	ought	ought	ought	nd			ned
larf	laugh	laugh	laugh	le			lle
ort	aught	aught	aught	kils			cles
low	lough	lough	lough	Vil			Vle
n	kn			Vils			Vles
yn			gn	Vm			Vmb
iy	igh	igh	igh	mps		mpts	mpts
ne			ghne	ms		mps	
nes			gnes	mt		mpt	mpt
n			gn	mees			mys
nV			gnV	mis	mys	mys	mys
				mee			my

syllable segmentation rules added awareness of multiple words, ensuring that the first character of each word was the start of a new syllable. The following demonstrates a SAPS-L transformation. 'dancin with the chicken slats' becomes 'DanSin With The ChiKen Slats'. A capital letter indicates the start of a syllable. The weights used for the SAPS-L matching algorithm were: 6 for matching capitals, −2 for mismatching capitals, 1 for matching lower case and white space, −1 for mismatching lower case and white space, −4 for a mismatch of both case and letter. Penalties for insertions and deletions were −3 for capitals and −1 for other characters.

5   Analysis of Results

The query set was processed to retrieve the 100 most relevant results for each query using each of the three techniques in turn. All candidates selected by two or more assessors were used in the relevance set to determine the effectiveness of the results. The average precision of the results was computed and the mean of these values is shown in Table 2.

Table 2. MAP calculated with the most relevant 15 results and also all 100 results

Technique	Mean Average Precision	
	15 results	100 results
Editex	0.541	0.552
Edit Distance	0.549	0.559
SAPS-L	0.507	0.519

While the MAP of edit distance is marginally better than that of Editex, it was not found to be statistically significant and cannot be interpreted as evidence that edit distance provides better results than Editex. The t-test was used to determine if the average precision values were likely to have been drawn from different populations, indicating that the difference in the mean was significant. The p-value returned by the t-test was far too high to be conclusive. This indicates that further work is required to determine the benefits of one technique over the other.

SAPS-L's slightly disappointing MAP was contributed to by a number of factors. The first was the presence of compound words in the text. Compound words are single words made up of two separate words. While these are sometimes hyphenated, the more common word pairs are usually written as a single word. When two words are put together as a compound word, it is difficult to generate a phonetic version of the word. This is because letters which can usually be expected to form a particular sound, no longer do so. An example of this is 'uphill', 'ph' in almost all cases would normally form a 'f' sound. But here it does not. These compound words do not react well to phonetic transformations. A second issue is that often SAPS-L would return results that appeared to have a similar rhythm to the query but did not sound similar. For longer queries it seems reasonable to expect that there would not be many results available that would have both similar rhythm and sound to the query.

Reciprocal rank was computed without the use of the relevance judgements supplied by the assessors. It is a different method of evaluation that is suited to information retrieval tasks where there is assumed to be a single correct answer. For this task we used the data from the misheard lyrics website where the queries were sourced. Each of the misheard lyrics that we have used as queries was submitted along with the correct lyric and that correct lyric was taken to be the single correct answer for this evaluation measure. The rank of a result is equal to the location of the correct answer in the list of results. If the correct answer was returned first, that is the technique considered it the most relevant, the rank is 1 and the reciprocal rank is 1/1. Whereas if the correct answer had been returned 4th, the rank would be 4 and the reciprocal rank would be 1/4. The mean reciprocal rank found in Table 3 is the mean taken over each of the 50 queries in the query set.

Table 3. Mean Reciprocal Rank calculated with the most relevant 15 results and also all 100 results

Technique	Mean Reciprocal Rank	
	15 results	100 results
Editex	0.713	0.715
Edit Distance	0.707	0.710
SAPS-L	0.749	0.750

The greater effectiveness of SAPS-L on this measure does not imply greater effectiveness, as the number of reciprocal ranks computed is below the recommended minimum for this measurement and the t-test showed no statistical significance in the results. However, further experimentation could indicate greater potential for this technique. The results from edit distance and Editex closely resemble those from the MAP evaluation.

We also examined whether the matching techniques were disproportionately successful for longer queries, indicating that the surrounding possibly correct words were responsible for successful retrieval. Table 4 shows the correlation coefficients for query lengths of each matching technique with each evaluation measure. While there is a weak correlation with query length, it can be seen from the graphs that short queries are often successful. There is also no evidence that one technique is better than the others for short queries.

The phonetic transformation carried out in the first stage of SAPS-L processing had the effect of wiping out some of the nuance of the sound of the words.

Table 4. Correlation coefficients for query length versus evaluation measure

	Average Precision	Reciprocal Rank
Editex	0.282	0.418
Edit Distance	0.328	0.439
SAPS-L	0.284	0.436

These nuances can, and often are, exploited by vocalists to produce phonetic sounds far removed from the expected phonetic sound of the written word. Emphasis can be placed much more heavily and on unusual phonemes in the word. That emphasis can rarely be predicted by viewing the lyrics in written form. This suggests that this type of algorithm could be challenging to fine tune for better results. A related issue is that the phonetic substitutions used in SAPS-L appear to have been developed for British English (for example, 'ought' being substituted with 'ort'), whereas the majority of pop songs are sung with American pronunciation regardless of the origin of the singer.

6   Conclusion

We have investigated the effectiveness of some techniques for retrieving misheard lyrics in a music information retrieval system. The measures used to evaluate the effectiveness of the techniques did not reveal any one clear leader. However SAPS-L showed promise and it is hoped that greater effectiveness can be achieved by ironing out the issues hampering its effectiveness. Further experimentation will determine the gains that can be made in SAPS-L's retrieval effectiveness.

The experiment indicated several areas of importance for further exploration of this problem. Of primary importance is improving the quality of the collection set that contains the complete lyrics possibly by retrieving better data from multiple sources using the method outlined by Knees et al. [7] to produce the most correct form of the lyric. The aim would be to reduce noise and minimise phonetic-unfriendly elements such as abbreviations, acronyms, unhyphenated compound words, and spelling errors. Of secondary importance is an increase in the number of queries evaluated. The query data source used for this study yielded 4124 queries, an amount more than sufficient for statistically significant results. However, we have clearly shown that songs can be retrieved with queries consisting of misheard lyrics, and that some variation of approximate pattern matching can be used for this purpose.

References

1. Bello, J.P., Chew, E. (eds.): International Conference on Music Information Retrieval, September 2008, vol. 9 (2008)
2. Büttcher, S., Clarke, C.L.A., Yeung, P.C.K., Soboroff, I.: Reliable information retrieval evaluation with incomplete and biased judgements. In: SIGIR 2007: Proceedings of the 30th annual international ACM SIGIR conference on Research and development in information retrieval, pp. 63–70. ACM, New York (2007)
3. Gong, R., Chan, T.K.Y.: Syllable alignment: A novel model for phonetic string search. IEICE - Trans. Inf. Syst. E89-D(1), 332–339 (2006)
4. Harman, D.: Overview of the second text retrieval conference (TREC-2). Information Processing and Management 31(3), 271–289 (1995)
5. Hosoya, T., Suzuki, M., Ito, A., Makino, S.: Lyrics recognition from a singing voice based on finite state automaton for music information retrieval. In: Buyoli, C.L., Loureiro, R. (eds.) International Conference on Music Information Retrieval, October 2004, vol. 5, pp. 532–535 (2004)

6. Kleedorfer, F., Knees, P., Pohle, T.: Oh oh oh whoah! towards automatic topic detection in song lyrics. In: Bello, Chew (eds.) [1], pp. 287–292
7. Knees, P., Schedl, M., Widmer, G.: Multiple lyrics alignment: Automatic retrieval of song lyrics. In: Proceedings of 6th International Conference on Music Information Retrieval (ISMIR 2005), London, UK, pp. 564–569 (2005)
8. Kondrak, G., Dorr, B.: Identification of confusable drug names: A new approach and evaluation methodology. In: Proceedings of 21st International Conference on Computer Linguistics (COLING 2004), Geneva, Switzerland, pp. 952–958 (2004)
9. Logan, B., Kositsky, A., Moreno, P.: Semantic analysis of song lyrics. In: IEEE International Conference on Multimedia and Expo (ICME), pp. 827–830. IEEE, Los Alamitos (2004)
10. Manning, C.D., Schutze, H.: Foundations of statistical natural language processing. MIT Press, Cambridge (1999)
11. Mayer, R., Neumayer, R., Rauber, A.: Rhyme and style features for musical genre categorisation by song lyrics. In: Bello, Chew (eds.) [1], pp. 337–342
12. Mellody, M., Bartsch, M.A., Wakefield, G.H.: Analysis of vowels in sung queries for a music information retrieval system. Journal of Intelligent Information Systems 21(1), 35–52 (2003)
13. Müller, M., Kurth, F., Damm, D., Fremerey, C., Clausen, M.: Lyrics-based audio retrieval and multimodal navigation in music collections. In: Kovács, L., Fuhr, N., Meghini, C. (eds.) ECDL 2007. LNCS, vol. 4675, pp. 112–123. Springer, Heidelberg (2007)
14. Neumayer, R., Rauber, A.: Integration of text and audio features for genre classification in music information retrieval. In: Amati, G., Carpineto, C., Romano, G. (eds.) ECiR 2007. LNCS, vol. 4425, pp. 724–727. Springer, Heidelberg (2007)
15. Skiena, S.: The Algorithm Design Manual. Springer, Heidelberg (1998)
16. Suzuki, M., Hosoya, T., Ito, A., Makino, S.: Music information retrieval from a singing voice using lyrics and melody information. EURASIP Journal on Advances in Signal Processing 2007 (2007)
17. Yaguchi, Y., Oka, R.: Song wave retrieval based on frame-wise phoneme recognition. In: Lee, G.G., Yamada, A., Meng, H., Myaeng, S.-H. (eds.) AIRS 2005. LNCS, vol. 3689, pp. 503–509. Springer, Heidelberg (2005)
18. Zobel, J., Dart, P.: Phonetic string matching: Lessons from information retrieval. In: Proceedings of the ACM-SIGIR Conference on Research and Development in Information Retrieval, Zurich, Switzerland, pp. 166–173 (1996)

Selecting Effective Terms for Query Formulation

Chia-Jung Lee, Yi-Chun Lin, Ruey-Cheng Chen, and Pu-Jen Cheng

Department of Computer Science and Information Engineering National Taiwan University
{r97037,r95021,pjcheng}@csie.ntu.edu.tw
cobain@turing.csie.ntu.edu.tw

Abstract. It is difficult for users to formulate appropriate queries for search. In this paper, we propose an approach to query term selection by measuring the effectiveness of a query term in IR systems based on its linguistic and statistical properties in document collections. Two query formulation algorithms are presented for improving IR performance. Experiments on NTCIR-4 and NTCIR-5 ad-hoc IR tasks demonstrate that the algorithms can significantly improve the retrieval performance by 9.2% averagely, compared to the performance of the original queries given in the benchmarks.

Keywords: Query Formulation, Query Term Selection.

1 Introduction

Users are often supposed to give effective queries so that the return of an information retrieval (IR) system is anticipated to cater to their information needs. One major challenge they face is what terms should be generated when formulating the queries. The general assumption of previous work [14] is that nouns or noun phrases are more informative than other parts of speech (POS), and longer queries could provide more information about the underlying information need. However, are the query terms that the users believe to be well-performing really effective in IR?

Consider the following description of the information need of a user, which is an example description query in NTCIR-4: *Find articles containing the reasons for NBA Star Michael Jordan's retirement and what effect it had on the Chicago Bulls.* Removing stop words is a common way to form a query such as "*contain, reason, NBA Star, Michael Jordan, retirement, effect, had, Chicago Bulls*", which scores a mean average precision (MAP) of 0.1914. It appears obviously that terms *contain* and *had* carry relatively less information about the topic. Thus, we take merely nouns into account and generate another query, "*reason, NBA Star, Michael Jordan, retirement, effect, Chicago Bulls*", which achieves a better MAP of 0.2095. When carefully analyzing these terms, one could find that the meaning of *Michael Jordan* is more precise than that of *NBA Star*, and hence we improve MAP by 14% by removing *NBA Star*. Yet interestingly, the performance of removing *Michael Jordan* is not as worse as we think it would be. This might be resulted from that *Michael Jordan* is a famous *NBA*

G.G. Lee et al. (Eds.): AIRS 2009, LNCS 5839, pp. 168–180, 2009.

Star in *Chicago Bulls*. However, what if other terms such as *reason* and *effect* are excluded? There is no explicit clue to help users determine what terms are effective in an IR system, especially when they lack experience of searching documents in a specific domain. Without comprehensively understanding the document collection to be retrieved, it is difficult for users to generate appropriate queries.

As the effectiveness of a term in IR depends on not only how much information it carries in a query (subjectivity from users) but also what documents there are in a collection (objectivity from corpora), it is, therefore, important to measure the effectiveness of query terms in an automatic way. Such measurement is useful in selection of effective and ineffective query terms, which can benefit many IR applications such as query formulation and query expansion.

Conventional methods of retrieval models, query reformulation and expansion [13] attempt to learn a weight for each query term, which in some sense corresponds to the importance of the query term. Unfortunately, such methods could not explain what properties make a query term effective for search. Our work resembles some previous works with the aim of selecting effective terms. [1,3] focus on discovering key concepts from noun phrases in verbose queries with different weightings. Our work focuses on how to formulate appropriate queries by selecting effective terms or dropping ineffective ones. No weight assignments are needed and thus conventional retrieval models could be easily incorporated. [4] uses a supervised learning method for selecting good expansion terms from a number of candidate terms generated by pseudo-relevance feedback technique. However, we differ in that, (1) [4] selects specific features so as to emphasize more on the relation between original query and expansion terms without consideration of linguistic features, and (2) our approach does not introduce extra terms for query formulation. Similarly, [10] attempts to predict which words in query should be deleted based on query logs. Moreover, a number of works [2,5,6,7,9,15,16,18,19,20] pay attention to predict the quality or difficulty of queries, and [11,12] try to find optimal sub-queries by using maximum spanning tree with mutual information as the weight of each edge. However, their focus is to evaluate performance of *a* whole query whereas we consider units at the level of terms.

Given a set of possible query terms that a user may use to search documents relevant to a topic, the goal of this paper is to formulate appropriate queries by selecting effective terms from the set. Since exhaustively examining all candidate subsets is not feasible in a large scale, we reduce the problem to a simplified one that iteratively selects effective query terms from the set. We are interested in realizing (1) what characteristic of a query term makes it effective or ineffective in search, and (2) whether or not the effective query terms (if we are able to predict) can improve IR performance. We propose an approach to automatically measure the effectiveness of query terms in IR, wherein a regression model learned from training data is applied to conduct the prediction of term effectiveness of testing data. Based on the measurement, two algorithms are presented, which formulate queries by selecting effective terms and dropping ineffective terms from the given set, respectively.

The merit of our approach is that we consider various aspects that may influence retrieval performance, including linguistic properties of a query term and statistical relationships between terms in a document collection such as co-occurrence and context dependency. Their impacts on IR have been carefully examined. Moreover, we have conducted extensive experiments on NTCIR-4 and NTCIR-5 ad-hoc IR tasks to evaluate the performance of the proposed approach. Based on term effectiveness prediction and two query formulation algorithms, our method significantly improve MAP by 9.2% on average, compared to the performance of the original queries given in the benchmarks.

In the rest of this paper, we describe the proposed approach to term selection and query formulation in Section 2. The experimental results of retrieval performance are presented in Sections 3. Finally, in Section 4, we give our discussion and conclusions.

2   Term Selection Approach for Query Formulation

2.1   Problem Specification

When a user desires to retrieve information from document repositories to know more about a topic, many possible terms may come into her mind to form various queries. We call such set of the possible terms *query term space* $T=\{t_1, ..., t_n\}$. A query typically consists of a subset of T. Each query term $t_i \in T$ is expected to convey some information about the user's information need. It is, therefore, reasonable to assume that each query term will have different degree of effectiveness in documents retrieval. Suppose Q denotes all subsets of T, that is, $Q=Power\ Set(T)$ and $|Q|=2^n$. The problem is to choose the best subset Δq^* among all candidates Q such that the performance gain between the retrieval performance of T and Δq ($\Delta q \in Q$) is maximized:

$$\Delta q^* = argmax_{\Delta q \in Q}\{(pf(T) - pf(\Delta q))/pf(T)\}. \qquad (1)$$

where $pf(x)$ denotes a function measuring retrieval performance with x as the query. The higher the score $pf(x)$ is, the better the retrieval performance can be achieved.

An intuitive way to solve the problem is to exhaustively examine all candidate subset members in Q and design a method to decide which the best Δq^* is. However, since an exhaustive search is not appropriate for applications in a large scale, we reduce the problem to a simplified one that chooses the most effective query term t_i ($t_i \in T$) such that the performance gain between T and $T-\{t_i\}$ is maximized:

$$t_i^* = argmax_{t_i \in T}\{(pf(T) - pf(T - \{t_i\}))/pf(T)\}. \qquad (2)$$

Once the best t_i^* is selected, Δq^* could be approximated by iteratively selecting effective terms from T. Similarly, the simplified problem could be to choose the most ineffective terms from T such that the performance gain is minimized. Then Δq^* will be approximated by iteratively removing ineffective or noisy terms from T.

Our goals are: (1) to find a function $r: T \rightarrow R$, which ranks $\{t_1, ..., t_n\}$ based on their effectiveness in performance gain (MAP is used for the performance measurement in this paper), where the effective terms are selected as candidate query terms, and (2) to formulate a query from the candidates selected by function r.

2.2   Effective Term Selection

To rank term t_i in a given query term space T based on function r, we use a regression model to compute r directly, which predicts a real value from some observed features of t_i. The regression function $r\colon T \to R$ is generated by learning from each t_i with the examples in form of $<f(t_i),(pf(T) - pf(T - \{t_i\}))/pf(T)>$ for all queries in the training corpus, where $f(t_i)$ is the feature vector of t_i, which will be described in Section 2.4.

The regression model we adopt is Support Vector Regression (SVR), which is a regression analysis technique based on SVM [17]. The aim of SVR is to find the most appropriate hyperplane $\mathbf{w}$ which is able to predict the distribution of data points accurately. Thus, r can be interpreted as a function that seeks the least dissimilarity between ground truth $y_i = (pf(T) - pf(T - \{t_i\}))/pf(T)$ and predicted value $r(t_i)$, and r is required to be in the form of $w\,f(t_i)+b$. Finding function r is therefore equivalent to solving the convex optimization problem:

$$Min_{w,b,\xi_{i,1},\xi_{i,2}}\ \ \frac{1}{2}\,\|\mathbf{w}\|^2 + C\sum_i (\xi_{i,1} + \xi_{i,2}). \tag{3}$$

subject to:

$$\forall\, t_i \in T \qquad\qquad y_i \quad (\mathbf{w}\, f(t_i)+b) \geq \varepsilon + \xi_{i,1} \tag{4}$$

$$\forall\, i\colon\ \xi_{i,1},\xi_{i,2} \geq 0 \qquad\qquad (\mathbf{w}\, f(t_i)+b) \quad y_i \geq \varepsilon + \xi_{i,2}. \tag{5}$$

where C determines the tradeoff between the flatness of r and the amount up to which deviations larger than ε are tolerated, ε is the maximum acceptable difference between the predicted and actual values we wish to maintain, and $\xi_{i,1}$ and $\xi_{i,2}$ are slack variables that cope with otherwise infeasible constraints of the optimization problem. We use the SVR implementation of LIBSVM [8] to solve the optimization problem.

Ranking terms in query term space $T=\{t_1, ..., t_n\}$ according to their effectiveness is then equivalent to applying regression function to each t_i; hence, we are able to sort terms $t_i \in T$ into an ordering sequence of effectiveness or ineffectiveness by $r(t_i)$.

2.3   Generation and Reduction

Algorithms *Generation* and *Reduction* formulate queries by greedily selecting effective terms or dropping ineffective terms from space T based on function r.

When formulating a query from query term space T, the Generation algorithm computes a measure of effectiveness $r(t_i)$ for each term $t_i \in T$, includes the most effective term t_i^* and repeats the process until k terms are chosen (where k is a empirical value given by users). Note that T is changed during the selection process, and thus statistical features should be re-estimated according to new T. The selection of the best candidate term ensures that the current selected term t_i^* is the most informative one among those that are not selected yet.

Compared to generation, the Reduction algorithm always selects the most ineffective term from current T in each iteration. Since users may introduce noisy terms in query term space T, Reduction aims to remove such ineffective terms and will repeat the process until $|T|-k$ terms are chosen.

Algorithm Generation	Algorithm Reduction
Input: $T=\{t_1,t_2,...,t_n\}$ (query term space)	**Input:** $T=\{t_1,t_2,...,t_n\}$ (query term space)
k (# of terms to be selected)	k (# of terms to be selected)
$\Delta q \leftarrow \{\ \}$	$\Delta q \leftarrow \{\ t_1,t_2,...,t_n\ \}$
for i = 1 **to** k **do**	**for** i = 1 **to** n-k **do**
$t_i^* \leftarrow argmax_{t_i \in T}\{r(t_i)\}$	$t_i^* \leftarrow argmin_{t_i \in T}\{r(t_i)\}$
$\Delta q \leftarrow \Delta q \cup \{t_i^*\}$	$\Delta q \leftarrow \Delta q - \{t_i^*\}$
$T \leftarrow T - \{t_i^*\}$	$T \leftarrow T - \{t_i^*\}$
end	**end**
Output Δq	**Output** Δq

Fig. 1. The Generation Algorithm and the Reduction Algorithm

2.4   Features Used for Term Selection

Linguistic and statistical features provide important clues for selection of good query terms from viewpoints of users and collections, and we use them to train function r.

Linguistic Features: Terms with certain linguistic properties are often viewed semantics-bearing and informative for search. Linguistic features of query terms are mainly inclusive of parts of speech (POS) and named entities (NE). In our experiment, the POS features comprise noun, verb, adjective, and adverb, the NE features include person names, locations, organizations, and time, and other linguistic features contain acronym, size (i.e., number of words in a term) and phrase, all of which have shown their importance in many IR applications. The values of these linguistic features are binary except the size feature. POS and NE are labeled manually for high quality of training data, and can be tagged automatically for purpose of efficiency alternatively.

Statistical Features: Statistical features of term t_i refer to the statistical information about the term in a document collection. This information could be about the term itself such as term frequency (TF) and inverse document frequency (IDF), or the relationship between the term and other terms in space T. We present two methods for estimating such term relationship. The first method depends on co-occurrences of terms t_i and t_j ($t_j \in T$, $t_i \neq t_j$) and co-occurrences of terms t_i and T-$\{t_i\}$ in the document collection. The former is called *term-term co-occur feature* while the latter is called *term-topic co-occur feature*. The second method extracts so-called context vectors as features from the search results of t_i, t_j, and T-$\{t_i\}$, respectively. The *term-term context feature* computes the similarity between the context vectors of t_i and t_j while the *term-topic context feature* computes the similarity between context vectors of t_i and T-$\{t_i\}$.

   Term-term & term-topic co-occur features: The features are used to measure whether query term t_i itself could be replaced with another term t_j (or remaining terms T-$\{t_i\}$) in

T and how much the intension is. The term without substitutes is supposed to be important in T. Point-wise mutual information (PMI), Chi-square statistics (X^2), and log-
likelihood ratio (LLR) are used to measure co-occurrences between t_i and Z, which is
either t_j or T-$\{t_i\}$ in this paper. Suppose that N is the number of documents in the collection, a is the number of documents containing both t_i and Z, denoted as a = #d(t_i,Z).
Similarly, we denote b = #d(t_i,~Z) c = #d(~t_i,Z) and d = #d(~t_i,~Z) i.e., $Z=N-a-b-c$.

PMI is a measure of how much term t_i tells us about Z.

$$PMI(t_i, Z) = \log[p(t_i, Z)/p(t_i)p(Z)] \approx \log[a \times N/(a + b)(a + c)] \tag{6}$$

X^2 compares the observed frequencies with frequencies expected for independence.

$$\chi^2(t_i, Z) = [N \times (a \times d - b \times c)^2]/[(a + b)(a + c)(b + d)(c + d)] \tag{7}$$

LLR is a statistical test for making a decision between two hypotheses of dependency
or independency based on the value of this ratio.

$$-2 \log LLR(t_i, Z) =$$

$$a \log \frac{a \times N}{(a + b)(a + c)} + b \log \frac{b \times N}{(a + b)(b + d)} + c \log \frac{c \times N}{(c + d)(a + c)} + d \log \frac{d \times N}{(c + d)(b + d)} \tag{8}$$

We make use of average, minimum, and maximum metrics to diagnose term-term co-
occur features over all possible pairs of (t_i, t_j), for any $t_j \neq t_i$:

$$f_{avg}^X(t_i) = \frac{1}{|T|} \sum_{\forall t_j \in T, t_i \neq t_j} X(t_i, t_j), \tag{9}$$

$$f_{max}^X(t_i) = \max_{\forall t_j \in T, t_i \neq t_j} X(t_i, t_j) \ and \ f_{min}^X(t_i) = \min_{\forall t_j \in T, t_i \neq t_j} X(t_i, t_j) \tag{10}$$

where X is *PMI, LLR* or X^2. Moreover, given $T=\{t_1, ..., t_n\}$ as a training query term
space, we sort all terms t_i according to their $f_{avg}^X(t_i)$, $f_{max}^X(t_i)$, or $f_{min}^X(t_i)$, and their
rankings varied from 1 to n are treated the additional features.

The *term-topic co-occur features* are nearly identical to the *term-term co-occur
features* with an exception that *term-topic co-occur features* are used in measuring the
relationship between t_i and query topic T-$\{t_i\}$. The co-occur features can be quickly
computed from the indices of IR systems with caches.

Term-term & term-topic context features: The co-occurrence features are reliable for
estimating the relationship between high-frequency query terms. Unfortunately, term
t_i is probably not co-occurring with T-$\{t_i\}$ in the document collection at all. The context features are hence helpful for low-frequency query terms that share common
contexts in search results. More specifically, we generate the context vectors from the
search results of t_i and t_j (or T-$\{t_i\}$), respectively. The context vector is composed of a
list of pairs <document ID, relevance score>, which can be obtained from the search
results returned by IR systems. The relationship between t_i and t_j (or T-$\{t_i\}$) is captured by the cosine similarity between their context vectors. Note that to extract the

context features, we are required to retrieve documents. The retrieval performance may affect the quality of the context features and the process is time-consuming.

3   Experiments

3.1   Experiment Settings

We conduct extensive experiments on NTCIR-4 and NTCIR-5 English-English ad-hoc IR tasks. Table 1 shows the statistics of the data collections. We evaluate our methods with description queries, whose average length is 14.9 query terms. Both queries and documents are stemmed with the Porter stemmer and stop words are removed. The remaining query terms for each query topic form a query term space T. Three retrieval models, the vector space model (TFIDF), the language model (Indri) and the probabilistic model (Okapi), are constructed using Lemur Toolkit [21], for examining the robustness of our methods across different frameworks. MAP is used as evaluation metric for top 1000 documents retrieved. To ensure the quality of the training dataset, we remove the poorly-performing queries whose average precision is below 0.02. As different retrieval models have different MAP on the same queries, there are different numbers of training and test instances in different models. We up-sample the positive instances by repeating them up to the same number as the negative ones. Table 2 summarizes the settings for training instances.

Table 1. Adopted dataset after data clean. Number of each setting is shown in each row for NTCIR-4 and NTCIR-5

Table 2. Number of training instances. (x : y) shows the number of positive (x) and negative (y) MAP gain instances, respectively

	NTCIR-4	NTCIR-5		Indri	TFIDF	Okapi
	<desc>	<desc>	Original	674(156:518)	702(222:480)	687(224:463)
#(query topics)	58	47	Upsample	1036(518:518)	960(480:480)	926(463:463)
#(distinct terms)	865	623	Train	828(414:414)	768(384:384)	740(370:370)
#(terms/query)	14.9	13.2	Test	208(104:104)	192(96:96)	186 (93:93)

3.2   Performance of Regression Function

We use 5-fold cross validation for training and testing our regression function r. To avoid inside test due to up-sampling, we ensure that all the instances in the training set are different from those of the test set. The R^2 statistics ($R^2 \in [0, 1]$) is used to evaluate the prediction accuracy of our regression function r:

$$R^2 = \frac{\Sigma_i (y_i - \hat{y}_i)^2}{\Sigma_i (y_i - \bar{y})^2} , \qquad (11)$$

where R^2 explains the variation between true label $y_i = (pf(T) - pf(T - \{t_i\}))/pf(T)$ and fit value $\hat{y}_i = wf(t_i) + b$ for each testing query term $t_i \in T$, as explained in Section 2.2. $\bar{y}$ is the mean of the ground truth.

Table 3 shows the R^2 values of different combinations of features over different retrieval models, where two other features are taken into account for comparison. Content load (*Cl*) [14] gives unequal importance to words with different POS. Our modified content load (m-Cl) sets weight of a noun as 1 and the weights of adjectives, verbs, and participles as 0.147 for IR. Our m-SCS extends the simplified clarity score (*SCS*) [9] as a feature by calculating the relative entropy between query terms and collection language models (unigram distributions).

It can be seen that our function *r* is quite independent of retrieval models. The performance of the statistical features is better than that of the linguistic features because the statistical features reflect the statistical relationship between query terms in the document collections. Combining both outperforms each one, which reveals both features are complementary. The improvement by m-Cl and m-SCS is not clear due to their similarity to the other features. Combining all features achieves the best R^2 value 0.945 in average, which guarantees us a large portion of explainable variation in *y* and hence our regression model *r* is reliable.

Table 3. R^2 of regression model *r* with multiple combinations of training features. L: linguistic features; C1: co-occurrence features; C2: context features

Performance of	One Group of Features			Two Groups of Features			Three	Four (3+1)		All
Regression Model *r*	L	C1	C2	L&C1	L&C2	C1&C2	L&C1 &C2	m-Cl	m-SCS	
R^2 Indri	0.120	0.145	0.106	0.752	0.469	0.285	0.975	0.976	0.975	0.976
TFIDF	0.265	0.525	0.767	0.809	0.857	0.896	0.932	0.932	0.932	0.932
Okapi	0.217	0.499	0.715	0.780	0.791	0.910	0.925	0.926	0.925	0.926
Avg.	0.201	0.390	0.529	0.781	0.706	0.697	0.944	0.945	0.944	0.945

3.3   Correlation between Feature and MAP

Yet another interesting aspect of this study is to find out a set of key features that play important roles in document retrieval, that is, the set of features that explain most of the variance of function *r*.   This task can usually be done in ways fully-addressed in regression diagnostics and subset selection, each with varying degrees of complexity. One common method is to apply correlation analysis over the response and each predictor, and look for highly-correlated predictor-response pairs.

Three standard correlation coefficients are involved, including Pearson's product-moment correlation coefficient, Kendall's tau, and Spearman's rho.   The results are given in Fig. 2, where x-coordinate denotes features and y-coordinate denotes the value of correlation coefficient. From Fig. 2, two context features, "cosine" and "cosineinc", are found to be positively- and highly-correlated ($\rho > 0.5$) with MAP, under Pearson's coefficient.   The correlation between the term-term context feature (cosine) and MAP even climbs up to 0.8. For any query term, high context feature value indicates high deviation in the result set caused by removal of the term from the query topic. The findings suggest that the drastic changes incurred in document ranking by removal of a term can be a good predictor. The tradeoff is the high cost in feature computation because a retrieval processing is required. The co-occurrence features such as PMI, LLR, and χ^2 also behave obviously correlated to MAP. The minimum value of LLR correlates more strongly to MAP than the maximum one does, which means that the independence between query terms is a useful feature.

In the linguistic side, we find that two features "size" and "phrase" show positive, medium-degree correlation ($0.3 < \rho < 0.5$) with MAP. Intuitively, a longer term might naturally be more useful as a query term than a shorter one is; this may not always be the case, but generally it is believed a shorter term is less informative due to the ambiguity it encompasses. The same rationale also applies to "phrase", because terms of noun phrases usually refer to a real-world event, such as "911 attack" and "4th of July", which might turn out to be the key of the topic.

We also notice that some features, such as "noun" and "verb", pose positive influence to MAP than others do, which shows high concordance to a common thought in NLP that nouns and verbs are more informative than other type of words. To our surprises, NE features such as "person", "geo", "org" and "time" do not show as high concordance as the others. This might be resulted from that the training data is not sufficient enough. Features "idf" and "m-SCS" whose correlation is highly notable have positive impacts. It supports that the statistical features have higher correlation values than the linguistics ones.

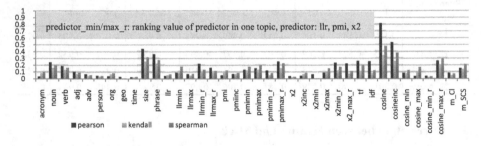

Fig. 2. Three correlation values between features and MAP on Okapi retrieval model

3.4   Evaluation on Information Retrieval

In this section, we devise experiments for testing the proposed query formulation algorithms. The benchmark collections are NTCIR-4 and NTCIR-5. The experiments can be divided into two parts: the first part is a 5-fold cross-validation on NTCIR-4 dataset, and in the second part we train the models on NTCIR-4 and test them on NTCIR-5. As both parts differ only in assignment of the training/test data, we will stick with the details for the first half (cross-validation) in the following text.

The result is given in Table 4. Evaluation results on NTCIR-4 and NTCIR-5 are presented in the upper- and lower-half of the table, respectively. We offer two baseline methods in the experiments: "BL1" puts together all the query terms into one query string, while "BL2" only consider nouns as query terms since nouns are claimed to be more informative in several previous works. Besides, the upper bound

UB is presented in the benchmark: for each topic, we permute all sub queries and discover the sub-query with the highest MAP. As term selection can also be treated as a classification problem, we use the same features of our regression function r to train two SVM classifiers, Gen-C and Red-C. Gen-C selects terms classified as "effective" while Red-C removes terms classified as "ineffective". Gen-R and Red-R denote our Generation and Reduction algorithms, respectively. The retrieval results are presented in terms of MAP. Gain ratios in MAP with respect to the two baseline methods are given in average results. We use two-tailed t-distribution in the significance test for each method (against the BL1) by viewing AP values obtained in all query session as data points, with $p<0.01$ marked ** and $p<0.05$ marked *.

Table 4. MAP of baseline and multiple proposed methods on NTCIR-4 <desc> regression model. (+x, +y) shows the improvement percentage of MAP corresponding to BL1 and BL2. TFIDF and Okapi models have PRF involved, Indri model does not. Best MAP of each retrieval model is marked **bold** for both collections.

Settings	Method	Indri	TFIDF	Okapi	Avg.
NTCIR-4	UB	0.2233	0.3052	0.3234	0.2839
<desc>	BL1	0.1742	0.2660	0.2718	0.2373
Queries	BL2	0.1773	0.2622	0.2603	0.2332
	Gen-C	0.1949**	0.2823**	**0.2946****	0.2572(+8.38%,+10.2%)
	Gen-R	0.1954**	**0.2861****	0.2875*	0.2563(+8.00%,+9.90%)
	Red-C	0.1911**	0.2755**	0.2854**	0.2506(+5.60%,+7.46%)
	Red-R	**0.1974****	0.2773**	0.2797	0.2514(+5.94%,+7.80%)
NTCIR-5	UB	0.1883	0.2245	0.2420	0.2182
<desc>	BL1	0.1523	0.1988	0.1997	0.1836
Queries	BL2	0.1543	0.2035	0.1969	0.1849
	Gen-C	0.1699**	0.2117*	0.2213*	0.2009(+9.42%,+8.65%)
	Gen-R	0.1712**	**0.2221***	**0.2232***	0.2055(+11.9%,+11.1%)
	Red-C	0.1645**	0.2194*	0.2084	0.1974(+7.51%,+6.76%)
	Red-R	**0.1749****	0.2034**	0.2160*	0.1981(+7.89%,+7.13%)

From Table 4, the MAP difference between two baseline methods is small. This might be because some nouns are still noisy for IR. The four generation and reduction methods significantly outperform the baseline methods. We improve the baseline methods by 5.60% to 11.9% in the cross-validation runs and on NTCIR-5 data. This result shows the robustness and reliability of the proposed algorithms. Furthermore, all the methods show significant improvements when applied to certain retrieval models, such as Indri and TFIDF; performance gain with Okapi model is less significant on NTCIR-5 data, especially when reduction algorithm is called for. The regression methods generally achieve better MAP than the classification methods. This is because the regression methods always select the most informative terms or drop the most ineffective terms among those that are not selected yet. The encouraging evaluation results show that, despite the additional costs on iterative processing, the performance of the proposed algorithms is effective across different benchmark collections, and based on a query term space T, the algorithms are capable of suggesting better ways to form a query.

We further investigate the impact of various ranking schemes based on our proposed algorithms. The ranking scheme in the Generation algorithm (or the Reduction algorithm) refers to an internal ranking mechanism that decides which term shall be included in (or discarded away). Three types of ranking schemes are tested based on our regression function r. "max-order" always returns the term that is most likely to contribute relevance to a query topic, "min-order" returns the term that is most likely to bring in noise, and "random-order" returns a randomly-chosen term.

Figure 3 shows the MAP curve for each scheme by connecting the dots at (1, $MAP^{(1)}$), ... , (n, $MAP^{(n)}$), where $MAP^{(i)}$ is the MAP obtained at iteration i. It tells that the performance curves in the generation process share an interesting tendency: the curves keep going up in first few iterations, while after the maximum (locally to each method) is reached, they begin to go down rapidly. The findings might informally establish the validity of our assumption that a longer query topic might encompass more noise terms. The same "up-and-down" pattern does not look so obvious in the reduction process; however, if we take the derivative of the curve at each iteration i (i.e., the performance gain/loss ratio), we might find it resembles the pattern we have discovered. We may also find that, in the generation process, different ranking schemes come with varying degrees of MAP gains. The ranking scheme "max-order" constantly provides the largest performance boost, as opposed to the other two schemes. In the reduction process, "max-order" also offers the most drastically performance drop than the other two schemes do. Generally, in the generation process, the best MAP value for each setting might take place somewhere between iteration n/2 to 2n/3, given n is the size of the query topic.

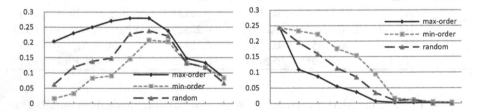

Fig. 3. MAP curves based on regression model for description queries of NTCIR-4 on TFIDF model, each with three selection order. X coordinate is # of query terms; Y coordinate is MAP.

4   Discussions and Conclusions

In this paper, we propose an approach to measure and predict the impact of query terms, based on the discovery of linguistic, co-occurrence, and contextual features, which are analyzed by their correlation with MAP. Experimental results show that our query formulation approach significantly improves retrieval performance.

The proposed method is robust and the experimental results are consistent on different retrieval models and document collections. In addition, an important aspect of this paper is that we are able to capture certain characteristics of query terms that are highly effective for IR. Aside from intuitive ideas that informative terms are often lengthy and tagged nouns as their POS category, we have found that the statistical features are more likely to decide the effectiveness of query terms than linguistics ones do. We also observe that context features are mostly correlated to MAP and thus are most powerful for term difficulty prediction. However, such post-retrieval features require much higher cost than the pre-retrieval features, in terms of time and space.

The proposed approach actually selects local optimal query term during each iteration of generation or reduction. The reason for this greedy algorithm is that it is inappropriate to exhaustively enumerate all sub-queries for online applications such as search engines. Further, it is challenging to automatically determine the value of parameter k in our algorithms, which is selected to optimize the MAP of each query topic. Also, when applying our approach to web applications, we need web corpus to calculate the statistical features for training models.

References

1. Allan, J., Callan, J., Croft, W.B., Ballesteros, L., Broglio, J., Xu, J., Shu, H.: INQUERY at TREC-5. In: Fifth Text REtrieval Conference (TREC-5), pp. 119–132 (1997)
2. Amati, G., Carpineto, C., Romano, G.: Query difficulty, robustness, and selective application of query expansion. In: McDonald, S., Tait, J.I. (eds.) ECIR 2004. LNCS, vol. 2997, pp. 127–137. Springer, Heidelberg (2004)
3. Bendersky, M., Croft, W.B.: Discovering key concepts in verbose queries. In: 31st annual international ACM SIGIR, pp. 491–498 (2008)
4. Cao, G., Nie, J.Y., Gao, J.F., Robertson, S.: Selecting good expansion terms for pseudo-relevance feedback. In: 31st annual international ACM SIGIR, pp. 243–250 (2008)
5. Carmel, D., Yom-Tov, E., Soboroff, I.: SIGIR Workshop Report: Predicting Query Difficulty - Methods and Applications. In: Workshop Session: SIGIR, pp. 25–28 (2005)
6. Carmel, D., Yom-Tov, E., Darlow, A., Pelleg, D.: What makes a query difficult? In: 29th annual international ACM SIGIR, pp. 390–397 (2006)
7. Carmel, D., Farchi, E., Petruschka, Y., Soffer, A.: Automatic query refinement using lexical affinities with maximal information gain. In: 25th annual international ACM SIGIR, pp. 283–290 (2002)
8. Chang, C.C., Lin, C.J.: LIBSVM (2001),
   http://www.csie.ntu.edu.tw/~cjlin/libsvm
9. He, B., Ounis, I.: Inferring query performance using pre-retrieval predictors. In: 11th International Conference of String Processing and Information Retrieval, pp. 43–54 (2004)
10. Jones, R., Fain, D.C.: Query word deletion prediction. In: 26th annual international ACM SIGIR, pp. 435–436 (2003)
11. Kumaran, G., Allan, J.: Effective and efficient user interaction for long queries. In: 31st annual international ACM SIGIR, pp. 11–18 (2008)
12. Kumaran, G., Allan, J.: Adapting information retrieval systems to user queries. In: Information Processing and Management, pp. 1838–1862 (2008)
13. Kwok, K.L.: A new method of weighting query terms for ad-hoc retrieval. In: 19th annual international ACM SIGIR, pp. 187–195 (1996)
14. Lioma, C., Ounis, I.: Examining the content load of part of speech blocks for information retrieval. In: COLING/ACL 2006 Main Conference Poster Sessions (2006)

15. Mandl, T., Womser-Hacker, C.: Linguistic and statistical analysis of the CLEF topics. In: Third Workshop of the Cross-Language Evaluation Forum CLEF (2002)
16. Mothe, J., Tanguy, L.: ACM SIGIR 2005 Workshop on Predicting Query Difficulty - Methods and Applications (2005)
17. Vapnik, V.N.: Statistical Learning Theory. John Wiley & Sons, Chichester (1998)
18. Yom-Tov, E., Fine, S., Carmel, D., Darlow, A., Amitay, E.: Juru at TREC 2004: Experiments with prediction of query difficulty. In: 13th Text Retrieval Conference (2004)
19. Zhou, Y., Croft, W.B.: Query performance prediction in Web search environments. In: 30th Annual International ACM SIGIR Conference, pp. 543–550 (2007)
20. Zhou, Y., Croft, W.B.: Ranking Robustness: A novel framework to predict query performance. In: 15th ACM international conference on Information and knowledge management, pp. 567–574 (2006)
21. The Lemur Toolkit:
    http://www.lemurproject.org/

Discovering Volatile Events in Your Neighborhood: Local-Area Topic Extraction from Blog Entries

Masayuki Okamoto and Masaaki Kikuchi

Corporate R&D Center, Toshiba Corporation
1 Komukai Toshiba-cho, Saiwai-ku, Kawasaki 212-8582, Japan
masayuki4.okamoto@toshiba.co.jp

Abstract. This paper presents a method for the detection of occasional or volatile local events using topic extraction technologies. This is a new application of topic extraction technologies that has not been addressed in general location-based services. A two-level hierarchical clustering method was applied to topics and their transitions using time-series blog entries collected with search queries including place names. According to experiments using 764 events from 37 locations in Tokyo and its vicinity, our method achieved 77.0% event findability. It was found that the number of blog entries in urban areas was sufficient for the extraction of topics, and the proposed method could extract typical volatile events, such as performances of music groups, and places of interest, such as popular restaurants.

Keywords: Hot topic extraction, hierarchical clustering, locality.

1   Introduction

Along with the spread of mobile terminals, such as cellular phones and mobile personal computers, there has been growth in location-based services (LBSs) [16], such as car navigation systems, personal navigation systems, and location-based recommendation. Such LBSs mainly deal with static POI (point of interest) data and usually do not pay attention to occasional or volatile events, such as performances of singers on street corners, and new topical spots. Usually, it is difficult to find out about such events, and even if one does, it could be after the event.

The features of volatile events include the following: one or more words about an event or topic are mentioned in blog entries frequently in a short period, and these words correlate with a specific place name. For example:

- *I enjoyed XX at the new YY restaurant near the ZZ station.*
- *XX event at YY was very exciting!*

With regard to the first feature, many techniques related to hot topic extraction such as a timeline analysis-based method [1] or the burst detection method [10], which

G.G. Lee et al. (Eds.): AIRS 2009, LNCS 5839, pp. 181–192, 2009.
© Springer-Verlag Berlin Heidelberg 2009

detects the increase in the number of documents following the occurrence of an event [7], is often applied to news documents or blog entries. However, the number of web-pages that mention a particular local event is often too small for the successful extraction of topics.

With regard to the second feature, one of the issues concerns determining the type of location name that would be useful. In urban areas in Japan, many places are near one or more train stations, and station names are possible keys for accessing local events. Moreover, the locality of each event, the fact that topic terms appear only in a few specific locations, can be used as well as the topicality of each event.

As one solution, this study proposes a topic extraction method that finds new or volatile events from time-series text data obtained from blog searches carried out using a query including station names.

Although there have been studies on clustering and topic detection for webpages or blog entries, transitory location-based events have not been dealt with. The purpose of this paper is to investigate this issue. In this study, a topic extraction method based on a two-level topic clustering technique for time-series documents [9], which was applied to Japanese EPG (electronic program guide) data in the previous report, is used. Moreover, the techniques of named entity recognition and locality calculation are introduced in this paper.

The remainder of this paper is organized as follows. Section 2 shows the flow of the local-area topic extraction method. Section 3 reports on experiments for evaluating our local-area topic extraction method. Section 4 shows related works.

2   Local-Area Topic Extraction

We define the problem of local event detection as a variation of the hot topic extraction task. In this section, the extraction method is introduced.

2.1   Overview

Figure 1 shows the process flow of the proposed topic extraction method.

First, blog entries are collected in advance. Entries are collected hourly in the RSS (rich site summary) format with a common blog search engine using query words including geographic names such as 'Tokyo' or 'Akihabara.' Then, the prefiltering module removes entries including NG terms usually used for advertising or offensive to public order and morals.

Second, document vectors are generated from the collected entries using morphological analysis, named entity recognition, and IDF-based weighting function, and topics are extracted using the hierarchical clustering technique. An agglomerative approach [8] in which the Euclidean distance is used as the distance metric is employed. Then, the topic words for each topic cluster are extracted using the C-value technique [6].

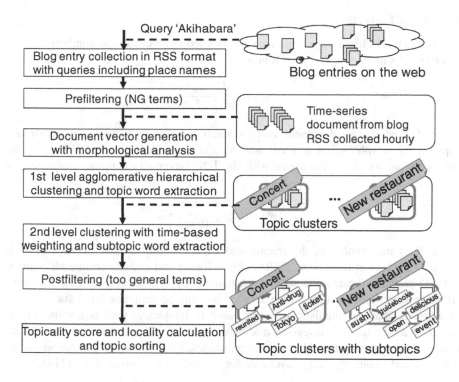

Fig. 1. Process flow of topic and event extraction in the two-level hierarchical clustering method

Third, second-level clustering is performed and subtopic clusters are extracted. A subtopic means a secondary important event or a shorter-term event in the topic. In this process, the time difference is additionally considered as a distance metric for detecting events within the same period. The subtopic words for each subtopic cluster are then extracted with the C-value technique, and the postfiltering process is performed to remove entries irrelevant to the topic, which are sometimes extracted when the topic word used is too general.

Finally, the topicality score is calculated for each topic. The topicality score is defined as the ratio of the short-term (2 days in this paper) average frequency of a topic to the long-term (3 days in this paper) average frequency of the same topic. If the Z test shows that the topicality score is significantly large for a topic, the topic is considered to be a hot topic. The locality score is also calculated for each topic. The locality score is defined as the number of locations at which a topic appears on the same day. A large locality score means that the topic appears in many locations and the topic is not a *local* event. Thus, events that share the same topic word are removed in this process.

Using the above method, the hot topics for each area are extracted. Figure 2 shows examples of extracted topics. This style is compact and applicable for a small display on a mobile terminal.

2.2   Topic Extraction

In the first-level clustering, cosine measure is used to calculate the similarity s_{ab} between two documents a and b as follows:

$$s_{ab} = \frac{d_a \cdot d_b}{\|d_a\|\|d_b\|}$$

where d_a and d_b are document vectors. Each term w in a document vector d is acquired by morphological analysis and named entity recognition (NER), and the corresponding weight is calculated with the IDF (inversed document frequency) as follows:

$$\mathrm{idf}(w) = \log_{10} \frac{N}{\mathrm{df}(w)} + 1$$

where N is the number of documents and $\mathrm{df}(w)$ is the number of documents in which w appears at least once. NER seeks to locate and classify atomic elements in text into predefined categories such as the names of persons, organizations, locations, expressions of time, quantities, monetary values, and percentages. NER systems have been created that use linguistic grammar-based techniques as well as statistical models [2]. We currently use linguistic rule-based techniques with over one hundred generic named entity classes covering person names, place names, organization names, numbers, and so on, originally developed for a question answering system [14].

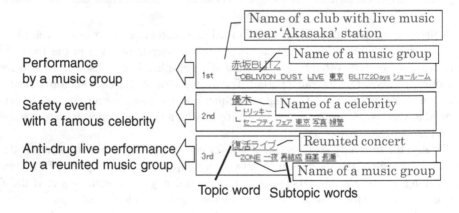

Fig. 2. Examples of extracted topics. Each topic includes a topic word and subtopic words.

Topics are extracted with the agglomerative clustering process by folding two clusters with the largest similarity score until the largest similarity score becomes lower than a threshold θ_T. When two clusters are folded, document vectors composing clusters are added as follows:

$$d_{ab} = d_a + d_b \ .$$

2.3   Topic Word Extraction

A topic word is extracted from each topic cluster with a modified version of the C-value method [6]. The C-value of a candidate collocation cw is calculated as follows:

$$\text{C-value}(cw) = (\text{length}(cw) - 1)\left(\text{n}(cw) - \frac{\text{t}(cw)}{\text{c}(cw)}\right)$$

1where $\text{length}(cw)$ is the number of characters, $\text{n}(cw)$ is the number of times cw appears, $\text{t}(cw)$ is the total frequency of cw in longer candidate collocations, and $\text{c}(cw)$ is the number of those candidate collocations. In this paper, the number of characters is used for $\text{length}(cw)$ because we put higher priority on a long word than a collocation with short words, though $\text{length}(cw)$ is the number of words in the original version of C-value. Finally, a collocation with the highest C-value is selected as the topic word.

2.4   Subtopic Extraction

For the topic extraction described in Section 2.2, subtopics are extracted with the second-level clustering. In this process, a time-based similarity function is additionally introduced. A decaying function $W(a, b)$ for the similarity between clusters a and b is as follows:

$$W(a, b) = \exp(-\alpha(t_a + t_b)^2)$$

where t_a and t_b are the average times when the clusters a and b occurred respectively, and α is a constant value. Finally, similarity $\hat{s}_{ab}$ between the clusters a and b is

$$\hat{s}_{ab} = W(a, b) \cdot s_{ab} .$$

In addition, the IDF score for each word is recalculated on the condition that documents in the cluster are considered as the document set. Through this process, the weight for topic word decreases and the other words' weight increases. Thus, subtopic words are expected to be extracted. As described in Section 2.2, subtopics are extracted with the agglomerative clustering process by folding two clusters with the largest similarity score until the largest similarity score $\hat{s}$ becomes lower than a threshold θ_E, and then subtopic words are extracted by the same process described in Section 2.3.

2.5   Topicality and Locality Calculation

Hot topics appear more frequently in the nearest days. For example, a hot topic appears more frequently in the nearest three days than in the nearest seven days. Therefore, we need to determine whether or not the number of short-term occurrences is significantly larger than the number of long-term occurrences.

We assume that the long-term occurrence follows the rectangular distribution and determine the topicality by testing the null hypothesis 'the short-term occurrence also follows the rectangular distribution in the same way as the long-term occurrence' with Z test.

If the probability of occurrence of a document follows rectangular distribution, we can assume that the distribution of the number of documents that occurred in the latest n days v follows a binomial distribution with the occurrence probability $p = n/N$ and the trial number of times u. It is known that a probability function $Pr(v) =_u C_v p v (1 - p)^{u-v}$ can be approximated by a normal distribution $(up, up(1 - p))$. Thus, the value of Z with the number of observed documents in the latest n days v_0 is calculated with

$$Z = \frac{v_0 - up}{\sqrt{(up(1-p))}} .$$

We used Z score for the rank of each topic, i.e., the higher Z score a topic has, the higher the topic is ranked. In this paper, we used one day for the short term and three days for the long term.

For finding a local event, locality is also important. When a hot topic appears only in a place, it will be a local event, while a hot topic which appears in many places is not a local event but a pervasive event. In fact, events that share the same topic word are removed in this process.

Finally, topics sorted by topicality score are shown as Figure 2.

3 Experiments

We investigated to what degree extracted topics are appropriate as local events.

3.1 Topic Extraction Test Collection from Local Blogs

Before the experiment, we investigated what location is applicable. The number of relevant topics that could be extracted from local blogs was investigated. First, blogs were collected in advance for 856 train stations in Tokyo and Kanagawa in the first half of December 2008. Then, some of the stations were selected and topics were extracted to investigate the types of events that could be extracted.

Figure 3 shows the distribution of the number of blog entries. From this data it was found that there were 95 stations with 100 or more blog entries per day on average, a number considered sufficient to extract topics. Other stations had fewer entries, but it was still possible to extract small events that were referred to by a number of blog entries at around the same time. Therefore, the data in Figure 3 indicates that the number of entries is sufficient for detecting transitory events in large urban areas.

We introduce two kinds of location sets from top-rated locations: the vague location set and the non-vague location set.

The *vague location* is a location whose name is used not only for the local area but also for the wider area. For example, the word 'Tokyo' may be used to indicate the station name (Tokyo Station), the city itself, or the greater Tokyo area, which includes

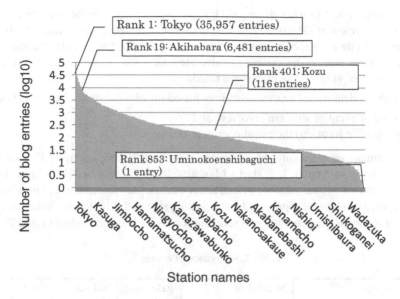

Station names

Fig. 3. Distribution of blog entries for 856 stations in Tokyo and Kanagawa in the first half of December 2008. One or more topics are extracted from 853 stations.

parts of the surrounding prefectures. Another example is the word 'Kawasaki.' This word may be used not only for the station name or the city itself but also for the last name of a person.

The *non-vague location* is a location whose name is used almost only for a local area. For example, the word 'Akihabara' may be used only to indicate Akihabara Station and its vicinity.

We chose 14 locations for the vague location set and 23 locations for the non-vague location set from locations with 10,000 or more blog entries in the first half of December 2008 (Table 1).

Table 1. Selected locations

Vague locations	Akasaka, Fuchu, Ginza, Kamakura, Kawasaki, Machida, Meguro, Nihonbashi, Ryogoku, Shibuya, Shinagawa, Tokyo, Ueno, Yokohama
Non-vague locations	Akihabara, Asakusa, Chigasaki, Daiba, Daikanyama, Ebisu, Harajuku, Ikebukuro, Jiyugaoka, Kagurazaka, Kichijoji, Korakuen, Minatomirai, Odawara, Omotesando, Roppongi, Shimokitazawa, Shinbashi, Shinjuku, Shiodome, Tsukiji, Yokosuka, Yurakucho

3.2 Extraction Accuracy

We investigated the extent to which extracted topics are recognized as local events.

Two subjects rated to what degree each topic seems to be a local event according to three levels: relevant (2); found relevant after checking the original documents (1); irrelevant (0). The relevance a user recognizes is calculated with only the rating 2, whereas the relevance of the system is calculated with the ratings 2 and 1. For each location-date pair, at most five events are rated.

Each subject determined relevance ratings based on whether or not he found both

- what kind of event or who are involved, and
- where (specific location) the event occurs or occurred.

Table 2 summarizes the results of our experiments for 764 topics from 37 locations for one week (from March 3, 2009 to March 9, 2009). In Table 2, the 'number of events' column means the number of extracted events, the 'relevance' column means the percentage of relevant topics (at least one subject rated 2 or 1), the 'relevance for user' column means the percentage of relevant topics (at least one subject rated 2).

Table 2. Relevance of events

Location	Number of evaluated events	Relevance	Relevance for user
Vague location	289	174 (60.2%)	101 (34.9%)
Non-vague location	475	335 (70.5%)	199 (41.9%)
Total	764	509 (66.6%)	300 (39.3%)

From Table 2, we found there is a large difference in relevance score between non-vague locations (70.5%) and vague locations (60.2%). It is because of the ambiguity of place names from vague locations, as we mentioned in Section 3.1. We also found there is a large difference between 'relevance' score and 'relevance for user' score. The reason is that some cue terms, such as building names or street names, are not extracted as a topic word or a subtopic word because these terms are written in various representations and it makes C-value of these terms smaller.

Table 3. Findability of events

Location	Number of locations	Findability of events
Vague location	78	58 (74.4%)
Non-vague location	135	106 (78.5%)
Total	213	164 (77.0%)

Table 3 also summarizes the results of our experiments from the findability of events. In Table 3, the 'findability of events' column means the percentage for which a user recognized at least one topic as a local event (ratings 2) for each set of 5 events.

Table 3 shows that at least one subject found one or more events for 77.0% of sets of 5 events. This means that if we develop an event-recommendation application, a user can find some events by 77% for each location.

We also found the relevance score is much lower than we had expected because our previous report using EPG showed that 94.3% of topics are relevant [9]. This is because of the difference of document well-formedness between EPG documents and blog RSS documents.

Figure 4(a) shows the relevance for non-vague locations and Figure 4(b) shows the relevance for vague locations. From Figure 4, most non-vague locations achieved 0.6 or more relevance, while only 4 vague locations achieved 0.6 or more relevance.

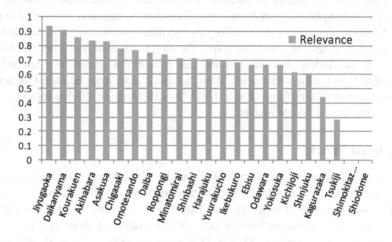

(a) Non-vague locations

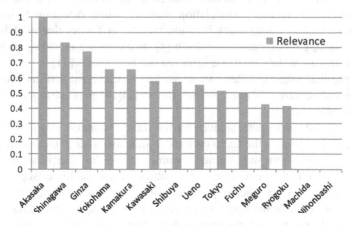

(b) Vague locations

Fig. 4. Relevance for each location

3.3   Feature of Extracted Events

Investigation of these stations revealed the following examples of typical events or spots:

- events held near a station (e.g., performances, such as a music group),
- Hokkaido souvenir festival (temporary souvenir shops),
- Ome marathon (sports event),
- Tokyo Auto Salon 2009 (a motor show), and
- Kenya Hara design event (an artist-related event in an art museum).

In addition to the above, restaurants on which bloggers wrote their impressions and incidents broadcast on news websites were also extracted.

However, problems related to place names were also revealed. One problem is the ambiguity of some place names from vague locations as mentioned in the above sections. Another problem is in the handling of the date information. The date information is useful for detecting events on a specific day, but it also functions as a hub and results in unrelated topics becoming connected. In addition, some real estate information was also found. This information, though it may sometimes be useful, is not used for extracting events. Our future research efforts will focus on solving these problems and improving the extraction results.

4   Related Works

There are many previous works on conventional topic detection and tracking (TDT) [1]. Hot topic extraction is a kind of topic detection task to find a topic that appears frequently over a period of time [3], and this technology is based on term-weighting theories, topicality calculation theories, and sentence modeling theories [5].

For term-weighting theories, TF-IDF [15] has been commonly used, and recently other methods such as TF*PDF [3] that focus on some feature of topic extraction have been proposed. For topicality calculation theories, burst detection techniques are widely used [10]. Another approach is to use the Aging Theory that models a topic's life cycle of birth, growth, decay, and death [4]. For sentence modeling theories, vector space models are widely used and achieved the best results [11].

In the topic detection task, there are some key characteristics of an event such as where the event occurred. Named entities (NEs) [2][14] play an important role and were successfully used [10].

Though we used common techniques for hot topic extraction, two new features are shown in this paper: one is the focus on local events using blog documents instead of extracting news topic from well-formed documents, and the other is the use of a two-level clustering method to extract detailed subtopics and subtopic words. For using location, a spatiotemporal theme pattern mining method is proposed [12]. This approach focuses on how the hotness of a given topic changes spatiotemporally from weblog data collected with a topic term such as 'Hurricane Katrina' with a probabilistic approach, where our approach is based on a clustering approach.

Moreover, the focus of this paper is not only on extracting local events but also on presenting the topic in a set of terms for mobile use.

There are also previous works on information presentation for mobile terminals with a small display, e.g., text summarization [13] and search result summarization [17]. We believe that the topic expression with topic words and subtopic words, described in this paper will be another successful presentation style for mobile terminals.

5   Conclusions

This paper introduced the possibility of topic extraction technologies for local-area event detection and new LBS applications. The hierarchical clustering method was used to extract topics and their transitions from blog entries collected with search queries including place names. An empirical investigation was also conducted into the types of event topics that could be extracted from blog entries collected using the names of 856 locations in the search queries. Moreover, according to an evaluation measuring relevance for 764 topics from 37 locations, we found the features of extracted topics. From these investigations, we confirmed that the proposed method was successful in extracting typical volatile events that would be difficult to extract with common web search engines or the burst detection method.

Our future work will include: improvement of filtering quality, classification of event types, and development of applications for mobile devices with geographic information systems.

References

1. Allan, J.: Topic Detection and Tracking: Event-Based Information Organization. Kluwer Academic Publication, Dordrecht (2002)
2. Borthwick, A.: A Maximum Entropy Approach to Named Entity Recognition. PhD thesis, New York Univeristy (1999)
3. Bun, K.K., Ishizuka, M.: Topic Extraction from News Archive Using TF*PDF Algorithm. In: Proceedings of International Conference on Web Information Systems Engineering (WISE 2002), pp. 73–82 (2002)
4. Chen, C.C., Chen, Y.T., Sun, Y., Chen, M.C.: Life Cycle Modeling of News Events Using Aging Theory. In: Lavrač, N., Gamberger, D., Todorovski, L., Blockeel, H. (eds.) ECML 2003. LNCS (LNAI), vol. 2837, pp. 47–59. Springer, Heidelberg (2003)
5. Chen, K.-Y., Luesukprasert, L., Chou, S.T.: Hot Topic Extraction Based on Timeline Analysis and Multidimensional Sentence Modeling. IEEE Transactions on Knowledge and Data Engineering 19(8), 1016–1025 (2007)
6. Frantsi, K., Ananiadou, S.: Extracting Nested Collocations. In: Proceedings of International Conference on Computational Linguistics (COLING 1996), pp. 41–46 (1996)
7. Fujiki, T., Nanno, T., Suzuki, M., Okumura, M.: Identification of Bursts in a Document Stream. In: Proceedings of International Workshop on Knowledge Discovery in Data Streams (2004)
8. Kamvar, S., Klein, D., Manning, C.: Interpreting and Extending Classical Agglomerative Clustering Algorithms Using a Model-Based Approach. In: Proceedings of International Conference on Machine Learning (ICML 2002), pp. 283–290 (2002)

9. Kikuchi, M., Okamoto, M., Yamasaki, T.: Extraction of Topic Transition through Time Series Document based on Hierarchical Clustering. Journal of the DBSJ 7(1), 85–90 (2008)
10. Kleinberg, J.: Bursty and Hierarchical Structure in Streams. In: Proceedings of ACM SIGKDD International Conference on Knowledge Discovery and Data Mining (KDD 2002), pp. 91–101 (2002)
11. Kumaran, G., Allan, J.: Text Classification and Named Entities for New Event Detection. In: Proceedings of Annual International ACM SIGIR Conference on Research and Development in Information Retrieval (SIGIR 2004), pp. 297–304 (2004)
12. Mei, Q., Liu, C., Su, H., Zhai, C.: A Probabilistic Approach to Spatiotemporal Theme Pattern Mining on Weblogs. In: Proceedings of International World Wide Web Conference (WWW 2006), pp. 533–542 (2006)
13. Otterbacher, J., Radev, D., Kareem, O.: News to Go: Hierarchical Text Summarization for Mobile Devices. In: Proceedings of Annual International ACM SIGIR Conference on Research and Development in Information Retrieval (SIGIR 2006), pp. 589–596 (2006)
14. Sakai, T., Saito, Y., Ichimura, Y., Koyama, M., Kokubu, T., Manabe, T.: ASKMi: A Japanese question answering system based on semantic role analysis. In: Proceedings of Recherche d'Information Assistée par Ordinateur (RIAO 2004), pp. 215–231 (2004)
15. Salton, G., Yang, C.S.: On the Specification of Term Values in Automatic Indexing. J. Documentation, 351–372 (1973)
16. Schiller, J.H., Voisard, A.: Location-based Services. Morgan Kaufmann Publishers, San Francisco (2004)
17. Yasukawa, M., Yokoo, H.: Clustering Search Results for Mobile Terminals. In: Proceedings of Annual ACM SIGIR Conference on Information Retrieval (SIGIR 2008), p. 880 (2008)

A Unified Graph-Based Iterative Reinforcement Approach to Personalized Search

Yunping Huang, Le Sun, and Zhe Wang

Institute of Software, Chinese Academy of Sciences, Beijing, 100190, China
{yunping07,sunle,wangzhe07}@iscas.ac.cn

Abstract. General information retrieval systems do not perform well in satisfying users' individual information need. This paper proposes a novel graph-based approach based on the following three kinds of mutual reinforcement relationships: RR-Relationship (Relationship among search results), RT-Relationship (Relationship between search results and terms), TT-Relationship (Relationship among terms). Moreover , the implicit feedback information, such as query logs and immediately viewed documents, can be utilized by this graph-based model. Our approach produces better ranking results and a better query model mutually and iteratively. Then a greedy algorithm concerning the diversity of the search results is employed to select the recommended results. Based on this approach, we develop an intelligent client-side web search agent GBAIR, and web search based experiments show that the new approach can improve search accuracy over another personalized web search agent.

Keywords: Information Retrieval, Personalized Search, Graph-Based Model.

1 Introduction

General information retrieval systems do not perform well in satisfying users' individual information need [20]. Many existing retrieval systems fail to discern individuals' search goals since most queries (usually short, ambiguous and lacking discriminative terms in general) don't provide a complete specification of the user's information need. In order to overcome such problems, we need to exploit users' personalized information to accurately capture user's information need. There have been many attempts on this topic, such as user specification [7] and relevance feedback [15]. Both methods are effective. However, they require users' extra effort; users are usually reluctant to provide additional information [1].

Implicit Feedback is a method which does not need the user's extra effort. Implicit feedback information includes users' query logs, browsing history, users' client-side interaction records and other desktop information. Several previous studies [20, 16, 19, 12] have shown that implicit feedback can improve retrieval accuracy. In this paper, we utilize the implicit information, including short-term query logs and the immediately viewed documents (also named click-through data), which are the clicked search results in the same query.

G.G. Lee et al. (Eds.): AIRS 2009, LNCS 5839, pp. 193–204, 2009.

Besides the implicit feedback information, there are many other resources which can be used for query model construction and result re-ranking. The clustering hypothesis [24] implies the following: closely related documents should have similar scores when ranking the document, and closely related terms should have similar scores when constructing the query model, so the relationships among terms and the relationships among search results can be exploited to improve effectiveness. The TT-Relationship (Relationship among terms) has been used for query model construction in [5, 2]. The contextual terms can be mutually reinforced. For example, the weight of "computer" increases as many related contextual terms such as "hardware" and "software" co-occur. RR-Relationship (Relationship among search results) is also a useful resource to re-rank the search results [23]. The relevance can be propagated among the search results. It is especially effective in personalized search framework since the immediately viewed result can be used to vote for similar search results.

In the web, because there are many redundant and duplicate pages, ranking method based on the content may make the top results lack diversity, which will make the system fail to learn much from the feedback. Diversity is important for obtaining good feedback information, especially when there is not sufficient information for the search goals. Therefore, in our approach, when selecting the search result to be recommended, a greedy algorithm is used to penalize the search results overlap with other related search results, and the penalty degree is tuned based on the quantity of user's feedback information.

This paper aims to make contributions on the following aspects: (1) We exploit three kinds of mutual reinforcement relationships: RR-Relationship (Relationship among search results), RT-Relationship (Relationship between search results and terms), TT-Relationship (Relationship among terms); (2) Based on the three kinds of relationships, we utilize a graph-based model which uses the ideas of PageRank and HITS to produce better ranked results and better query model mutually and iteratively, and the query model is updated when there is new implicit feedback information which triggers the iterative algorithm; (3) In order to obtain good feedback information and avoid redundancy, a conformity penalty algorithm is used to select search results to recommend.

In our approach, RR-Relationship reflects the mutual reinforcement among search results, RT-Relationship reflects the mutual reinforcement between search results and terms, and TT-Relationship reflects the mutual reinforcement among terms. To the best of our knowledge, how to exploit and utilize the three kinds of relationships in a unified model has not been well addressed in previous works. Moreover, our approach produces better ranked results and a better query model mutually and iteratively. The experiments clearly show that all types of relationships can result in improvements on retrieval effectiveness.

The remainder of this paper is organized as follows. Section 2 describes previous work related to our approach. Section 3 describes our novel approach for personalized search. Section 4 provides the architecture of GBAIR system and some specific techniques used to implement the proposed ideas. Section 5 presents the details of the experiments and the evaluation results. Section 6 draws some conclusions of our work.

2 Related Works

In this section, we focus on the related work on personalized search using implicit feedback information, and the related graph-based ranking methods.

Many studies exploited query logs to capture users' information need. UCAIR [16], PAIR [12] and [19] all use query history (query terms and clickthough data) to capture user's interests. UCAIR uses the query history in the same search session. PAIR uses the search-related query history in the last 24 hours. [19] tries to mine the long-term search history. Both UCAIR and PAIR integrate the interaction information during the search process. [10, 14] utilize the server-side query logs to infer users' interest. The user's browsing history [13], web communities [18], user's client side interactions [3] are also exploited to improve retrieval performance. Some studies [20, 8] combined several kinds of implicit feedback information. Our approach also utilizes the query-related query logs and immediately viewed documents. The main difference is that such implicit feedback information is naturally integrated with the three kinds of relationships (RR, RT and TT) into a unified graph-based model.

In recent years, several graph-based ranking methods similar to HITS algorithm [11] or PageRank [4] have been proposed. [22] uses sentence-to-word relationships to rank the sentences and the words. [12] uses the word-to-result relationships to rank the results and words. [23] uses the affinity graph to compute the information richness of each search result. [21] utilizes sentence-to-sentence, word-to-word and sentence-to-word relationships to rank the sentences and words. The graph-based model in our work is partly inspired by [21]. However, they use it to select the sentences and then to form a document summarization while we focus on how to combine the relationships with the implicit feedback information (query logs and immediately viewed documents) to discern individuals' search goals by producing better ranking results and a better query model mutually and iteratively.

[12] is closely related to our work in that they also exploit immediately viewed documents and short-term history in query logs, and use a graph model to implement result re-ranking and query expansion. However, our work differs from that of [12] in three respects: (1) They do not utilize the relationship among results (RR-Relationship) and the relationship among terms (TT-Relationship), thus, their result re-ranking and query expansion are not as effective as ours; (2) In their work, the initial query model is not taken into consideration, the new query model are only relied on the scores of search results, which will make query language model deviate too much from its initial value; (3) They don't take the diversity of the recommended results into consideration.

3 Graph-Based Iterative Reinforcement Approach

3.1 Overview

In this study, we exploit three kinds of mutual reinforcement relationships:

(1). RR-Relationship: the relevance score of a search result would be influenced by the relevance score of the search result that links with it. A search result should be relevant if it is linked with many relevant search results, which reflects the mutual reinforcement among search results.

(2). RT-Relationship: a result should be relevant if it contains many relevant terms, and a term should be relevant if it occurs in many relevant results, which reflects the mutual reinforcement between search results and terms.

(3). TT-Relationship: a term should be relevant if it is linked with many other relevant terms, which reflects the mutual reinforcement among terms.

Through these mutual reinforcement relationships, the query model and the ranked search results can be mutually boosted. The ranked search results can be used to optimize the query model, and a better query model results in better ranked search results. Therefore, we can produce better ranking results and a better query model mutually and iteratively in a unified model. And the implicit feedback information can be utilized naturally to trigger the iterative algorithm by this unified model. Based on these mutual reinforcement relationships, we can leverage the ideas of PageRank and HITS to construct a graph to implement this model. We will illustrate how to construct these graphs in detail in the following section.

3.2 Graph Construction

The proposed approach uses PageRank-like model to reflect the relationships among terms and relationships among search results, and use HITS-like model to reflect the relationships between terms and search results.

We build the three sub-graphs based on the above relationships. To reflect the RR-Relationship, we build a directed graph $G_1 = (V_1, E_1)$, where the nodes V_1 correspond to the search results, and an edge " $p \leftrightarrow q \in E_1(p, q \in V_1)$ " is weighted by the content similarity between p and q.

To reflect the TT-Relationship, we build another directed graph $G_2 = (V_2, E_2)$, where the nodes V_2 correspond to the terms, and an edge " $p \to q \in E_2(p, q \in V_2)$ " is weighted by the word relationship between p and q.

To reflect the RT-Relationship, we build a third directed graph $G_3 = (V_1 \cup V_2, E_3)$, where an edge " $p \leftrightarrow q \in E_3(p \in V_1, q \in V_2)$ " is weighted by the importance of term p in the search result q.

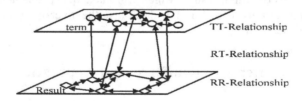

Fig. 1. Illustration of the Relationships

3.3 Parameters of the Graph

The initial value of the node will be discussed in section 4. Here we will describe the computation of the weights of the edges.

3.3.1 RR-Relationship

The content similarity between two search results is used to represent the relationships of them. The content similarity between two search results is simply computed with the cosine measure. We associate each edge between two search results with a weight as follows:

$$w(ri \rightarrow rj) = w(rj \rightarrow ri) = \cos(\vec{ri}, \vec{rj}) \tag{1}$$

3.3.2 RT-Relationship

If a term t_i occurs in a search result r_j, we create an edge between t_i and r_j, and the weight of the edge specifies the importance of t_i in r_j, which is computed as follows:

$$w(ti \rightarrow rj) = w(rj \rightarrow ti) = p(ti \mid rj) = tfij \,/\, |\,rj\,| \tag{2}$$

Where tf_{ij} represents the term frequency of t_i in search result r_j, and $|\,r_j\,|$ represents the length of r_j.

3.3.3 TT-Relationship

We use the following formulas to estimate the relationships between words:

$$w(ti \rightarrow tj) = P(tj \mid ti) = \delta * Pc(tj \mid ti) + (1-\delta) * Pcq(tj \mid ti)$$

$$Pc(tj \mid ti) = \frac{c(ti, tj \mid W, C)}{\sum_{t} c(t, ti \mid, W, C)} \qquad Pcq(tj \mid ti) = \frac{c(ti, tj \mid W, Cq)}{\sum_{t} c(t, ti \mid W, Cq)} \tag{3}$$

where $Pc(tj \mid ti)$ is the language model based on the predefined collection C, and $Pcq(tj \mid ti)$ is the language model based on the collection Cq constructed by the query's retrieved results, $\delta \in [0,1]$ controls the weights of the above two terms; $c(ti, tj \mid W, C)$ is the count of co-occurrences of t_i and t_j within the predefined window W in the collection C, $c(ti, tj \mid W, Cq)$ is the count of co-occurrences of t_i and t_j within the predefined window W in the collection Cq. $Pcq(tj \mid ti)$ can make the relationship among words more suitable for the query, as the user give new feedback information, the system can retrieved more new results, $Pcq(ti \mid tj)$ will be changed, and then the relationship model can be updated. The relationships between two words are asymmetrical.

3.4 Iterative Algorithm

After initialization, the iterative process of result score computation and query model optimization starts. We use $p(t \mid \theta_{iter})$ to represent the score of term t in the iterative algorithm, $p(t \mid \theta)$ to represent the query model, and $S(r)$ to represent the score of result r.

$$p(t_i \mid \theta_{iter})^{k+1} = \alpha \times \sum_{\forall j: r_j \rightarrow t_i} S(r_j)^k \frac{w(r_j \rightarrow t_i)}{\sum_{\forall n: r_j \rightarrow t_n} w(r_j \rightarrow t_n)} + \beta \times \sum_{\forall j: t_j \rightarrow t_i} p(t_j \mid \theta)^k \frac{w(t_j \rightarrow t_i)}{\sum_{\forall n: t_j \rightarrow t_n} w(t_j \rightarrow t_n)} \tag{4}$$

where $p(t_i \mid \theta_{iter})^{k+1}$ is the value of $p(t_i \mid \theta_{iter})$ after (k+1)-th iteration; $S(r_j)^k$ is the value of the search result r_j after k-th iteration. The first term implies that the weight of the term relies on the search results linked with it, and the second term implies that the weights of the term rely on the terms linked with it.

$$S(r_j)^{k+1} = \alpha \times \sum_{\forall i: t_i \to r_j} p(t_i \mid \theta)^k \frac{w(t_i \to r_j)}{\sum_{\forall n: t_i \to r_n} w(t_i \to r_n)} + \beta \times \sum_{\forall i: r_i \to r_j} S(r_i)^k \frac{w(r_i \to r_j)}{\sum_{\forall n: r_i \to r_n} w(r_i \to r_n)} \qquad (5)$$

where $S(r_j)^{k+1}$ is the value of $S(r_j)$ after (k+1)-th iteration; $p(t_i \mid \theta_{iter})^k$ is the value of $p(t_i \mid \theta_{iter})$ after k-th iteration; The first term implies that the score of the search result relies on the terms linked with it, the second term implies that the score of the search result relies on the search results linked with it.

α and β specify the relative contributions, and we have $\alpha+\beta=1$. After re-computation, $p(t \mid \theta_{iter})$ and $S(r)$ are normalized after each iteration as follows:

$$p(t_i \mid \theta_{iter}) = p(t_i \mid \theta_{iter}) / \sum_{tk} p(t_k \mid \theta_{iter}) \qquad (6)$$

$$S(r_j) = S(r_j) / \sum_{rk} S(r_k) \qquad (7)$$

The query model will be updated after each iteration:

$$p(t \mid \theta) = \lambda \times p(t \mid \theta) + (1-\lambda) \times p(t \mid \theta_{iter}) \qquad (8)$$

where $\lambda \in [0,1]$ is a parameter to control the weight on the current query model. In order to compute the score of the search results and the terms, the following steps are repeated until convergence or the iterative time reaches a predefined threshold.

1. Apply (4) and (5) to compute the value of $p(t \mid \theta_{iter})^{k+1}$ and $S(rj)^{k+1}$

2. Apply (6) and (7) to normalize the scores of the terms and the scores of the search results.

3. Apply (8) to update the query model.

The algorithm converges when the changes of $p(t \mid \theta)$ and the scores of the results are smaller than some predefined threshold θ (e.g. 10^{-6}), the changes c is computed as the following formula:

$$c = \sum_{t} (p(t \mid \theta)^{k+1} - p(t \mid \theta)^k)^2 + \sum_{r} (S(r)^{k+1} - S(r)^k)^2 \qquad (9)$$

3.5 Greedy Algorithm for Conformity Penalty

After the iterative algorithm, the search results are ranked by the relevance, however, the top search result may be very similar and redundant. In order to get rich feedback information, the top search results may need to keep diversity. In our algorithm, conformity penalty is imposed through a greedy algorithm similar to [23], which is applied to select n search results which are recommended to the users, and it is similar to the MMR[6] criterion, the algorithm goes as follows:

1. Input the search results and their scores A={(r$_i$,S(r$_i$))} computed by the iterative algorithm, the number of recommended results n.
2. Select the search result r$_i$ which has the highest score, n=n-1

   For each search result r$_j$,j≠i:

$$S(r_j) = S(r_j) - x * 2^{-c} * sim(r_j, r_i) * S(r_i)$$

3. Go to step 2 and iterate until n = 0.

It decreases the ranking score of the result that is similar to the results which have been selected, $x * 2^{-c}$ is the penalty degree factor, $x > 0$, and c is the times that the user click the result, when the user click more result, the system get more feedback information, the search goal becomes clearer, the conformity penalty should be smaller. After the greedy algorithm, the top n unseen search results are recommended to the user, here n is a predefined number (i.e. n=3). Also, the top m (i.e.50%) terms with highest weights in the query model are remained.

4 Implementation

4.1 System Architecture

In this section, we present a client-side web search agent GBAIR, which is an IE plug-in based on the popular Web search engine Google. As showed in Figure 2, GBAIR has three main modules: (1) Result retrieval module retrieves results from search engine; (2) The user modeling module captures user's implicit feedback information such as history information and any clicked search results, then updates the query model and triggers the iterative algorithm; (3) Iterative Algorithm module implements the graph-based iterative algorithm described in section 3, this module updates the query model and re-ranks the unseen search results.

4.2 Trigger Iterative Algorithm through Implicit Feedback Information

Each query log contains query text, query time and the corresponding clicked search results (consist of URL, title and snippet). We judge whether a query log is related to the current query according to the similarity between the query log and the current query text. When computing the similarity, the query log and the current query text are represented as a vector. If the similarity between the two vectors exceeds a predefine threshold, the query log will be considered to be related. When there are related query logs, the system utilized a method similar to PAIR [12] to extract representative terms from query logs, and then use the Fixed Coefficient Interpolation method of [17] to build initial query model.

Immediately clickthrough document is important implicit feedback information we utilize. When the user click a result to view, the system utilized a method similar to PAIR [12] to extract representative terms extracted from immediately viewed documents. After extracting the representative terms, we apply Bayesian estimation method used in [17] to update the query model.

When the system captures new implicit feedback information, the iterative algorithm is triggered. The initial score of a term can be obtained from the query model, and the initial score of a search result can be set as its content similarity with the immediately viewed result when the iterative algorithm is triggered by the immediately viewed result, otherwise, the initial scores of a search results are set to be equal.

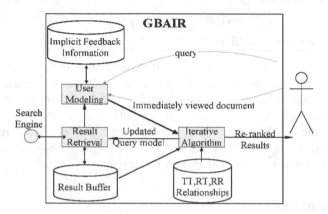

Fig. 2. The architecture of GBAIR

5   Experiments

5.1   Experiment Setup

We evaluate seven systems in our experiments: Google, PAIR[12](a system which utilizes RT-Relationship), GBAIR BASE (the basic GBAIR system which do not use the RR-Relationship and TT-Relationship, the conformity penalty is imposed, RT-Relationship is used), BASE+RR (GBAIR system which use the RR-Relationship beyond GBAIR BASE), BASE+TT (GBAIR system which use the TT-Relationship beyond GBAIR BASE), BASE+TT+RR (GBAIR system which use the TT-Relationship and RR-Relationship beyond GBAIR BASE), BASE+TT+RR-DIV (BASE+TT+RR system which do not impose conformity penalty). PAIR [12] is selected as our baseline model, PAIR extracts representative terms from Query Logs and immediately viewed documents for result re-ranking and query expansion, it can be considered as a variation of pseudo-relevance feedback (PRF).

GBAIR is based on the results from Google: when the user input a query, the system fetches the top N (N=30) results from Google, and adds them into a pool; when new expanded terms are obtained, the system will also fetched some new results and adds them to the pool. The process is the same as that of PAIR.

It is a challenge to quantitatively evaluate the potential performance improvement of the proposed approach over other systems in an unbiased way [16, 12]. Here, we adopt a similar quantitative evaluation to that of [16] and [12] to evaluate our system GBAIR. We recruited 8 students to participate in our experiments. Our experiments contain the following steps: First, we gave each participant a query set which contains HTRDP 2005 and 2004 topics[1] and some frequent queries from a commercial search engine, each participant browsed the query set and selected the queries related to their everyday activities, or they constructed queries themselves. The frequent query list was used to help the user to form query. Each query was submitted to the seven systems. During each query process, the participants click to view some search results, just as in normal web search. Then, at the end of each query, the 30 top ranked search results from these seven different systems were randomly and anonymously mixed together so that every participant would not know where a search result comes from. For every search result, the participants gave a rating ranging from 0 to 2, dividing the relevant results in two categories, 1: relevant and 2: highly relevant. At last, we respectively measure the precision and NDCG (Normalized discount cumulative gain) [9] at top 5, 10, 20 and 30 documents of these systems. In the experiment, each participant submitted 5~10 queries. The NDCG value of a ranking list at position n is calculated as follow:

$$N(n) = Z_n \sum_{j=1}^{n} \frac{2^{r(j)} - 1}{\log(1+j)}$$

where r(j) is the rating of the j-th document in the list, and the normalization constant Zn is chosen so that the perfect list gets a NDCG score of 1.

5.2 Results and Analysis

65 queries are submitted to the seven systems, most of the query selected by the participants is informational, 3711 documents were evaluated by the user during the experiment, and the average evaluated documents for each query is 57; 645 documents are judged as relevant (relevant rating is 1 or 2) from Google search results. The corresponding numbers of relevant documents from PAIR, "GBAIR BASE", "BASE+TT", "BASE+RR", "BASE+RR+TT", "BASE+RR+TT-DIV" are: 827, 825, 835, 852, 879 and 834 respectively. Table 1 shows the average precision of these seven systems at top n results among the 65 queries, and Table 2 shows the NDCG at top n results.

As we can see, all the versions of GBAIR perform better than PAIR; "BASE+RR+TT" system performs best. Comparing "GBAIR BASE" with PAIR, we can see "GBAIR BASE" performs better than PAIR. This indicates that integrating the initial query model can achieve the better performance. However, if the initial query model is not accurate, this will result in worse performance. For example, if query logs considered to be related to the current query deviate from the information need, the original query model may be not accuracy, the performance of "GBAIR BASE" decreases more than PAIR.

The performance of "BASE+TT" is better than that of "GBAIR BASE". This shows that reinforcement relationships among terms are helpful in improving retrieval effectiveness. One explanation is that the approach can mine the term related the user's interest. For example, a term may not appear in the user's query, but it is supported by many related terms. Through the reinforcement relationship, this term

---

[1] HTRDP Evaluation. http://www.863data.org.cn/

will gain a high weight, thus, an accurate query model can be constructed, and then the search results can be improved.

"BASE+RR" outperforms "GBAIR BASE". This indicates the reinforcement relationships among search results are effective. A possible explanation is that a relevant search result may be not relevant to the query because the terms in the relevant search result may mismatch with the terms in the query. However, if it is strongly linked with many relevant search results, this result has a high probability to gain a high score through the reinforcement relationships among search results. When a search result is clicked, the relevance can be transmitted to the similar search results. Therefore using the reinforcement relationships can better rank the search results.

"BASE+RR+TT" system can greatly improve the retrieval effectiveness compared to "BASE+RR" and "BASE+TT". This indicates that although a single factor can work, integrating the three kinds of reinforcement relationships can produce an even better performance. The effects of the factors can be mutually boosted.

Table 1. Average precision@n (PAIR is selected as the baseline model, Google just as a reference system, "+" indicates the improvement over the baseline, ** and * mean significant changes in t-test with respect to the baseline, at the level of p<0.01 and p<0.05 respectively

System	P@5	P@10	P@20	P@30
Google	0.498	0.478	0.375	0.331
PAIR(RT)	0.541	0.493	0.458	0.424
GBAIR BASE(RT)	0.572 (+5.68%)**	0.525 (+6.23%)**	0.475 (+3.87%)**	0.423 (-0.24%)
BASE +TT	0.588 (+8.52%)**	0.555 (+12.46%)**	0.489 (+6.89%)**	0.428 (+0.97%)*
BASE +RR	0.60 (+10.80%)**	0.542 (+9.66%)**	0.478 (+4.37%)**	0.437 (+3.02%)**
BASE+ RR+TT	0.618 (+14.20%)**	0.586 (+18.69%)**	0.506 (+10.59%)**	0.451 (+6.29%)**
BASE+ RR+TT-DIV	0.578 (+6.82%)**	0.534 (+8.10%)**	0.471 (+2.86%)**	0.428 (+0.85%)*

Table 2. Average NDCG@n

System	NDCG@5	NDCG@10	NDCG@20	NDCG@30
Google	0.431	0.427	0.420	0.437
PAIR(RT)	0.465	0.468	0.497	0.543
GBAIR BASE(RT)	0.496 (+6.48%)**	0.502 (+7.16%)**	0.534 (+7.45%)**	0.564 (+3.80%)*
BASE +TT	0.514 (+10.34%)**	0.528 (+12.82%)**	0.549 (+10.48%)**	0.577 (+6.27%)**
BASE +RR	0.515 (+10.73%)**	0.516 (+10.21%)**	0.536 (+7.92%)**	0.578 (+6.21%)**
BASE +RR+TT	0.541 (+16.33%)**	0.554 (+18.25%)**	0.568 (+14.28%)**	0.594 (+9.31%)**
BASE+ RR+TT-DIV	0.517 (+11.00%)**	0.515 (+10.06%)**	0.533 (+7.26%)**	0.566 (+4.31%)**

"BASE+RR+TT" has better performance than "BASE+RR+TT-DIV". The greedy algorithm concerning the diversity is effective. For one thing, keeping the diversity of the top result will make the probability that the relevant result appears in the top results increased; for the other thing, keeping the diversity can avoid redundancy to some degree, the user can see more different relevant results, and the user will click more results(the user are not likely to click result which is very similar to those he/she has clicked, the average number of clickthrough results for all queries of "BASE+RR+TT" and "BASE+RR+TT-DIV" respectively is 2.2 and 1.9), which can give the system more feedback information. When the query is difficult, much of the search results are not relevant, "BASE+RR+TT" can outperform "BASE+RR+TT-DIV" more.

5.3 Efficiency of GBAIR

The approach we proposed is efficient. The relationship among words is computed offline. The average iteration times of our approach in the experiment are 9, and the response time of the proposed approach (exclusive the time for fetching results from Google) is imperceptible for users (usually less than 0.1s on a desktop).

6 Conclusions

In this paper, we studied how to exploit implicit feedback information and the three kinds of mutual reinforcement relationships to improve retrieval effectiveness for personalized search. Unlike most previous work, we utilize a novel graph-based iterative algorithm that can make use of query logs, immediately viewed documents, and the three kinds of reinforcement relationships (RR-Relationship, RT-Relationship, TT-Relationship) to update the query model and re-rank the search results mutually and iteratively. Experiments clearly show that our approach is both effective and efficient.

The work can be further improved on several aspects, such as exploiting other types of personalized information, studying the effects of the parameters used in the approach, etc.

Acknowledgements

This work was supported by the National Science Foundation of China (60736044, 60773027), as well as 863 Hi-Tech Research and Development Program of China (2006AA010108, 2008AA01Z145).

References

1. Anick, P.: Using terminological feedback for Web search refinement: a log-based study. In: Proceedings of WWW, pp. 89–95 (2004)
2. Bai, J., Nie, J., Bouchard, H., Cao, G.: Using Query Contexts in Information Retrieval. In: Proceedings of SIGIR, pp. 15–22 (2007)

3. Bharat, K.: SearchPad: Explicit capture of search context to support Web search. Computer Networks 33(1-6), 493–501 (2000)
4. Brin, S., Page, L.: The anatomy of a large-scale hypertextual Web search engine. In: Proceedings of WWW (1998)
5. Cao, G., Nie, J., Bai, J.: Integrating word relationships into language models. In: Proceedings of SIGIR, pp. 298–305 (2005)
6. Carbonell, J., Goldstein, J.: The use of MMR, diversity-based reranking for reordering documents and producing summaries. In: Proceedings of SIGIR, pp. 335–336 (1998)
7. Chirita, P., Nejdl, W., Paiu, R., Kohlschütter, C.: Using ODP metadata to personalize search. In: Proceedings of SIGIR, pp. 178–185 (2005)
8. Chirita, P., Firan, C., Nejdl, W.: Personalized Query Expansion for the Web. In: Proceedings of SIGIR, pp. 7–14 (2007)
9. Jarvelin, K., Kekalainen, J.: Cumulated gain-based evaluation of IR techniques. ACM Transactions on Information Systems (2002)
10. Joachims, T.: Optimizing search engines using clickthrough data. In: Proceedings of SIGKDD, pp. 133–142 (2002)
11. Kleinberg, J.M.: Authoritative sources in a hyperlinked environment. Journal of the ACM 46(5), 604–632 (1999)
12. Lv, Y., Sun, L., et al.: An Iterative Implicit Feedback Approach to Personalized Search. In: The Proceedings of the COLING/ACL, pp. 585–592 (2006)
13. Morita, M., Shinoda, Y.: Information filtering based on user behavior analysis and best match text retrieval. In: Proceedings of SIGIR, pp. 272–281 (1994)
14. Qiu, F., Cho, J.: Automatic identification of user interest for personalized search. In: WWW 2006, pp. 727–736 (2006)
15. Salton, G., Buckley, C.: Improving retrieval performance by relevance feedback. Journal of the American Society for Information Science 41(4), 288–297 (1990)
16. Shen, X., Tan, B., Zhai, C.: Implicit User Modeling for Personalized Search. In: Proceedings of CIKM, pp. 824–831 (2005)
17. Shen, X., Tan, B., Zhai, C.: Context Sensitive Information Retrieval Using Implicit Feedback. In: Proceedings of SIGIR, pp. 43–50 (2005)
18. Sugiyama, K., Hatano, K., Yoshikawa, M.: Adaptive Web search based on user profile constructed without any effort from user. In: Proceedings of WWW, pp. 675–684 (2004)
19. Tan, B., Shen, X., Zhai, C.: Mining Long-term Search History to Improve Search Accuracy. In: SIGKDD, pp. 718–723 (2006)
20. Teevan, J., Dumais, S.T., Horvitz, E.: Personalizing search via automated analysis of interests and activities. In: Proceedings of SIGIR, pp. 449–456 (2005)
21. Wan, X., Yang, J., Xiao, J.: Towards an Iterative Reinforcement Approach for Simultaneous Document Summarization and Keyword Extraction. In: Proceedings of ACL, pp. 552–559 (2007)
22. Zha, H.Y.: Generic summarization and key phrase extraction using mutual reinforcement principle and sentence clustering. In: Proceedings of SIGIR, pp. 113–120 (2002)
23. Zhang, B., Li, H., Liu, Y., Ji, L., Xi, W., Fan, W., Chen, Z., Ma, W.Y.: Improving web search results using affinity graph. In: Proceedings of SIGIR (2005)
24. Zhou, D., Weston, J., Gretton, A., Bousquet, O., Scholkopf, B.: Ranking on data manifolds. In: Advances in Neural Information Processing Systems, pp. 169–176 (2004)

Exploiting Sentence-Level Features
for Near-Duplicate Document Detection

Jenq-Haur Wang[1] and Hung-Chi Chang[2]

[1] National Taipei University of Technology, Taiwan
[2] Academia Sinica, Taiwan
jhwang@csie.ntut.edu.tw, hungchi@iis.sinica.edu.tw

Abstract. Digital documents are easy to copy. How to effectively detect possible near-duplicate copies is critical in Web search. Conventional copy detection approaches such as document fingerprinting and bag-of-word similarity target at different levels of granularity in document features, from word n-grams to whole documents. In this paper, we focus on the *mutual-inclusive* type of near-duplicates where only partial overlap among documents makes them similar. We propose using a simple and compact sentence-level feature, *the sequence of sentence lengths*, for near-duplicate copy detection. Various configurations of sentence-level and word-level algorithms are evaluated. The experimental results show that sentence-level algorithms achieved higher efficiency with comparable precision and recall rates.

Keywords: Near-duplicate, sentence-level copy detection, mutual inclusive.

1 Introduction

Digital documents are easy to copy at very low costs. With the proliferation of Web community tools such as *blog* and *wiki* that simplify Web publishing, users (or *bloggers*) can easily post their opinions on news events or topics. For common users without suitable tools and training in citations, they might simply organize their favorite articles by "copy-and-paste" part of the document and adding their own comments. From the content manager's point of view, these documents are not well-organized. First, in the absence of explicit citation links, it's not easy to find relations among documents without further analysis. Second, it's a waste of storage to host duplicate fragments in separate bloggers' articles. It would be useful if there are effective ways of identifying partially duplicated documents and issuing early warning for possible citations or plagiarism.

Conventional copy detection methods span a wide range of spectrum in terms of what we need to detect. Since exact copies among whole documents can be simply detected with checksums, most research focuses on detecting partial copies or *near-duplicates* that are almost identical. The simplest approach models the process of copying as a series of edit operations such as insertion, deletion, and substitution, and *edit distance* [9] among documents is calculated to measure their similarity. This is useful only if the difference is small. Some approaches focus on measuring document

G.G. Lee et al. (Eds.): AIRS 2009, LNCS 5839, pp. 205–217, 2009.
© Springer-Verlag Berlin Heidelberg 2009

relationships such as containment and resemblance [3] using simple sketches of documents called *shingles*, which are basically word n-grams. The more shingles two documents share, the more likely they are near-duplicates. Storing and matching all word n-grams takes too much space and time, which people try to reduce by *document fingerprinting* approaches. Hashing word n-grams can save the space, but takes extra time for hash function computation; while selecting subsets of word n-grams as the signature can reduce computation, but details might be lost. Other approaches utilize *information retrieval* techniques to identify "relevant" documents based on "bag-of-word" similarity. We are more interested in partial copies among documents, namely, the *mutual-inclusive* type of near-duplicates, as shown in Fig. 1.

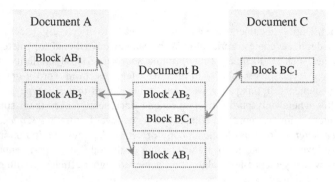

Fig. 1. The *mutual-inclusive* type of near-duplicates shows partial content sharing

As shown in Fig. 1, three documents A, B, and C are mostly different, except for the only identical parts that are shared among them, i.e., blocks AB_1 and AB_2 between documents A and B, and block BC_1 between documents B and C. These shared blocks might start at any location in different orders for different documents holding the content. In this paper, we focus on detecting this type of "near-duplicates", without differentiating the original and the copy. This definition is different from ordinary ones such as Broder [3] and Heintze [11]. First, documents could share partial content, which is not necessarily a containment relationship. Second, they are not "roughly the same" in the sense of resemblance [3] because the percentage of overlap is not large enough. They might only contain small "key" portions of overlap that make them "very similar". Third, it's also different from IR approaches using bag-of-word similarity. On the one hand, documents sharing similar words are not necessarily near-duplicates. For example, news articles talking about the same people on the same topic might relate to different events. On the other hand, near-duplicates might only share a small portion of contents with different word distributions. Finally, computing word-level similarity is time-consuming. In practice, we need an efficient way of near-duplicate detection.

To effectively identify the mutual-inclusive type of near-duplicates, the following issues have to be addressed. First, it is important to measure the degree of duplication with suitable similarity criteria. Second, the efficiency of document matching highly depends on the similarity criteria. To address these issues, we could exploit different granularity levels of document features: from characters, words, phrases, sentences, paragraphs, to the whole document. At both extremes, it's either too rough to represent the whole document as a single feature, or too detailed to count the number of different characters as in edit distance approaches. Also, matching the distributions of word frequencies as in IR techniques might not help. Users usually copy-and-paste in larger blocks than word n-grams and matching large number of word n-grams takes higher costs in space and time. Selecting subsets of word n-grams could reduce the costs, but the detection robustness might be significantly influenced. To avoid unnecessary details while capturing enough useful information for near-duplicate detection, we propose to measure the degree of duplication at sentence-level.

The major contribution of this paper is to provide a comprehensive empirical study of sentence-level detection algorithms. First, we propose a simple but effective sentence-level feature, *the sequence of sentence lengths*, for near-duplicate copy detection. Instead of comparing the sentences themselves, we only count the number of words in sentences with a sliding window. It is a very rough document overview which is independent of specific linguistic content. Also, we exploit sentence-level features in document fingerprinting techniques, and compare their effectiveness and efficiency with conventional word-level features. As the experimental results demonstrate, our proposed approach can achieve comparable precision and recall rates as the full fingerprinting and shingling methods more efficiently. Also, our approach takes significantly less time and storage for indexing. We also discuss the possible applications and limitations to our approach.

The rest of the paper is organized as follows. Section 2 lists some related works. In Section 3, the methods of the proposed approach are illustrated. Section 4 details the experiments and discussions. In Section 5, we list the conclusions.

2 Related Work

The field of document copy detection has received much attention thanks to many popular applications, for example, spam site detection using "phrase"-level replication detection techniques [10], duplicate Web page detection [12] and removal in Web search engines, versioned and plagiarized document detection [13] in digital libraries, and duplicate document removal in databases [2, 21]. Conventionally, there are several general approaches to near-duplicate document copy detection: edit distance, document fingerprinting, shingling, and bag-of-word similarity. The edit distance [9] between two documents can be measured by the minimum number of edit operations to make two documents identical. Broder [3] measured the concept of "roughly the same" and "roughly contained" by resemblance and containment metrics. Relatively small sketches or *shingles* of documents which can be computed fairly fast were proposed. The document fingerprinting approaches by Manber [15], Heintze [11], and Manku et al. [16] detected almost-identical documents with hash-value based signatures. Earlier copy detection systems such as SCAM [21] and COPS [2] are some examples. The bag-of-words approach identified similar documents according to the distribution of word frequencies. Charikar [7] developed a locality sensitive hashing approach based on random projection of words. Henzinger [12] evaluated two state-of-the-art algorithms, that is, Broder [3] and Charikar [7], for detecting near-duplicate

Web pages. I-Match [8] used collection statistics such as inverse document frequency for duplicate detection.

Recent advances showed rising interests in developing new algorithms for near-duplicate copy detection. For example, Yang and Callan [24] applied an instance-level constrained clustering approach that incorporates document attributes and content structure into the clustering process to form near-duplicate clusters. Huffman et al. [14] improved the recall while maintaining high precision by combining multiple signals among results for the same query. Xiao et al. [23] proposed new filtering techniques by exploiting the ordering information in prefix filtering. SpotSigs [22] combined stopword antecedents with short chains of adjacent content terms.

Web community applications such as blogs also raised new challenges to the conventional approaches. People can easily create more partial copies, which near-duplicate detection techniques might not directly apply. Therefore, more research focuses on "intermediate-level" or partial copy detection. For example, Metzler et al. [17], calculated sentence-level similarities by comparing the word overlap, TF-IDF, and relative frequency measures between sentences. Then, document similarity is estimated by combining sentence-level similarity. The ideas of local text reuse detection [20] and accurate discovery of co-derivative documents [1] are similar to the mutual-inclusive type of near-duplicates, but the methods and the unit of document comparison are different. It is similar to the idea of super-shingles in Broder [4], without shingling twice.

3   The Proposed Approach

To compare different features in various applications, we exploit a flexible system architecture [5, 6] as illustrated in Fig. 2. First, we can search for possible near-duplicate candidates of a given query, or within a document collection. Second, we can easily compare different similarity features. Third, it can be applied to various types of media such as images or videos if we replace the corresponding feature extraction modules. Here we focus on finding near-duplicate candidates of a given textual query document using sentence-level features.

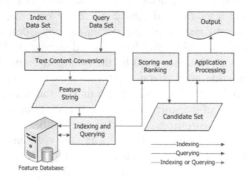

Fig. 2. System architecture of the proposed approach

There are four main functional modules in the architecture: text content conversion, indexing and querying, scoring and ranking, and application processing. First, each document in the index data set is converted into a *feature string* by the algorithm used in the *text content conversion* module. Then, *feature vectors* are extracted from the feature string by the *indexing and querying* module and stored in the feature database, which contain the signatures for identifying documents.

In a query session, the same procedure as described above is applied to the query to obtain the feature string, extract the feature vectors, and query the feature database. Then, document scores are calculated and highly-ranked documents are included in the candidate set by the *scoring and ranking* module, which are the potential near-duplicates of the query. Other post-processing steps can be further applied depending on the *application processing* module.

3.1 Sentence-Level Features

Before extracting sentence-level features from documents, the pre-defined *delimiter set* [5, 6] decides the sentence boundaries. Basically, sentences are separated by characters in the delimiter set. Since delimiters influence what sentences would be and also the quality of subsequent functions, they should be chosen carefully. After the sentence boundaries are determined, we extract two different sentence-level features: *sentence-level fingerprints* and *sentence lengths*.

Sentence-level Fingerprints (SL-FP). One of the most popular strategies for implementing fingerprinting schemes is to extract word *n*-grams as the fingerprints. For example, in the Full-Fingerprinting (Full-FP) scheme, every word *n*-gram is extracted and hashed by standard hash algorithms, such as Rabin [4] or NIST's SHA-1 [18]. Shingling methods (SG) are similar to Full-FP except that only selected *n*-grams are hashed. Instead of using word *n*-grams, sentence-level fingerprints are built with each sentence employed as the unit of decomposition. The feature string is a sequence of sentence fingerprints, and each feature vector is simply one hashed sentence.

Sentence Lengths. We extract a sequence of sentence lengths from the text as the feature string by the *sliding window* approach [5, 6]. Specifically, a fixed-length window of size *WS* slides through the feature string from which the feature vectors are extracted. The window moves forward a constant number of tokens (the step width) each time. The idea of dynamic jumping is to skip repeated patterns from consecutive windows [5]. The granularity of comparison depends on the window size and step width. On the one hand, we want to store more feature vectors to have higher chances of matching common feature vectors with near-duplicate documents. On the other hand, we want to minimize the number of feature vectors stored to save the space and time for matching. Thus, there's a tradeoff between accuracy and efficiency depending on window sizes and step widths [5].

For accuracy concerns, a larger window size matches longer substrings at a time, which will achieve higher precision but miss shorter substrings. A smaller window size will match more substrings, thus achieving higher recall rates with more false positives. With regards to efficiency, for a given window size *WS* and step width *SW*, the number of feature vectors extracted from document d_i in the data set D is:

$$N(d_i) = \left\lceil \frac{SN_{d_i} - WS + 1}{SW} \right\rceil \tag{1}$$

where SN_{di} is the number of sentences in document d_i. The space requirement for storing all documents in D is approximately proportional to:

$$Space(D) \propto \sum_{d_i \in D} \left\lceil \frac{SN_{d_i} - WS + 1}{SW} \right\rceil \times WS \approx \left\lceil \frac{WS}{SW} \right\rceil \times \sum_{d_i \in D} SN_{d_i} \tag{2}$$

In Equation (2), we can see that a smaller SW preserves more details of overlapping fragments which implies more feature vectors to be stored and higher recall rates. A smaller WS and a larger SW will produce a more compact index, which greatly reduces the comparison time as we will verify in our experiments.

3.2 Scoring and Ranking

After querying the feature database with feature vectors extracted from the query, we obtain a candidate set where each document contains at least one common feature vectors with the query. Next, we need to estimate the degree of similarity and determine the ranking among these documents. We denote the N ordered feature vectors of the query Q as:

$$V_Q = (V_{Q1} \ V_{Q2} \dots V_{QN})$$

where N is computed similarly as Equation (1) with Q in place of d_i, and V_{Qi} is the i-th feature vector. The candidate set C after searching all N feature vectors is:

$$C = \bigcup_i C_i$$

$$C_i = \{d_k : d_k \in D \text{ and } d_k \text{ contains } V_{Qi}\}$$

Let $|C_i|$ denote the number of documents that contains V_{Qi}, the vector W for the inverse document frequency of feature vectors in Q is then expressed as:

$$W = (w_1 \ w_2 \dots w_N), \text{ where } w_i = 1 / | C_i |, \text{ if } | C_i | \neq 0$$

$$w_i = 0, \text{ otherwise}$$

For each document d_k in C, we construct another binary vector:

$$D_k = (d_{k,1} \ d_{k,2} \dots d_{k,N}), \text{ where } d_{k,i} = 1, \text{ if } d_k \in C_i$$

$$d_{k,i} = 0, \text{ otherwise}$$

which indicates the co-occurrence between the document d_k and the query Q. Then the similarity score of d_k is calculated by:

$$score_{d_k} = (W \cdot D_k) \times \frac{\sum_{i=1}^{N} d_{k,i}}{\sum_{d_j \in C} \sum_{i=1}^{N} d_{j,i}} \tag{3}$$

where $(W.D_k)$ denotes the inner product of vectors W and D_k. The fractional part in Equation (3) is considered as the *match ratio* of d_k with Q, where the numerator is the *match count* between d_k and Q and the denominator is the *total match count* of all documents in C with Q. Since the score is the product of the overall feature weight and the match ratio of d_k, a document is more likely to contain duplicate content with the query if it has a higher score. The ranking among candidates can then be determined by sorting their scores. Equation (3) can be simplified by ignoring the denominator since it's the same for a given query.

4 Experiment

In our experiments, we selected six English document collections from the CLIR task in NTCIR workshop [19] as the *index data set* in Table 1.

Table 1. The index data set

Document collection	# of doc	Size (MB)
EIRB010 (Taiwan News and Chinatimes English News, 1998 - 1999)	10,204	24.6
Hong Kong Standard (1998 – 1999)	96,856	253.2
Korea Times (1998 - 1999)	19,599	55.9
Korea Times (2000 - 2001)	30,530	81.1
Mainichi Daily News (1998 – 1999)	12,723	33.3
Mainichi Daily News (2000 – 2001)	12,155	26.3
Total	182,067	474.4

Among the 182,067 documents in the *index data set*, 180 documents were randomly selected as the *query data set* which was equally divided into three subsets. For each document in the first subset, a selected number of beginning sentences (10 at most, and about 6 on average) were copied and concatenated into a single *query document*. For the second and third subsets, the middle and ending sentences were similarly copied and added respectively. The resulting query document is assumed to be the "near-duplicate" of each document in the query data set.

4.1 Data Analysis and Algorithm Configuration

With the delimiter set {. ; ! ?}, there are 12,892,057 sentences in the index data set and 70.81 sentences per document on average. The average sentence length is 5.74, which will be later used to determine the substring size for fingerprinting and shingling algorithms. The distribution of sentence lengths is illustrated in Fig. 3. Most sentences (97.46%) contain 1-20 words.

To compare the performance of sentence-level and word-level features, we evaluated the performance of the proposed approach and other word-level algorithms with various parameter configurations as follows:

Full Fingerprinting (Full-FP). Every substring of 6 words (or 6-grams) in the documents is extracted and hashed by SHA-1. The substring size 6 is determined by the average sentence length (5.74) of the index data set.

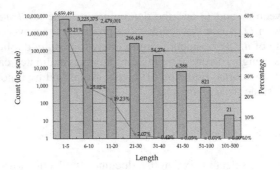

Fig. 3. The distribution of sentence lengths

Shingling (SG-63 and SG-66). It is similar to Full-FP except that one in every w 6-grams is hashed, where w is set to 3/6 for SG-63/SG-66, respectively.

Sentence Length. The algorithm as proposed in Section 3.1 with various parameter configurations $WS_iSW_j(wDJ)$ for window size i, step width j, with/without dynamic jumping. Most punctuation marks such as { , ; " . ? !} are included in the delimiter set.

Sentence-level Fingerprinting (SL-FP and SL-FP-SD). The algorithm as proposed in Section 3.1. SL-FP uses the same delimiter set as in the sentence length algorithm, while SL-FP-SD only includes { . ; ! ?} in the delimiter set.

The *collision rates* in the feature database are explored in Table 2. From indexing perspective, a *collision* occurs when documents contain an identical feature vector. The efficiency will suffer from high collision rates, since extra mechanisms are needed to handle the collisions in the database. This will be verified in Section 4.2.

Table 2. Collision rates under various algorithm configurations

Algorithm	Total number of collisions	Collision rate (%)
Full-FP	7,410,325	10.197
SG-63	1,874,838	7.701
SG-66	827,897	6.771
WS8SW1	390,021	6.837
WS8SW1wDJ	369,264	6.541
WS8SW2	189,105	6.336
WS16SW1	254,169	5.774
WS16SW1wDJ	245,607	5.650
WS16SW2	125,551	5.426
WS16SW4	**65,970**	**5.193**
SL-FP	1,438,232	21.143
SL-FP-SD	309,439	9.649

For sentence length algorithms, the collision rate for each *WS* setting is slightly reduced as *SW* increases, and it's also improved with the dynamic jumping (*DJ*) scheme enabled. Note that SL-FP has the highest collision rate among all algorithms. The reason might be having most punctuation marks in the delimiter set leads to higher chances of extracting common phrases.

4.2 Duplicate Detection

For each algorithm in Section 4.1 we apply the same feature extraction method at the indexing and query stages, except the following. At querying stage, full fingerprinting was applied to the query document for shingling (SG) algorithms, and non-jumping scheme $(SW = 1)$ was applied to sentence length algorithms. To illustrate how the symmetric procedure of feature extraction at the indexing and query stages influences the result, we evaluated another algorithm SG-63-Sym, which is identical to SG-63 except that the same shingling scheme was used at both stages.

Efficiency Measurements. We first explored the efficiency at the indexing stage for each algorithm in Table 3.

Table 3. Storage and efficiency measurements for indexing

Algorithm	Pre-processing time (sec.)	Converting time (sec.)	# of features (K)	Index size (MB)	Index time (sec.)
Full-FP	16.407	550	72,669	3,481	247,447
SG-63	17.226	241	24,344	1,185	32,350
SG-66	17.122	162	12,227	598	10,734
WS8SW1	17.638	5.336	5,705	188	2,034
WS8SW1wDJ	17.074	**4.709**	5,645	187	2,043
WS8SW2	16.410	5.536	2,984	100	474
WS16SW1	16.569	5.334	4,402	194	1,565
WS16SW1wDJ	16.713	5.119	4,347	191	1,568
WS16SW2	16.657	5.210	2,314	103	306
WS16SW4	**16.382**	5.286	**1,270**	**56**	**49**
SL-FP	17.724	299	6,802	305	4,018
SL-FP-SD	17.100	330	3,207	155	726

The pre-processing step involves HTML tag removal and punctuation mark recognition where all algorithms have similar performance. In converting documents to feature strings, the sentence length algorithm took the least time and size due to its simple computation and compact representation, while the fingerprinting algorithms used a time-consuming hash function (SHA-1). Word-level fingerprinting algorithms (Full-FP and SG) produce a larger index structure in terms of the number of features, index size, and access time than sentence-level algorithms as we expected in Section 4.1. Note that the size of index structure produced by the sentence-level fingerprinting is comparable with the sentence length algorithm. To summarize, sentence-level features, including sentence length and sentence fingerprints, are more compact and efficient than word-level features.

Effectiveness Measurements. We evaluated the effectiveness of different algorithms for searching as listed in Table 4.

As shown in Table 4, *valid result* is the number of documents in the candidate set, and *recognition* indicates how many "near-duplicate" documents can be found in the candidate set. The rank and the score of the *last correctly recognized* (denoted as *LCR*) document in the ranked list can help determine the appropriate threshold for scoring. The ratio of *recognition* to *valid result* can be regarded as the overall precision rate, whereas the ratio of *recognition* to *rank of LCR* reflects the ranking quality. For fingerprinting algorithms, although the overall precision rates are low, their ranking qualities are good. This indicates the proposed scoring function can assign appropriate weights to the candidates so that near-duplicates can be found in top ranks of the valid results. Note that SL-FP took almost 100 times longer than SL-FP-SD for searching. One possible reason could be its much higher collision rate as shown in Table 2.

Table 4. Performance for searching

Algorithm	Search time (sec.)	Valid result	Recog.	Rank of LCR	Score of LCR	Recog. / Valid result	Recog. / Rank of LCR
Full-FP	151	413	180	189	59.83	0.4358	0.9524
SG-63-Sym	1.032	222	72	133	1.00	0.3243	0.5414
SG-63	103	571	180	192	22.00	0.3152	0.9375
SG-66	1.344	524	180	193	11.00	0.3435	0.9326
WS8SW1	0.110	188	178	188	1.00	0.9468	0.9468
WS8SW1wDJ	0.125	188	178	188	1.00	0.9468	0.9468
WS8SW2	0.094	185	177	185	1.00	0.9568	0.9568
WS16SW1	0.094	146	145	146	1.00	**0.9932**	**0.9932**
WS16SW1wDJ	0.094	146	145	146	1.00	**0.9932**	**0.9932**
WS16SW2	0.094	144	142	144	1.00	0.9861	0.9861
WS16SW4	**0.062**	130	128	130	1.00	0.9846	0.9846
SL-FP	21.219	220	180	187	4.01	0.8182	0.9626
SL-FP-SD	0.203	189	180	189	1.00	0.9524	0.9524

To compare the effectiveness of each algorithm, we checked the top-k precision, recall and F_1 rates (for $k=5$ to 225, in increments of 5) using the test query. For better visual effects, partial results are shown in Fig. 4.

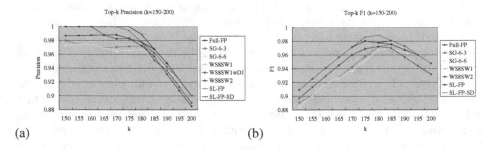

(a)   (b)

Fig. 4. The top-k precision and F_1 rates for $k=150\text{-}200$.

In Fig. 4(a), the top-k precision rates showed an earlier drop for shingling and sentence-length algorithms with larger window sizes than for sentence fingerprinting and sentence-length algorithms with smaller window sizes. All algorithms showed similar performance in top-k recall rates, which is not shown here. Most algorithms differ in top-k F_1 rates within the range of $k=150\text{-}200$, which we further explored as in Fig. 4(b). Specifically, Table 5 lists the top-180 precision, recall, and F_1 rates.

Table 5. Top-180 precision/recall/F1 rates

Algorithm	Top-180 recall	Top-180 precision	Top-180 F1
Full-FP	0.9778	0.9778	0.9778
SG-63-Sym	0.4	0.4	0.4
SG-63	0.9722	0.9722	0.9722
SG-66	0.9667	0.9667	0.9667
WS8SW1	0.9778	0.9778	0.9778
WS8SW1wDJ	0.9778	0.9778	0.9778
WS8SW2	0.9722	0.9722	0.9722
WS16SW1	0.8056	0.9797	0.8841
WS16SW1wDJ	0.8056	0.9797	0.8841
WS16SW2	0.7889	0.9861	0.8765
WS16SW4	0.7111	0.9846	0.8258
SL-FP	0.9778	0.9778	0.9778
SL-FP-SD	**0.9889**	**0.9889**	**0.9889**

As shown in Table 5, SL-FP-SD achieved the best effectiveness for searching among all algorithms. It also showed more efficient computation than word-level fin-gerprinting features for indexing in Table 3. Full-FP achieved high top-180 recall rates, but it takes much more space and time to keep details of the original contents. In sum-mary, SL-FP-SD achieves a good balance between efficiency and effectiveness.

4.3 Discussion

There are several potential applications of the proposed approach. For example, it can be used to detect quotations without explicit references on blog sites. When a blogger posts a new article, the system can check the submitted content against the document archive on the site. Another example is to detect the possible implicit links under the assumption that documents containing the same duplicate contents could be regarded relevant to each other.

There is one potential limitation to the proposed sentence-level approach: short documents are difficult to detect. Since features are extracted in sentence-level, short documents will have short sequence of features, which makes detection of possible copies more difficult. Other examples include poems or lyrics that have regular pat-terns and fixed sentence lengths. From sentence lengths alone, we cannot effectively distinguish among different documents. In this case, word-based approaches could be adopted to complement the sentence-level approach.

5   Conclusions

We proposed a sentence-level approach to the mutual-inclusive type of near-duplicate detection. A simple but effective feature, the sequence of sentence lengths, was pro-posed, which can be directly applied to languages with fixed sentence delimiters without knowledge of the specific language. We also compared sentence-level with word-level features for near-duplicate detection. The experimental results showed the potential of the effectiveness and efficiency for sentence-level features. In the future, we intend to investigate other sentence-level features for partial copy detection.

References

1. Bernstein, Y., Zobel, J.: Accurate Discovery of Co-derivative Documents via Duplicate Text Detection. Information Systems 31(7), 595–609 (2006)
2. Brin, S., Davis, J., Garcia-Molina, H.: Copy Detection Mechanisms for Digital Documents. In: The 1995 ACM International Conference on Management of Data (SIGMOD 1995), pp. 398–409 (1995)
3. Broder, A.: On the Resemblance and Containment of Documents. In: Compression and Complexity of Sequences, pp. 21–29 (1997)
4. Broder, A., Glassman, S., Manasse, M., Zweig, G.: Syntactic Clustering of the Web. In: The 6th International Conference on World Wide Web (WWW 1997), pp. 393–404 (1997)
5. Chang, H.C., Wang, J.H.: Organizing News Archives by Near-duplicate Copy Detection in Digital Libraries. In: Goh, D.H.-L., Cao, T.H., Sølvberg, I.T., Rasmussen, E. (eds.) ICADL 2007. LNCS, vol. 4822, pp. 410–419. Springer, Heidelberg (2007)
6. Chang, H.C., Wang, J.H., Chiu, C.Y.: Finding Event-Relevant Content from the Web Using a Near-duplicate Detection Approach. In: The 2007 IEEE/WIC/ACM International Conference on Web Intelligence (WI 2007), pp. 291–294 (2007)
7. Charikar, M.S.: Similarity Estimation Techniques from Rounding Algorithms. In: The 34th Annual ACM Symposium on Theory of Computing (STOC 2002), pp. 380–388 (2002)
8. Chowdhury, A., Frieder, O., Grossman, D., McCabe, M.C.: Collection Statistics for Fast Duplicate Document Detection. ACM Transactions on Information Systems (TOIS) 20(2), 171–191 (2002)
9. Damerau, F.J.: A Technique for Computer Detection and Correction of Spelling Errors. Communications of the ACM 7(3), 171–176 (1964)
10. Fetterly, D., Manasse, M., Najork, M.: Detecting Phrase-level Duplication on the World Wide Web. In: The 28th Annual International ACM SIGIR Conference on Research and Development in Information Retrieval (SIGIR 2005), pp. 170–177 (2005)
11. Heintze, N.: Scalable Document Fingerprinting. In: The 2nd USENIX Workshop on Electronic Commerce (1996)
12. Henzinger, M.: Finding Near-duplicate Web Pages: A Large-scale Evaluation of Algorithms. In: The 29th Annual International ACM SIGIR Conference on Research and Development in Information Retrieval (SIGIR 2006), pp. 284–291 (2006)
13. Hoad, T.C., Zobel, J.: Methods for Identifying Versioned and Plagiarized Documents. Journal of the American Society for Information Science and Technology 54(3), 203–215 (2003)
14. Huffman, S.B., Lehman, A.R., Stolboushkin, A.P., Wong-Toi, H., Yang, F., Roehrig, H.: Multiple-signal Duplicate Detection for Search Evaluation. In: The 30th Annual International ACM SIGIR Conference on Research and Development in Information Retrieval (SIGIR 2007), pp. 223–230 (2007)
15. Manber, U.: Finding Similar Files in a Large File System. In: USENIX Winter Technical Conference, pp. 1–10 (1994)
16. Manku, G.S., Jain, A., Sarma, A.D.: Detecting Near-duplicates for Web Crawling. In: The 16th International Conference on World Wide Web (WWW 2007), pp. 141–150 (2007)
17. Metzler, D., Bernstein, Y., Croft, W.B., Moffat, A., Zobel, J.: Similarity Measures for Tracking Information Flow. In: The 14th ACM Conference on Information and Knowledge Management (CIKM 2005), pp. 517–524 (2005)
18. NIST. Secure hash standard. Federal Information Processing Standards, FIPS 180-1 (1995)

19. NTCIR (NII Test Collection for IR Systems) project,
    http://research.nii.ac.jp/ntcir/ (accessed on January 23, 2009)
20. Seo, J., Croft, W.B.: Local Text Reuse Detection. In: The 31st Annual International ACM SIGIR Conference on Research and Development in Information Retrieval (SIGIR 2008), pp. 571–578 (2008)
21. Shivakumar, N., Garcia-Molina, H.: SCAM: A Copy Detection Mechanism for Digital Documents. In: International Conference on Theory and Practice of Digital Libraries (1995)
22. Theobald, M., Siddharth, J., Paepcke, A.: SpotSigs: Robust and Efficient Near Duplicate Detection in Large Web Collections. In: The 31st Annual International ACM SIGIR Conference on Research and Development in Information Retrieval (SIGIR 2008), pp. 563–570 (2008)
23. Xiao, C., Wang, W., Lin, X., Yu, J.X.: Efficient Similarity Joins for Near Duplicate Detection. In: The 17th International Conference on World Wide Web (WWW 2008), pp. 131–140 (2008)
24. Yang, H., Callan, J.: Near-duplicate Detection by Instance-level Constrained Clustering. In: The 29th Annual International ACM SIGIR Conference on Research and Development in Information Retrieval (SIGIR 2006), pp. 421–428 (2006)

Language Models of Collaborative Filtering

Jun Wang

Department of Computer Science, University College London
Malet Place, London, WC1E 6BT, UK
jun_wang@acm.org

Abstract. Collaborative filtering is a major technique to make personalized recommendations about information items (movies, books, webpages etc) to individual users. In the literature, a common research
objective is to predict unknown ratings of items for a user, on the condition that the user has explicitly rated a certain amount of items. Nevertheless, in many practical situations, we may only have *implicit* evidence
of user preferences, such as "playback times of a music file" or "visiting
frequency of a web-site". Most importantly, a more practical view of the
recommendation task is to directly generate a top-N ranked list of items
that the user is most likely to like.

In this paper, we take these two concerns into account. Item ranking in
recommender systems is considered as a task highly related to document
ranking in text retrieval. Firstly, two practical item scoring functions
are derived by adopting the generative language modelling approach of
text retrieval. Secondly, to address the uncertainty associated with the
score estimation, we introduce a *risk-averse* model that penalizes the
less reliable scores. Our experiments on real data sets demonstrate that
significant performance gains have been achieved.

1 Introduction

The Digital Revolution on information storage and transmission increases the
amount of information that we deal with in our daily lives. Although we enjoy
the entertainment and convenience brought to us by such a variety of sources,
the volume of information is increasing far more quickly than our ability to digest
it. For instance, the Internet has become the most significant media source and
is growing at an exponential speed. But the user ability of obtaining useful
information in the Internet grows slowly. Tools that support for the effective
retrieval of relevant information are still primitive – most information retrieval
systems heavily rely on textual queries of users to identify their information needs
[13]. Queries constructed by keywords only are, however, not powerful enough to
express the needs of a particular user both semantically and contextually. To see
this, consider the following common search scenario: "find movies showing this
weekend in nearby cinemas that I most likely to like." Such a user information
need requires the retrieval system at least to be able to capture user interest
("most likely to like"). Unfortunately, most existing retrieval models and search
technologies are incapable of achieving such a realistic retrieval goal, because

G.G. Lee et al. (Eds.): AIRS 2009, LNCS 5839, pp. 218–229, 2009.

they only focus on building the *correspondence* between textual queries and documents and lack mechanisms to model individual users who issue queries. Hence, it is essential to accurately model various user information needs beyond queries. With the recent advances in Human-Computer Interfacing and sensor technologies that make use of cameras, motion detectors, voice captures, GPSs etc., we have witnessed a research transition from information (document) centric computing into user centric computing; consider for example user profiling, that attempts to broadly understand users' various interests, intentions etc. on the basis of the recorded human-computer interactions.

Also, large amounts of information exist in a dynamic form. To process streams of incoming data, we need an information system that can play a more active role during the information seeking process. Therefore, information filtering systems arise [2]. In contrast to most retrieval systems that passively wait for user queries to respond accordingly, they aim to actively filter out, refine and systematically represent the relevant information and intuitively ignore superfluous computations on redundant data.

Combination of these two demands has created increasing interests in building a recommender system that can steer users towards their personal interests and actively filter relevant information items on the users' behalf. As one of the dominant techniques, collaborative filtering has appeared in the domain of Information Retrieval (IR) and Human-Computer Interaction (HCI) [8]. They attempt to filter information items such as books, CDs, DVDs, movies, TV programs, and electronics, based on a history of the user's likes and dislikes. Examples include the Amazon's book recommendation engine (amazon.com) and the Netflix DVD recommendation engine (netflix.com). We believe recommender systems will eventually support companies to realize a shift from offering mass products and services to offering customized goods and services that efficiently satisfy desires and needs of individual users.

In this paper, we would like to emphasize the following two crucial observations:

1. Although collaborative filtering exists in various forms in practice, its purposes can be generally regarded as "item ranking" and "rating prediction". They are illustrated in Fig 1. The rating prediction (see Fig 1 (a) and (b)) aims at predicting an unknown rating of an item for the user, with the requirement that the user has to explicitly rate a certain amount of items. This type of recommendation has been widely conceived and well studied in the research literature, since the pioneering work on the MovieLens systems (http://movielens.umn.edu); from the early work on filtering netnews [15] and the movie recommender systems ([9]) to the latest Netflix competition (http://www.netflixprize.com/), most approaches accept by default that the rating prediction is the underlying task for recommender systems. However, in many practical systems such as Amazon (http://amazon.com) and Last.Fm (http://last.fm), it is sometimes more favorable to formulate collaborative filtering as an item ranking problem, because we often face the situation where our ultimate task is to generate the top-N list of the end user's most favorite items (see Fig 1 (c) and (d)).

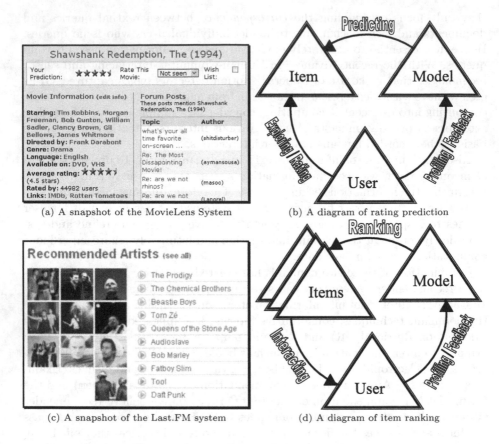

Fig. 1. The Two Forms of Recommendation

2. User profiles can be explicitly obtained by asking users to rate items that they know. However these explicit ratings are hard to gather in a real system [5]. It is highly desirable to infer user preferences from implicit observations of user interactions with a system. These implicit interest functions usually generate frequency-counted profiles, like "playback times of a music file", or "visiting frequency of a web-site" etc. So far, academic research into frequency-counted user profiles for collaborative filtering has been limited. A large body of research work for collaborative filtering by default focuses on rating-based user profiles [1,9,10,16,21].

This motivated us to conduct a formal study on probabilistic item ranking for collaborative filtering. The remainder of the paper is organized as follows. We first describe related work, and then establish the generative language model for collaborative filtering. After that, we extend the model by considering the uncertainty of the estimation. Finally, we provide an empirical evaluation of the recommendation performance, and conclude our work.

2 Related Work

In the memory-based approaches, all rating examples are stored *as-is* into memory (in contrast to learning an abstraction), forming a heuristic implementation of the "Word of Mouth" phenomenon. In the rating prediction phase, similar users or (and) items are sorted based on the memorized ratings. Relying on the ratings of these similar users or (and) items, a prediction of an item rating for a test user can be generated. Examples of memory-based collaborative filtering include user-based methods [3,9,15], item-based methods [7,16] and unified methods [19]. The advantage of the memory-based methods over their model-based alternatives is that less parameters have to be tuned; however, the data sparsity problem is not handled in a principled manner.

In the model-based approaches, training examples are used to generate an "abstraction" (model) that is able to predict the ratings for items that a test user has not rated before. In this regard, many probabilistic models have been proposed. For example, to consider user correlation, [14] proposed a method called personality diagnosis (PD), treating each user as a separate cluster and assuming a Gaussian noise applied to all ratings. It computes the probability that a test user is of the same "personality type" as other users and, in turn, the probability of his or her rating to a test item can be predicted. On the other hand, to model item correlation, [3] utilizes a Bayesian Network model, in which the conditional probabilities between items are maintained. Some researchers have tried mixture models, explicitly assuming some hidden variables embedded in the rating data. Examples include the aspect models [10,12], the cluster model [3] and the latent factor model [4]. These methods require some assumptions about the underlying data structures and the resulting 'compact' models solve the data sparsity problem to a certain extent. However, the need to tune an often significant number of parameters has prevented these methods from practical usage. For instance, in the aspect models [10,12], an EM iteration (called "fold-in") is usually required to find both the hidden user clusters or/and hidden item clusters for any new user.

Memory-based approaches are commonly used for rating prediction, but they can be easily extended for the purpose of item ranking. For instance, a ranking score for a target item can be calculated by a summation over its similarity towards other items that the target user liked (i.e. in the user preference list). Taking this item-based view, we formally have the following basic ranking score:

$$o_u(i) \equiv \sum_{i' \in L_u} s_I(i', i) \tag{1}$$

where u and i denote the target user and item respectively, and $i' \in L_u$ denotes any item in the preference list of user u. S_I is the similarity measure between two items, and in practice cosine similarity and Pearson's correlation are generally employed. To specifically target the item ranking problem, researchers in [7] proposed an alternative, TFxIDF-like similarity measure, which is shown as follows:

$$s_I(i', i) = \frac{Freq(i', i)}{Freq(i') \times Freq(i)^\alpha} \qquad (2)$$

where $Freq$ denotes the frequency counts of an item $Freq(i')$ or co-occurrence counts for two items $Freq(i', i)$. α is a free parameter, taking a value between 0 and 1. On the basis of empirical observations, they also introduced two normalization methods to further improve the ranking. In our previous work, we have introduced the concept of relevance into collaborative filtering [17]. Items can be then ranked by estimating the probability of the relevance between users (preferences) and items [18,20]. In this paper, we take another angle, considering a generative process between items and users.

3   A Statistic Language Model for Collaborative Filtering

Collaborative filtering aims at finding information items that a user is most likely to like, given his or her preference. To achieve this, we could formally measure how probable an item (denoted as i) is to be suggested to a given user (denoted as u): $p(i|u)$, and then rank items accordingly:

$$\begin{aligned}
o_u(i) \equiv p(i|u) &= \frac{p(u|i)p(i)}{p(u)} \\
&\propto_i \log p(u|i) + \log p(i) - \log p(u) \\
&\propto_i \log p(u|i) + \log p(i)
\end{aligned} \qquad (3)$$

where $\log p(u)$ can be removed since it is independent of the target item i. The item ranking has two parts: its likelihood towards the user preference $p(u|i)$ and its popularity $p(i)$. The probability $p(i)$ can be easily estimated by counting the frequency from the collection.

To estimate the likelihood $p(u|i)$, we follow the argument of the language model of information retrieval. In the language modelling approach of information retrieval [6], one needs to assess how probable a query q would be generated from a document *language model* θ_d, and then rank each of the documents d in the collection on the basis of the generative probability $p(q|\theta_d)$. Similarly, in collaborative filtering, we first choose an *optimal* generative model θ_i for each candidate item i; it captures the underlying distribution of users (or user preferences) who liked the item. Probability $p(u|\theta_i)$ is then used to estimate how probable a user preference (as a query) is to be generated by that model. Replacing it into Eq. (3) gives

$$p(i|u) \propto_i \log p(u|\theta_i) + \log p(i) \qquad (4)$$

By doing this, we relate the language modelling of text retrieval and the collaborative filtering modelling at a probabilistic level. Yet, at the feature representation level they are quite apart from each other, as their input data and purposes are completely different. Consequently, applying the text retrieval model to collaborative filtering is not trivial. The difficulty lies in the fact that in text retrieval both queries and documents are represented by texts, which provide an important information channel to link queries (user needs) and documents. Due to the lack of relevance observations, the language models in text retrieval shift their focus from directly estimating the correspondence (relevance) between user needs (queries) and documents to estimating word statistics in the documents and/or queries and then building up the link through these statistics. Conversely, in collaborative filtering, in most cases, we do not have such extra information to relate user preferences and information items. Instead, recorded in the system are only user preferences, which are thought of as indirect observations of the relevance between a user interest and an information item. Thus, the central question in modelling collaborative filtering is how to relate users and items through this usually very sparse user-item matrix, where its elements record the frequency counts, like "playback times of a music file", or "visiting frequency of a web-site" etc.

The estimation of the likelihood $p(u|\theta_i)$ depends on the representation of the user preference. From the data stored in the user-item matrix, if we use a set of items $i' \in L_u$ to present user u, and assume that each item i' in the user preference L_u is independently generated, we have

$$p(i|u) \propto_i \sum_{i' \in L_u} \log p(i'|\theta_i) + \log p(i) \tag{5}$$

In text retrieval, the interpretation of the likelihood function $p(t|\theta_d)$ is relatively straightforward as both queries and documents are represented by the same set of features, i.e., words. Subsequently, θ_d is estimated conveniently by looking at the words occur in document d. By contrast, the interpretation of the likelihood $p(i'|\theta_i)$ in Eq. (5), which links the target item i to another item i' in the target user's preference, is slightly different; it measures how probable an item i would be generated from a user preference where an item i occurs. It is estimated by considering the following two steps: 1) aggregating the user preferences in which item i occurs, and 2) from them, calculating how frequent item i' is also present - the Maximum Likelihood Estimate (MLE) would be $\hat{\theta}_i = p(i'|\hat{\theta}_i) \equiv \frac{c(i',i)}{c(i)}$, where $c(i',i)$ denotes the number of user preferences where both items i and i' occur, and $c(i)$ denotes the number of user preferences where item i occurs. The hat on $\hat{\theta}_i$ indicates that it is an estimated value.

Like text retrieval, due to the sparsity of the data, only considering the co-occurrence statistics is unreliable. One can smooth the estimate from the collection statistics; using the linear smoothing method [22], we have the following ranking formula:

$$o_u(i) \equiv \sum_{i' \in L_u} \ln \left(\lambda P(i'|i) + (1-\lambda)P(i') \right) + \ln P(i)$$

$$\equiv \sum_{i' \in L_u} \ln \left(\lambda \frac{c(i',i)}{c(i)} + (1-\lambda) \frac{c(i')}{\sum_{i'} c(i')} \right) + \ln \frac{c(i)}{\sum_{i'} c(i')} \tag{6}$$

where the ranking score of a target item i is essentially a combination of its popularity (expressed by the prior probability $P(i)$) and its co-occurrence with the items $i' \in L_u$ in the preference list of the target user (expressed by the conditional probability $P(i'|i)$. $\lambda \in [0, 1]$ is used as a linear smoothing parameter to further smooth the conditional probability from a background model ($P(i')$). $\sum_{i'} c(i') = \sum_i c(i)$ denotes the number of user preferences in the collection.

Alternatively, one can apply the Bayes-smoothing technique [22] to smooth the estimation. More formally, we have:

$$\hat{\theta}_i = p(i'|\hat{\theta}_i) \equiv \frac{c(i', i) + \mu \cdot p(i')}{c(i) + \mu} \tag{7}$$

where μ is the smoothing parameter and $p(i') \equiv \frac{c(i')}{\sum_{i'} c(i')}$. Replacing Eq. (7) into Eq. (5) results in the following Bayes-smoothing-based ranking formula:

$$o_u(i) \equiv \sum_{i' \in L_u} \ln\left(\frac{c(i', i) + \mu \cdot \frac{c(i')}{\sum_{i'} c(i')}}{c(i) + \mu}\right) + \ln\frac{c(i)}{\sum_i c(i)} \tag{8}$$

In summary, we have derived two ranking formulae in Eq. (6) and Eq. (8), respectively, by following the school of thinking in the language modelling approaches of text retrieval. The two scoring functions are item-based as the scoring relies on the co-occurrence statistics between items. It is worth noticing that a parallel user-based method can similarly be derived by considering the co-occurrence between users.

4   Risk-Aware Ranking for Collaborative Filtering

As described, the classic language modelling approaches, thus including Eq. (6) and Eq. (8), consider the model parameters as unknown fixed constants, and apply point estimate such as the MLE, the linear-smoothing, or the Bayes-smoothing technique. The main drawback of this approach is that exact measures of the uncertainty associated with the estimation are not handled in a principled manner. As a result, unreliably-estimated items may be ranked highly in the ranked list, reducing the retrieval performance of the top-N returned items.

To model the uncertainty of the estimate, we follow the Bayesian viewpoint, considering parameter θ_i itself has a probability distribution associated with it. We propose to use variance $Var(\theta_i)$ to summarize the uncertainty, inspired by the risk-aware language models introduced in [23]. A large variance indicates that the estimate is unreliable and its rank score should be penalized accordingly. Based on this, we have the following formula:

$$\hat{\theta}_i \equiv Mean(\theta_i) - \frac{b}{2}Var(\theta_i) \tag{9}$$

where $Mean(\theta_i)$ is the mean of θ_i while $Var(\theta_i)$ denotes its variance. $b > 1$, and it is a parameter that adjusts the risk preference and can be tuned from data.

Table 1. Comparison with the other approaches. Precision is reported in the Last.FM data set. The best results are in bold type. A Wilcoxon signed-rank test is conducted and the significant ones (P-value < 0.05) over SuggestLib are marked as *.

	Top-1	Top-3	Top-10
LM-LS	0.572	0.507	0.416
LM-BS	**0.585***	**0.535***	**0.456***
SuggestLib	0.547	0.509	0.421

(a) User Profile Length 5

	Top-1	Top-3	Top-10
LM-LS	0.673	0.617	**0.517***
LM-BS	**0.674***	**0.620***	**0.517***
SuggestLib	0.664	0.604	0.503

(b) User Profile Length 10

	Top-1	Top-3	Top-10
LM-LS	0.669	0.645	0.555
LM-BS	**0.761***	**0.684***	**0.568***
SuggestLib	0.736	0.665	0.553

(c) User Profile Length 15

Here the user preference data is assumed to follow Multinomial distribution, and the conjugate prior is Dirichlet distribution. Thus, the mean and variance are obtained as follows:

$$Mean(\theta_i) = \frac{c(i',i)}{c(i)}, Var(\theta_i) = \frac{c(i',i)\Big(c(i) - c(i',i)\Big)}{c(i)^2\Big(c(i)+1\Big)} \tag{10}$$

5   Experiments

5.1   Data Sets and Experiment Protocols

The standard data sets used in the evaluation of collaborative filtering algorithms (i.e. MovieLens and Netflix) are rating-based, which are not suitable for testing our method using implicit, frequency-counted user profiles. This paper adopts two implicit user profile data sets.

The first data set comes from a well known social music web site: Last.FM. It was collected from the play-lists of the users in the community by using a plug-in in the users' media players (for instance, Winamp, iTunes, XMMS etc). Plug-ins send the title (song name and artist name) of every song users play to the Last.FM server, which updates the user's musical profile with the new song. For our experiments, the triple {userID, artistID, Freq} is used.

The second data set was collected from one well-known collaborative tagging Web site, del.icio.us. Unlike other studies focusing on directly recommending contents (Web sites), here we intend to find relevance tags on the basis of user profiles as this is a crucial step in such systems. For instance, the tag suggestion is

Table 2. Comparison with the other approaches. Precision is reported in the Del.icio.us data set. The best results are in bold type. A Wilcoxon signed-rank test is conducted and the significant ones (P-value < 0.05) over SuggestLib are marked as *.

	Top-1	Top-3	Top-10
LM-LS	**0.306***	**0.253***	**0.208***
LM-BS	0.253	0.227	0.173
SuggestLib	0.168	0.141	0.107

(a) User Profile Length 5

	Top-1	Top-3	Top-10
LM-LS	**0.325***	**0.256***	**0.207***
LM-BS	0.248	0.226	0.175
SuggestLib	0.224	0.199	0.150

(b) User Profile Length 10

	Top-1	Top-3	Top-10
LM-LS	**0.322***	**0.261***	**0.211***
LM-BS	0.256	0.231	0.177
SuggestLib	0.271	0.230	0.171

(c) User Profile Length 15

needed in helping users assigning tags to new contents, and it is also useful when constructing a personalized "tag cloud" for the purpose of exploratory search . The Web site has been crawled between May and October 2006. We collected a number of the most popular tags, found which users were using these tags, and then downloaded the whole profiles of these users. We extracted the triples {userID, tagID, Freq} from each of the user profiles. User IDs are randomly generated to keep the users anonymous.

For 5-fold cross-validation, we randomly divided this data set into a training set (80% of the users) and a test set (20% of the users). Results are obtains by averaging 5 different runs (sampling of training/test set). The training set was used to estimate the model. The test set was used for evaluating the accuracy of the recommendations on the new users, whose user profiles are not in the training set. For each test user, 5, 10, or 15 items of a test user were put into the user profile list. The remaining items were used to test the recommendations. Our experiments here consider the *recommendation precision*, which measures the proportion of recommended items that are ground truth items.

5.2 Performance

The Language Models. We choose a state-of-the-art item ranking algorithm [7] discussed in Section 2 as our strong baseline. We adopt their implementation, the top-N suggest recommendation library[1], which is denoted as SuggestLib. The proposed language modelling approach of collaborative filtering in Eq. (6) is denoted as LM-LS while its variant using the Bayes' smoothing given in Eq. (8) is denoted as LM-BS. The optimal parameters are tuned by applying cross-validation.

---

[1] http://glaros.dtc.umn.edu/gkhome/suggest/overview

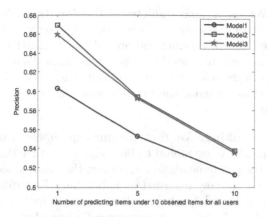

Fig. 2. Recommendation Precision in Last.FM data set. Model 1: the linear smoothing method; Model 2: risk-aware linear smoothing method; and Model 3: risk-aware Maximum Likelihood Estimate method.

Table 3. Recommendation Precision in Last.FM. Model 1: the linear smoothing method; Model 2: risk-aware linear smoothing model ; and Model 3: risk-aware Maximum Likelihood Estimate model. A Wilcoxon signed-rank test is conducted and the significant ones (P-value < 0.05) over Model 1 the linear smoothing model are marked as *.

Observed items	Method	Parameter	Top-10	Top-5	Top-1
5	①	λ=0.9949	0.4696	0.5081	0.5988
	②	λ=0.989 b=22.9	0.4888*	0.5281*	0.6279*
	③	b =130.8	0.4786	0.5243	0.5925
10	①	λ=0.99718	0.5131	0.5530	0.6029
	②	λ=0.9969 b=16	0.5378*	0.5938*	0.6694*
	③	b =170.8	0.5356	0.5925	0.6590
15	①	λ=0.998363	0.5632	0.6146	0.6757
	②	λ=0.99712 b=17.01	0.5969*	0.6674*	0.7526*
	③	b =190.8	0.5784	0.6482	0.7339

The results are shown in Table 1 and 2. From the tables, we can see that our ranking methods, derived from the language models of text retrieval, performs consistently better than the heuristic ranking method, SuggestLib, over all the configurations. A Wilcoxon signed-rank test [11] is done to verify the significance. We believe that the effectiveness of our methods is due to the fact that the models naturally integrate frequency counts and probability estimation into the ranking formula. For the two smoothing methods, we have obtained a mixed result - in the Last.FM data, the Bayes smoothing method outperforms the linear smoothing, while in the `del.icio.us` data, the linear smoothing is better than the Bayes smoothing method.

The Risk-aware Ranking. We continue our experiment with the risk-aware model given in Eq. (9). As we intend to investigate whether the added variance bit could improve the recommendation accuracy, the linear smoothing approach (denoted as Model 1) is now regarded as a baseline, Two different risk-aware models are evaluated: Model 2, using the linear smoothing to estimate the mean ($Mean(\theta_i)$), and Model 3, using the maximum likelihood to estimate the mean ($Mean(\theta_i)$). The results, under the three configurations top-1, top-10, and top-15, are shown in Fig. 2 and Table 3. We can see that Model 2 and Model 3 significantly outperform Model 1 in all configurations. Model 2 is slightly better than Model 3, implying that, the variance plays a more critical role than the smoothing from the collection. And, even without smoothing from the collection, the risk-model that considers the variance only provides a robust scoring function.

6   Conclusions

In this paper, we have presented a novel statistic model for item ranking in collaborative filtering. It is inspired by the widely-adopted language models of text retrieval. To consider the uncertainty of the parameter estimation and reflect it during the ranking, we then presented a risk-averse ranking model by considering the variance of the parameters. The experiments on two real data sets have shown that the significance of our approaches.

One of the assumptions in our model is that the items in users' profiles are independent of each other. This is unrealistic in practice. In the future, we intend to explore this dependence. It is also of great interest to study the method of combing content descriptions under the proposed framework.

Acknowledgement

The experiment of the risk-aware model was conducted by Mofei Han when he was working on his MSc thesis under the supervision of the author.

References

1. Adomavicius, G., Tuzhilin, A.: Toward the next generation of recommender systems: A survey of the state-of-the-art and possible extensions. IEEE Transactions on Knowledge and Data Engineering 17(6), 734–749 (2005)
2. Belkin, N.J., Croft, W.B.: Information filtering and information retrieval: two sides of the same coin? Commun. ACM 35(12), 29–38 (1992)

3. Breese, J., Heckerman, D., Kadie, C.: Empirical analysis of predictive algorithms for collaborative filtering. In: Proceedings of the 14th Annual Conference on Uncertainty in Artificial Intelligence, UAI 1998 (1998)
4. Canny, J.: Collaborative filtering with privacy via factor analysis. In: SIGIR 2002 (2002)
5. Claypool, M., Le, P., Wased, M., Brown, D.: Implicit interest indicators. In: IUI 2001 (2001)
6. Croft, B.W., Lafferty, J.: Language Modeling for Information Retrieval. Springer, Heidelberg (2003)
7. Deshpande, M., Karypis, G.: Item-based top-N recommendation algorithms. ACM Trans. Inf. Syst. 22(1), 143–177 (2004)
8. Goldberg, D., Nichols, D., Oki, B.M., Terry, D.: Using collaborative filtering to weave an information tapestry. Commun. ACM 35(12), 61–70 (1992)
9. Herlocker, J.L., Konstan, J.A., Borchers, A., Riedl, J.: An algorithmic framework for performing collaborative filtering. In: SIGIR 1999 (1999)
10. Hofmann, T.: Latent semantic models for collaborative filtering. ACM Trans. Info. Syst. 22(1), 89–115 (2004)
11. Hull, D.: Using statistical testing in the evaluation of retrieval experiments. In: SIGIR 1993 (1993)
12. Jin, R., Si, L., Zhai, C.: A study of mixture models for collaborative filtering. Inf. Retr. 9(3), 357–382 (2006)
13. Manning, C.D., Raghavan, P., Schtze, H.: Introduction to Information Retrieval. Cambridge University Press, Cambridge (2008)
14. Pennock, D.M., Horvitz, E., Lawrence, S., Giles, C.L.: Collaborative filtering by personality diagnosis: A hybrid memory and model-based approach. In: UAI 2000 (2000)
15. Resnick, P., Iacovou, N., Suchak, M., Bergstrom, P., Riedl, J.: Grouplens: an open architecture for collaborative filtering of netnews. In: CSCW 1994 (1994)
16. Sarwar, B., Karypis, G., Konstan, J., Reidl, J.: Item-based collaborative filtering recommendation algorithms. In: WWW 2001 (2001)
17. Wang, J.: Relevance Models for Collaborative Filtering. Delft University of Technology (2008),
http://web4.cs.ucl.ac.uk/staff/jun.wang/papers/phdthesis.pdf, ISBN 978-90-9022932-4
18. Wang, J., de Vries, A.P., Reinders, M.J.: A user-item relevance model for log-based collaborative filtering. In: Lalmas, M., MacFarlane, A., Rüger, S.M., Tombros, A., Tsikrika, T., Yavlinsky, A. (eds.) ECIR 2006. LNCS, vol. 3936, pp. 37–48. Springer, Heidelberg (2006)
19. Wang, J., de Vries, A.P., Reinders, M.J.: Unified relevance models for rating prediction in collaborative filtering. ACM Trans. on Information System, TOIS (2008)
20. Wang, J., Roberston, S.E., de Vries, A.P., Reinders, M.J.T.: Probabilistic relevance models for collaborative filtering. Journal of Information Retrieval (2008)
21. Xue, G.-R., Lin, C., Yang, Q., Xi, W., Zeng, H.-J., Yu, Y., Chen, Z.: Scalable collaborative filtering using cluster-based smoothing. In: SIGIR 2005 (2005)
22. Zhai, C., Lafferty, J.: A study of smoothing methods for language models applied to ad hoc information retrieval. In: SIGIR 2001 (2001)
23. Zhu, J., Wang, J., Taylor, M., Cox, I.: Risky business: Modeling and exploiting uncertainty in information retrieval. In: SIGIR 2009 (2009)

Efficient Text Classification Using Term Projection

Yabin Zheng, Zhiyuan Liu, Shaohua Teng, and Maosong Sun

State Key Laboratory on Intelligent Technology and Systems,
Tsinghua National Laboratory for Information Science and Technology,
Department of Computer Science and Technology, Tsinghua University,
Beijing 100084, China
yabin.zheng@gmail.com, liuliudong@gmail.com,
tengshaohua@gmail.com, sms@mail.tsinghua.edu.cn

Abstract. In this paper, we propose an efficient text classification method using term projection. Firstly, we use a modified χ^2 statistic to project terms into predefined categories, which is more efficient compared to other clustering methods. Afterwards, we utilize the generated clusters as features to represent the documents. The classification is then performed in a rule-based manner or via SVM. Experiment results show that our modified χ^2 statistic feature selection method outperforms traditional χ^2 statistic especially at lower dimensionalities. And our method is also more efficient than Latent Semantic Analysis (LSA) on homogeneous dataset. Meanwhile, we can reduce the feature dimensionality by three orders of magnitude to save training and testing cost, and maintain comparable accuracy. Moreover, we could use a small training set to gain an approximately 4.3% improvement on heterogeneous dataset as compared to traditional method, which indicates that our method has better generalization capability.

Keywords: Text classification, χ^2 statistic, Term projection, Cluster-based classification.

1 Introduction

Text classification [1, 2] is a fundamental task in text mining research area. The goal of text classification is to assign documents with pre-defined semantically meaningful labels. Traditional text classification methods always follow a supervised learning strategy: use some labeled training set and machine learning technologies, like Naïve Bayesian, KNN, and SVM [3, 4] to build a model, and then classify the documents in test set. In general, those algorithms have demonstrated reasonable performance.

Standard representation of text uses bag-of-word (BOW) model, with each term corresponds to a dimension. It is obvious that BOW representation will bring sparse and noisy problems, especially when the training set is relatively small. Moreover, this will also lead to curse of dimensionality issue, which is tough and common in text classification.

A sophisticated methodology to reduce feature dimensionality is feature selection [5], such as χ^2 statistic, mutual information and information gain. In [6], they show that χ^2 statistic has better performance on Chinese text dataset when the dimensionality is

G.G. Lee et al. (Eds.): AIRS 2009, LNCS 5839, pp. 230–241, 2009.

relatively high. We did some modification on χ^2 statistic feature selection method. After feature selection, the next procedure is cluster-based text classification. The goal is to group the similar terms into clusters, in [7, 8, 9, 10], they do term clustering according to the distribution of terms with different clustering algorithm applied. Then they utilize the generated clusters as features to represent the documents for classification.

In this paper, we follow the similar procedure described above. First, we found that traditional χ^2 statistic method doesn't take term frequency into account. However, we argue that term frequency indeed shows relationship between terms and categories. The more a term emerges in a corresponding category, the stronger their relationship is. With this tiny modification, we get better performance on both English and Chinese dataset on varied dimensionality.

Second, we do term projection according to the modified χ^2 statistics, which can be considered as a rule-based clustering algorithm. The advantage of our projection method is that no additional computational cost is needed. In [7, 8, 9, 10], they have tried diverse term clustering algorithms, which bring different degrees of computational costs. Moreover, they have to determine the number of clustering result, which is difficult for different datasets.

After term projection step, we utilize the generated clusters as features to represent the documents. The benefit of using clusters as features include: (a) make most use of semantic meanings of terms. We can group similar terms into the same cluster, and condense the feature space to reduce the sparse problem to a certain extent, (b) efficient classification speed. We can reduce the feature dimensionality by three orders of magnitude, from 60,000 to 55 in Chinese dataset, while still and maintain comparable accuracy as compared to LSA. Besides, the classification speed can be greatly accelerated as a result of much smaller feature size, (c) small and better generalization classification model. With the feature dimensionality greatly reduced, we need less parameter to determine the model. Furthermore, experiment result shows that this model also has better generalization capability, (d) a complement to feature selection. Feature selection aims at removing noisy features, while term clustering is good at decreasing redundant features by putting them together. In practice, we generally do feature selection first to keep meaningful features, and then condense the feature space by clustering.

Our contributions in this paper include: (a) we modify the traditional χ^2 statistic, to the best of our knowledge, no one has take term frequency into consideration when using χ^2 statistic to do feature selection. This modification improves the performance, especially at lower dimensionalities, (b) rule-based term projection algorithm, we project the terms according to the modified χ^2 statistics, which is quite efficient and practical, (c) we reduce the feature dimensionality by three orders of magnitude, and still maintain comparable accuracy in comparison with LSA on homogeneous dataset. This indicates that our method require less computational cost to achieve the same performance, (d) we have observed some improvement both on classification accuracy and speed on heterogeneous dataset using a small training corpus.

The rest of the paper is organized as follows. In Section 2, we will review some related works that using term clustering technology for text classification as well as transfer learning that used to solve the heterogeneous dataset problem. Then we introduce our term projection method in section 3. After that, projection based classification algorithms are discussed in section 4, using a rule-based manner or via SVM. Experiment results and discussions are shows in section 5. Section 6 concludes the whole paper and gives some future works.

2 Related Work

Pereira *et al.* [11] firstly proposed the distributional clustering scheme of English words in 1993, followed by a group of researchers in text classification area [7, 8, 9, 10], to establish a more sophisticated text representation than bag-of-words model via term clusters.

Baker and McCallum [7] apply term clustering according to the distributions of class labels associated with them. Then use these learned clusters to represent the documents in a new reduced feature space. They get only a slight decrease in accuracy; however, the cluster-based representation is significantly more efficient than BOW model.

Bekkerman *et al.* [8, 9] follow the similar idea. They introduced a new information bottleneck method to generate cluster-based representations of documents. What's more, combined with SVM, their experiment result outperforms other methods in both accuracy and efficiency. The shortcoming of all mentioned work is that they spend extra computational cost on term clustering, and some parameters like number of clusters should be determined. Unfortunately, this is always tough for different applications. In this paper, we use modified χ^2 statistic to do term projection, which can be obtained straightforwardly from the feature selection step, with no additional computational cost introduced.

On the other hand, traditional text classification strategies always make a basic assumption: the training and test set are sampling from the same distribution. However, this assumption may be violated in reality. For example, it is not reasonable to assume that web-pages on the internet are homogeneous because they change frequently. New terms emerge; old terms disappear; identical terms have different meanings. Recently, transfer learning [12, 13] is designed to solve this problem. Transfer learning is the application of skills and knowledge learned in one context being applied in another context. In this paper, we make use of cluster-based representations of documents to alleviate this heterogeneous problem. We gain improvement on both classification accuracy and speed, which give evidence of better generalization ability of our method.

3 Term Projection

In this section, we will introduce our term projection algorithm, which is significantly more efficient than other clustering algorithms. First, we present the modified χ^2 statistic formula, with term frequency taken into consideration. We also give some explanation and benefit of doing this. We also use the modified χ^2 statistic to do feature selection in the following experiment. Second, we straightforwardly utilize modified χ^2 statistic to do term projection, which is quite efficient.

3.1 Modified χ^2 Statistic

Yang *et al.* [5] has investigated several feature selections for text classification. They found that information gain and χ^2 statistic is most effective on English text dataset among five feature selection methods. In fact, χ^2 statistic measures the lack of independence between term t and class label c. We first review the traditional χ^2 statistic formula as follows.

Using a two-way contingency table of term t and class label c, we can found four elements in the table, where A is the number of times that both t and c occur, B is the number of times that only t occurs, C is the number of times that only c occurs, D is the number of times that neither c nor t occurs. N is the number of documents in the training set. Statistics are performed at document level. The formula is defined to be:

$$\chi^2(t,c) = \frac{N \times (AD - CB)^2}{(A+C) \times (B+D) \times (A+B) \times (C+D)}$$
(1)

Ideally, t and c always occur or disappear together, which means that t and c have a strong relationship. Then once a document contains term t, maybe we are confident enough to classify it to class c. On the other hand, if t and c are completely independent, then we get a value of zero in formula (1). We computed the χ^2 statistic between a particular term t and all the class labels in the training set. We use the maximum value to represent the final score of term t, which is known as χ^2_{max}. The formula is defined as: (m is the number of classes in the training set)

$$\chi^2_{max}(t) = \max_{1 \leq i \leq m} \chi^2(t, c_i)$$
(2)

We also record the χ^2_{total} value for future use, which is defined as:

$$\chi^2_{total}(t) = \sum_{i=1}^{m} \chi^2(t, c_i)$$
(3)

We argue that traditional χ^2 statistic ignores the term frequency information. For example, a document d which belongs to class c contains two terms, $t1$ and $t2$. Suppose that $t1$ occurs only one time in d, while $t2$ occurs 1000 times in d, which is much more frequent than $t1$. It is reasonable to consider that $t2$ has a stronger relationship with c than $t1$ has. But this term frequency information is not revealed in formula (1). Traditional χ^2 statistic makes statistics at the document level, which assumes $t1$ and $t2$ have the same relation with c.

We perform statistic at the term level to consider the missing term frequency information discussed above. Use the same annotations before; now A is the total frequency of t occurs in class c, B is the total frequency of t occurs in other classes except c, C is the total frequencies of other terms occur in c, D is the total frequencies of all terms (term t not included) outside class c, N is the total frequency of all terms in training set. Experiment results on both English and Chinese datasets show that our modified χ^2 statistic has better performance, especially at lower dimensionalities. Actually, we obtain 18.4% improvement on Chinese dataset at 200 dimensions.

3.2 Projection by χ^2 Statistic

As stated above, we use modified χ^2 statistic and χ^2_{max} to do feature selection. In traditional methods, they sort the terms according to their χ^2_{max} values, and choose top T candidates. However, χ^2_{max} values have natural semantic meanings. For example, in our experiment dataset, the term "导演(director)" gains χ^2_{max} in class "电影(movie)", which gives evidence that "导演" has the strongest semantic relationship with "电影" among all the classes. So, we have every reason to use χ^2_{max} information to do semantic term projection.

Furthermore, we record the χ^2_{max} values of each term and the corresponding class that it gains χ^2_{max} value. Then we make semantic matching between terms and classes, which can be considered as term projection (clustering) procedure straightforwardly after feature selection. Similar terms are projected to the same cluster.

The benefit of our proposed projection method is: (a) make most use of semantic meanings of the dataset. Dataset is usually labeled by human, which is in high quality. We use exactly the same taxonomy of training set to do projection, (b) no clustering parameters introduced. Other clustering algorithms always require extra parameters, such as clustering numbers, iteration convergence control parameters. It is always difficult to set those extra parameters for various dataset, (c) our method is significantly more efficient. Unlike other clustering algorithms, all those projections are generated straightforwardly after previous feature selection step without any extra computational cost brought about.

Meanwhile, we do some post-processing jobs to reduce noise. Certain terms may appear uniformly in classes, such as some stop words. It can be projected to every class on different datasets. Consider classifying documents into classes by individual sport (like basketball, football, volleyball). It is suitable to project term "coacher" to either class. To solve this problem, we only keep the terms whose χ^2_{max} value makes up at least $\lambda\%$ of their χ^2_{total} value. λ is set as 50 in this paper.

4 Cluster-Based Classification

In this section, we will show how to utilize the generated clusters as features to represent the documents. We reduce the feature dimensionality by three orders of magnitude using this more sophisticated text representation. The direct benefit of this approach is that much more efficient classification speed. For practical applications, we always desire splendid processing speed as the documents on the internet accumulate exponentially. Besides, we also gain better generalization performance using this representation on heterogeneous dataset.

In subsection 4.2, we proposed two strategies to do classification task using cluster-based representation. The former performs in a rule-based manner, which classifies the document based on the values on individual cluster features; the latter takes advantage of classification power of SVM. In our experiment, the latter method achieves better performance with a little more computational cost. While in practice, we are free to choose either method under different situation.

4.1 Cluster-Based Representation

First, we will introduce how to use clusters to represent documents in detail. As discussed before, this representation is more sophisticated than traditional BOW model with semantic meanings of terms taken into consideration. Similar terms are projected to the identical cluster. We project the documents from previous feature space to the newly created feature space, in which each dimension corresponds to a cluster.

Then, we elaborate the concept of discriminability and a straight metric of this concept [14]. Discriminability measures how unbalanced is the distribution of term among the classes. A term is said to be have high discriminability if it appears significantly more frequent in one class c than others. Once this term appears for quite a lot of times in one document, it is reasonable to infer that this document belongs to class c. Forman [15] proposed a straight metric named *probability ratio* (for brief, we use PR instead hereafter) to measure the discriminability of a term, where df means number of documents:

$$PR(t,c) = \frac{P(t \mid c_+)}{P(t \mid c_-)} = \frac{df(t,c_+)/df(c_+)}{df(t,c_-)/df(c_-)} \tag{4}$$

Like χ^2 statistic discussed before, we use the maximum value to represent the final discriminability of term t, which is denoted as $\mathrm{PR_{max}}$.

$$PR_{\max}(t) = \max_{1 \leq i \leq m} PR(t,c_i) \tag{5}$$

Suppose document d contains three terms, t_1, t_2 and t_3, with occurrences of n_1, n_2 and n_3, respectively. For simplicity, we only label d as positive or negative, which is a binary classification problem. Assume that t_1 and t_2 are projected to positive class c_1 and t_3 is projected to negative class c_2 in term projection step. Moreover, suppose t_1 and t_2 have better discriminability ($\mathrm{PR_{max}}(t_1) > \mathrm{PR_{max}}(t_3)$, $\mathrm{PR_{max}}(t_2) > \mathrm{PR_{max}}(t_3)$) as well as more occurrences in d than t_3 ($n_1 > n_3$, $n_2 > n_3$). Then, we represent d in the new 2-dimensional feature space, the first dimension corresponds to c_1 and the second dimension corresponds to c_2. Corresponding feature weighting is performed as:

$$weight_1 = \log(n_1 + 1) \times \log(PR_{\max}(t_1)) + \log(n_2 + 1) \times \log(PR_{\max}(t_2))$$
$$weight_2 = \log(n_3 + 1) \times \log(PR_{\max}(t_3)) \tag{6}$$

The weighting schema is similar to traditional TF*IDF, with discriminability taken into consideration. As a result, $weight_i$ indicates the possibility that d belongs to class c_i. Therefore, we tend to label d as positive according to above information.

As we can see, feature dimensionality is greatly reduced using this representation, which is only related to the number of classes in the training set. Straightforwardly, our method has significantly more efficient training and classification speed, especially when the dataset contains hundreds of thousands of class labels.

4.2 Classification

In the last subsection, we have demonstrated have to use the clusters to represent the documents. The dimensionality of the new feature space is exactly the number of classes in training set, with each dimension corresponds to a class.

In fact, the value of $weight_i$ in the new representation of document d implies the probability that d belongs to $class_i$. The naive and intuitive idea is classifying d to the class in which d obtains maximum weight. Based on this idea, the classification can be performed in a rule-based manner, which is quite efficient. At the meantime, experiment results prove this method maintain comparable accuracy. In addition, we can exploit the classification power of SVM, which is considered as a powerful tool for machine learning task especially text classification. We apply the same cluster-based representations both to training and test sets, then use training set to build a SVM model, which is used to classify documents in the test set.

5 Experiment

5.1 Experimental Setting

We carry out experiments both on Chinese and English datasets. 20Newsgroups [16] is a widely used English document collection. We choose this collection as a secondary validation case for modified χ^2 statistic.

For Chinese document collection, we involve two datasets. One is the electronic version of Chinese Encyclopedia (**CE**). This collection contains 55 categories and 71669 single-labeled documents (9:1 split to training and test set). This collection is homogeneous. The other is Chinese Web Documents (**CWD**) collection. It has the same taxonomy as CE, including 24016 single-labeled documents. The distributions of two Chinese text collections are diverse though under the same taxonomy, which reflects the heterogeneous problem.

Libsvm [17] with linear kernel is used as our SVM classifier. Previous work [6] shows that Chinese character bigram has better performance than Chinese word unit at higher dimensionality. Besides, we don't have to consider Chinese word segmentation problem. We use bigram as our term unit. Finally, Micro-average F1-Measure is adopted as performance evaluation metric.

In addition, "traditional method" used hereafter follows a straightforward strategy: use traditional χ^2 statistic with various dimension cutoff values to do feature selection, and then use Libsvm to train a SVM model, finally, classify the documents in test set.

5.2 Modified χ^2 Statistic

In this subsection, we will first illustrate that modified χ^2 statistic that takes term frequency into account indeed improves accuracy especially at lower dimensionalities. We use SVM as classifier in this experiment, with different χ^2 statistic methods applied.

Experiment result on CE is shown in Fig.1. X-axis represents the dimension cutoff value, and Y-axis means the corresponding F1 value. It is remarkable that we gain 17% improvement at dimensionality of 100, from 35.1% to 52.1% and 18.4% improvement at 200, from 44.4% to 62.8%. Furthermore, we can promote the performance on various dimensionalities to a certain extent.

To verify the effectiveness of our method further, we also carry out the same experiment on 20NG dataset. Result is shown in Fig.2, which is similar to CE. This shows that our proposed modified χ^2 statistic approach has good generalization performance. On the other hand, we can infer that term frequency is helpful information for feature selection. In this context, we adopt the modified χ^2 statistic as our feature selection method.

In addition, we also obtain greater promotion on Chinese dataset compared to English dataset, which indicates that our method is more appropriate for Chinese text classification. Therefore, we use CE (for homogenous case) and CWD (for heterogeneous case) collections in the following experiments.

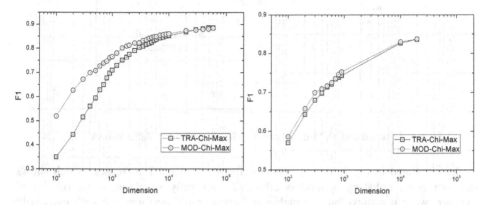

Fig. 1. Modified χ^2 statistic on CE   **Fig. 2.** Modified χ^2 statistic on 20NG

5.3 Homogeneous Dataset

We first focus on homogeneous case, which means that training and test set are sampling from the same distribution. We split CE dataset to training and test set according to the proportion of 9:1. With 64529 documents used as training set and 7140 used as test set.

Follow the instructions described in section 3 and 4. We first use the modified χ^2 statistic to do feature selection and term projection. Two parameters introduced in this step, one is the dimension cutoff value T, which determines the selected feature size, and the other is λ, which remove the noisy term whose χ^2_{max} value is relatively small compared to its χ^2_{total} value. We set λ to 50 in the following experiments. In other words, we only keep the terms whose χ^2_{max} value makes up at least 50% of their χ^2_{total} value. We have done different trials with various values of T. In fact, we are able to utilize all the terms in the training set, as we only need to record the projection result of them. While in traditional method, each term corresponds to a dimension, it's impossible to handle the dimension at that level.

Then we use the term projection information to represent the documents in training and test set, following the procedure in subsection 4.1. Using this cluster-based representation, we can extremely reduce the feature dimensionality to the number of classes in training set. In other words, we use vectors in a 55-dimensional feature space to represent the documents in CE dataset. We can apply rule-based or SVM-based algorithms to do classification discussed in subsection 4.1. The results of both methods are shown in Fig.3 along with different values of T.

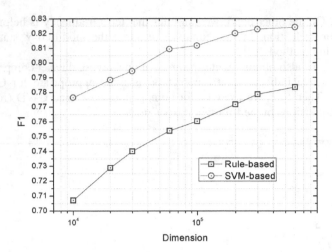

Fig. 3. Rule-based and SVM-based results with different T on homogeneous datasets

As illustrated in Fig.3, we get better performance when T increases, which shows that our noise reducing method is effective. Generally speaking, using rule-based manner, we can achieve an acceptable and comparable performance with extremely efficient classification speed. Furthermore, we can an approximately 5% improvement by SVM compared to rule-based method with a little more computational cost. We gain the best performance when T reaches 600,000, which indicates that we almost use all the useful terms in the training set, this is impossible for traditional method because of dimensionality curse problem.

Rule-based method gets F1-value of 78.3%, a comparable result with traditional method with dimension cutoff value of 2,000. While SVM-based method gains F1-value of 82.4%. Using the traditional method, we gains the same performance with dimension cutoff value of 4,000. In other words, we can get an acceptable and comparable result with less training efforts and greater classification efficiency. The training and test time are shown in Table 1, in a PC with Intel Core2 2.10GHz CPU and 3G memory.

As we can see clearly from the table, using SVM-based method, we can save training time by 75 percent, as well as test time by 50 percent without any lost in F1. Furthermore, rule-based method does not require model training process, and the classification step is really efficient. In fact, we spend only about 2 percent test time of traditional method, but gain a small improvement.

Table 1. Training and test time on homogeneous datasets

Algorithms	Dimensionality	Training Time	Test Time	F1
Traditional Method	4,000	430.676s	139.873s	82.4%
SVM-based	**55**	**101.95s**	**66.795s**	82.4%
Traditional Method	2,000	247.805s	91.957s	77.9%
Rule-based	**55**	**0s**	**2.238s**	**78.3%**

We also do comparisons between Latent Semantic Analysis (LSA) and our method. As we all known that LSA is time consuming and computational intractable, which is not suitable for practical application. In a PC with Intel Core2 2.10GHz CPU and 3G memory, it takes about an hour to do singular value decomposition (SVD) when the dimensionality is reduced to 200. We also perform similar experiments on dimensionalities of 55 and 100. Results are shown in Table 2. On the contrary, we obtain some improvement both on classification accuracy and speed compared with LSA. Rule-based and SVM-based methods get F1-value of 78.3% and 82.4% with dimensionality reduced to 55, which gains improvements of 2.4% and 6.5% compared to LSA with the same dimensionality.

Table 2. Comparisons with LSA

Algorithms	Dimensionality	F1
LSA	55	75.9%
LSA	100	78.5%
LSA	200	80.9%
Rule-based	55	78.3%
SVM-based	55	**82.4%**

5.4 Heterogeneous Dataset

To verify the generalization performance of our proposed methods, we carry out similar experiments on heterogeneous datasets. As shown before, distributions of CE and CWD datasets are distinct. CE stands for a more constant distribution, while CWD reflects the characteristic of web documents which change from time to time. We use the smaller portion of CE as our training set in the following experiment, which contains 7140 documents. CWD with 24016 documents is used as test set.

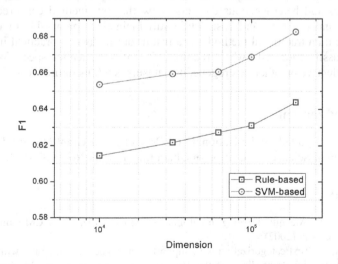

Fig. 4. Rule-based and SVM-based results with different T on heterogeneous datasets

We follow the same steps in previous experiment. Results of rule-based and SVM-based methods on heterogeneous datasets are shown in Fig. 4. F1-value of both methods increases along with larger value of T. Overall, performance of SVM-based method exceeds rule-based method by 3%. Both methods reach peak performance when T is 200,000, with F1-values of 64.4% and 68.3%.

We also compare our methods with traditional one. Traditional method gains best F1-value of 64% at dimensionality of 60,000. Both of our methods get better performance with less training and test time. SVM-based method uses about 2 percent training time and 7 percent test time of traditional method, but gains improvement of 4.3%. Besides, Rule-based method ignores training process and uses only about 1 percent test time of traditional method, while still maintains comparable performance.

Table 3. Training and test time on heterogeneous datasets

Algorithms	Dimensionality	Training Time	Test Time	F1
Traditional Method	60,000	179.51s	462.875s	64%
SVM-based	**55**	4.065s	31.678s	**68.3%**
Rule-based	**55**	**0s**	**5.938s**	64.4%

6 Conclusion

In this paper, we proposed an efficient text classification method based on term projection. First, we show that our modified χ^2 statistic promotes the performance especially at lower dimensionalities. Then, we project the terms to appropriate classes using the modified χ^2 statistic, this can make most use of semantic meanings of terms. We also use a more sophisticated cluster-based text representation to reduce the feature dimensionality by three orders of magnitude. Finally, Rule-based and SVM-based methods are adopted to do classification. Experiment results on both homogeneous and heterogeneous datasets show that our method can greatly reduce the training and test time and cost, while still maintains comparable or even better performance than traditional method. As a result, our method is practical in the large-scale text classification tasks which require efficient classification speed. Whether our method is effective on other heterogeneous datasets is left as future work.

Acknowledgement

This work is supported by the National 863 Project under Grant No. 2007AA01Z148 and the National Science Foundation of China under Grant No. 60873174.

References

1. Sebastiani, F.: Machine learning in automated text categorization. ACM Computing Surveys 34(1), 1–47 (2002)
2. Joachims, T.: Text categorization with support vector machines: learning with many relevant features. In: Proceedings of the 10th European Conference on Machine Learning, pp. 137–142 (1998)

3. Duda, R.O., Hart, P.E., Stork, D.G.: Pattern classification, 2nd edn. Wiley-Interscience, New York (2000)
4. Yang, Y.M., Liu, X.: A re-examination of text categorization methods. In: Proceedings of the 22nd Annual International Conference on Research and Development in Information Retrieval, pp. 42–49 (1999)
5. Yang, Y.M., Pedersen, J.O.: A comparative study on feature selection in text categorization. In: Proceedings of 14th International Conference on Machine Learning, pp. 412–420 (1997)
6. Li, J.Y., Sun, M.S., Zhang, X.: A comparison and semi-quantitative analysis of words and character-bigrams as features in Chinese text categorization. In: Proceedings of COLING-ACL 2006, pp. 545–552 (2006)
7. Baker, L.D., McCallum, A.K.: Distributional clustering of words for text classification. In: Proceedings of 21st ACM International Conference on Research and Development in Information Retrieval, pp. 96–103 (1998)
8. Bekkerman, R., El-Yaniv, R., Tishby, N., Winter, Y.: On feature distributional clustering for text categorization. In: Proceedings of 24th ACM International Conference on Research and Development in Information Retrieval, pp. 146–153 (2001)
9. Bekkerman, R., El-Yaniv, R., Tishby, N., Winter, Y.: Distributional word clusters vs. words for text categorization. Journal of Machine Learning Research 3, 1183–1208 (2003)
10. Chen, W.L., Chang, X.Z., Wang, H.Z., Zhu, J.B., Yao, T.S.: Automatic word clustering for text categorization using global information. In: First Asia Information Retrieval Symposium, pp. 1–6 (2004)
11. Pereira, F., Tishby, N., Lee, L.: Distributional clustering of English words. In: Proceedings of the 31st Annual Meeting of the Association for Computational Linguistics, pp. 183–190 (1993)
12. Ling, X., Dai, W.Y., Jiang, Y., Xue, G.R., Yang, Q., Yu, Y.: Can Chinese Web Pages be Classified with English Data Source? In: Proceedings of the 17th international conference on World Wide Web (2008)
13. Dai, W.Y., Xue, G.R., Yang, Q., Yu, Y.: Transferring Naive Bayes Classifiers for Text Classification. In: Proceedings of the 22nd AAAI Conference on Artificial Intelligence (2007)
14. Li, J.Y., Sun, M.S.: Scalable term selection for text categorization. In: Proceedings of EMNLP 2007, pp. 774–782 (2007)
15. Forman, G.: An extensive empirical study of feature selection metrics for text classification. Journal of Machine Learning Research 3, 1289–1305 (2003)
16. Rennie, J.: 20Newsgroups dataset,
    http://people.csail.mit.edu/jrennie/20Newsgroups/
17. Chang, C.-C., Lin, C.-J.: LIBSVM: a library for support vector machines (2001),
    http://www.csie.ntu.edu.tw/cjlin/libsvm

IPHITS: An Incremental Latent Topic Model for Link Structure

Huifang Ma, Weizhong Zhao, Zhixin Li, and Zhongzhi Shi

Key Lab of Intelligent Information Processing, Institute of Computing Technology,
Chinese Academy of Sciences, 100190 Beijing, China
Graduate University of the Chinese Academy of Sciences, 100049 Beijing China
mahf@ics.ict.ac.cn
http://www.intsci.ac.cn/users/mahuifang/index.html

Abstract. The structure of linked documents is dynamic and keeps on changing. Even though different methods have been proposed to exploit the link structure in identifying hubs and authorities in a set of linked documents, no existing approach can effectively deal with its changing situation. This paper explores changes in linked documents and proposes an incremental link probabilistic framework, which we call IPHITS. The model deals with online document streams in a faster, scalable way and uses a novel link updating technique that can cope with dynamic changes. Experimental results on two different sources of online information demonstrate the time saving strength of our method. Besides, we make analysis of the stable rankings under small perturbations to the linkage patterns.

Keywords: link analysis, incremental learning, PHITS, IPHITS.

1 Introduction

Link-based document ranking algorithms concern about link mining techniques that can identify "authoritative" or "prestigious" documents from set of hyperlinked webpages or from other citation data within a certain structure. These methods always try to find out relationships among documents or authors and documents. There exist a variety of algorithms that rank documents in a social network according to criteria that reflect structural properties of the network. Madadhain et al. [14] grouped these algorithms in three major paradigms: betweenness centrality [3], eigenvector centrality [20] and voltage-based rankers [22]. A comprehensive overview on link analysis can be found in [10] and [2], which made a taxonomy of common link mining task. Link analysis encompasses a wide range of tasks and we only focus on the core issues addressed by a majority of ongoing research in the field of link-based document ranking.

PageRank [17] and HITS [13] algorithms are the most famous approaches in link-based document ranking. Both of them are eigenvector methods for identifying "authoritative" or "influential" articles, given hyperlink or citation information. There are many extensions on these methods. Bharat and Henzinger [1] and

G.G. Lee et al. (Eds.): AIRS 2009, LNCS 5839, pp. 242–253, 2009.
© Springer-Verlag Berlin Heidelberg 2009

Chakrabarti et al. [4] made modifications to HITS to exploit webpage content for weighing pages and links based on relevance. Cohn and Chang [6] introduced PHITS as a probabilistic analogue of the HITS algorithm, attempting to explain the link structure in terms of a set of latent factors. Ding et al. [8] introduced a unified framework encompassing both PageRank and HITS and presented several new ranking algorithms. Cohn and Hofmann [7] constructed Link-PLSI, a joint latent probabilistic model to integrate content and connectivity together while Doan et al. [9] presented probabilistic models inspired by HITS and PageRank, which incorporate both content and link structure. Xu [23] proposed a ranking algorithm trying to introduce the content information into link-based methods as implicit links. Matthew Richardson et al. [19] achieved a novel ranking algorithm using features that are independent of the link structure of the Web. Ng et al. [16] analyzed the stability of PageRank and HITS to small perturbations in the link structure and presented modifications to HITS that yield more stable rankings. O'Madadhain and Smyth [14] and O'Madadhain et al. [15] proposed a framework for ranking algorithms that respect event sequences and provided a natural way of tracking changes in ranking over time.

The dramatic growth and changes of link information, however, exhibit dynamic patterns of the importance of linked documents, as references and links can be changed or become inaccessible. As the amount of data grows and the number of sources expands, techniques which can track the changes in documents rank over time become a hot issue. As a naive approach to catch the update of links, static ranking methods can be applied to data streams over various time intervals, which means to re-run the batch algorithm from scratch on all existing data each time a new data comes in. But it is computationally expensive for use during link development. Another obvious weakness is that changes to the links themselves can not be automatically updated with the content of stable ranking results maintained after re-running of the batch algorithm.

In this paper, we develop a novel incremental probabilistic model, which models hyperlinked data by sequentially folding in the corresponding new documents and citations. Initially, IPHITS uses the traditional PHITS technique to recognize links and stores them into the PHITS model. For the arriving of new linked data and removal of out of date data, IPHITS analyzes the changes and the new structure of linking repository and then updates the corresponding links. Since the method does not need to recalculate the probabilities completely whenever it receives new linked data, it is very fast computationally.

The key contributions of this paper include a novel link updating technique using our novel incremental PHITS algorithm (IPHITS) that can cope with link changes. Importantly, with the great reduction in time complexity, the model allows latent topics of citations to be automatically maintained. Before we describe our new model, we summarize the main notations used in this paper in Table 1.

The rest of this paper is organized as follows: In Section 2, we introduce the PHITS model and its principles. In Section 3, we give detailed information on our proposed IPHITS algorithm. Section 4 considers our test corpora, the

Table 1. Notations of some frequently occurring variables

D	Citing document set
C	Cited document set
K	Number of latent topics
d_i	A citing document
c_r	A cited document
z_k	A latent topic
$A_{r,i}$	The count of cited document c_r from citing document d_i

performance measures, together with the experiment results. We conclude and discuss future work in Section 5.

Note that in the rest of the paper, we use the terms "citation" and "hyperlink" interchangeably. Likewise, the term "citing" is synonymous to "linking" and so is "cited" to "linked".

2   PHITS Model

PHITS [6] model is based on a two way factor analysis that is in most respects identical to the topic model used by Hofmann [11]. The model attempts to explain two sets of observables (citing documents and cited documents) in terms of a small number of common but unobserved variables. Under the PHITS assumptions, a document is modeled as a mixture of latent topics that generates citations. A representation of the model in terms of graphical model is depicted in Fig 1. We can see that PHITS introduces a hidden variable $z_k (k \in 1, \ldots, K)$ in the generative process of each cited document $c_r (r \in 1, \ldots, R)$ in a document $d_i (i \in 1, \ldots, N)$. Given the unobservable variable z_k, each c_r is independent of the document d_i it comes from and the latent topic can be identified within individual research communities.

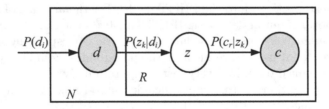

Fig. 1. Graph Model of PHITS

Essentially, we can describe the model as a generative process: a document d_i is generated with some probability $P(d_i)$, a latent topic z_k associated with documents and citations is chosen probabilistically so that their association can be represented as conditional probabilities $P(c_r|z_k)$ and $P(z_k|d_i)$. The joint model for predicting citations in documents is defined as:

$$P(c_r, d_i) = P(d_i)P(c_r|d_i), \tag{1}$$

$$P(c_r|d_i) = \sum_k P(c_r|z_k)P(z_k|d_i), \tag{2}$$

where c_r and d_i both refer to document in the document set and they may be identical. They are kept separate notationally to reinforce different roles they play in the model, c_r is conveyed by being cited and d_i is conveyed by citing [11].

An EM algorithm is used to compute the parameters $P(c_r|z_k)$ and $P(z_k|d_i)$ through maximizing the following log-likelihood function of the observed data:

$$\mathcal{L} = \sum_i \sum_r A_{r,i} log P(c_r, d_i). \tag{3}$$

The steps of the EM algorithm are described as follows:

E-step: The conditional distribution $P(z_k|d_i, c_r)$ are computed from the previous estimate value of the parameters $P(c_r|z_k)$ and $P(z_k|d_i)$:

$$P(z_k|d_i, c_r) = \frac{P(c_r|z_k)P(z_k|d_i)}{\sum_k P(c_r|z_k)P(z_k|d_i)}. \tag{4}$$

M-step: The parameters $P(c_r|z_k)$ and $P(z_k|d_i)$ are updated with the new expected values and $P(z_k|d_i, c_r)$:

$$P(c_r|z_k) = \frac{\sum_i A_{r,i}P(z_k|d_i, c_r)}{\sum_i \sum_r A_{r,i}P(z_k|d_i, c_r)}, \tag{5}$$

$$P(z_k|d_i) = \frac{\sum_r A_{r,i}P(z_k|d_i, c_r)}{\sum_k \sum_r A_{r,i}P(z_k|d_i, c_r)}. \tag{6}$$

In order to make full use of existing information, an incremental learning algorithm is presented that can efficiently model the incoming information. Instead of modeling documents all at once, we start with a given document set and add new documents or delete documents at each stage while preserving document coherence. The basic idea of our updating algorithm is straightforward: the PHITS algorithm is performed once on the initial linked-documents at the beginning. When a set of new documents introduce new citations, a cycle should be created for folding documents and citations and the model is then updated during the cycle.

Chou et al. [5] proposed an Incremental PLSI (IPLSI), aiming to address the problem of online event detection, this model captures the basic concept of incremental learning for PLSI and offers an excellent foundation on which to build our model. Our technique is computationally similar to the IPLSI procedure. However, what we are trying to do is, conceptually speaking, very different.

3   Incremental PHITS for Link Update

In this section, for the update of links when new links are added or deleted, we develop an incremental approach to PHITS technique. The PHITS algorithm is executed once on the initial documents. Then, for each new adding or deleting of linked data, we need to adjust link-topic probabilities at the lowest cost. On one hand, we need to update citation-topic probabilities for out of date

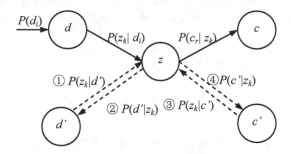

Fig. 2. Illustration of IPHITS Model

links; on the other, new citation-topic and new document-topic probabilities should be adjusted corresponding. We take advantage of folding-in method, a partial version of EM algorithm, to update the unknown parameters with the known parameters kept fixed so as to maximize the likelihood with respect to the previously trained parameters.

Fig. 2 is an illustration of sequences for updating related information of IPHITS, where d' and c' indicate new documents and new citations respectively. As the figure shows, new documents should first be folded in with old links fixed, and then $P(d'|z_k)$ are calculated which sets a foundation for folding in new links. In this way, $P(c_{all}|z_k)$ and $P(z_k|d_{all})$ are updated as better initial values for the final EM algorithm and this guarantees a faster convergence. (Note that d_{all} is a final document in the entire document set, and so is c_{all}). Algorithm 1 gives a detailed illustration of Incremental PHITS algorithm.

In Step 2, preprocessing is done as the first phase for the incremental learning, involving elimination of out-of-date documents and hyperlinks. Documents not used anymore should be discarded and hyperlinks related to these documents should be removed as well. The model can not be augmented directly, as the basic principle of probability that the total probability will be equal to one should be observed, the remaining parameters need to be renormalized proportionally. Note that $P_0(c_r|z_k)$ stand for the probabilities of the remaining terms and citations.

In Step 4, we fold in documents with old links kept fixed since old links are well trained and the arriving documents contain old links. In contrast, old documents convey no corresponding information to aid the folding in of hyperlinks.

It is a pity that $P(c_{new}|z)$ can not be folded in using $P(z|d_{new})$ directly. That is because the sum of all probabilities of citations in old citation sets

Algorithm 1. Incremental PHITS

Input: Set of new documents D_{new}, set of new citations C_{new}, $P(z_k|d_i)$ for each latent topic z_k and document d_i, $P(c_r|z_k)$ for each citation c_r and latent topic z_k

Output: New updated $P(z|d_{all})$ for all documents, updated $P(c_{all}|z)$ for all citations

1. Execute PHITS;
2. For all the remaining citations do

$$P(c_r|z_k) = \frac{P_0(c_r|z_k)}{\sum_{r'} P_0(c_{r'}|z_k)};$$

3. End For
4. While not convergent do

Randomize $P(z_k|d_{new})$ and ensure $\sum_k P(z_k|d_{new}) = 1$;
For each new document d_{new} in D_{new} do

$$P(z_k|d_{new}, c_r) = \frac{P(c_r|z_k)P(z_k|d_{new})}{\sum_{k'} P(c_r|z_{k'})P(z_{k'}|d_{new})};$$

End For
For each new document d_{new} in D_{new} and each latent topics z_k do

$$P(z_k|d_{new}) = \frac{\sum_{r \in d_{new}} A_{r,new} P(z_k|d_{new}, c_r)}{\sum_{k'} \sum_{r' \in d_{new}} A_{r',new} P(z_{k'}|d_{new})};$$

End For
5. End While
6. For each new document d_{new} in D_{new}

$$P(d_{new}|z_k) = \frac{\sum_{r \in d_{new}} A_{r,new} P(z_k|d_{new}, c_r)}{\sum_{d \in D_{new}} \sum_{r' \in d} A_{r',new} P(z_k|d_{new}, c_{r'})};$$

7. End For
8. While not convergent do

Randomize $P(z_k|c_{new})$ and ensure $\sum_k P(z_k|c_{new}) = 1$;
For each new document c_{new} in C_{new} do

$$P(z_k|d_{new}, c_{new}) = \frac{P(z_k|c_{new})P(d_{new}|z_k)}{\sum_{k'} P(z_{k'}|c_{new})P(d_{new}|z_{k'})};$$

End For
For each new document c_{new} in C_{new} and each latent topics z_k do

$$P(z_k|c_{new}) = \frac{\sum_{d_i \in D_{new}} A_{new,i} P(z_k|d_i, c_{new})}{\sum_{k'} \sum_{d \in D_{new}} A_{r',new} P(z_{k'}|d_{new})};$$

End For
9. End While
10. For each citation c_{all} in the entire C_{all}

$$P(c_{all}|z_k) = \frac{\sum_{d_i \in D_{all}} A_{all,i} P(z_k|d_i, c_{all})}{\sum_{d_{i'} \in D_{all}} \sum_{c_r \in d} A_{ri'} P(z_k|d_{i'}, c_r)};$$

11. End For
12. Execute PHITS;

under z already equals to one, which means $P(c_r|z_k)$ have been well trained and normalized. If we randomize and normalize all $P(c_{new}|z_k)$ when new documents arrive, the sum of the probabilities of all citations under z will be larger than one. This restriction makes it inapplicable to update new terms and citations directly. To avoid this, we first derive $P(d_{new}|z_k)$ in Step 6.

In Step 8, we develop a mechanism for new citations update, which can satisfy the basic of requirement of probabilities of latent topics under new citations equal to one. $P(z_k|c_{new})$ are randomly initialized and normalized. We then update $P(z_k|c_{new})$ with $P(d_{new}|z_k)$ fixed.

In Step 10, our method deals with issues of how to get the final normalized $P(c_{all}|z_k)$ by means of adjusting $P(c_r|z_k)$. For new citations c_{new}, $P(z_k|d_i, c_j)$ are calculated according to Step 8 while for old terms and citations, we use Eq.(4) to get $P(z_k|d_i, c_r)$.

Finally, $P(c_{all}|z_k)$ and $P(z_k|d_i)$ are adopted as better initial value to execute the original EM algorithm for updating the model. As new documents arrive and old documents disappear, the above algorithm can preserve the probability and continuity of the latent parameters during each revision of the model in a fast way.

4   Experiments

This section reports on the empirical evaluation on the performance and usefulness of our approach. We design two experiments to test the viability of the model: time expenditure and analysis of the stability of the algorithm.

4.1   Data Description and Baseline Representation

The performance of our model is evaluated using two different types of linked data: scientific literature from Citeseer which is connected with citations, Wikipedia Webpages dataset containing hyperlinks. We adjust the link structure to include the incoming links and outgoing links only within each corpus, and then take advantage of these dataset for our model construction with adding new documents and citations and deleting out of date information.

The Citeseer data can be obtained from Citeseer collection that was made publicly available by Lise Getoor's research group at University of Maryland [21]. There are altogether 3312 documents using abstract, title and citation information in the corpus. The Citeseer dataset only includes articles that cite or are cited by at least two other documents. Thereafter the corpus size is limited to 1168 documents, of which only 168 documents have both incoming and outgoing links.

The dataset of Wikipedia Webpages is downloaded from Wikipedia by crawling within the Wikipedia domain, starting from the "Artificial Intelligence" Wikipedia page and the dataset is composed of 6324 documents and 16387 links.

We compare our IPHITS with the following methods:

• Naive IPHITS. For every advance of the new documents, the EM algorithm uses new random initial setting to re-estimate all the parameters of the PHITS algorithm.

• PageRank. The outgoing and incoming links are counted and the implementation of PageRank used in the experiments comes from [17].

• HITS. A n-by-n adjacency matrix is constructed, whose (i, j)-entry is 1 if there is a link from i to j, and 0 otherwise. An iterative process is used to identify the principal eigenvector and principal community of the matrix.

4.2 Experiment on Time Cost

We perform this experiment aiming at evaluating time efficiency of IPHITS in comparison with Naive IPHITS. In order to observe the correlation between the number of topics k and the time saved by our model, we examine the impact of different numbers of latent variables and run IPHITS on the two datasets with different k. For each k, we run these two algorithms on the subset of each database consisting of 90% of the entire documents respectively. We then randomly delete 10% subset of the documents and add the same amount of data. Table 2 gives a detailed illustration on the total time and the number of iterations required to achieve convergence (The total time of IPHITS is divided into two parts: Link-PLSI time and folding time).

Table 2. Execution time (in seconds) of Naive IPHITS and IPHITS

	Wiki				Citeseer			
	Naive IPHITS		IPHITS		Naive IPHITS		IPHITS	
k	Ave Iter	Total Time	Ave Iter	Total Time	Ave Iter	Total Time	Ave Iter	Total Time
10	32.13	7131	3.42	752	38.42	1872	2.87	142
15	30.47	7421	4.32	796	40.34	2354	3.43	183
20	28.54	8996	3.67	848	42.76	2843	2.92	242
25	25.62	9443	4.45	932	39.52	3084	2.69	305
30	24.38	13542	3.71	974	38.41	3465	2.96	326

Note: Aver Iter stands for Average Iterations; k indicates number of latent topics

As seen in Table 2, the IPHITS method can save a large amount of time. In general, the computation time of the Naive IPHITS approach is 10 times longer than that of the our algorithm. With $k = 30$ on Wiki dataset, IPHITS can reduce the time cost by 14 times. The reason is that the Naive IPHITS approach uses new random initial settings to re-estimate all relevant parameters of EM algorithm each time and requires a large number of iterations to converge to a different local optimum while IPHITS has preserved a better starting point and can therefore converge much faster. The larger the dataset is, the more time our model can save. Furthermore, when k increases, time cost increases as well, these results are consistent with our intuition.

4.3   Stability of the Algorithm

This experiment aims to analysis the sensitivity of the rank orderings of different algorithms using different datasets. We believe that small changes in the dataset should have a small effect on the output of a stable algorithm.

Specifically, we run the Naive IPHITS and IPHITS on the subset of the Wiki database consisting of 514 Artificial Intelligence webpages, and examined the list of papers that they considered "influential". To evaluate the stability of these two algorithms, we also constructed 20 perturbed versions of the datasets (which are denoted as wiki-1, wiki-2, etc), each of which containing a randomly deleted 10% subset of the documents and added of the same amount of webpages. Note that we keep the top ranking 10 documents in each dataset, assuming that a robust and stable model is not sensitive to the perturbation of the link structure. Likewise, we develop a similar experiment on Citeseer dataset. In this experiment, the number of latent topics is set to 5.

Due to the limitation of the paper, we only present the most authoritive webpages generated by IPHITS and Naive IPHITS under the topic "artificial intelligence" along with its probability in Table 3 and Table 4 respectively. The leftmost column is the authority ranking obtained by analyzing the initial set of Wiki dataset, the 5 rightmost columns demonstrate the ranks on some perturbed datasets. We can see substantial variation across the different datasets.

We found that results obtained from Naive IPHITS are not consistent to perturbation of these datasets. And under this perturbation it sometimes ignores some webpages indeed "influential" and returns different webpages as top-ranked documents while our method almost always returns similar results as its pervious outcome. This qualitative evaluation reveals the stable property of our model to perturbation.

Table 3. Naive IPHITS results on Wiki for the topic of Artificial Intelligence

1	Artificial intelligence(0.1214)	2	1	2	1	3
2	Machine learning(0.0929)	1	3	3	5	2
3	Computer science(0.0643)	8	12	19	6	8
4	Natural language processing(0.0483)	12	5	5	21	19
5	Data mining(0.0397)	44	17	117	13	58
6	Speech recognition(0.0376)	11	37	160	29	39
7	Knowledge representation(0.0367)	292	22	170	321	301
8	Information retrieval(0.0353)	173	9	98	112	37
9	Information(0.0351))	18	18	11	13	34
10	Cyc (0.0329)	11	15	8	27	13

Table 4. IPHITS results on Wiki for the topic of Artificial Intelligence

1	Artificial intelligence(0.1214)	1	1	1	1	1
2	Machine learning(0.0929)	2	2	2	2	2
3	Computer science(0.0643)	3	4	3	3	3
4	Natural language processing(0.0483)	4	3	4	5	4
5	Data mining(0.0397)	5	5	10	4	6
6	Speech recognition(0.0376)	7	6	5	6	5
7	Knowledge representation(0.0367)	6	7	6	8	8
8	Information retrieval(0.0353)	8	9	8	7	9
9	Information(0.0351))	9	8	9	9	7
10	Cyc (0.0329)	10	10	7	10	10

In terms of quantitive evaluation, the difference between two rank orderings is defined as the mean absolute difference in rank ordering between pair of individual datasets, borrowed from [14]:

$$d(M_j, M_k) = \frac{\sum_{d \in D} |O_{M_j}(d) - O_{M_k}(d)|}{|D|}, \tag{7}$$

where M_j and M_k represent two different models on different dataset, and $O_{M_j}(d)$ denotes the index of the rank assigned to d.

For a given document set D, $d(M_j, M_k)$ obtains its maximum value when M_j produces a reserve ordering of M_k:

$$d_{max}(M_j, M_k) = \frac{\sum_{k=1}^{\frac{|D|}{2}} (2k - 1)}{|D|} = \frac{|D|}{2}. \tag{8}$$

Table 5 shows the results of cross-comparison of four ranking models over variation on different datasets. We can see that IPHITS and PageRank are much more stable than that of the Naive IPHITS and HITS on different datasets.

Similar with [16], we count the number of top 10 pages under each topic which drop or rise in ranking drastically in each dataset and made 5 trials on each dataset. Since we perturb the dataset by deletions and insertions, large rises in ranking the algorithms and large drops in ranking are all interesting to observe. Table 6 shows the average percentage of different algorithm that suffer rank drops(rises). The Percentage of rank drop(rise) is defined as the percentage of the number of top 10 pages that drop(rise) below(above) rank 20 and the total number of times of trails. It is clear that our model rank drop(rise) only changes slightly compared with Naive IPHITS. The reason is that the latent variables generated by the Naive IPHITS are discontinuous, whereas our algorithm maintains good continuity in the content of latent variables. The ranking for our algorithm also appears less stable on Wiki than on Citeseer, which indicates the different characteristic of cited papers and hyperlinked webpages. In the Citeseer experiments, removing a document from the dataset does not

Table 5. Comparison of four algorithms on different datasets

(a) Naive IPHITS results on Wiki

DataSet	Wiki-1	Wiki-2	Wiki-3	Wiki-4
Wiki-1	0.00	26.34	29.18	28.47
Wiki-2	26.34	0.00	24.33	30.18
Wiki-3	29.18	24.33	0.00	26.45
Wiki-4	28.47	30.18	26.45	0.00

(b) IPHITS results on Wiki

DataSet	Wiki-1	Wiki-2	Wiki-3	Wiki-4
Wiki-1	0.00	1.30	1.82	1.64
Wiki-2	1.30	0.00	2.17	2.45
Wiki-3	1.82	2.17	0.00	0.43
Wiki-4	1.64	2.45	0.43	0.00

(c) HITS results on Wiki

DataSet	Wiki-1	Wiki-2	Wiki-3	Wiki-4
Wiki-1	0.00	34.18	38.47	29.65
Wiki-2	34.18	0.00	35.43	33.56
Wiki-3	38.47	35.43	0.00	29.54
Wiki-4	29.65	33.56	29.54	0.00

(d) PageRank results on Wiki

DataSet	Wiki-1	Wiki-2	Wiki-3	Wiki-4
Wiki-1	0.00	2.32	2.17	2.45
Wiki-2	2.32	0.00	1.13	2.18
Wiki-3	2.17	1.13	0.00	1.73
Wiki-4	2.45	2.18	1.73	0.00

Table 6. Percentage of rank rise, drop

Data	Naive IPHITS	IPHITS	HITS	PageRank
Wiki	27.36%, 23.26%	16.36%, 13.26%	22.34%, 20.26%	17.34%, 15.26%
Citeseer	25.44%, 21.26%	12.88%, 11.38%	21.16%, 22.18%	15.92%, 13.48%

remove any of the papers cited by or citing while in the Wiki case, each deleted webpage also removes its surrounding link structure. So do new adding of papers or webpages.

5   Conclusions and Future Work

Many emerging applications require linked documents to be repeatedly updated. Such documents include cited papers, webpages, and shared community resources such as Wikipedia. In order to accommodate dynamically changing topics, efficient incremental algorithms need to be developed. In this paper, we have developed an incremental technique to effectively update the hyperlinked information dynamically. The novelty of our model is the ability to identify hidden topics while reducing the amount of computations and maintaining the latent topic from one time period to another incrementally. When tested on a corpus of Wikipedia articles and Citeseer papers, our model performs much faster than the batched methods. Extending the model to the unified probabilistic model of the content and connections of linked documents is our future work.

Acknowledgments. This work is supported by the National Basic Research Priorities Programme (No. 2007CB311004), 863 National High-Tech Program (No.2007AA01Z132)and the National Science Foundation of China (60775035).

References

1. Bharat, K., Henzinger, M.R.: Improved algorithms for topic distillation in a hyperlinked environment. In: 21st annual international ACM SIGIR Conference on Research and Development in Information Retrieval, Melbourne, Australia, pp. 104–111 (1998)
2. Borodin, A., Roberts, G.O., Rosenthal, J.S., Tsaparas, P.: Link Analysis Ranking: Algorithms, Theory, and Experiments. ACM Transactions on Internet Technology 5(1), 231–297 (2005)
3. Brandes, U.: A faster algorithm for betweenness centrality. Journal of Mathematical Sociology 25(2), 163–177 (2001)
4. Chakrabarti, S., Dom, B., Gibson, D., Kleinberg, J., Raghavan, P., Rajagopalan, S.: Automatic resource list compilation by analyzing hyperlink structure and associated text. In: 7th International World Wide Web Conference, Brisbane, Austrilia, pp. 65–74 (1998)
5. Chou, T.C., Chen, M.C.: Using incremental PLSA for threshold resilient online event analysis. IEEE Trans. Knowledge and Data Engineering 20(3), 289–299 (2008)

6. Cohn, D., Chang, H.: Learning to probabilistically identify authoritative documents. In: 7th International Conference on Machine Learning, Austin, Texas, pp. 167–174 (2000)
7. Cohn, D., Hofmann, T.: The missing link - a probabilistic model of document content and hypertext connectivity. Neural Information Processing Systems 13 (2001)
8. Ding, C., He, X., Husbands, P., Zha, H., Simon, H.D.: PageRank, HITS and a unified framework for link analysis. In: 25th annual international ACM SIGIR Conference on Research and Development in Information Retrieval, Tampere, Finland, pp. 353–354 (2002)
9. Doan, A., Domingos, P., Halevy, A.Y.: Learning to match the schemas of data sources: A multistrategy approach. Machine Learning 50(3), 279–301 (2003)
10. Getoor, L., Diehl, C.P.: Link mining: a survey. ACM SIGKDD Explorations Newsletter 7(2), 2–12 (2005)
11. Hofmann, T.: Unsupervised learning by probabilistic latent semantic analysis. Maching Learning 42(1), 177–196 (2001)
12. Jeh, G., Widom, J.: Scaling personalized web search. In: 12th International World Wide Web Conference, Budapest, Hungary, pp. 271–279 (2003)
13. Kleinberg, J.: Authoritative sources in a hyperlinked environment. Journal of the ACM 46(5), 604–632 (1999)
14. Madadhain, J.O'., Hutchins, J., Smyth, P.: Prediction and ranking algorithms for even-based network data. SIGKDD Explorations 7(2) (2005)
15. Madadhain, J.O'., Smyth, P.: EventRank: A framework for ranking time-varying networks. In: 3rd KDD Workshop on Link Discovery LinkKDD, Issues, Approaches and Applications, Chicago, Illinois, pp. 9–16 (2005)
16. Ng, A.Y., Zheng, A.X., Jordan, M.I.: Link analysis, eigenvectors and stability. In: 17th International Joint Conference on Artificial Intelligence, Seattle, USA, pp. 903–910 (2001)
17. Page, L., Brin, S., Motwani, R., Winograd, T.: The PageRank citation ranking: bringing order to the web. Technical report, Stanford University (1998)
18. Richardson, M., Domingos, P.: The intelligent surfer: probabilistic combination of link and content information in PageRank. Advances Neural Information Processing Systems 14 (2002)
19. Richardson, M., Prakash, A., Brill, E.: Beyond PageRank: machine learning for static ranking. In: 15th International World Wide Web Conference, Edinburth, Scotland, pp. 707–715 (2006)
20. Seeley, J.: The net of reciprocal influence: A problem in treating sociometric data. Canadian Journal of Psychology 3, 234–240 (1949)
21. http://www.cs.umd.edu/~sen/lbc-proj/LBC.html
22. Wu, F., Huberman, B.: Discovering communities in linear time: A physics approach. Europhysics Letters 38, 331–338 (2004)
23. Xu, G.: Building implicit links from content for forum search. In: 29th annual international ACM SIGIR conference on Research and development in information retrieval, Seattle, Washington, pp. 300–207 (2006)

Supervised Dual-PLSA for Personalized SMS Filtering

Wei-ran Xu[1], Dong-xin Liu[1], Jun Guo[1], Yi-chao Cai[1], and Ri-le Hu[2]

[1] School of Information and Communication Engineering, Beijing University of Posts and Telecommunications, Beijing, 100876, China
{xuweiran,guojun}@bupt.edu.cn, dongdongdedipan@gmail.com,
caiyichaobupt@163.com
[2] Nokia Research Center, Beijing, 100176, China
rile.hu@nokia.com

Abstract. Because users hardly have patience of affording enough labeled data, personalized filter is expected to converge much faster. Topic model based dimension reduction can minimize the structural risk with limited training data. In this paper, we propose a novel supervised dual-PLSA which estimate topics with many kinds of observable data, i.e. labeled and unlabeled documents, supervised information about topics. c-w PLSA model is first proposed, in which word and class are observable variables and topic is latent. Then, two generative models, c-w PLSA and typical PLSA, are combined to share observable variables in order to utilize other observed data. Furthermore, supervised information about topic is employed. This is supervised dual-PLSA. Experiments show the dual-PLSA has a very fast convergence. Within 100 gold standard feedback, dual-PLSA's cumulative error rate drops to 9%. Its total error rate is 6.94%, which is the lowest among all the filters.

Keywords: Spam Filtering, Personalized Filtering, Probabilistic Latent semantic Analysis, Latent Dirichlet Allocation, dual-PLSA.

1 Introduction

Text Categorization (TC) is one of the most important task in the information system. Sebastiani believes that TC is the meeting point of machine learning and information retrieval (IR), the "mother" of all disciplines concerned with automated content-based document management [1].

TC can be defined as the task of determining an assignment of a tag from $C=\{c_1,...,c_K\}$, a set of pre-defined categories, to each document in $D=\{d_1,...,d_L\}$, a set of documents. In traditional supervised learning framework, all the users share one classifier, and the classification task is to train a classification model on a labeled training data, and then, use the learned model to classify a test data set.

Short message (SMS) filtering is to distinguish between spam and ham for any individual user. Different user may have different definition of spam messages, which implies that some users' spam message may be classified as ham by others. For example, although some group-sending-SMS, such as weather forecast, financial information, news report, etc., may be regarded as spam by user A, but on the

G.G. Lee et al. (Eds.): AIRS 2009, LNCS 5839, pp. 254–264, 2009.

contrary user B may have an urgent need for them. It is necessary to train a specific filter for each user.

Personalized SMS filtering task is similar to TREC Spam Track [15]: there is a set of chronologically ordered SMS, $D=\{d_1,...,d_L\}$; D are presented one at a time to the filter, which yields a binary judgment (spam or ham); the user will immediately feedback a gold standard, and the filter should be updated at once.

Adaptive filtering has been widely discussed. In this paper, we face a new challenge that the filter is expected to converge much faster than ever before, e.g. within 10 labeled SMSs or less. It seems impossible for filter, but it is reasonable for users. Users hardly have patience of affording enough labeled data and waiting for filter's learning.

The dimension of the feature space can be defined as the number of different words (based on our statistics, there are 27,054 different words in 200,000 pieces of SMS) in the vector space model. And therefore, even if a linear filter is adopted, it still has more than 20,000 parameters to be estimated, which is difficult for training sample size to meet this requirement. According to the statistical learning theory [6], dimension reduction can minimize the structural risk with a limited size of training set. In this paper, latent topics are employed as feature space to minimize dimensions. We focus on utilizing the topic models to achieve knowledge extraction, knowledge representation and knowledge utilization.

The rest of the paper is organized as follows. In Section 2, we give a brief review of related work. In Section 3, the key issues are analyzed, and c-w PLSA model is introduced. In Section 4, dual-PLSA is presented to solve the problem of insufficient labeled samples. In section 5, supervised dual-PLSA is proposed to deal with practical problems. In section 6, experiments results and the analysis are shown.

2 Related Work

A recent trend in dimensionality reduction is using the probabilistic models. These models, which include generative topological mapping, factor analysis, independent component analysis and probabilistic latent semantic analysis(PLSA), are generally specified in terms of an underlying independence assumption or low-rank assumption. The models are generally fit with maximum likelihood, although Bayesian methods are sometimes used [9].

In LSA [2], the singular value decomposition (SVD) is used to extract the latent semantic features, which form a low dimensional vector space. The shortcoming of LSA is the lack of statistical interpretation. By using the conditional probability to interpret the semantic feature (latent topic) z_i, Hofmann presented PLSA (Fig. 1) [3]. Blei added a Dirichlet prior assumption and obtained Latent Dirichlet Allocation (LDA) generative model (Fig. 2) [4]. In particular, LDA is a Bayesian model in the spirit of PLSA that models each data point as a collection of draws from a mixture model in which each mixture component is known as a topic [5]. Typical LSA, PLSA and LDA are unsupervised methods, in which the documents' categories are not taken into account. Those methods can be employed in document classification. Blei obtained a generative model for classification by using one LDA module for each class [5]. And he believed the LDA also can be applied in the discriminative framework [5].

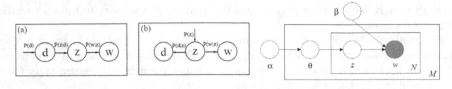

Fig. 1. PLSA   **Fig. 2.** LDA

Currently, many supervised topic models have been presented. Blei added LDA a response variable to get supervised latent Dirichlet allocation (sLDA), which was used to predict movie ratings and web page popularity (Fig. 3) [8]. A discriminative variation on LDA named DiscLDA employs a class-dependent linear transformation on the topic mixture proportions (Fig. 4) [9]. And Xue proposed topic-bridged PLSA for cross-domain text classification [7]. Some other methods also make use of category information to obtain topics or clusters [10] [11] [13]. One limitation of these models is that a topic often belongs to only one category, which means the conditional probability of category c_k and topic z_i, $P(c_k|z_i)$, has a binary value (1 or 0), such as in [4], [7], and [10],etc.. Another limitation is that topics are always regarded as unobservable, which means no supervised information about topics is employed. Sometimes topics are observable or semi-observable, and information from these observed topics can greatly improve the performance of filters. For example, [14] presented a semi-supervised topic model, which takes advantage of the high readability of the expert review to structure the unorganized ordinary opinions.

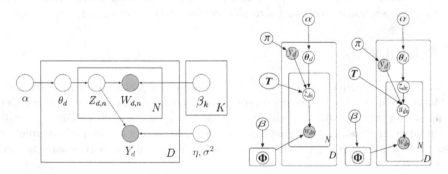

Fig. 3. sLDA   **Fig. 4.** DiscLDA

In this paper, we find that there are only two key conditional probabilities for TC: $P(W|Z)$ and $P(Z|C)$, and they are regarded as unknown and nonbinary. Z is the latent topic vector, C is the category tag vector, and W is word item vector. c-w PLSA model is first proposed to extract topics with sufficient labeled SMSs. In c-w PLSA, C and W are observable variables, and Z is latent. Labeled data are needed to estimate parameters. Unfortunately, we can't afford enough training data. A reasonable assumption is that more observed data is helpful for learning latent topic. Therefore, two generative models, c-w PLSA and typical PLSA, are combined to share observable variables in order to utilize other observed data. This is the dual-PLSA, which has an additional observable variable D. Furthermore, observable information about Z is employed to acquire more useful topic model. Since Z is no longer latent, this model is called supervised dual-PLSA.

3 c-w PLSA Model

Assume that category set is $C=\{c_1=\text{spam}, c_2=\text{ham}\}$, and word set is $W=\{w_1, ..., w_M\}$. Each SMS is represented as a "bag-of-words" vector. The posterior probability of an input SMS d_l is:

$$P(c_k \mid d_l) = P(d_l \mid c_k)P(c_k)/P(d_l) = \frac{P(c_k)}{P(d_l)} \prod_{j=1}^{n(d_l)} P(w_j \mid c_k) \quad (1)$$

$n(d_l)$ is the total word number of d_l. Assume it has N latent topic, $Z=\{z_1,...,z_N\}$, then:

$$P(w_j \mid c_k) = \sum_{i=1}^{N} P(w_j \mid z_i)P(z_i \mid c_k) \quad (2)$$

From (1) and (2) we can get:

$$P(c_k \mid d_l) = \frac{P(c_k)}{P(d_l)} \prod_{j=1}^{n(d_l)} \sum_{i=1}^{N} P(w_j \mid z_i)P(z_i \mid c_k) \quad (3)$$

(3) means three kinds of information are needed to filter d_l: what words are contained in d_l, $P(w_j|z_i)$ and $P(z_i|c_k)$. And only $P(w_j|z_i)$ and $P(z_i|c_k)$ are need to be learned. Therefore, we present c-w PLSA model shown in expression (4) and figure 5. C and W are observable variables, and Z is latent. Labeled data is needed to estimate $P(W|Z)$. Unfortunately, we can't afford enough training data, and then the dual-PLSA model is presented.

$$P(c_k, w_j) = P(c_k)P(w_j \mid c_k) = P(c_k)\sum_{i=1}^{N} P(w_j \mid z_i)P(z_i \mid c_k) \quad (4)$$

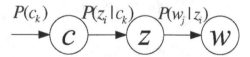

Fig. 5. c-w PLSA model. C and W are observable variables, and Z is latent.

4 Dual-PLSA Model

The c-w PLSA is similar with the typical PLSA (shown in figure 1), and they share the same conditional probability, $P(w_j|z_i)$. However, the c-w PLSA needs plenty of labeled data. In our problem, labeled data denoted as D^l is deficient, while unlabeled data denoted as D^u is abundant. Because typical PLSA model is suitable for unsupervised learning, we employ a PLSA model, named d-w PLSA, to extract latent topic from unlabeled data. And we use the c-w PLSA to estimate $P(z_i|c_k)$ and to optimize $P(w_j|z_i)$. This combined model is the dual-PLSA, shown in Figure 6(a). Suppose d_l, c_k, and w_j are statistically independent on condition that z_i is given, the dual-PLSA model can be equivalent to Figure 6(b). However, whether dual-PLSA is a generative model is not our key problem. In this model, C, D and W are all observable variables. Since c-w PLSA and typical PLSA both have the same parameter $P(w_j|z_i)$, they can share the observed data to estimate this parameter.

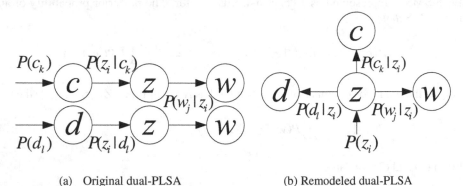

(a)   Original dual-PLSA   (b) Remodeled dual-PLSA

Fig. 6. dual-PLSA model. Suppose d_l, c_k, and w_j are statistically independent, given z_i, model (a) and model (b) are equivalent. In this model, C, D and W are all observable variables.

Now the co-occurrence of (d_l,w_j) can be counted from D^u; the co-occurrence of (c_k,w_j) can be counted from D^l; and the parameters to be estimated are $P(w_j|z_i)$ and $P(z_i|c_k)$. In this model, the hidden variables include latent topics Z and the category tags of all the SMSs in D^u. EM algorithm is employed to resolve this problem. $P^u(w_j|z_i)$ and $P^l(w_j|z_i)$ obtained from typical PLSA and c-w PLSA respectively are combined with a changeable radio to acquire a more accurate $P(w_j|z_i)$.

E-step is to update $P(z_i|d_l,w_j)$, $P(z_i|c_k,w_j)$ and $P(c_k|d_l)$ via (3), (5) and (6).

$$P(z_i \mid d_l, w_j) = \frac{P(w_j \mid z_i)P(z_i \mid d_l)}{\sum_{i'=1}^{N} P(w_j \mid z_{i'})P(z_{i'} \mid d_l)} \tag{5}$$

$$P(z_i \mid c_k, w_j) = \frac{P(w_j \mid z_i)P(z_i \mid c_k)}{\sum_{i'=1}^{N} P(w_j \mid z_{i'})P(z_{i'} \mid c_k)} \tag{6}$$

M-step is to update $P(w_j|z_i)$, $P(z_i|c_k)$ and $P(z_i|d_l)$ via (7), (8) and (11). $P(w_j|z_i)$ can be estimated by both d-w PLSA and c-w PLSA. $P^u(w_j|z_i)$, obtained from d-w PLSA, can't reflect the category information; while c-w PLSA can't acquire enough labeled data to obtain $P^l(w_j|z_i)$. So we weight sum $P^u(w_j|z_i)$ and $P^l(w_j|z_i)$ to get $P(w_j|z_i)$ via (11). In the beginning of EM iteration, $P^l(w_j|z_i)$ comes from unreliable $P(c_k|d_l)$, so the parameter a

should be setting approximated to 1. With the iteration time going up, a should decrease. In the end, a should be approximate to 0. $n(d_l, w_j)$ is counted from D^u, and $n(c_k, w_j)$ is counted from D^u and D^l.

$$P(z_i \mid d_l) = \frac{\sum_{j=1}^{M} n(d_l, w_j) P(z_i \mid d_l, w_j)}{n(d_l)} \tag{7}$$

$$P(z_i \mid c_k) = \frac{\sum_{j=1}^{M} n(c_k, w_j) P(z_i \mid c_k, w_j)}{n(c_k)} \tag{8}$$

$$P^u(w_j \mid z_i) = \frac{\sum_{l=1}^{L} n(d_l, w_j) P(z_i \mid d_l, w_j)}{\sum_{j=1}^{M} \sum_{l=1}^{L} n(d_l, w_j) P(z_i \mid d_l, w_j)} \tag{9}$$

$$P^l(w_j \mid z_i) = \frac{\sum_{k=1}^{K} n(c_k, w_j) P(z_i \mid c_k, w_j)}{\sum_{j=1}^{M} \sum_{k=1}^{K} n(c_k, w_j) P(z_i \mid c_k, w_j)} \tag{10}$$

$$P(w_j \mid z_i) = a \cdot P^u(w_j \mid z_i) + (1-a) \cdot P^l(w_j \mid z_i) \tag{11}$$

5   Supervised Topic Learning Dual-PLSA

In the above mentioned models, supervised information about topics is still not employed. Actually supervised information is indispensable to make the filter converged as quickly as possible. To acquire supervised information, some SMS taxonomies with labeled data can be utilized. For example, from a taxonomy we pick out categories which are related with spam filtering, and consider them as topics. Then the samples in the set D^l, which consists of SMSs labeled with these categories, are regarded as the observed values of topics. D^l can be used to estimate $P(w_j|z_i)$ directly when the following two conditions in formula (2) are satisfied:

(i) z_i and $z_j (i \neq j)$ are mutually exclusive;

(ii) $\sum_i P(z_i) = 1$.

Therefore, Algorithm 1, shown in table 1, is proposed. Initial value $P_0(w_j|z_i)$ is obtained from D^l via (11) before the EM iteration of dual-PLSA model.

$$P_0(w_j \mid z_i) = \frac{1 + n^t(w_j, z_i)}{M + \sum_{j'=1}^{M} n^t(w_{j'}, z_i)} \tag{12}$$

Table 1. Algorithm 1

Algorithm 1 Supervised dual-PLSA
Input: D^u, D^l, and D^t
Output: $P(w_j|z_i)$, $P(z_i|c_k)$
1. Set topic number N, and initialize $P(w_j|z_i)$ via (11). D^l is an empty set.
2. EM algorithm to estimate $P(w_j|z_i)$ and $P(z_i|d_l)$. That is to iterate (5)(7)(9).
3. for each SMS d_l do
4.   Calculate $P(c_k|d_l)$ by (3) and predict the category of d_l (spam or ham).
5.   Obtain the gold standard of d_l, and add d_l to D^l.
6.   The EM iteration of dual-PLSA to update $P(w_j|z_i)$ and $P(z_i|c_k)$. That is to iterate (3),(5)-(11).
7. end for

Algorithm 1 can't guarantee to be processed in real-time, then we simplify it and get algorithm 2. $P(w_j|z_i)$ is estimated only from D^t. And (13) is used to initialize $P_0(z_i|c_k)$ before the EM iteration.

$$P_0(z_i \mid c_k) = \frac{1 + n^l(c_k, z_i)}{N + \sum_{i'=1}^{N} n^l(c_k, z_{i'})} \tag{13}$$

Table 2. Algorithm 2

Algorithm 2 Supervised supervised dual-PLSA
Input: D^u, D^l, and D^t
Output: $P(w_j|z_i)$, $P(z_i|c_k)$
1. Set topic number N, and initialize $P(w_j|z_i)$ by (11). D^l is an empty set.
2. EM algorithm to estimate $P(w_j|z_i)$ and $P(z_i|d_l)$. That is to iterate (5)(7)(9).
3. for each SMS d_l do
4.   Calculate $P(c_k|d_l)$ by (3) and predict the category of d_l (spam or ham).
5.   Obtain the gold standard of d_l, and add d_l to D^l.
6.   Initialize $P(z_i|c_k)$ by (13)
7.   The EM iteration to update $P(z_i|c_k)$. That is to iterate (6)(8).
8. end for

6 Experiments

6.1 Datasets

Datasets include 4 SMS sets, named D^u, $D^{l\text{-}train}$, $D^{l\text{-}test}$, and D^t. D^u is unlabeled set, which contains 715,589 SMSs. Both $D^{l\text{-}train}$ is the set labeled with "spam" and $D^{l\text{-}test}$ is labeled with "ham". $D^{l\text{-}train}$ is used for training PLSA, so it needs plenty of samples. $D^{l\text{-}test}$ is a small set which is used for testing the convergence rate. The sample numbers of $D^{l\text{-}train}$ and $D^{l\text{-}test}$ are 310,000 and 10,000. D^t, built for supervised topic learning, is labeled with 5 tags (working, living, illegal ads, legitimate notification, and others). D^t has 91,403 samples.

Table 3. Datasets

Data	Description	Tags	Size
D^u	unlabeled set	NONE	715,589
$D^{l\text{-}train}$	for training c-w PLSA	spam, ham	310,000
$D^{l\text{-}test}$	for testing the convergence rate	spam, ham	10,000
D^t	for training z-w PLSA	working, living, illegal ads, legitimate notification, and others	91,403

6.2 Algorithms Explanation

Two different types of algorithms are employed for comparison with proposed c-w PLSA model and adaptive supervised dual-PLSA. The first one is Naïve Bayesian classifier (NBC). Before tested with $D^{l\text{-}test}$, no training is given to NBC, which implies all the filtering knowledge is acquired by adaptive learning from $D^{l\text{-}test}$. The second method is LDA discriminative approach with five topics, which is trained with D^u. When tested by $D^{l\text{-}test}$, a NBC is introduced to estimate $P(z_i|c_k)$ by adaptive learning. c-w PLSA model is trained with $D^{l\text{-}train}$. Adaptive supervised dual-PLSA is trained with D^t to obtain $P(w_j|z_i)$, and it estimate $P(z_i|c_k)$ by the adaptive learning in table 2.

Table 4. Compared Algorithms

Algorithm	Training Data	Testing Data	Adaptive Learning From Testing Data
NBC	NONE	$D^{l\text{-}test}$	Yes
LDA discriminative approach	D^u	$D^{l\text{-}test}$	Yes
c-w PLSA model	$D^{l\text{-}train}$	$D^{l\text{-}test}$	No
supervised dual-PLSA	D^t	$D^{l\text{-}test}$	Yes

6.3 Evaluation Metric

There exist many evaluation metrics for measuring the classification performance. In this paper, we employ the error rate, cumulative error rate and instant error rate for comparing different algorithms. The error rate is defined as: $E=|\{d|d\in D^{l\text{-}test} \wedge T(d)\neq L(d)\}|/|D^{l\text{-}test}|$. T is a function which maps from document d to its true class label $c = T(d)$, and L is the function which maps from document d to its prediction label $c = L(d)$ by the filter. The cumulative error rate is a function of processed messages number. Assume n is the processed messages number, and $D^{l\text{-}test}(n)$ is the processed messages set. The cumulative error rate is defined as: $E_c(n)=|\{d|d\in D^{l\text{-}test}(n) \wedge T(d)\neq L(d)\}|/|D^{l\text{-}test}(n)|$. The instant error rate is approximated by: $E_I(n)=|\{d|d\in D_n^{l\text{-}test} \wedge T(d)\neq L(d)\}|/|D_n^{l\text{-}test}|$; $D_n^{l\text{-}test}=\{ d_{n-Nw}, d_{n-Nw+1}, ..., d_n\}$; and Nw is set to 1000 here, which is the neighborhood size of the nth SMS.

6.4 Overall Performance and Conclusion

The experimental results are listed in table 5~7 and figure 7~9. To our surprise, neither c-w PLSA or LDA has better performance than NBC. c-w PLSA's error rate is 10.53%, though trained previously with 310,000 ham/spam labeled samples. By adaptive learning with only 2,000 gold standard feedback, other filters' cumulative error rates become lower than c-w PLSA. The reason maybe lies in table 5. From this table we can see that topic 4, with a poor discriminability, is the main component of both spam and ham. By checking $P(W|z_4)$, we find topic 4 includes many high-frequency words of $D^{l\text{-}train}$. Most of those words co-occur in spam and ham, yet a relatively considerable quantity of discriminable words lie either in spam or in ham. Then these words are wasted by topic 4. LDA is also not as good as we expect. Its error rate is 9.83%.The key reason of bad performance from c-w PLSA and LDA is that: topics in them are served to generate documents other than distinguish categories. Therefore, supervised information about topics is indispensable.

Table 5. $P(z_i|c_k)$ in c-w PLSA (c_1 =spam, c_2 =ham)

	z_1	z_2	z_3	z_4	z_5
c_1	0.0255	0.0119	0.0145	0.9300	0.0180
c_2	0.0116	0.0245	0.0194	0.9281	0.0163

Table 6. Error Rate

Algorithm	Testing Data	error rate
NBC	$D^{l\text{-}test}$	7.94%
LDA	$D^{l\text{-}test}$	9.83%
c-w PLSA	$D^{l\text{-}test}$	10.53%
dual-PLSA	$D^{l\text{-}test}$	6.94%

Table 7. Cumulative Error Rate

feedback	NBC	LDA	c-w PLSA	d-PLSA
100	25.51%	30.00%	7.00%	9.00%
200	17.26%	23.00%	10.50%	10.00%
300	13.49%	17.33%	11.33%	8.67%
400	13.35%	16.00%	11.00%	8.00%
500	11.21%	13.60%	11.00%	8.60%
1000	10.07%	10.30%	10.90%	8.00%
2000	9.50%	9.55%	9.90%	7.55%
3000	9.19%	9.73%	10.53%	7.60%
4000	8.76%	9.30%	10.20%	7.05%
5000	8.53%	9.64%	10.04%	6.86%

Table 7 and figure 8 show that dual-PLSA converges very fast. Within 100 gold standard feedback, the cumulative error rate drops to 9%. Its total error rate is 6.94%, which is the lowest. However, NBC needs 4,000 feedback to reduce the cumulative error rate to below 9%. And its total error rate is 7.94%. The cumulative error rate of NBC keeps slow descending in the whole test process.

The dual-PLSA's outstanding performance is due to supervised information about topics. Though it only has 5 topics, the meanings of those topics are determined by labeled data other than by unsupervised methods like LDA. And the real value of $P(z_i|c_k)$ are approximate to 1 or 0, which allow these probabilities to be estimated accurately and easily.

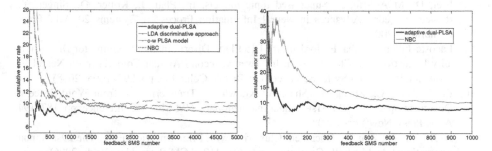

Fig. 7. Cumulative Error Rate(5000 feedback)   **Fig. 8.** Cumulative Error Rate(1000 feedback)

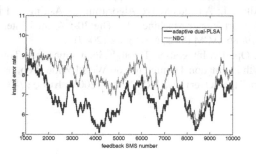

Fig. 9. Instant error rate

Acknowledgments. This work was supported by the national High-tech Research and Development Plan of China under grant No. 2007AA01Z417 and the 111 Project of China under grant No. B08004. It was also supported by Nokia Research Center (China). We thank the anonymous reviewers for their useful comments.

References

1. Sebastiani, F.: Machine Learning in Automated Text Categorization. ACM computing Surveys, 11–12, 32–33 (2002)
2. Deerwester, S., Dumais, S.T., Furnas, G.W., Landauer, T.K., Harshman, R.: Indexing By Latent Semantic Analysis. Journal of the American Society for Information Science and Technology, 391–407 (1990)
3. Hofmann, T.: Probabilistic Latent Semantic Indexing. In: The 22st Annual International ACM SIGIR Conference, pp. 50–57. ACM Press, New York (1999)

4. Lee, D.D., Sebastian Seung, H.: Learning the parts of objects by non-negative matrix factorization. Nature, 788–791 (1999)
5. Blei, D.M., Ng, A.Y., Jordan, M.I.: Latent Dirichlet allocation. Journal of Machine Learning Research, 993–1022 (2003)
6. Vapnik, V.N.: The Nature of Statistical Learning Theory. Springer, Heidelberg (2000)
7. Xue, G.-R., Dai, W., Yang, Q., Yu, Y.: Topic-bridged PLSA for cross-domain text classification. In: The 31st Annual International ACM SIGIR Conference, pp. 627–634. ACM Press, New York (2008)
8. Blei, D., McAuliffe, J.: Supervised topic models. In: Platt, J., Koller, D., Singer, Y., Roweis, S. (eds.) Advances in Neural Information Processing Systems 20. MIT Press, Cambridge (2008)
9. Lacoste-Julien, S., Sha, F., Jordan, M.: DiscLDA: Discriminative Learning for Dimensionality Reduction and Classification. In: Twenty-Second Annual Conference on Neural Information Processing Systems, Vancouver, British Columbia, pp. 897–1005 (2008)
10. Li, T., Ding, C., Zhang, Y., Shao, B.: Knowledge Transformation from Word Space to Document Space. In: The 31st Annual International ACM SIGIR Conference, pp. 187–194. ACM Press, New York (2008)
11. Ji, X., Xu, W., Zhu, S.: Document Clustering with Prior Knowledge. In: The 29st Annual International ACM SIGIR Conference, pp. 405–412. ACM Press, New York (2006)
12. Sindhwani, V., Keerthi, S.S.: Large Scale Semi-supervised Linear SVMs. In: The 29st Annual International ACM SIGIR Conference, pp. 477–484. ACM Press, New York (2006)
13. Raghavan, H., Allan, J.: An InterActive Algorithm for Asking and Incorporating Feature Feedback into Support Vector Machines. In: The 30st Annual International ACM SIGIR Conference, pp. 79–86. ACM Press, New York (2007)
14. Lu, Y., Zhai, C.: Opinion Integration Through Semi-supervised Topic Modeling. In: Proceeding of the 17th international conference on World Wide Web, pp. 121–130 (2008)
15. TREC 2007 Spam Track Overview, TREC 2007 (2007), http://trec.nist.gov

Enabling Effective User Interactions in Content-Based Image Retrieval

Haiming Liu[1], Srđan Zagorac[1], Victoria Uren[1,*], Dawei Song[2],
and Stefan Rüger[1]

[1] Knowledge Media Institute, The Open University, Milton Keynes, MK7 6AA, UK
[2] School of Computing, The Robert Gordon University, Aberdeen, AB25 1HG, UK
{h.liu,s.zagorac,s.rueger}@open.ac.uk
v.uren@dcs.shef.ac.uk, d.song@rgu.ac.uk

Abstract. This paper presents an interactive content-based image retrieval framework—uInteract, for delivering a novel four-factor user interaction model visually. The four-factor user interaction model is an
interactive relevance feedback mechanism that we proposed, aiming to
improve the interaction between users and the CBIR system and in turn
users overall search experience. In this paper, we present how the framework is developed to deliver the four-factor user interaction model, and
how the visual interface is designed to support user interaction activities. From our preliminary user evaluation result on the ease of use and
usefulness of the proposed framework, we have learnt what the users like
about the framework and the aspects we could improve in future studies.
Whilst the framework is developed for our research purposes, we believe
the functionalities could be adapted to any content-based image search
framework.

Keywords: uInteract, four-factor user interaction model, content-based
image retrieval.

1 Introduction

Content-based image retrieval (CBIR) has been researched for decades, but it
has yet to be widely applied in common Web search engines. In our view, one of
the reasons for this is that CBIR is normally performed by computing the dissimilarity, e.g., Euclidean distance, between objects and queries based on their
multidimensional feature vectors in content feature spaces, for example, colour,
texture and structure features. There is a well-known gap, called the "semantic
gap", between low-level feature of an image and its high-level meaning to a user.
Another reason that CBIR is not yet widely used is that most existing CBIR systems are designed principally for evaluating search accuracy. Less attention has
been paid to designing interactive visual systems that support users in grasping
how feedback algorithms work and how they can be manipulated.

---

* Present address: Department of Computer Science, Regent Court, 211 Portobello,
  University of Sheffield, Sheffield, S1 4DP United Kingdom.

G.G. Lee et al. (Eds.): AIRS 2009, LNCS 5839, pp. 265–276, 2009.
© Springer-Verlag Berlin Heidelberg 2009

One way to bridge this gap and increase the use of CBIR systems is to make the CBIR system more human-centric. A human-centric system should deliver a user-oriented search making the user feel that they, rather than the system, are driving the search process. In [1], Bates addressed two issues for search system design: "(1) the degree of user vs. system involvement in the search, and (2) the size, or chunking, of activities; that is, how much and what type of activity the user should be able to direct the system to do at once." To investigate the first issue, we had developed an interactive relevance feedback (RF) mechanism named four-factor user interaction model in our early research, which aims to improve the interaction between users and the system as well as to improve the search accuracy. According to the results of our simulated experiments, the model can improve the search accuracy in some circumstances. However, we are not able to do user evaluation on the ease of use and usefulness of the interactive functionalities without an effect visual search interface.

In terms of the second issue, White, et al. in [13] has also addressed "When providing new search functionality, system designers must decide how the new functionality should be offered to users. One major choice is between (a) offering automatic features that require little human input but give little human control; or (b) interactive features which allow human control over how the feature is used, but often give little guidance over how the feature should be best used." One question arises here for our study: How should the functionalities be presented visually to the user by the interface to enable users to directly control the model in a effective way?

In this paper, we introduce a novel CBIR framework, which delivers a four-factor user interaction model we have developed aiming at providing a user-oriented search platform. The framework provides functionalities that support the user's interactive search process and allow the user to control all four factors in our model. The design of an innovative CBIR search interface is the main focus of the paper.

2   Background and Motivation

This section will review the related work and explain our motivation on resolving the interaction issue (the degree of search control deployed to users and system) and the design issue (the best way to deliver the framework functionalities to users through interface) from two aspects, namely: the user interaction models and interactive search interface.

2.1   User Interaction Models

In this section, we review and analyze a number of existing UI models that inspired our four-factor user interaction model.

In [10] Spink, et al. proposed the three-dimensional spatial model to support the user interactive search for text retrieval. The model emphasizes that partial relevance is as important as binary relevance / non-relevance and indeed it can be more important to the inexperienced user. The three dimensions are: levels of relevance, regions of relevance and time. The levels of relevance indicates why a particular document is relevant to the user. The regions of relevance indicates how relevant the document is. And the third dimension - time - captures information seeking stage and successive searches. We consider that their model is a useful foundation from which to develop further detailed user interaction models and techniques for CBIR.

Other research has tended to focus more on a single dimension, such as time. For example, Campbell in [2] proposed the Ostensive Model that indicates the degree of relevance relative to when a user selected the evidence from the results set. The model includes four relevance profiles: increasing, decreasing, flat and current profiles. Later, Urban, et al. applied the increasing profile to CBIR [12]. Their preliminary study showed that the system based on the Ostensive Model was preferred by users over traditional CBIR search engines.

Ruthven, et al. [9] adapted and combined two dimensions from Spink, et al. three-dimensional spatial model, namely: regions of relevance and time to text retrieval. Their experimental results showed that combining partial and time relevance criteria does help the interaction between the user and the system. It will be interesting to see how the combined model performs in our CBIR framework.

Most of these models were tested and applied in text retrieval, we were motivated to investigate what the outcome would be were we to adapt combined three-dimensional spatial model with the Ostensive Model together, and, further, to add another factor - frequency - to the combination, into CBIR. Therefore, we developed a new model for CBIR, namely 'four-factor user interaction model', which includes relevance region, relevance level, time and frequency [6]. Our hypothesis is that the four-factor user interaction model will provide enhanced search experience in terms of the level of interaction between the system and users and the search accuracy. However, without a visual search interface, we are not able to test the interaction aspects. This motivated us to develop a novel interactive CBIR framework and an innovative interactive user interface in order to enable effective user interactions and evaluate the model with real, as opposed to simulated, users.

2.2   Interactive Search Interface

When providing new search functionality, we should decide how the new functionality should be delivered to users [1, 13]. In this section we investigate a number of search interfaces in order to explain why we developed the search interface in the way we did.

Flexible Image Retrieval Engine (FIRE) [3] is one tool that allows users to provide non-relevant feedback from the result set. The research in [4, 7, 8] also usefully referred to the importance of providing both negative and positive examples as feedback. In addition, from the result of our simulated experiments, we found that limiting user's selection of non-relevant feedback to the poorest matches in the results list will improve search accuracy, but we realized this is not going to be intuitive to users. Therefore, we are encouraged to design the system to enable users to provide the negative examples from the worst matches in a natural way.

Urban, et al. developed an image search system based on the Ostensive Model [12]. Like FIRE, this is a browsing based search system, which uses a dynamic tree view to display the query path and results, thus enabling users to re-use their previous queries at a later stage. Whilst the query path functionality is useful, the user display becomes overly crowded even after a relatively small number of iterations. This limitation would become even more evident were the system to allow the user to provide negative as well as positive examples. Why not then harness the benefits of the query path functionality but in a search-based system, which separates query and results and applies the linear display to both queries and results?

Later, Urban, et al. in [11] presented another system—Effective Group Organization (EGO), which is as a personalized image search and management tool that allows the user to search and group the results. The user's groupings are then used to influence the outcome of the results of the next search iteration. This system supports long-term user and search activity by capturing the user's personalized grouping history, allowing users to break and re-commence later without the need to re-create their search groupings from scratch. From this study, we can see that providing personalized user search history can improve the interaction between the system and users.

In [5] Hopfgartner, et al. defined explicit and implicit feedback. They concluded that explicit feedback is given actively and consciously by the user to instruct the system what to do. Whereas, implicit feedback is inferred by the system from the what the user has done unconsciously and here the system assumes or infers the user's purpose. In other words, explicit feedback means the user is actively controlling the search process and implicit feedback means the system is exercising control over the search process. Their simulated user study results showed that combining implicit RF with explicit RF may provide better search results than explicit RF by itself. We are then encouraged to combine the implicit and explicit RF in our system.

In the following sections, we will present our proposed interactive CBIR framework, namely uInteract, which will implement the ideas we have developed to overcome the shortcomings of the related work and to apply the inspiration from the related work. Table 1 shows how the related work maps to the features of

Table 1. How the related work maps to the features of uInteract

Feature	Deselaers et al.	Urban et al.	Urban et al.(EGO)	Hopfgartner et al.	uInteract
Search-based system	No	No	Yes	No	Yes
Providing positive feedback	Yes	Yes	Yes	Yes	Yes
Providing negative feedback	Yes	No	No	Yes	Yes
Range of (non)relevant level	No	No	No	No	Yes
Query history functionality	No	Yes	No	No	Yes
Showing negative result	No	No	No	No	Yes

the uInteract (note that in this paper we only compare the CBIR features and ignore the textual search features). Moreover, the next sections will describe how we developed uInteract to deliver our four-factor user interaction model.

3   uInteract - An Interactive CBIR Framework

Our framework consists of two logically different modules, an interaction model based frontend and a backend search engine with a fusion controller supported by the interaction model. In addition, the login functionality of the framework supports personalized long-term searching.

3.1   Four-Factor User Interaction Model

The four factors taken into account in the model are relevance region, relevance level, time and frequency. The relevance region comprises two parts: relevant (positive) evidence and non-relevant (negative) evidence. The relevance level here is a quantitative level, which indicates how relevant/non-relevant the evidence is, and differs from the original qualitative level presented in the work of Spink, et al. We adapted the Ostensive Model to the time factor to indicate the degree of relevance/non-relevance relative to when the evidence was selected, which is different from the original OM that only applies to positive evidence. The new factor - frequency - captures the number of appearances of an image in the user selected evidence both for positive and negative evidence separately.

The weight scheme of the model is given by W_{opf}, which is the multiplication of W_o and W_p and W_f [6]. W_o is the ostensive weight, which can be different depending on the profile of the Ostensive Model: e.g., increasing profile (the earlier selected images deemed less important), decreasing profile (the earlier selected images deemed more important), flat profile (all the selected images treated equally) and current profile (only the latest selected images considered). The value of W_o relies on the number of search iterations when the query was applied; W_p is the partial weight, which is the range of relevance/non-relevance level based on the relevance and non-relevance regions. The value of W_p depends on the score of the positive and negative feedback provided by the users; W_f is the frequency weight, which is the number of times an image appears (frequency) in the query across all the iterations. The value of W_f is the frequency that the image is taken as a query image.

3.2   Two Fusion Approaches

In many CBIR applications, it is important to take more than one image as a query and so a fusion approach is needed to merge the results of each query image. We used two separate fusion approaches to support two different relevance feedback scenarios: Firstly, the vector space model (VSM) [8] approach was deployed for positive query images only. By combining this with the four-factor user interaction model the approach is represented by:

$$D = \sum_i (D_{ij}/(W_{opf})), \qquad (1)$$

where the D is the overall dissimilarity value between a query (containing a number of positive examples) and an object image. It is computed as the sum of the dissimilarity values between all the example images in the query and the object image. D_{ij} is the original distance between the feature vectors for the query image i and the object image j. W_{opf} is the weight scheme of the four-factor user interaction model (see section 3.1).

Secondly, because the VSM in [8] only uses positive RF, we applied k-Nearest Neighbours (k-NN) for both positive and negative RF [8]. Here, by taking into account the weighting scheme of our four-factor user interaction model, k-NN is given by:

$$D = \frac{\sum_{i \in N}(D_{ij}/W_{opf} + \varepsilon)^{-1}}{\sum_{i \in P}(D_{ij}/W_{opf} + \varepsilon)^{-1} + \varepsilon}, \qquad (2)$$

where D is the dissimilarity value between an object image and the example images (positive and negative) in a query. ε is a small positive number to avoid division by zero. N and P denote the sets of positive and negative images in the query.

4   uInteract - The Interface

In our view a appropriate interface is vital to allow our new interaction CBIR framework to fully function, because the interface is the communication platform between the system and user. We will outline our developed interface and describe how it underpins the four-factor interaction model.

Table 2. Which parts of the interface support the four-factor user interaction model

Factor	Functionality
Relevance region	Positive and negative feedback in [2] and [7]
Relevance level	score in [3]
Time	Positive and negative query history in [8] and [9]
Frequency	Positive and negative query history in [8] and [9]

The search interface (see Figure 1) takes on a simple search-based grid style so that the user does not need to learn the new visual layout before they start a search. Different colour backgrounds have been applied to the different panels which is aimed at supporting user navigation and appreciation of the differences between the panels. Each panel provides a different level of interaction to the user, where some of the four factors are controlled indirectly and others more directly. Table 2 shows how the interface supports each of the four factors (note: the numbers on the table indicate the functionalities on the screen shot). The rest of this section describes the features of those panels.

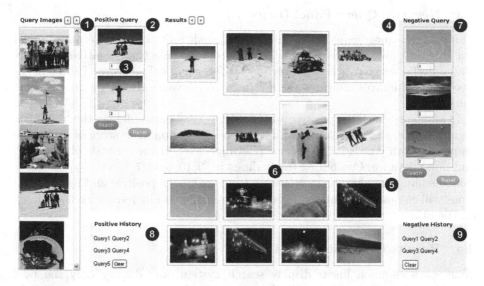

Fig. 1. The uInteract interface. Key: [1] The browsing based query images where the initial query is selected; the initial query images go into [2] as positive query to start a search; users can score (integer 1-20, bigger is better) the selected images in [3] with their preference; [4] and [5] the search result shows the best matches and worst matches to the query respectively; [6] a horizontal line divides the two parts of result visually; [7] negative query examples that users selected from previous results; [8] positive query history records the positive queries was used before; [9] negative query history records the negative queries from previous search.

4.1 Query Image Browsing Panel (Region 1)

The query image panel is a browsing panel. The user browses the query panel and selects one or more images from the provided query images as an initial query image(s) prior to starting the search.

4.2 Positive Query Panel (Region 2)

The positive query panel contains images that the user considers are good positive examples of what they are searching for. Users can provide as many images as they want as positive queries. These images can be selected from the query images, the search results or a combination of both. Users are also able to eliminate positive examples by simply clicking on them.

After the user selects positive images, the system automatically gives their importance score by their display order. If the user is not happy with the default score, he can re-score the importance of the images by changing the number (integer 1-20, bigger is better) in the text box underneath each image. This functionality delivers the 'relevance level' factor. The intention of the design is to provide users an explicit control to the importance level of query image examples.

4.3   Negative Query Panel (Region 7)

The negative query panel has similar functionality to the positive query panel but this time for negative queries. The only difference is that negative examples may only be selected from the previous search results. The score of these negative example images indicates the level of non-relevance (integer 1-20, bigger is worse).

In summary, both the positive and negative query panels deliver the 'relevance region' factor, such as relevant and non-relevant region. The score of image examples in both panels indicates the 'relevance level' factor—a scale of relevance and non-relevance. Combining the findings in [9, 10] and [7, 8], our hypothesis is that blending the non-binary relevance level with both positive and negative regions will enhance user interaction on the one hand and increase search accuracy on the other.

4.4   Results Panel (Regions 4 and 5)

Whereas a common linear display search system may display only the best matching results, our system displays both the best and poorest matches. In our view this added functionality allows users to gain a better understanding of the data set they are searching. By seeing both good and bad results, the user can gain better understanding of the data they are searching. Additionally, for experienced users, the extreme results can aid their special search purposes, for instance, when a user searches for two extremely different colour images, say one pink and one blue.

Furthermore, users can indicate positive examples from the good matches and negative examples from the poorest matches by selecting them with a single mouse click. The selected images will appear automatically in either the positive or negative query panels. According to our simulated experimental results, taking the worse matches as negative query examples outperforms the query example from good matches. Therefore, we designed the interface to support the search mechanism by showing the poorest as well as the best matches. Users will need some training on the way that the interface works. We assume that the users will be able to search naturally after a couple of search iterations although this functionality is not intuitive to start with.

To aid navigation, we have inserted a horizontal line between the good and bad results to clearly divide the two.

4.5   Positive History Panel (Region 8)

This is an important feature of our search system. This panel records the user's earlier positive queries used during previous search iterations. This enables the user to go back and reuse a previous query if required. This might be needed, for instance, if the user got lost during the search process.

In addition, this panel delivers two important factors to our four-factor user interaction model: Firstly, the 'time' factor which is computed by the Ostensive Model and takes a search iteration as a time unit. Secondly, the 'frequency' factor that judges the importance of an image by reference to how many time the image was used as a query.

These two factors are fully controlled by the system, and all previous queries will be taken into account in the final weighting scheme.

4.6 Negative History Panel (Region 9)

This panel is similar to the positive history panel but instead records the negative queries selected from each search iteration. The negative query history is introduced together with the negative query as two of the new features of our search interface. The introduction of query history functionality has been encouraged in [2,12] and we would like to investigate the effects on user interaction and search accuracy by adding the negative factor.

4.7 Summary

In summary, the key features of the proposed interface are:

(1) Users can provide both positive and negative examples to a search query, and further expand or reformulate the query. This is a way to deliver the 'relevance region' factor.

(2) By allowing the user to override the automatically generated score of positive and negative query images, we are enabling the user to directly influence the importance level of the feedback. The 'relevance level' factor is generated by the score functionality.

(3) The display of the results in the interface takes a search-based linear display format but with the addition of showing not only the best matches but also the worst matches. This functionality aims to enable users to control the model directly in a natural way.

(4) The query history not only provides users with the ability to reuse their previous queries, but also enables them to expand future search queries by taking previous queries into account. The positive and negative history panels together with the current query feed the 'time' and 'frequency' factor of our four-factor user interaction model.

5 Evaluation

The goal of our evaluation is testing whether users can adapt to the new search strategy easily and whether users find the uInteract system useful.

5.1 Evaluation Baseline

The baseline systems we used to compare with the uInteract system are: baseline1 - a typical Relevance Feedback (RF) mechanism, where users are allowed to give positive RF from search results through a simplified interface; baseline2 - a system based on Urban, et al. model [12] provides positive query history functionality which is in addition to baseline1; baseline3 - a system based on Ruthven, et al. model [9] enhances baseline2 by adding partial relevance (we call it **importance score** in this paper) functionality.

5.2   Evaluation Setup

A total of 17 subjects, who are a mixture of males and females, undergraduate and postgraduate students, and academic staff from a variety of departments with different age and levels of image search experience, participated in the evaluation. Subjects can be classified into two groups - younger/older and inexperienced/experienced - based on age and image search experience respectively. We take 26 as the age cut off point. Everybody older then 26 is in the older group, otherwise they are in the younger group. We consider that people are experienced users if they search images at least once a week, and otherwise they are inexperienced users.

The subjects attempted four different complexity level search tasks on the four systems randomly in a random order (limited five minutes for each task) and provided feedback on their experience through questionnaires and comments made during informal discussions. The complexity of the tasks is based on the task description. Task one provides both search topic and example images, so we consider it is the easiest task. Task two gives example images without a topic description. Task three has only a topic but no image examples. Task four described a broad search scenario without any clear topic and image examples, so it is the most complex task in our view. The questionnaires used five point Likert scales. The questions are about the general feeling, ease of use, novelty, search result, search performance and satisfaction with the systems, etc.

5.3   Evaluation Results

The following preliminary results were obtained based on the ANOVA analysis (with $\alpha = 0.05$) of questionnaires and informal feedbacks from the subjects.

(1) The uInteract system gain higher novelty score than the baseline1 (p-value$\leq$0.05). The search result of uInteract showed better result than baseline1 (p-value$\leq$0.05). This result is promising, which encouraged us to do further detailed analysis on other data we obtained from the evaluation, such as, videos and actual image search results.

(2) Task factor had very strong impact on users' responses to most of the questions. Most of the subjects thought that task two and task four, and task three and task four were significantly different (p-value$\leq$0.05). One interesting observation is that the subjects liked to give higher scores to the questions when they performed easier tasks. Due to the tasks strong impact on the results, the score of different systems will become less sensitive.

(3) The person factor is another important factor which affects some of the question results. Different subjects seemed to have very different opinions on scoring the questions, but what is the trend on this factor? In an effort to find the trend, we took the age and image search experience into account. The statistical results showed that older subjects and experienced subjects feel more comfortable with using the systems than younger and inexperienced subjects.

(4) The older subjects thought the systems helped them better understand the quality of the results they could get from the collection data more than younger

subjects, and the score is estimated to increase by 0.05 per year as age increases. There is a clear evidence that age and image search experience affect system satisfaction, and the score is estimated to be higher for older and experienced users. There is also clear evidence that age and image search experience affect feeling in control using the system and search result satisfaction, but the score is estimated to be higher for younger and inexperienced users.

(5) The image search experience of the subjects also impacted on the opinions of query history usefulness, negative query ease of use and negative result usefulness. The experienced subjects gave higher score on these questions than inexperienced subjects (p-value$\leq$0.05).

(6) While the subjects though the uInterace framework was useful, they felt the ease of use of the functionalities can be improved by making the negative result optional, so that subjects can see and use it when they need to do, and using drag and drop to higher or lower position to indicate the importance of the query images in the query panel, enabling them to show the images in every query when the mouse is over to the query history labels, etc.

6   Conclusions and Future Work

This work was undertaken to achieve three objectives: (a) to deliver an effective interactive CBIR framework, in particular through a novel four-factor user interaction model, (b) to design the interaction activities of the interface to enable users to directly control the model in a natural way, and (c) to test the easy of use and usefulness of the new search functionalities through a use study.

The uInterace framework was built on our observations about the insufficiencies of previous research and inspirations from the literature. In order to evaluate the interactive framework, we have done a user evaluation on the ease of use and usefulness of the framework. The preliminary evaluation result showed, while the framework was considered as easy of use and useful in general, there are still places to improve and more data to analyze.

Our next step is to extract some factors from the captured video of the evaluation, such as, how many queries did the subjects use to complete the tasks, how many pages did the subjects look through the results, etc. Furthermore, we will also look into the image result data the subjects selected for completing each tasks, and analyze the difference on the results of the evaluated systems.

Acknowledgments

This work was partially supported by the PHAROS project sponsored by the Commission of the European Communities as part of the Information Society Technologies programme under grant number IST-FP6-45035, and AutoAdapt project funded by the UK's Engineering and Physical Sciences Research Council, grant number EP/F035705/1.

References

1. Bates, M.J.: Where should the person stop and the information search interface start? Information Processing and Management 26(5), 575–591 (1990)
2. Campbell, I.: Interactive evaluation of the ostensive model using a new test collection of images with multiple relevance assessments. Journal of Information Retrieval 2(1) (2000)
3. Deselaers, T., Keysers, D., Ney, H.: Fire – flexible image retrieval engine: Imageclef 2004 evaluation. In: Peters, C., Clough, P., Gonzalo, J., Jones, G.J.F., Kluck, M., Magnini, B. (eds.) CLEF 2004. LNCS, vol. 3491, pp. 688–698. Springer, Heidelberg (2005)
4. Heesch, D., Rüger, S.: Performance boosting with three mouse clicks-relevance feedback for CBIR. In: Proceeding of the European Conference on IR Research 2003 (2003)
5. Hopfgartner, F., Urban, J., Villa, R., Jose, J.: Simulated testing of an adaptive multimedia information retrieval system. In: Proceeding of Content-Based Multimedia Indexing (CBMI), pp. 328–335 (2007)
6. Liu, H., Uren, V., Song, D., Rüger, S.: A four-factor user interaction model for content-based image retrieval. In: Proceeding of the 2nd international conference on the theory of information retrieval, ICTIR (2009)
7. Müller, H., Müller, W., Marchand-Maillet, S., Pun, T.: Strategies for positive and negative relevance feedback in image retrieval. In: Proceedings of the International Conference on Pattern Recognition (ICPR 2000), Barcelona, Spain, September 2000, vol. 1, pp. 1043–1046 (2000)
8. Pickering, M.J., Rüger, S.: Evaluation of key frame-based retrieval techniques for video. Computer Vision and Image Understanding 92(2-3), 217–235 (2003)
9. Ruthven, I., Lalmas, M., van Rijsbergen, K.: Incorporating user search behaviour into relevance feedback. Journal of the American Society for Information Science and Technology 54(6), 528–548 (2003)
10. Spink, A., Greisdorf, H., Bateman, J.: From highly relevant to not relevant: examining different regions of relevance. Information Processing Management 34(5), 599–621 (1998)
11. Urban, J., Jose, J.M.: Ego: A personalized multimedia management and retrieval tool. International Journal of Intelligent Systems 21, 725–745 (2006)
12. Urban, J., Jose, J.M., van Rijsbergen, K.: An adaptive technique for content-based image retrieval. Multimedia Tools and Applications 31, 1–28 (2006)
13. White, R.W., Ruthven, I.: A study of interface support mechanisms for interactive information retrieval. Journal of the American Society for Information Science and Technology (2006)

Improving Text Rankers by Term Locality Contexts

Rey-Long Liu and Zong-Xing Lin

Department of Medical Informatics
Tzu Chi University
Hualien, Taiwan, R.O.C.
rlliutcu@mail.tcu.edu.tw

Abstract. When ranking texts retrieved for a query, semantics of each term t in the texts is a fundamental basis. The semantics often depends on locality context (neighboring) terms of t in the texts. In this paper, we present a technique CTFA4TR that improves text rankers by encoding the term locality contexts to the assessment of term frequency (TF) of each term in the texts. Results of the TF assessment may be directly used to improve various kinds of text rankers, without calling for any revisions to algorithms and development processes of the rankers. Moreover, CTFA4TR is efficient to conduct the TF assessment online, and neither training process nor training data is required. Empirical evaluation shows that CTFA4TR significantly improves various kinds of text rankers. The contributions are of practical significance, since many text rankers were developed, and if they consider TF in ranking, CTFA4TR may be used to enhance their performance, without incurring any cost to them.

Keywords: Text Ranking, Term Locality Context, Term Frequency Assessment.

1 Introduction

Ranking of those texts that are retrieved by natural language queries is essential. It helps users to easily access those texts that are relevant to the queries, and hence significantly reduces the gap between users and the huge information space. Main difficulties of text ranking lie on identifying semantics of each term, which is often diverse and depends on different contexts of discussion.

In this paper, we explore how contextual information may be encoded to improve existing ranking techniques so that more information relevant to natural language queries may be ranked higher for users to access. More specifically, we focus on context-based assessment of the *term frequency* (TF) of each term in a document[1]. Since TF is a fundamental component of many ranking techniques, the refined TF assessment may be used to improve the ranking techniques.

Therefore, we develop a context-based TF assessment technique CTFA4TR (*Context-based TF Assessment for Text Ranking*). To assess TF of a term t, CTFA4TR employs locality contexts of t, based on the observation that semantics of a term t in a

---

[1] TF of a term t in a document d is the times of occurrences of t in d.

G.G. Lee et al. (Eds.): AIRS 2009, LNCS 5839, pp. 277–288, 2009.
© Springer-Verlag Berlin Heidelberg 2009

document d often depends on locality context (neighboring) terms of t in d. We will show that, by encoding the locality contexts into TF assessment, many ranking techniques are more capable of ranking relevant information higher.

In the next section, we discuss related studies. CTFA4TR is then presented in Section 3. To evaluate CTFA4TR, experiments are conducted on OHSUMED [6], which is a popular collection of medical queries and literature (Section 4). CTFA4TR is efficient online, and the experimental results show that it may significantly improve many ranking techniques under various environmental settings.

2 Related Work

To rank texts with respect to a query, systems often assign a score to each text. Therefore, many scoring methods were developed. They often estimated how the text matched the query, and hence each method was also treated as a *feature* of the text (for a list of popular features, the reader is referred to [9]). Since each of the features has its relative strengths and weaknesses, many techniques were developed to integrate the features in the hope to achieve better ranking. The problem thus turned to *learning to rank* – the system actually learns to properly integrate the features to produce better ranking. Typical methods included support vector machines (RankingSVM [8] and its variants such as [3]), boosting [8][17], neural networks [4], association rules [15], and genetic algorithms [18].

In contrast to the previous studies, we explore how the features (scoring methods) may be improved, so that those ranking techniques that employ the features may be improved as well. We employ term locality contexts to improve the features. Actually, contributions of term contexts to the retrieval of information was noted in previous studies. For example, considerations of adjacent terms (in fixed order [16] or whatever order [13]), nearby terms (e.g. in a locality distance smaller than 3 to 5 words [1][5]), and distances between each pair of query terms [2][11][14] were shown to be helpful.

In contrast to the previous approaches, we focus on encoding term contexts to refine TF assessment so that many text rankers may be improved, without needing to revise the algorithms and training processes of the rankers. Although some previous approaches employed term proximity (distances between pairs of query terms) to improve text rankers [2][11][14], they only focused on very few basic scoring methods (e.g., BM25 [12]) by directly adding a proximity-based score to the scores from the methods. Their applicability to other scoring methods deserve more exploration, since scoring methods may produce scores in different scales, making it difficult to set a universal weight for the proximity-based score. The scale problem is more complicated for those ranking techniques that integrate multiple scoring methods (features).

3 Context-Based Term Frequency Assessment

Fig. 1 illustrates an overview of CTFA4TR, which is proposed to improve text rankers by context-based TF assessment, without requiring any change to the rankers. The basic idea is that, for a term t appearing in both a query q and a text d, its semantics in q is more possible to be similar to its semantics in d if a higher percentage of terms in q appear around t (in d). In that case, TF of t in d should be amplified in order to direct the ranking techniques to promoting the score of d with respect to q.

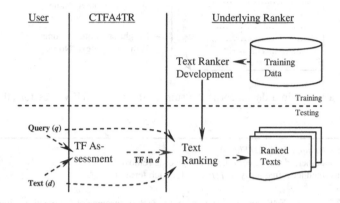

Fig. 1. Overview of CTFA4TR

Given a document d and a query q, CTFA4TR assesses TF of each term t in d with respect to q. As illustrated in Fig. 2, each occurrence of t in d may have its term contexts, including a *neighbor context* and a *window context*. As defined in Fig. 3, for each occurrence of t in d, CTFA4TR computes a weighted occurrence value (WeightedOcc) based on the contexts. If a term t in document d is not in query q, its WeightedOcc is simply 1 (as done by traditional TF assessment); otherwise CTFA4TR increments its WeightedOcc value by a linear combination of its neighbor context score and window context score (Equation B in Fig. 3). Refined TF of t is simply the sum of the weighted occurrence values for all occurrences of t in d (Equation A in Fig. 3).

3.1 Neighbor Context Score

For a query term t occurring at position k in document d, its neighbor context terms are those *distinct* query terms that *consecutively* appear around position k in d. For the example in Fig. 2, two terms (w_1 and w_2) serve as neighbor context terms for t at position t_1 if they are distinct terms in query q. Similarly, at position t_2, four terms (x_1~x_4) serve as neighbor context terms for t if they are distinct terms in query q. Certainly, some occurrences of t might have no neighbor terms (e.g., position t_3 in Fig. 2).

For a query term t occurring at position k in document d, CTFA4TR computes its neighbor context score by considering *percentage* of distinct query terms that serve as neighbor context terms (including t per se) around position k in d (ref., Equation C in Fig. 3). Obviously, the larger the percentage is, the larger the neighbor context score at position k should be. In that case, CTFA4TR increases the weighted occurrence value of t at position k, which in turn amplifies TF of t in order to direct various kinds of ranking techniques to promoting the score of d with respect to q.

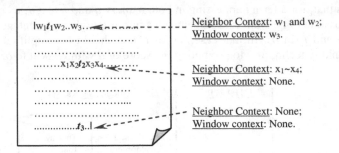

Fig. 2. TF of a term t in a document is the sum of the refined TF values for all occurrences $(t_1{\sim}t_3)$ of t

---

Procedure: *ComputeRefinedTF(d, q)*
Given: A document d and a query q, with stopwords removed;
Return: Refined term frequency $RTF(t, d, q)$ for each term t in d;
Equation:

    (A) $RTF(t, d, q) = \Sigma_k WeightedOcc(t, k, d, q)$, for each occurrence of t at position k in d.

    (B) *WeightedOcc* $(t, k, d, q) =$
        1, if t is not in query q;
        $1 + \alpha{\times}NeighborContext(k, d, q) + (1{-}\alpha){\times}WindowContext(k, d, q)$, otherwise.

    (C) *NeighborContext*$(k, d, q) =$
        0, if $N_q=1$;
        $Log_2 B_k / Log_2 N_q$, otherwise; where
        N_q = Number of distinct terms in q,
        B_k = Number of distinct terms in q that *consecutively* appear around position k in d.

    (D) *WindowContext*$(k, d, q) =$
        0, if $N_q=1$;
        1, if $N_q=B_k$;
        $(W_k{-}B_k) / (N_q{-}B_k)$, otherwise; where
        W_k = Number of distinct terms in q that appear in the term window [Max($1, k{-}2{\times}|q|$),
            Min($|d|, k{+}2{\times}|q|$)] in d.

Fig. 3. Refining TF assessment by term locality

3.2 Window Context Score

For a query term t occurring at position k in document d, its window context terms are those *distinct* query terms that (1) are not considered as neighbor context terms at position k, and (2) appear in a term window whose center lies at position k. For the example in Fig. 2, one term (w_3) serves as the window context term for t at position t_1 if it is a query term that appears in the term window at position t_1 and not considered as a neighbor context term at position t_1. Certainly, some occurrences of t might have no window context terms (e.g., positions t_2 and t_3 in Fig. 2).

For a query term t occurring at position k in document d, CTFA4TR computes its window context score by considering *percentage* of distinct query terms that serve as window context terms around position k in d (ref. Equation D in Fig. 3). The size of the window depends on the number of terms in the query–a longer query leads to a larger term window.

Therefore, both neighbor context scores and window context scores are computed by considering percentage of distinct query terms serving as contexts. CTFA4TR actually conducts *query-dependent* TF assessment—each term in a document may get different TF values with respect to different queries. Moreover, the relative weight of the neighbor context (i.e., α in Equation B) is set to 0.4, aiming to give a little bit more preference to the window context, since it governs the completeness of semantics of query q appearing in the term window.

It is interesting to note that, CTFA4TR mainly considers locality of query terms in the document being ranked. It is thus quite efficient to conduct TF assessment online[2], without requiring any training processes and data. Moreover, the TF assessment may be directly input to various kinds of ranking techniques, without requiring any revisions to the algorithms and development processes of the ranking techniques.

4 Empirical Evaluation

CTFA4TR is applied to several ranking techniques. To measure the contributions of CTFA4TR, experiments are designed and summarized in Table 1. Experimental results show that it may significantly improve the ranking techniques on ranking texts.

4.1 Experimental Data

Experimental data is from OHSUMED [6], which is a popular database of 348,566 medical references. OHSUMED contains 106 queries that are descriptions of medical information requests. It also provides 16,140 query-reference pairs that indicate the relevance of a reference to a query: *definitively relevant*, *possibly relevant*, and *not relevant*. The relevance information helps to evaluate contributions of CTFA4TR.

When ranking the references with respect to queries, we consider titles and abstracts of the references (i.e., the .T and .W fields in OHSUMED) and information requests of the queries (i.e., the .W field in OHSUMED). A few steps are conducted to preprocess the data, including removing stopwords, removing non-alphanumeric characters, changing all characters into lower case, and removing 's' if it is the last character of a term. Moreover, those references that do not have abstracts in OH-SUMED are not employed in training and testing if ranking is based on abstracts of the references.

To conduct 4-fold cross validation, we evenly partition the 16,140 pairs into four parts. Each experiment is conducted four times so that each part is used for testing only once (and the other three parts are for training). Average performance in the 4-fold experimentation is then reported. Moreover, to conduct more complete evaluation, we set up two kinds of partitions, which are named *CrossVaidation1* and *CrossValidation2*. Results on both partitions are reported separately.

---

[2] Suppose query terms appear p times in document d, CTFA4TR only checks neighbor contexts and window contexts for the p occurrences, and hence is more efficient than those previous approaches (e.g., [2][11][14]) that computed the distance between each pair of occurrences.

Table 1. Summary of experiment design

Aspect	Setting
Source of data	16140 query-reference pairs in OHSUMED
Cross validation	*CrossValidation1* and *CrossValidation2*: Two splits, with each partitioning the 16140 pairs into 4 folds for cross validation
Underlying rankers	(1) Individual ranking methods: 5 methods that consider TF in ranking in Table 2, with each applied to abstracts of references in OHSUMED (2) Integrative ranking techniques: 　(2A) *RankingSVM* that integrates 15 methods (5 methods that consider TF in ranking in Table 2, with each applied to 3 parts of each reference: title, abstract, and title+abstract) 　(2B) *RankingSVM* that integrates 21 methods (7 methods in Table 2, with each applied to 3 parts of each reference: title, abstract, and title+abstract)
Evaluation criteria	(1) *Mean Average Precision* (MAP) (2) *Normalized Discounted Cumulative Gain at x* (NDCG@x) where x=1~10

4.2 Underlying Rankers

Many basic ranking methods were implemented and tested in previous studies (e.g., [3][9][10]). For each document being ranked with respect to a query, each method produced a score, and hence each method was also treated as a *feature* of the document with respect to the query. In the experiment, we test the same set of features in [3], which is defined in Table 2. When performing ranking, the features are applied to the abstracts of the references in OHSUMED. Among the 7 features in Table 2, there are 5 features that consider TF of each term in the document being ranked (i.e., features L1, L4, L5, L6, and L7), and hence CTFA4TR is applicable to the 5 features. We aim at exploring the extent to which CTFA4TR improves them individually.

Moreover, as noted in Section 2, there were also techniques that aimed at achieving better ranking performance by integrating individual features. Among the integration techniques, we employed RankingSVM [8], which has been one of the most important techniques routinely tested in many previous studies (e.g., [3] [9] [15] [17] [18]). It models the ranking problem as a classification problem whose input is an instance pair and output may be one of two categories: correctly ranked and incorrectly ranked. We employ SVM^{light} to implement RankingSVM[3].

To conduct more complete evaluation, two feature sets are constructed for RankingSVM. The first feature set is derived by the 5 features that consider TF (recall L1, L4, L5, L6, and L7 in Table 2). Following [9], we apply the 5 features to 3 parts of each reference in OHSUMED—title, abstract, and both title and abstract. Therefore, there are 15 features (=5×3) in the feature set. On the other hand, the second feature set is built by applying all the 7 features in Table 2 to the 3 parts of each reference, and hence the second feature set contains 21 features (=7×3).

4.3 Evaluation Criteria

Two evaluation criteria are employed to measure the contributions of CTFA4TR: *mean average precision* (MAP) and *normalized discount cumulative gain at x*

---

[3] SVM^{light} is available at http://svmlight.joachims.org

Table 2. Individual ranking methods

Basic ranking methods (features)	Whether TF is considered
L1: $\displaystyle\sum_{q_i \in q \cap d} \log_2(c(q_i, d) + 1)$	Yes
L2: $\displaystyle\sum_{q_i \in q \cap d} \log_2(\frac{|C|}{c(q_i, C)} + 1)$	No
L3: $\displaystyle\sum_{q_i \in q \cap d} \log_2(idf(q_i) + 1)$	No
L4: $\displaystyle\sum_{q_i \in q \cap d} \log_2(\frac{c(q_i, d)}{|d|} + 1)$	Yes
L5: $\displaystyle\sum_{q_i \in q \cap d} \log_2(\frac{c(q_i, d)}{|d|} \cdot idf(q_i) + 1)$	Yes
L6: $\displaystyle\sum_{q_i \in q \cap d} \log_2(\frac{c(q_i, d)}{|d|} \cdot \frac{|C|}{c(q_i, C)} + 1)$	Yes
L7: $\log_2(BM25score)$, where $BM25score =$ $$\sum_{i=1}^{n} idf(q_i) \cdot \frac{c(q_i, d) \cdot (k_1 + 1)}{c(q_i, d) + k_1 \cdot (1 - b + b \cdot \frac{|d|}{avgdl})}$$ where $k_1=2$, $b=0.75$, and $avgdl$ is average document length [4]	Yes

Note: q_i is a term appearing in both the query and the document; $c(q_i, d)$ is the times q_i appearing in document d (i.e. term frequency, TF); $idf(q_i)$ is the inverse document frequency of q_i; $|d|$ is the length of d (i.e., number of terms in d); C is the set of training documents; $c(q_i, C)$ is the times q_i appearing in C; $|C|$ is number of terms in C.

(NDCG@x) [7], which were commonly employed in previous studies in text ranking. MAP is defined to be

$$MAP = \frac{\sum_{i=1}^{106} P(i)}{106}, \quad P(i) = \frac{\sum_{j=1}^{k} \frac{j}{Doc_i(j)}}{k} \tag{1}$$

where k is number of relevant documents for the ith query, and $Doc_i(j)$ is the number of documents whose ranks are higher than or equal to that of the jth relevant document for the ith query. That is, $P(i)$ is actually the average precision (AP) of the ith query, and MAP is simply the average of the AP values of the 106 queries. On the other hand, NDCG@x is defined to be

$$NDCG @ x = \frac{\sum_{i=1}^{106} N_i(x)}{106}, \quad N_i(x) = Z_x \sum_{j=1}^{x} \frac{2^{r_i(j)} - 1}{\log_2(1 + j)} \tag{2}$$

where $r_i(j)$ is the relevance level of the text whose rank is j with respect to the ith query, and Z_x is set to some value so that a perfect ranking for the ith query gets an

---

[4] When compared with BM25 defined in [12], the score computation excludes the effect of term frequency in each query, since all queries in OHSUMED are quite short. Moreover, it employs traditional IDF(t_i), i.e., $(1+N)/(1+n_i)$, instead of $(0.5+N-n_i)/(0.5+n_i)$, where N is total number of training documents and n_i is number of training documents containing t_i. We avoid the possible problem of negative logarithm values, which might occur when N-$n_i < n_i$.

$N_i(x)$ value of 1.0. That is, $N_i(x)$ actually measures the relevance of top-x documents with respect to the ith query, with larger weights assigned to the relevance of higher-ranked documents. NDCG@x is simply the average of the $N_i(x)$ values of the 106 queries. In the experiment, we report experimental results when x ranges from 1 to 10.

Also note that MAP only considers binary relevance levels (i.e., relevent and non-relevant), while NDCG@x may consider more relevancy levels. Therefore, following many previous studies (e.g., [9]), when computing MAP, only those references that are judged to be "definitively relevant" to a query are relevant to the query (i.e., both "possibly relevant" and "not relevant" are treated as "non-relevant"). When computing NDCG@x, those references that are judaged to be "definitely relevant," "possibly relevant," and "not relevant" are given scores of 2 , 1, and 0, respectively.

4.4   Result and Discussion

We separately discuss contributions of CTFA4TR to individual ranking methods (features) and integrative ranking techniques (i.e., RankingSVM).

4.4.1   Contributions to Individual Ranking methods

Fig. 4 illustrates contributions of CTFA4TR to the 5 basic ranking methods that consider TF. L1 ranks documents by considering TF of each term. CTFA4TR successfully improves L1 in *all* evaluation criteria (MAP and NDCG@1~NDCG@10) under *both* cross validations (i.e., CrossValidation1 and CrossValidation2). We also conduct a significance test to verify the significance of the improvement (two-tailed, paired t-test with 95% confidence level), and those improvements that are statistically significant are marked with '•'. The results show that, in addition to the large improvements on NDCG@1, CTFA4TR contributes statistically significant improvements to L1 on NDCG@6, NDCG@7, and MAP. The improvements in the two cross validations are quite similar.

L4 ranks documents by considering normalized TF (or the probability of a term occurring in the document being ranked). Again, CTFA4TR successfully improves L4 in *all* evaluation criteria under *both* cross validations, and the improvements in the two cross validations are quite similar. It is interesting to note that, L4 performs worse than L1, indicating that the normalization conducted by L4 is not helpful. In this case, CTFA4TR contributes much more statistically significant improvements (NDCG4~10 and MAP).

L5 ranks documents by considering both normalized TF (as L4 does) and IDF (inverse document frequency). Again, CTFA4TR successfully improves L5 in *all* evaluation criteria under *both* cross validations, while the improvements in the two cross validations are somewhat different. It is also interesting to note that, L5 performs better than L4, indicating that the IDF component of L5 is helpful. On the other hand, performance of L5 is more stable than that of L1, especially in NDCG@7~10 in which CTFA4TR contributes more statistically significant improvements.

L6 ranks documents by considering both normalized TF (as L4 and L5 do) and a component functioning like IDF (focusing on times of occurrences of terms in the training corpus). In this case, CTFA4TR improves L6 and provides several statistically significant improvements to L6, although the performance differences in many criteria are not large, which may be attributed to the fact that the component functioning like IDF ($|C|/c(q_i, C)$) tends to have a larger value range than IDF, and hence has a more dominant effect in ranking. It is also interesting to note that, L6 performs better than L5, especially in NDCG@1~3, and CTFA4TR contributes statistically significant improvements to L6 in NDCG@2 and MAP.

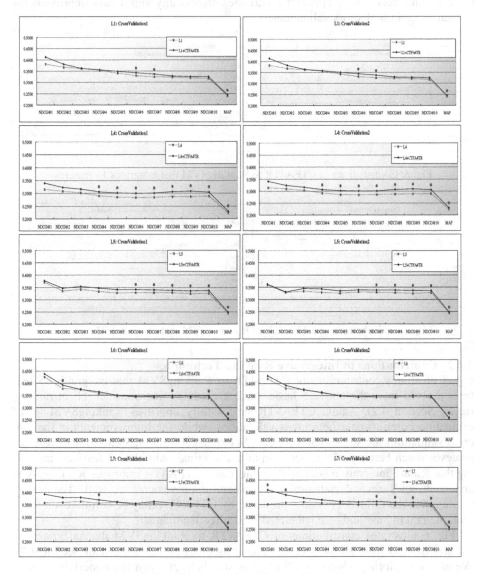

Fig. 4. Contributions to individual ranking methods

L7 ranks documents by BM25 scores. Again, CTFA4TR successfully improves L7 in *all* evaluation criteria under *both* cross validations, while the improvements in the two cross validations are somewhat different. It is also interesting to note that, L7 has the most stable performance than other methods, however, it performs worse than L6 in NDCG@1~3. CTFA4TR contributes statistically significant improvements to L7 in many evaluation criteria, including NDCG@1~2.

Therefore, the overall results justify the contribution and applicability of CTFA4TR to the basic ranking methods. It is also interesting to note that, for *all* the 5 basic ranking methods, CTFA4TR contributes statistically significant improvements to their MAP in both cross validations.

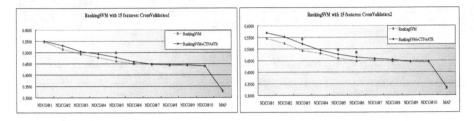

Fig. 5. Contributions of CTFA4TR to RankingSVM that integrates 15 features

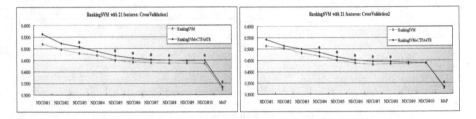

Fig. 6. Contributions of CTFA4TR to RankingSVM that integrates 21 features

4.4.2 Contributions to Integrative Ranking Techniques

Fig. 5 illustrates contributions of CTFA4TR to RankingSVM that integrates 15 features derived from the above 5 basic ranking methods that consider TF in ranking (recall Section 4.2). Again, CTFA4TR successfully improves RankingSVM in *all* evaluation criteria under *both* cross validations, while the improvements in the two cross validations are somewhat different. As expected, RankingSVM with 15 features achieves much better performance than all the original individual 5 features, indicating that feature integration is helpful. In this case, CTFA4TR improves RankingSVM and contributes several statistically significant improvements as well. The results indicate that the refined TF values produced by CTFA4TR may be helpful even when multiple ranking methods are integrated in a collective manner.

Fig. 6 illustrates contributions of CTFA4TR to RankingSVM that integrates 21 features derived from all the 7 basic ranking methods in Table 2 (recall Section 4.2). We aim at exploring whether CTFA4TR is still helpful when the underlying ranker integrates several features that do *not* consider TF in ranking (i.e., L2 and L3 in Table 2). Interestingly, the results show that CTFA4TR successfully improves RankingSVM in *all* evaluation criteria under *both* cross validations as well, and the improvements in the two cross validations are somewhat different. It is also interesting to note that RankingSVM with 21 features achieves much better performance than all the original individual 5 features, however, it performs worse than RankingSVM with 15 features in NDCG@1~3, indicating that adding more features is not necessarily helpful. In this case, CTFA4TR contributes more statistically significant improvements to RankingSVM with 21 features, and makes this version of RankingSVM able to achieve similar performance as RankingSVM with 15 features.

5 Conclusion

Text ranking is essential for the access of relevant information. In this paper, we explore how various kinds of text rankers may be improved by considering term locality contexts in the documents being ranked. We propose a technique CTFA4TR that recognizes term locality contexts by considering *completeness* of query terms appearing in both *neighbor contexts* and *window contexts* of each term in the documents being ranked. To improve various kinds of text rankers by the term locality contexts, the technique identifies *term frequency* (TF) as the target into which the term locality contexts are encoded. Given that TF is one of the most common components considered by text rankers, CTFA4TR is applicable to many text rankers. Moreover, CTFA4TR is efficient online without requiring any training. Empirical evaluation justifies the contributions of CTFA4TR, which successfully improves various kinds of ranking methods both *individually* and *collectively* (integrated with machine learning approaches) under different experimental settings. The contributions are of practical significance, since many text ranking techniques were developed, and if they consider TF in ranking, CTFA4TR may be used to enhance them without incurring any cost to them, since neither their algorithms nor training processes need to be changed.

Acknowledgments. This research was supported by the National Science Council of the Republic of China under the grant NSC 96-2221-E-320-001-MY3.

References

1. Alvarez, C., Langlais, P., Nie, J.-Y.: Word Pairs in Language Modeling for Information Retrieval. In: Proceedings of RIAO (Recherche d'Information Assistée par Ordinateur), pp. 686–705. University of Avignon (Vaucluse), France (2004)
2. Büttcher, S., Clarke, C.L.A., Lushman, B.: Term Proximity Scoring for Ad-Hoc Retrieval on Very Large Text Collections. In: Proceedings of the 29th annual international ACM SIGIR conference on research and development in information retrieval, Seattle, USA, pp. 621–622 (2006)
3. Cao, Y., Xu, J., Liu, T.-Y., Li, H., Huang, Y., Hon, H.-W.: Adapting Ranking SVM to Document Retrieval. In: Proceedings of the 29th annual international ACM SIGIR conference on research and development in information retrieval, Seattle, Washington, pp. 186–193 (2006)
4. Cao, Z., Qin, T., Li, T.-Y., Tsai, M.-F., Li, H.: Learning to Rank: From Pairwise Approach to Listwise Approach. In: Proceedings of the 24th International Conference on Machine Learning, Corvallis, OR, pp. 129–136 (2007)

5. Gao, J., Nie, J.-Y., Wu, G., Cao, G.: Dependence Language Model for Information Retrieval. In: Proceedings of the 27th annual international ACM SIGIR conference on research and development in information retrieval, Sheffield South Yorkshire, UK, pp. 170–177 (2004)
6. Hersh, W.: OHSUMED: An Interactive Retrieval Evaluation and New Large Test Collection for Research. In: Proceedings of the 17th annual international ACM SIGIR conference on research and development in information retrieval, pp. 192–201 (1994)
7. Järvelin, K., Kekäläinen, J.: IR Evaluation Methods for Retrieving Highly Relevant Documents. In: Proceedings of the 23rd annual international ACM SIGIR conference on Research and development in information retrieval, pp. 41–48 (2000)
8. Joachims, T.: Optimizing Search Engines using Clickthrough Data. In: Proceedings of ACM SIGKDD, Edmonton, Alberta, Canada, pp. 133–142 (2002)
9. Liu, T.-Y., Xu, J., Qin, T., Xiong, W., Li, H.: LETOR: Benchmark Dataset for Research on Learning to Rank for Information Retrieval. In: Proceedings of ACM SIGIR 2007 Workshop on Learning to Rank for Information Retrieval, pp. 3–10 (2007)
10. Nallapati, R.: Discriminative Models for Information Retrieval. In: Proceedings of SIGIR, July 25–29, pp. 64–71 (2004)
11. Rasolofo, Y., Savoy, J.: Term Proximity Scoring for Keyword-Based Retrieval Systems. In: Sebastiani, F. (ed.) ECIR 2003. LNCS, vol. 2633, pp. 207–218. Springer, Heidelberg (2003)
12. Robertson, S.E., Walker, S., Jone, S., Beaulieu, M., Gatford, M.: Okapi at TREC-3. In: Proceedings of the 3rd Text REtrieval Conference, Gaithersburg, USA (1994)
13. Srikanth, M., Srihari, R.: Biterm Language Models for Document Retrieval. In: Proceedings of the 25th annual international ACM SIGIR conference on research and development in information retrieval. Tampere, Finland (2002)
14. Tao, T., Zhai, C.: An Exploration of Proximity Measures in Information Retrieval. In: Proceedings of the 30th annual international ACM SIGIR conference on Research and development in information retrieval, Amsterdam, The Netherlands, pp. 23–27 (2007)
15. Veloso, A., Almeida, H.M., Gonçalves, M., Meira Jr., W.: Learning to Rank at Query-Time using Association Rules. In: Proceedings of the 31rd annual international ACM SIGIR conference on research and development in information retrieval, Singapore, pp. 267--274 (2008)
16. Wang, X., McCallum, A., Wei, X.: Topical N-grams: Phrase and Topic Discovery, with an Application to Information Retrieval. In: Proceedings of the IEEE 7th International Conference on Data Mining, Omaha NE, USA, pp. 697–702 (2007)
17. Xu, J., Li, H.: AdaRank: A Boosting Algorithm for Information Retrieval. In: Proceedings of the 30th annual international ACM SIGIR conference on research and development in information retrieval, Amsterdam, Netherlands, pp. 391–398 (2007)
18. Yeh, J.-Y., Lin, J.-Y., Ke, H.-R., Yang, W.-P.: Learning to Rank for Information Retrieval Using Genetic Programming. In: Proceedings of the 30th annual international ACM SIGIR conference on research and development in information retrieval, Amsterdam, Netherlands (2007)

Mutual Screening Graph Algorithm: A New Bootstrapping Algorithm for Lexical Acquisition

Yuhan Zhang and Yanquan Zhou

Research Center of Intelligence Science and Technology,
P.O. BOX #310 XITUCHENG ROAD #10
Beijing University of Posts and Telecommunications, Beijing, 100876, China
zhangyuhan1985@gmail.com, zhouyanquan@gmail.com

Abstract. Bootstrapping is a weakly supervised algorithm that has been the focus of attention in many Information Extraction(IE) and Natural Language Processing(NLP) fields, especially in learning semantic lexicons. In this paper, we propose a new bootstrapping algorithm called Mutual Screening Graph Algorithm (MSGA) to learn semantic lexicons. The approach uses only unannotated corpus and a few of seed words to learn new words for each semantic category. By changing the format of extracted patterns and the method for scoring patterns and words, we improve the former bootstrapping algorithm. We also evaluate the semantic lexicons produced by MSGA with previous bootstrapping algorithm Basilisk [1] and GMR (Graph Mutual Reinforcement based Bootstrapping) [4]. Experiments have shown that MSGA can outperform those approaches.

Keywords: Lexical acquisition, Bootstrapping, MSGA.

1 Introduction

In recent years, Bootstrapping methods [1, 2, 4] have attracted great attention in many IE and NLP fields and achieved good results. Although supervised methods usually could get better results than weakly supervised methods, they are highly constrained by annotated corpus and transplantation from a field to another often costs approximately the same workload as we do in the initial field. As a method of the weakly supervised approach, Bootstrapping uses only a small set of seed words and a lot of unannotated corpus to build semantic lexicon. This feature makes bootstrapping a useful approach in many IE [4, 5] and NLP fields, such as exploiting role-identifying nouns [3].

Semantic lexicons have been proved to be useful for many IE and NLP fields. Knowing the semantic classes of words (e.g., "Benz" is a brand of automobile) can be extremely valuable for many tasks, including question answer [15, 16], IE [17] and so on. Although some semantic dictionaries do exist (e.g. Word-Net), they are rarely complete, especially for large open classes (e.g., classes of people and objects) and rapidly changing categories (e.g., computer technology). It is reported that for every five terms generated by their semantic lexicon learner three were not present in Word-Net. Automatic semantic lexicon acquisition could be used to enhance existing

G.G. Lee et al. (Eds.): AIRS 2009, LNCS 5839, pp. 289–299, 2009.

resources such as Word-Net, or to produce semantic lexicons for specialized categories of domains. In recent years, several algorithms have been proposed to automatically learn semantic lexicons using supervised methods [6, 7], and weakly unsupervised methods [1, 4]. As a weakly-supervised method requires little manually-labeled training data, it has received more and more attention [8-10].

In this paper, a weakly-supervised bootstrapping algorithm called MSGA-Bootstrapping (Mutual Screening Graph Algorithm Based Bootstrapping) which automatically generates semantic lexicons is developed. Like other bootstrapping methods, MSGA-Bootstrapping begins with unlabeled corpus and a few seed words, and it will automatically iterate to learn lexicons. After analyzing the procedures of Basilisk [1] and GMR [4], we found that they both have a number of areas for improvement, such as the pattern format and the method of scoring patterns and words. Therefore, we develop a new format for extracted patterns and a more reasonable method for scoring patterns and words. We also add a new process to filter words without any field-related information. From the experimental results, lexicons acquired by MSGA-Bootstrapping (more details in Section 4) are more precise than those of Basilisk and GMR-Bootstrapping.

The reminder of the paper is organized as follows. In section 2, we introduce Basilisk and GMR-Bootstrapping, and point out the problem occurred by using these algorithms and explain why these problems occur. In Section 3, we introduce our MSGA-Bootstrapping and its new feature. In Section 4, experiments are presented to show the improvements, and the experimental results are discussed. Conclusions are given in Section 5.

2   Related Work

Many research works have been done on Bootstrapping in recent years. Among them, Basilisk [1] and its improved algorithm GMR-Bootstrapping [4] get most attention.

2.1   Basilisk

Michael Thelen and Ellen Riloff developed a weakly-supervised bootstrapping algorithm called Basilisk (Bootstrapping Approach to Semantic Lexicon Induction using Semantic Knowledge) [1], which automatically generates semantic lexicons. The input to Basilisk is an unannotated corpus and a few manually-defined seed words for each semantic category, as well.

The bootstrapping process begins by selecting a subset of the extraction patterns with the AutoSlog system [11]. Then it scores each pattern using the RlogF metric [11]. The top N extraction patterns are put into a pattern pool. Then Basilisk collects all noun phrases extracted by patterns in the pattern pool and puts the result into the candidate pool. The top M words are added to the lexicon, and the process starts over again.

In Basilisk, a pattern receives a high score if a high percentage of its extractions are category members. Or if a moderate percentage of its extractions are category members and it extracts a lot of them. A word receives a high score if it is extracted by patterns that also have a tendency to extract known category members.

Basilisk gives us a good method to learn Semantic Lexicon at a less cost than other previous methods. However, after analyzing the procedure of Basilisk we found that the method treats every word extracted by patterns equal, and it does not distinguish how many times the word has been extracted and how many patterns have extracted it. In other word, Basilisk evaluates a word without any quality and quantity information on the patterns that extract it. Suppose that word α has been extracted 10 times by three different candidate patterns in our corpus, while word β is extracted only once by one pattern. In Basilisk, word α and word β are equal. However, since word α is more popular than word β in this corpus, it should have a higher score than word α. Basilisk also has this problem in scoring patterns.

2.2 GMR-Bootstrapping

Hany Hassan, Ahmed Hassan and Ossama Emam proposed a new method based on bootstrapping called GMR-Bootstrapping (Graph Mutual Reinforcement based Bootstrapping) [4]. Different with Basilisk, GMR did not evaluate every word in the lexicon equal; every word in the lexicon has a score that represents its precision of belonging to this lexicon. When scoring patterns, the score of words will affect the result for making patterns, and patterns that have higher scores can be easily chosen as pattern candidates. So does the processing of extracting words.

GMR-Bootstrapping introduces a good idea that it assigns every word (or pattern) a score which represents the precision how this word (pattern) should be added to this words candidate (patterns candidate). When we use these words (patterns) to score a new pattern (word), it can provide the quality information of words (patterns). From the experimental result given by the author, we saw that GMR got a better result than Basilisk for adding quality information of words/patterns.

However, there is still some information we should add to in our effort to make the result more accurate. For example, we assume that pattern γ is a perfect pattern suitable for extracting words from the corpus, and word α has been extracted 10 times by pattern γ while word β has been extracted by this pattern only once. In GMR-Bootstrapping these two words will be evaluated equal, even though intuitively it seems that word α should get a higher score than word β. This paper focuses on how to improve this bootstrapping algorithm to get a more accurate result, and the key improvement of MSGA to the previous method.

3   MSGA-Bootstrapping

MSGA is a weakly supervised machine learning method. Same as other bootstrapping methods, the input of MSGA-Bootstrapping are a large set of unlabeled data and a small set of labeled data. Then MSGA-Bootstrapping iteratively trains and evaluates a classifier in order to improve its performance.

For lexical acquisition, we need a lot of unannotated corpus and a few seed words. Customarily, we select seed words by sorting the frequency of words in corpus and manually select some words which we think have more information on this field. Step 1, MSGA-Bootstrapping selects a subset of the extraction patterns that tend to extract the seed words. We then add these patterns in a set called the pattern pool. Step 2 we score every pattern in the pattern pool and select the best m patterns and put them in the pattern candidate pool. Step 3, we use patterns in the pattern candidate pool to extract words and put them in the word pool. Step 4, we sort the words in the word candidate pool and choose the best n words that do not belong to the common word dictionary. Then, we use the words in the word candidate pool to select patterns, and the whole process starts over again. After several rounds of iteration, the word candidate pool will expand and eventually it will become the semantic lexicons we need. Figure 1 shows the process of MSGA-Bootstrapping.

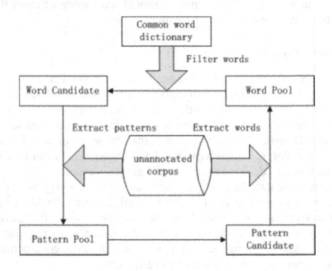

Fig. 1. MSGA algorithm

Figure 2 shows the MSGA-Bootstrapping process.

```
Words frequency statistics
Build up Common word dictionary
Word Candidate = {seed words}
i:=0

Bootstrapping
1. Extract patterns and put them in pattern pool
2. Score all extracted patterns in pattern pool
3. Pattern Candidate = top rank 3+i patterns
4. Extract words and put them in word pool
5. Score all extracted words in word pool
6. Add top 5 words in Word Candidate
7. i:=i+1
8. Go to Step 1
```

Fig. 2. MSGA's bootstrapping algorithm

We make three major changes to improve Basilisk and GMR-Bootstrapping.

3.1 Pattern Formats

In order to find new lexicon entries, extraction patterns are used to provide contextual evidence that shows which semantic class a word belongs to. Basilisk and GMR-Bootstrapping use the AutoSlog system [12] to represent extraction patterns. AutoSlog's extraction patterns represent linguistic expressions that extract a head noun of noun phrases in one of the three syntactic roles: subject, direct object, or prepositional phrase object.

However, AutoSlog's extraction patterns only can provide the evidence of linguistic expressions and words extracted by it are head nouns of noun phrases. This feature make AutoSlog's extraction pattern limited in processing corpuses of some field, for words that belongs to the lexicon of this field can be used as an adjective or other part of speech in our corpus. In other words, AutoSlog's extraction pattern can work well in a few fields, but if we want make our algorithm processing available for more corpus, we need find out a new format of extracted patterns.

To solve this problem, we performed many experiments and finally choose the pos pattern as our pattern format. In every pattern, we only use pos tags and their combinations to locate words in the text and extract them into the word pool. For example, suppose that we get a pattern "(*)/NN */AD */VV", where the asterisk represents any word; the word in brackets is what we want to extract. Using this pattern, we use pos message instead of information of the special word to match. According to the combination and the sequence of pos tag, we can extract patterns and words without knowing the exact word. Another advantage of using this pattern format is, for we do not need to know the exact word, we can use pos pattern format to process corpuses of different languages while AutoSlog's extraction patterns only can deal with English corpus.

3.2 Common Word Dictionary

In our experiment, we find that many words without any field-related information appear in many semantic lexicons. After analyzing these words, we find they often have relatively high frequencies of appearance in more than one corpus. In order to solve this problem and exclude these words from our lexicon, we add a dictionary called Common Word Dictionary(CWD) in our MSGA-Bootstrapping. We build CWD with these steps:

Step 1. Calculate word frequency in every corpus.
Step 2. Select words that have relatively high frequency in most corpuses. In our experiment, we use 200 as a threshold of frequency and words whose frequency exceeds this threshold in more than 75% corpus will be added into the CWD.

When scoring the words in the word pool, we filter out words contained in the CWD and this process can make words without any field-related information to be excluded from our lexicons. Experiments show that this method can greatly improve the precision of the lexicon we create.

3.3 MSGA Scoring

MSGA is different with Basilisk and GMR-Bootstrapping in its method for scoring words and patterns. We assume patterns which match words that have a higher precision of belonging to the semantic lexicon tend to be important. For example, pattern γ can extract word α which has the possibility of 80% to belong the semantic and pattern δ can only extract word β with the possibility of 60%, then we said pattern γ should get a higher score than pattern δ in this iteration. And if a pattern matches a word which proved to belong to the semantic lexicons many times, this pattern tends to be more important in this field. For example, pattern γ and pattern δ also can extract word α which has been proved to belong the lexicon. Pattern γ can extract this word 4 times in the corpus while pattern δ can only extract this word once in the same corpus. Then we assumed pattern γ is more important in this field than pattern δ, and it should achieve a higher score than pattern δ. Similar as processing the word. By this method, we can use not only quality information but also quantity information on calculating the score of patterns and words.

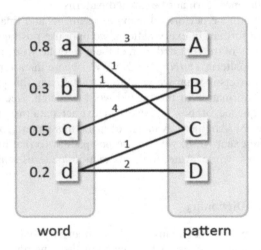

Fig. 3. MSGA scoring

Figure 3 shows an example of MSGA scoring, where left pool contains some words (such as a, b) and right pool contains some patterns (pattern A, B). The number beside words shows the score of these words, where the number of the line between a word and a pattern shows the times this word extract the pattern. With this figure, we can further explain the difference between MSGA and Basilisk and GMR, Basilisk only count on the information that whether exist a line between a word and a pattern. GMR involve the precision message of words in calculate the score of a pattern. MSGA use both quality information of words(score) and quantity information(times that words extract patterns) to calculate the score of patterns.

When got a lot of patterns that extract by the words in word candidate pool, we calculate these patterns with words' score of previous step. The score for each pattern is computed as:

$$F^{(i)}(p) = \sum_{w \in W(p)} sw^{(i-1)}(w)C(w) \tag{1}$$

Where $F(p)$ is a temporary variable used in calculating the score of a pattern; $W(p)$ is the set of words which extract this pattern; $sw(w)$ is the word's score which is calculated by previous iteration; $C(w)$ is the number of time that this word extracts the pattern; and i denotes the i-th iteration. Using this equation, we add both the quality information $sw(w)$ and the quantity information $C(w)$ in scoring patterns.

$$sp^{(i)}(p) = \frac{F^{(i)}(p)\log F^{(i)}(p)}{|W(p)|} \tag{2}$$

$F(p)$ is the result of Eq. (1). $|W(p)|$ is the number of words that extract this pattern. A pattern receives a high score if words that extract it many times have a high score (meaning that this word has a high possibility of belonging to the semantic lexicon) and they extract the patterns many times. We calculate $SP(p)$, which is the normalization factor:

$$SP^{(i)} = \sum_{W=1}^{|W|} \sum_{v=1}^{P(w)} C(v)sp^{(i)}(v) \tag{3}$$

Then the $sp(p)$ is normalized by the following equations:

$$sp^{(i)}(p) = \frac{sp^{(i)}(p)}{SP^{(i)}} \tag{4}$$

Where $sp(p)$ is the final score of the pattern. Then we sort the patterns in the pattern pool by the score calculated by Eq. (4), and add the top 3+j patterns to the pattern candidate. Note that j is the number of iterations. We use the value j to avoid our word candidate pool stagnant. For example, we assume that our MSGA performs perfectly, and we have three reliable patterns. Initially, they work well and give us a lot of correct words. As the iteration continues, however, the quality of words these patterns extract decreases. Why? Because the word set that these three patterns can extract becomes stagnant. In each round of iteration, we only get the top five words and the second top five words in the next round. For this reason, the pattern pool needs to be infused with new patterns so that more words become available for consideration.

Then we use these patterns to match the corpus again, and the words that have been matched are put into the word pool. We calculate the word score with the pattern that matches the words and the equation is:

$$sw^{(i)}(w) = \frac{\sum_{p \in P(w)} (C(p)\log(F^{(i)}(p)+1))}{|P(w)|} \tag{5}$$

Where $sw(w)$ is the final score of the word. $P(w)$ is the set which contains patterns match this word, and $F(p)$ is the score of the pattern calculated by Eq. (1). $C(p)$ is the number of time that this pattern matches the word in the corpus. $|P(w)|$ is the number denoting how many patterns in the pattern pool can match this word. A word receives a high score if patterns that match it many times have a high score (meaning these patterns have a high possibility on extracting words for this semantic lexicon) and match it many times.

At the end of this iteration, SW, which is the normalization factor, is calculated as

$$SW^{(i)} = \sum_{p=1}^{|P|} \sum_{u=1}^{W(p)} C(u) sw^{(i)}(u) \tag{6}$$

Then $sw(w)$ is normalized by the following equation:

$$sw^{(i)}(w) = \frac{sw^{(i)}(w)}{SW^{(i)}} \tag{7}$$

Where $sw(w)$ is the final score of the word. Sorting the words in word pool, we choose best 5 words and add them in a set called word candidate pool. Then the bootstrapping processing iterated.

4   Experiments

To compare the performance of MSGA-Bootstrapping with Basilisk and GMR-Bootstrapping, we design several experiments on official corpus of COAE 2008 (Chinese Opinion Analysis Evaluation) [13], which contains about 3000 texts (including both tests and training parts). All the words in the corpus are divided into four semantic categories: automobile, notebook, digital camera and cell phone. We compare MSGA-Bootstrapping with Basilisk and GMR-Bootstrapping on all of these fields.

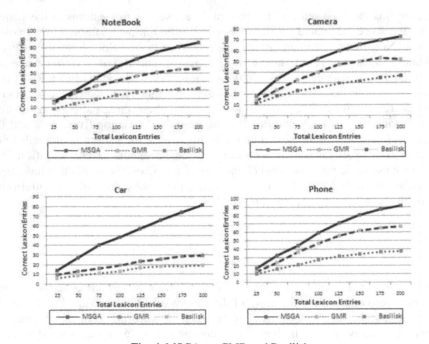

Fig. 4. MSGA vs. GMR and Basilisk

In order to ensure the fairness of the experiment, we add CWD also in MSGA, GMR and Basilisk. The results we show next are ruled out the influence of this processing, and they just show the improvement of MSGA-Bootstrapping to Basilisk and GMR-Bootstrapping on scoring patterns and words.

Figure 4 presents results of MSGA-Bootstrapping, Basilisk and GMR-Bootstrapping. Table 1 shows the results of some specific point of the experiment. For each category, five most frequent words that belong to the category are extracted as seed words. It is the same as the Thelen and Riloff (2002)'s method, and we ran the algorithm for N iterations, so 5*N words are extracted. The X axis shows the number of words extracted. The Y axis shows the number of correct ones. From the results we know that MSGA has a better performance on both precision and stability, contrasting with GMR and Basilisk.

More specifically, MSGA-Bootstrapping receives better results than GMR-Bootstrapping and Basilisk do in all four fields of our experiment. For precision, MSGA often gets a result more than 10 percent higher than GMR, and even more than 30 percent higher than that of Basilisk. For stability, we saw GMR and Basilisk obtain bad results on the corpus of Car, worse than their average results on the others corpuses. MSGA, on the other hand, retains a stable and excellent result on all corpuses.

By analyzing the experimental result, we found that the effect of segmentation directly restricted the operation of our algorithm results. Some words were wrongly segmented in the first step, so it was impossible for our algorithm to extract these words in semantic lexicons correctly, because we use pos tag as our patterns.

Table 1. Lexicon Results

Corpus	Total words	Basilisk	GMR Bootstrapping	MSGA Bootstrapping
Camera	50	18(36%)	24(48%)	**34(68%)**
	100	26(26%)	40(40%)	**53(53%)**
	150	32(21.3%)	50(33.3%)	**66(44%)**
	200	37(18.5%)	52(26%)	**73(36.5%)**
Notebook	50	14(28%)	27(54%)	**29(58%)**
	100	24(24%)	41(41%)	**57(57%)**
	150	29(19.3%)	51(34%)	**75(50%)**
	200	31(15.6%)	55(27.5%)	**86(43%)**
Car	50	9(18%)	13(26%)	**27(54%)**
	100	13(13%)	19(19%)	**48(48%)**
	150	18(12%)	25(16.7%)	**66(44%)**
	200	19(9.5%)	29(14.5%)	**81(40.5%)**
Phone	50	16(32%)	24(48%)	**32(64%)**
	100	27(27%)	47(47%)	**59(59%)**
	150	34(22.7%)	62(41.3%)	**80(53.3%)**
	200	37(18.5%)	67(33.5%)	**91(45.5%)**

5 Conclusion and Future Work

In this paper, we present an approach for lexical acquisition using a novel bootstrapping method called MSGA-Bootstrapping. We show the impact of both the quality information and quantity information of words and patterns when scoring the words and patterns created by them. Based on this idea, we improve the algorithm of scoring patterns and words so that it contains both quality information and quantity information. Experiments show that with this improvement MSGA-Bootstrapping significantly outperforms Basilisk and GMR-Bootstrapping in COAE corpus. By changing the pattern format, we also make our algorithm capable of processing corpuses of more languages, while Basilisk and GMR-Bootstrapping can only deal with English corpus. Furthermore, we add a new processing by using CWD in MSGA-Bootstrapping, which can prevent words without any field-related information from being added to semantic lexicons acquired by MSGA-Bootstrapping.

In the future, we will mainly improve our algorithm in two aspects: one is to add a step using frequency statistics to discover new words to solve the problem of error segment; another is to explore a new approach to use our algorithm as an unsupervised method that needs no field-related information to acquire semantic lexicons.

References

1. Thelen, M., Riloff, E.: A bootstrapping method for learning semantic lexicons using extraction pattern contexts. In: Proceedings of the ACL 2002 conference on Empirical methods in natural language processing, Philadelphia, USA, vol. 10, pp. 214–221 (2002)
2. Blum, A., Mitchell, T.: Combining labeled and unlabeled data with co-training. In: Proceedings of the eleventh annual conference on Computational learning theory, Madison, Wisconsin, United States, pp. 92–100 (1998)
3. Phillips, W., Riloff, E.: Exploiting Role-Identifying Nouns and Expressions for Information Extraction. In: 2007 Proceedings of Recent Advances in Natural Language Processing, RANLP 2007 (2007)
4. Hassan, H., Hassan, A., Emam, O.: Unsupervised Information Extraction Approach Using Graph Mutual. In: Proceedings of the 2006 Conference on Empirical Methods in Natural Language Processing (EMNLP 2006), pp. 501–508 (2006)
5. Patwardhan, S., Riloff, E.: Learning Domain-Specific Information Extraction Patterns from the Web. In: Proceedings of the Workshop on Information Extraction Beyond The Document, pp. 66–73 (2006)
6. Florian, R., Hassan, H., Ittycheriah, A., Jing, H., Kambhatla, N., Luo, X., Nicolov, N., Roukos, S.: A statistical model for multilingual entity detection and tracking. In: HLT-NAACL 2004: Main Proceedings, pp. 1–8 (2004)
7. Kambhatla, N.: Combining lexical, syntactic, and semantic features with maximum entropy models for in-formation extraction. In: The Companion Volume to the Proceedings of 42st Annual Meeting of the Association for Computational Linguistics, Barcelona, Spain, pp. 178–181 (2004)
8. Collins, M., Singer, Y.: Unsupervised models for named entity classification. In: Proceedings of the Joint SIGDAT Conference on Empirical Methods in Natural Language Processing and Very Large Corpora. University of Maryland, MD (1999)

9. Etzioni, O., Cafarella, M., Downey, D., Popescu, A., Shaked, S.T.: Unsupervised named-entity extraction from the web: An experimental study. Artificial Intelligence 165(1), 91–134 (2005)
10. Riloff, E., Wiebe, J., Wilson, T.: Learning subjective nouns using extraction pattern bootstrapping. In: Proceedings of the seventh conference on Natural language learning at HLT-NAACL 2003, Edmonton, Canada, vol. 4, pp. 25–32 (2003)
11. Riloff, E.: Automatically generating extraction patterns from untagged text. pattern bootstrapping. In: Proceedings of the Thirteenth National Conference on Artificial Intelligence, Portland, Oregon, pp. 1044–1049 (1996)
12. Riloff, E., Phillips, W.: An Introduction to the Sundance and AutoSlog Systems (2004)
13. COAE Proceedings: COAE proceedings. In: Proceedings of Chinese Opinion Analysis Evaluation 2008, COAE 2008 (2008)
14. Riloff, E., Jones, R.: Learning dictionaries for information extraction by multi-level bootstrapping. In: Proceedings of the 16th National Conference on Artificial Intelligence, Orlando, USA, pp. 474–479 (1999)
15. Hirschman, L., Light, M., Breck, E., Burger, J.D.: Deep read: A reading comprehension system. University of Maryland, United States (1999)
16. Moldovan, D., Harabagiu, S., Pasca, M., Mihalcea, R., Goodrum, R., Girju, R., Rus, V.: Lasso: A tool for surfing the answer net. In: Proceedings of the Eighth Text REtrieval Conference, TREC-8 (1999)
17. Riloff, E., Schmelzenbach, M.: An empirical approach to conceptual case frame acquisition. In: Proceedings of the Sixth Workshop on Very large Corpora, Montreal, Canada (August 1998)

Web Image Retrieval for Abstract Queries Using Text and Image Information

Kazutaka Shimada, Suguru Ishikawa, and Tsutomu Endo

Department of Artificial Intelligence, Kyushu Institute of Technology
680-4 Iizuka Fukuoka 820-8502 Japan
{shimada,s_ishikawa,endo}@pluto.ai.kyutech.ac.jp

Abstract. In this paper, we propose a method for image retrieval on the web. In this task, we focus on abstract words that do not directly link to images that we want. For example, a user might use a query "summer" to retrieve images of "fireworks" or "a white sand beach with the sea". In this case retrieval systems need to infer direct words for the images from the abstract query of the user. In our method, we extract related words about a query from the web first. Second, we retrieve images from the web by using the extracted words. Then, a user selects relevant images from the retrieved images. Next, the system computes a similarity between selected images and other images and ranks the images on the basis of the similarity. We use the Earth Mover's Distance as the similarity. The experimental result shows the effectiveness of our method that uses text and image information for the image retrieval process.

Keywords: Image retrieval, Abstract queries, Text and image, Feedback.

1 Introduction

As the World Wide Web rapidly grows, a huge number of online documents are easily accessible on the web. Finding information relevant to user needs has become increasingly important. There are many images on the web. Web image retrieval systems are one of the most significant tasks in web applications. Most of existing web image retrieval systems, such as Google image search[1], utilize keywords extracted from text areas surrounding images. The method is effective for concrete queries.

In this paper, we focus on abstract queries. For example, a user inputs "summer" as a query to a search engine. The query " summer", however, is ambiguous because it is an abstract query. The target images that related to the abstract query "summer" depend on the user. Moreover, the concrete words related to the abstract query do not always exist in text areas surrounding the images. For example, assume that a user wants images about "a white sand beach with the sea" and uses "summer" as a query. Existing systems can not often retrieve images that the user wants. If a web image retrieval system can connect the word "summer" with concrete words related to it, it is useful.

---

[1] http://images.google.com/

G.G. Lee et al. (Eds.): AIRS 2009, LNCS 5839, pp. 300–309, 2009.

Kato et al. [1] have proposed a web image search method for an abstract query. They used social tagging information in Flickr[2] for the image search method. Freng and Lapata [2] have reported a automatic tagging system for images in BBC news. The system tagged appropriate keywords extracted from the caption of an image, the headline of the news and so on. These methods were useful. However, they did not treate any information in images such as color. Many researchers have proposed content-based image retrieval techniques[3,4]. Sezaki and Kise [5] have proposed a recommending system about a tagging process for images. They used the co-occurrence of tags and a similarity between images. However, the method needed social tag information of images. Barthel et al. [6] have reported an image search system on the web. First, the method retrieved images from the web on the basis of a keyword-based search process. Next, it sorted the images according to their visual similarity. However, they did not argue abstract queries.

In this paper, we propose a web image retrieval system for abstract queries. We focus on both of text and image information. First, we detect concrete words associated with an abstract query by using results from a text-based web search engine. Next, we search and sort the images extracted by the text-based web search engine by using a similarity between images and user's feedback. We use the Earth Mover's Distance as the similarity.

2   Problems

In this section, we discuss search results concerning concrete queries and abstract queries on an existing system. We compared 6 concrete queries and 6 abstract queries in terms of precision rates in the top 20 and top 50 search results. Table 1 and Table 2 show the precision rates[3]. We evaluated the correctness of the extracted images from the existing system subjectively. The precision rates concerning concrete queries such as "Rose" and "Snow" were high. However, the precision rates concerning abstract queries such as "Summer" were generally low as compared with concrete queries. The reasons of the low precision rates are as follows:

1. a concrete query does not always appear in the text surrounding a target image.
2. a concrete query frequently appear in the text surrounding non-target images.

An example of the 1st reason is a target image about "a white sand beach with the sea" with an abstract query "Summer". The same tendency is shown in other abstract queries such as "Spring" and "Autumn". Moreover, the precision rate of a concrete query occasionally becomes low. See the case of "moon" in Table 1. The word "moon" is expressed "tsuki" in Japanese. The word "tsuki" contains two meanings; "tsuki as moon" and "tsuki as month". The ambiguity is the 2nd reason of the low precision rate.

---

[2] http://www.flickr.com/
[3] Actually, we used Japanese queries in this preliminary experiment.

Table 1. The precision rates of an existing system for concrete queries

Queries	Top 20	Top 50
Moon	0.25	0.14
Rose	0.95	0.92
Snow	0.90	0.70
Strawberry	0.80	0.76
Earphones	0.70	0.74
Pencil	0.85	0.60

Table 2. The precision rates of an existing system for abstract queries

Queries	Top 20	Top 50
Spring	0.10	0.04
Summer	0.30	0.12
Autumn	0.15	0.08
Winter	0.65	0.34
Japan	0.30	0.18
Beautiful	0.30	0.22

To solve the 1st problem, we extract words related to an abstract query from the web. For example, we detect concrete words, such as "sea", "beach" and "fireworks", concerning "Summer". For the 2nd problem, we apply a feedback process from a user into our retrieval method.

3   The Proposed Method

In this section, we explain our proposed method using text and image information. Also we apply user's feedback to the method. Figure 1 shows the outline of the proposed method.

Our method consists of three processes; (1) text processing, (2) feedback processing and (3) image processing. If a user inputs an abstract query, the system extracts concrete words associated with the query. Then, it searches images with the extracted words and displays the images as initial outputs. Next, the user selects images that he/she wants, namely relevant or positive examples. On the basis of the selected images, our system computes a similarity between images and sorts the images by using the similarity. The user iterates these processes until the system outputs images that he/she wants. The following subsections describe each process in our method.

3.1   Text Processing

In the text processing, our method detects related words concerning a query and searches images on the basis of the words. For the related word extraction, we

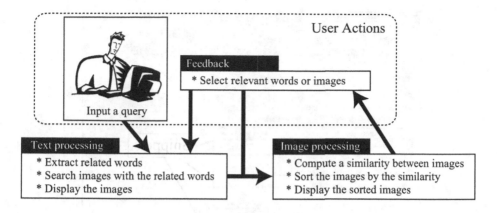

Fig. 1. The outline of our method

use snippets in search results. A snippet is a summary that is displayed in search results from a search engine. Figure 2 shows an example. We use Yahoo! Web search API[4] as the search engine in this process

For the related word extraction, it is not suitable to extract related words by using the original query that a user inputs. In the situation, the search engine usually outputs pages that contain the origin or the explanation of the query word. For example, one snippet of the word "spring" is "one of the four temperate seasons. Spring marks the transition from winter into summer." This snippet does not contain effective information for our system because we need concrete words to associate "Spring" such as "Cherry Blossom".

To solve this problem, we add the phrase "to ieba" to the initial query. In other words, we use the phrase "[Query] to ieba ([Query] is associated with)" as the new query. Using this phrase leads to reduction of the number of pages that we do not need in this process. First, we divide sentences in snippets of top 30 pages into words by using the Japanese morphological analyzer ChaSen[5]. Next, we compute frequency of each noun in the results. Finally, we employ high frequency words (top 15 words) as related words of the query.

Our method retrieves images by the initial query and the extracted words. Then, it displays top 10 images extracted by each query-set (the initial query and an extracted word) for the feedback process mentioned below. We also use Yahoo! Image search API for retrieving images in this process.

3.2 Feedback

Feedback methods are effective for the image retrieval systems [6,7]. We also apply user's feedback to our system to retrieve images that users want. Figure 3 shows the interface of our system.

---

[4] http://www.yahoo.co.jp/
[5] http://chasen-legacy.sourceforge.jp/

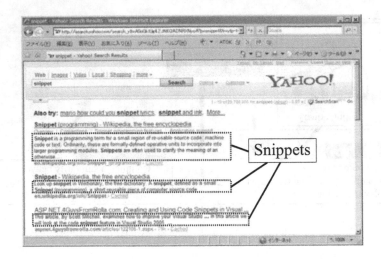

Fig. 2. An example of snippets on a search engine

As the 1st step, our system displays the images retrieved by each query-set in the previous subsection. In Figure 3, the initial query is "spring". In this case we obtained 15 related words (including the initial query itself) such as "Cherry" and "Flower". Each line with a related word in the figure denotes a kind of cluster of images that are retrieved with the related word. In our system, a user can select the related words and the images with a check box in the list. On the basis of the selected elements, our system retrieve images again from the web. If images are selected by a user, it computes a similarity between the selected images and other images retrieved and sorts the images by using the similarity. The similarity calculation is described in the next subsection.

3.3 Image Processing

Image retrieval systems based on keywords only do not always refine the search results. For example they can not distinguish between a image of a mouse as animals and a image of a mouse as devices. One solution to the problem is to apply a visual similarity measure[6]. In our system, we use a visual similarity measure. The inputs of this image processing are the images that a user selects in the feedback process described in the previous subsection. If a user select images in the list on our system, it computes a similarity. The features for the similarity calculation are a color signature which is a pair consisting of a color p_i and the ratio r_i of the color in all pixels. To compute a distance between color signatures, we use the Earth Mover's Distance (EMD), which has been used by Rubner et al for a metric for image retrieval [8]. In our method, we use the $L * a * b*$ color space and the Euclidean distance for the EMD calculation. Here

---

[6] http://similar-images.googlelabs.com/

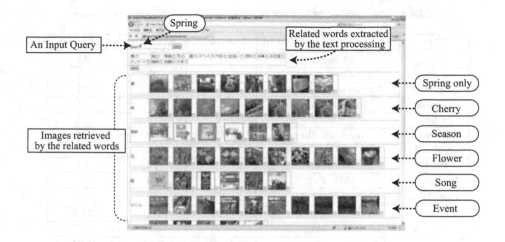

An Input Query

Spring

Related words extracted
by the text processing

Spring only

Cherry

Season

Flower

Song

Event

Images retrieved
by the related words

Fig. 3. The interface of our system

assume that the color signature consists of $P = \{(p_1, r_{p1}), (p_2, r_{p2}), ..., (p_m, r_{pm})\}$ and $Q = \{(q_1, r_{q1}), (q_2, r_{q2}), ..., (q_m, r_{qm})\}$ for image P and Q. Also the distance between elements of the color signature is $d_{ij} = d(p_i, q_j), (1 \leq i \leq m, 1 \leq j \leq n)$. In this situation, the EMD is computed by the following equation.

$$EMD(P, Q) = \frac{\sum_{i=1}^{m} \sum_{j=1}^{n} f_{ij} d_{ij}}{\sum_{i=1}^{m} \sum_{j=1}^{n} f_{ij}} \tag{1}$$

where f_{ij} is the optimal solution of Eq. (1) under the following conditions.

$$f_{ij} \geq 0 \quad (1 \leq i \leq m, 1 \leq j \leq n)$$

$$f_{ij} \leq r_{pi} \quad (1 \leq i \leq m)$$

$$f_{ij} \leq r_{qj} \quad (1 \leq j \leq n)$$

$$\sum_{i=1}^{m} \sum_{j=1}^{n} f_{ij} = min(\sum_{i=1}^{m} r_{pi}, \sum_{j=1}^{n} r_{qj})$$

In our system, we compute the EMD between a selected image and top 50 images from Yahoo Image search API. Our system displays top 20 images in ascending order of the EMD value.

4   Experiment

We compared an existing web image search system with our method. We used the Yahoo! image search API as the existing system. The criteria of evaluation in

Table 3. The number of correct images extracted and precision rates

Que	An example of Ext word	Que only	Ext only	Que+Ext
Spring	Cherry	2	46	44
Summer	Fireworks	6	29	29
Autumn	Japanese persimmon	4	18	23
Winter	Mandarin orange	17	12	23
Fish	Saury	12	15	13
Shrine	Yasukuni Shrine	38	19	19
Animal	Dog	7	15	36
Fruit	Strawberry	13	14	14
Mountain	Mt. Fuji	16	39	37
Beautiful	Flower	10	26	35
# of correct images		125	233	**273**
# of extracted images		450	429	447
Precision		0.28	0.54	**0.61**

this experiment were (1) the precision rates of top 50 images[7] from each method and (2) appropriateness of related words extracted by our method.

For the 1st criterion (the precision rate), we prepared 10 abstract queries. We subjectively defined the target images, namely the correct images for the queries. If we can imagine a query from an image, we accept the image as the correct image for the query. For example, the correct images for "Summer" were images about "sea", "watermelon" and "fireworks". Table 3 shows the experimental result. In the table, "Que" and "Ext word" denote an initial abstract query and one example of extracted words from our method, respectively. "Que only" denotes a naive method with the initial query only, namely the existing web image search system with an abstract query. "Ext only" denotes a method with the extracted word. As compared with "Que only", the "Ext only" yielded high precision rate (0.28 vs. 0.54). This result shows that the words extracted by our method were suitable for image retrieval with abstract queries. "Que+Ext" in the table denotes a system that used the initial query and the extracted word. The combination of two words, namely the initial query and the extracted word, produced the best performance.

The sorting process based on interaction including the EMD was not applied to three methods, namely "Que", "Ext word" and "Que+Ext", in Table 3. Next we evaluated a method with the EMD. Table 4 shows the experimental result of the effectiveness of the EMD. n denotes the number of images extracted. In other words, the precision rate on $n = 1$ denotes the correctness of the 1st output. "Que+Ext+EMD" denotes a method with the EMD. By sorting the images with interaction including the EMD, the change of the precision rates was small even if the n became large, as compared with two method without the EMD. Our

---

[7] Note that the number of images obtained from some queries in this experiment was less than 50.

system boosted correct images in lower ranks by using image information, i.e., the EMD between images. However, the results about the query "beautiful" did not improve by using the EMD. The query "beautiful" is too abstract word for our system because it is an adjective. Adjectives such as "beautiful" are associated with many words. To solve this problem, we need to apply other features or a framework of KANSEI image retrieval to our method.

In general, synonyms are effective in information retrieval as query expansion. We also evaluated a system based on synonyms extracted from some dictionaries such as Nihongo-Goi-Taikei [9]. However, using synonyms from the dictionaries was not effective. The reason was that the synonyms from the dictionaries were not always concrete words. For example, the synonyms of "Summer" are "Spring", "Winter" and so on. These words are not suitable to retrieve images about "Summer". This result shows the effectiveness of our method based on related words extracted from the web.

Table 4. Precision of Top n

	n=1	n=3	n=5	n=10	n=20	n=30
Que only	0.70	0.80	0.70	0.58	0.42	0.33
Que+Ext	0.80	0.73	0.70	0.70	0.62	0.57*
Que+Ext+EMD	0.70	0.70	0.74	0.71	0.71	0.64

*Here, this value (0.57) is lower than the value in Table 3 (0.61). The reason was that queries that contained correct images in lower ranks, namely from the 31st to the 50th, existed for the method "Que+Ext".

Table 5. Evaluation of the extracted words as tags for the images

Target image	Query	Ext	A	B	C	D	Ave
Ichiro in a batter's box	Ichiro	hit	2	2	4	1	2.3
(a baseball player)		Mariners	4	4	3	4	3.8
Illuminated	Tokyo Tower	night	3	3	5	4	3.8
Tokyo Tower		Tokyo	2	5	3	4	3.5
Let'sNote	note PC	mobile	2	4	3	3	3.0
(a PC's name)	(Laptop PC)	Let's	3	4	2	2	2.8
Mouse (as devices)	Mouse	optical	3	4	3	4	3.5
Mouse	Nezumi	rat	4	5	4	4	4.3
(as animals)		rattus	2	3	2	3	2.5
HUB	habu	HUB	5	5	4	5	4.8
	(as devices)	switching	4	4	2	3	3.3
		LAN	2	4	3	3	3.0
P.flavoviridis	habu	mongoose	1	1	3	1	1.5
	(as animals)	snake	4	5	5	4	4.3
		poison	2	1	4	2	1.8
Ave			2.9	3.6	3.1	3.1	3.2

In this experiment, we evaluated our system with abstract queries only. One future work is to evaluate our system with both of concrete and abstract queries.

The 2nd criterion was appropriateness of related words extracted by our method. In other words, we evaluated our method as a tagging system for images on the web. We regarded related words extracted by our method as tags of images. The number of test subjects was 4. In this experiment, we defined target images of each query first. For instance, "Ichiro who is a baseball player standing in a batter's box" as a concrete image for "Ichiro" as an abstract query. Next we iterated the retrieval process until the system output images that we wants. The test subjects evaluated the image and extracted word pairs that the system output. The scores for the evaluation were as follows:

1: reject, 2: weak reject, 3: fair, 4: weak accept, 5: accept.

Table 5 shows the experimental result. "A" to "D" in the table are each test subject. The score on average was 3.2. The result was fair but not enough for a tagging system of images. We need to improve the related word extraction process for the tagging system of images.

5   Conclusions

In this paper, we proposed a web image retrieval system for abstract queries. Our method could retrieve images of "fireworks" and "a white sand beach with the sea" from the query "summer". We used both of text and image information for the system. First, Our method extracted concrete words associated with an abstract query by using results from a text-based web search engine. Next, It searched and sorted the images by using a similarity between images and user's feedback. We used the Earth Mover's Distance as the similarity.

In the experiment, we compared our method with a naive method based on an existing web search engine. As a result, our method was effective as compared with the naive method. We also evaluated our method as a tagging system for images. The evaluation score was 3.2 (fair) on average. Future work includes (1) evaluation of other abstract queries and a large-scale experiment and (2) improvement of the related word extraction process, especially as a tagging system of images.

References

1. Kato, M., Ohshima, H., Oyama, S., Tanaka, K.: "likely" image search: Web image search using term sets representing typical features extracted from social tagging information. In: Proceedings of Data Engineering Workshop (DEWS 2008), IEICE (2008)
2. Freng, Y., Lapata, M.: Automatic image annotation using auxiliary text information. In: Proceedings of ACL 2008: HLT, Columbus, Ohio, June 2008, pp. 272–280. Association for Computational Linguistics (2008)
3. Gudivada, V., Raghavan, V.: Content-based image retrieval-systems. IEEE Comput. 28(9), 18–22 (1995)

4. Kushima, K., Akama, H., Konya, S., Yamamuro, M.: Content based image retrieval techniques based on image features. Transactions of Information Processing Society of Japan 40(SIG3(TOD1)), 171–184 (1999)
5. Sezaki, N., Kise, K.: Tagging system using co-occurrence of tags and similar images. In: Proceedings of Data Engineering Workshop (DEWS 2008). IEICE (2008)
6. Barthel, K.U., Richter, S., Goyal, A., Fllmann, A.: Improved image retrieval using visual sorting and semi-automatic semantic categorization of images. In: MMIU 2008, VISIGRAPP 2008 (2008)
7. Rui, Y., Huang, T.S., Mehrotra, S.: Relevance feedback techniques in interactive content-based image retrieval. In: Proceedings of Storage and Retrieval of Image and Video Databases VI (SPIE), pp. 25–36 (1998)
8. Rubner, Y., Tomasi, C., Guibas, L.: The earth mover's distance as a metric for image retrieval. International Journal of Computer Vision 40(2), 99–121 (2000)
9. Ikehara, S., Miyazaki, M., Shirai, S., Yokoo, A., Nakaiwa, H., Ogura, K., Ooyama, Y., Hayashi, Y. (eds.): Goi-Taikei. A Japanese Lexicon. Iwanami Shoten (1997)

Question Answering Based on Answer Trustworthiness

Hyo-Jung Oh, Chung-Hee Lee, Yeo-Chan Yoon, and Myung-Gil Jang

Electronics and Telecommunications Research Institute (ETRI)
161 Gajeong-dong, Yuseong-gu, Daejeon,
305-700, Korea
{ohj,forever,ycyoon,mgjang}@etri.re.kr

Abstract. Nowadays, we are faced with finding "trustworthy" answers not only "relevant" answers. This paper proposes a QA model based on answer trustworthiness. Contrary to the past researches which focused simple trust factors of a document, we identified three different answer trustworthiness factors: 1) incorporating document quality at the document layer; 2) representing the authority and reputation of answer sources at the answer source layer; 3) verifying the answers by consulting various QA systems at the sub-QAs layer. In our experiments, the proposed method using all answer trustworthiness factors shows improvement: 237% (0.150 to 0.506 MRR) for answering effectiveness and 92% (28,993 to 2,293 min.) for indexing efficiency.

Keywords: Question answering, answer trustworthiness, document quality.

1 Introduction

We are now faced with finding "trustworthy" answers not only "relevant" answers. Although IR pioneered some of the approaches used on the web for locating relevant information sources, trust-based retrieval is a relatively recent focus in that area of research.

The main thrust of our talk will be based our experience in developing and applying an answer trustworthiness-based QA model. Contrary to the past researches which focused simple trust factors of a document, we identify three different answer trustworthiness factors. The following, we (1) review some past studies relative to identify trustworthiness of documents or answers; (2) illustrate an overview of the proposed QA model; (3) describe our three different answer trustworthiness factors in detail; (4) analyze the effect of the proposed QA model with several experiments; (5) finally conclude with a suggestion for possible future works.

2 Related Work

There has been extensive research to use "trust" in information retrieval or question answering area. Trust in information retrieval is motivated by the need for not just relevant documents, but high-quality documents as well.

G.G. Lee et al. (Eds.): AIRS 2009, LNCS 5839, pp. 310–317, 2009.
© Springer-Verlag Berlin Heidelberg 2009

QA systems have proven to be helpful to users because they can provide succinct answers that do not require users to wade through a large number of documents. As same with our model, Ko [1] applied a probabilistic graphical model for answer ranking, which estimates the joint probability of correctness of all answer candidates. However, they focused on just candidate answers' correlations, not their trustworthiness or quality.

Trust is also an important area in question answering, since contradictory answers can be obtained from diverse sources in answer to a question. JiWoon [2] extracted a set of features from a sample of answers in Naver, a Korean QA portal similar to Yahoo! Answers. They built a model for answer quality based on features derived from the particular answer being analyzed, such as answer length, number of points received, etc., as well as user features, such as fraction of best answers, number of answers given, etc. They applied the method to improve the quality of the retrieval service that is attached to a community-based question answering web site and achieved significant improvement in retrieval performance.

According to Su [3], the quality of answers in commercial QA portals was good on average, but the quality of specific answers varies significantly. Agichtein [4] investigated methods for exploiting the community feedback to automatically identify high quality content. They introduced a general classification framework for combining the evidence from different sources of information, that can be tuned automatically for a given social media type and quality definition. In particular, for the community QA domain, they showed that their system is able to separate high-quality items from the rest with accuracy close to that of humans.

As the latest research, Cruchet [5] presented a solution using an existing QA system and investigated whether the quality of the answers extracted depends on the quality of the health web pages analyzed. According to the result the trustworthiness of the database used influence the quality and accuracy of the answers retrieved by the HON question answering system.

Most of works were based on only document quality analysis and a few researchers tried to use the quality of answers in a collection of question and answer pairs in the retrieval process. They used simple trust factor for improving only the quality of document in IR or answers in QA field. On the other hand, we propose a QA model using three different answer trustworthiness factors: 1) in the document layer incorporating document quality; 2) in the answer source layer representing the authority and reputation of answer sources; 3) in the sub-QAs layer verifying the answers by consulting various QA systems.

3 System Overview

Fig. 1 illustrates our system overview, which consists of *Question Analysis*, *Answer Manager,* and *Multiple sub-QA modules* based on several *Knowledge bases* generated by *Answer Indexing* from heterogeneous *Sources*. In this section, we describe how to analyze user questions and how to select the best appropriate answer candidate for the given question. At the answer selection stage, three different answer trustworthiness factors, to be explained in the Section 4, are combined.

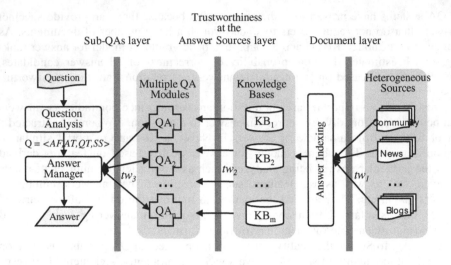

Fig. 1. System Overview of our QA model based on Answer Trustworthiness

A user question in the form of natural language is entered into the system and analyzed by the *Question Analysis* component that employs various linguistic analysis techniques such as POS tagging, chunking, answer type (AT) tagging [6], and some semantic analysis such as word sense disambiguation [7]. An internal question generated from the *Question Analysis* has the following form:

$$Q = <AF, AT, QT, SS> \tag{1}$$

where *AF* is the expected answer format, *AT* is the expected answer type, *QT* is the theme of the question, and *SS* is the information related to the expected answer source or QA module from which the answer is to be found [8].

Our QA framework using a learnable strategy makes use of a number of independent QA modules employing different answer finding methods. A strategy that determines the QA modules to be invoked when finding an answer is selected based on several factors such as the question's expected AF, AT, and SS corresponding to the expected answer sources [8].

The *Answer Manager* selects a strategy based on the internal query, translated by *Question Analysis* for the given question, invokes one or more QA modules depending on whether the answer confidence for each answer candidate is strong enough, and finalizes the answer by incorporating answers and their evidence values returned from multiple QA modules if necessary. For example of *"How to play Starcraft between 2 computers connected by a router?"*, the question should be analyzed as:

- Answer Format (AF): Descriptive
- Answer Type (AT): Method
- Question Theme (QT): target=Starcraft, focus=H/W problem
- Sub-QA Source (SS) = community-QA

then, the strategy for invoking community-QA is selected. When the candidate answer from the community-QA is higher than a threshold, the candidate can be served as the final answer to the user. If not, other QA modules are activated in sequence as in the strategy until the stopping condition is satisfied [8].

The goal of the *Answer Manager* is to determine which answer candidate has the maximum confidence value to support for the given question. Let Q and a be the user question and a candidate, respectively. Then the final answer a^* is selected by:

$$a^* = \arg \max C(a, Q) \tag{2}$$

where $C(a, Q)$ computes the confidence value between a and Q. As Q is translated into a set of internal query $<q_1, q_2, \ldots, q_N>$ (N is the number of the QA modules) for the chosen strategy by the *Question Analysis* module, $C(a, Q)$ is defined as follows:

$$C(a, Q) = \sum_i^N w_i \times S(a, q_i) \times (\alpha t w_{1i} + \beta t w_{2i} + \gamma t w_{3i}), \quad \alpha + \beta + \gamma = 1 \tag{3}$$

where $S(a, q_i)$ is the semantic distance between the answer candidate and the original question term, w_i is the weight assigned by the i-th QA module to the answer candidate a, and $tw_{(1to3)i}$ are three answer trustworthiness factors to be explained in the next Section. Semantic distance values can be computed with a lexical database [7] like Korean WordNet. When the original question looks for a "location," for example, it can be matched with "city," "province," or "country" with a decreasing values of semantic distance. Additional details can be found in [8].

4 Answer Trustworthiness Factors

We now focus on the task of determining trustworthiness of candidate answers and describe our overall approach to solving this problem. As shown in the Fig 1, we identify a set of trustworthiness factors for three different layers: document layer (Section 4.1), answer source layer (Section 4.2), and sub-QAs layer (Section 4.3). All these factors are used in *Answer Manager* as an input to the ranking equation (3) that can be tuned for confidence value for a particular answer candidate.

4.1 tw_1: Trustworthiness at the Document Layer

The first trustworthiness factor, tw_1, represents the intrinsic quality of answers and it is determined depending mainly on how the document's contents are *informative* and *readable*. We assumed that a document is informative when it is at least not a spam (not an advertisement), has originality (not just clipping), and has good quality with the big potential of containing answers. The informative documents can be detected by following processes: 1) spam filtering, 2) duplicate detection, and 3) document quality evaluation [9].

For spam filtering, we classify input documents into three categories - adult page, a spam, and ham (valid document) using SVM (Supported Vector Machine [10]) classification algorithm. Duplicated documents can be detected by breaking a document into sentences and comparing them with sentences of original documents. For efficiency of indexing process, duplicated documents should be filtered whereas "duplication" is an important clue for answer selection as the redundancy factor. To satisfy both sides, we only keep the original document, while remaining the number of duplication times.

Document quality is evaluated according to seven quality metrics reflecting *text-related* features [9], such as:

- Informativeness features: Document length, the # of words, the # of attached multimedia (such as image and video), Descriptive vocabularies
- Readability features: Lexical density, The # of frequency of internet words (such as emoticon), The # of frequency of vulgarism, and so on

After the document quality scores are normalized from 0 to 1, we cut off documents which have lower scores than a threshold at the *Answer Indexing* stage.

4.2 tw_2: Trustworthiness at the Answer Source Layer

The second trustworthiness factor, tw_2, represents the authority and reputation of answer sources. It is determined depending on how the answer source is *associate* and *reliable*. As shown in Fig 1, we collect documents from heterogeneous sources such as public media including news article, personalized media including blogs, internet community board articles, and commercial content providers (CPs). In usual, public media contents are more liable than blogs or community boards. Contrary to general expectations, in some cases, a particular community board might have a high reputation. For example, if a question is about "Starcraft", then the answer is trustworthy when it is extracted from the official user BBS developed by "Blizzard Entertainment" company, by which the game was created. It indicates that answer sources have different authority scores according to the question theme (QT), especially the question target. We built 1,000 <question, answer> pairs which are annotated the question target and the source for each pair. At the result, we set authority scores for 312 answer sources according to 137 high-frequently asked question targets.

To calculate the reputation score of answer sources, we also use four reputation metrics reflecting *non-textual* features such as:

- The publication date
- The # of reply
- The # of RSS feeds
- The # of user's recommendation

The final tw_2 can be calculated by authority scores and reputation scores in a ratio of five to three.

4.3 tw_3: Trustworthiness at the Sub-QAs Layer

The third trustworthiness factor, tw_3, represents the relevance to questions. It is determined depending on how the answering strategy is *appropriate* for the given question. As explain in Section 3, the strategy invokes the most suitable sub-QA module and attempts to verify the answers by consulting other modules. The performance of our QA depends heavily on the accuracy of the *Question Analysis* because a strategy is selected based on the sub-QA source (SS) of the analysis result, which determines the sub-QA module to be invoked first. Invoking more than one module can compensate any possible errors in the *Question Analysis* and in the answers from a sub-QA module. In other words, answers from the first sub-QA module are verified, and their confidence values are boosted if appropriate [8].

For strategy learning, we use 250 question/answer pairs of training data of various sorts in terms of sub-QA sources and difficulty levels, which are part of the entire set of 1,000 pairs used in the prevision section. We assume 250 pairs in the training set and four QA modules without loss of generality.

5 Empirical Results

To evaluate our proposed QA model, various information of answer sources should be collected including non-textual data such as content provider's name, publication date, and RSS feed tags. Therefore, we decide to develop a vertical QA focused on the "game" domain and analyzed the entire answer sources in advance. Nonetheless, it does not mean that our proposed methodologies are biased on a specific-domain.

For effectiveness comparisons, we employed a *mean reciprocal rank* (MRR, [11]). We also used *F-score*, which combines *precision* and *recall*, with the well-known "top-5" measure that considers whether a correct nugget is found in the top 5 answers. Like the TREC QA track [11], we have constructed various levels of question/answer types. We collected 5,028 questions from the Korean commercial web portal logs (Naver^{tm} Manual QA Service (http://kin.naver.com), Paran^{tm} Game IR Service (http://search.paran.com/pgame/)). Among them, a total of 250 question/answer pairs were used for building and tuning the system, and an additional 140 pairs were used for evaluation.

As shown in Fig 1, the first trustworthiness factor, tw_1, can be determined at answer indexing stage in document layer. If a document's tw_1 value exceeds a predetermined threshold, the document can be indexed as an answer candidate source. Otherwise, the document is considered as a noise then we ignored it. When the cut off threshold was 0.3, the total documents reduced by 96%; accordingly the indexing time also reduced by 92 % (28,993 to 2,293 min.)

Contrary to the first trustworthiness factor, the second and third factors, tw_2 and tw_3, can be determined at answering stage in real-time since they reflects relatedness with the given user question and candidate answers. To see the value of the impact of trustworthiness in answer selection, four cases were examined: (1) a traditional QA, (2) using only tw_1, (3) using tw_1 and tw_2, and (4) using all factors, the proposed model. For a traditional QA (the baseline), $C(a, Q)$ in equation (3) is simplified as follows:

$$C(a,Q) = \sum_{i}^{N} w_i \times S(a,q_i) \tag{4}$$

To archive the best performance for each case, we optimized not only the four different QA models but also individual sub-QA modules in Fig 1. The overall comparison shows in Table 1.

As shown in Table 2, the final MRR values for the traditional QA (baseline), using only tw_1, and using tw_1 and tw_2, and using all factors are, 0.150, 0.310, 0.385, and 0.506, respectively. Our proposed model using all factors also shows the highest overall performance in both F-score on Top1 (0.464) and Top5 (0.571).

Table 1. Overall Comparisons

	Top1		Top5		MRR	
	Precision	Improv.	Precision	Improv.	Precision	Improv.
Baseline	0.150		0.220		0.150	
tw_1	0.280	$86.7\%^1$	0.380	$72.7\%^1$	0.310	$106.7\%^1$
tw_1+tw_2	0.357	$27.5\%^2$ $138.0\%^1$	0.421	$10.8\%^2$ $91.4\%^1$	0.385	$24.2\%^2$ $156.7\%^1$
All $(tw_1+tw_2+tw_3)$	0.464	$30.0\%^3$ $65.7\%^2$ $209.3\%^1$	0.571	$35.6\%^3$ $50.3\%^2$ $159.5\%^1$	0.506	$31.4\%^3$ $63.2\%^2$ $237.3\%^1$

1: Improvement over the baseline; 2: over the using tw_1; 3: over the using tw_1+tw_2

The relative improvement shows the impact of each trustworthiness factor. In comparison with the baseline and using only tw_1, the improvement of MRR is up to 100% (0.150 to 0.310), whereas the tw_2 addition obtains only 24.2% (0.310 to 0.385) gains. By using all factors, 31.4% (0.385 to 0.506) gains can be achieved over the using tw_1 and tw_2. By reflecting this observation, the final QA model showed the best accuracy (0.506 MRR) when we set weight values, α, β, and γ, in Equation (3) respectively are 0.5, 0.2, and 0.3.

To sum up experimental results, the proposed method using all answer trustworthiness factors showed dramatic improvement: 237% (0.150 to 0.506 MRR) for answering effectiveness and 92% (28,993 to 2,293 min.) for indexing efficiency. These results indicate that by referring answer trustworthiness, we can not only save time but also provide better answers, while suppressing unreliable ones.

6   Conclusion

This paper proposed a QA model based on answer trustworthiness. Contrary to the past researches which focused simple trust factors of a document, we identified three different answer trustworthiness factors: 1) incorporating document quality at the document layer; 2) representing the authority and reputation of answer sources at the answer source layer; 3) verifying the answers by consulting various QA systems at the sub-QAs layer.

To prove the efficacy of the proposed model, we analyzed the impact of answer trustworthiness in indexing stage as well as the answering stage. In indexing, distilling unreliable documents which have lower trustworthiness value than 0.3 brings not only 96% reduced document size but also 92% speedy indexing time. To reveal trust effects on answering, we have conducted experiments for four different cases: (1) a traditional QA, (2) using only tw_1, (3) using tw_1 and tw_2, and (4) using all factors. The proposed method using all answer trustworthiness factors obtained an improvement over the traditional QA by 237% in effectiveness.

We plan to improve each answer trustworthiness calculation method extending various features. In addition, to overcome false-alarm of reliable document in the indexing process, a dynamic cut-off scheme will be developed.

Acknowledgement

This work was supported in part by the Korea Ministry of Knowledge Economy (MKE) under Grant No. 2008-S-020-02.

References

1. Ko, J.W., Si, L., Eric, N.: A Probabilistic Graphical Model for Joing Answer Ranking in Question Answering. In: The 30th annual international ACM SIGIR conference, pp. 343–350. ACM Press, New York (2007)
2. Jeon, J.W., Croft, W.B., et al.: A framework to predict the quality of answers with non-textual features. In: The 29th annual international ACM SIGIR conference, pp. 228–235. ACM Press, New York (2006)
3. Su, Q., Pavlov, D., Chow, J.-H., Baker, W.C.: Internet-scale collection of human-reviewed data. In: The 16th international conference on WWW conference, pp. 231–240. ACM Press, New York (2007)
4. Agichtein, E., Castillo, C., Donato, D.: Finding High-Quality Content in Social Media. In: Web Search and Data Mining (WSDM), pp. 183–194. ACM Press, Stanford (2008)
5. Cruchet, S., Gaudinat, A., Rindflesch, T., Boyer, C.: What about trust in the Question Answering world? In: AMIA 2009 Annual Symposium (2009)
6. Lee, C.K., Hwang, Y.G., Lim, S.J., et al.: Fine-Grained Named Entity Recognition Using Conditional Random Fields for Question Answering. In: Ng, H.T., Leong, M.-K., Kan, M.-Y., Ji, D. (eds.) AIRS 2006. LNCS, vol. 4182, pp. 581–587. Springer, Heidelberg (2006)
7. Choi, M.R., Hur, J., Jang, M.G.: Constructing Korean lexical concept network for encyclopedia question answering system. In: IEEE IECON 2004, pp. 3115–3119. IEEE Press, New York (2004)
8. Oh, H.J., Myaeng, S.H., Jang, M.G.: Strategy-driven Question Answering with Multiple Techniques. ETRI Journal 31(4) (2009)
9. Lee, H.G., Kim, M.J., Rim, H.C., et al.: Document Quality Evaluation for Question Answering System. In: 20th Conference of Hangul and Korean Information Processing (in Korean), pp. 176–181 (2008)
10. Lee, C.K., Jang, M.G.: Fast Training of Structured SVM Using Fixed-Threshold Sequential Minimal Optimization. ETRI Journal 31(2), 121–128 (2009)
11. Lin, J.: Is question answering better than information retrieval? A task-based evaluation framework for question series. In: HLT/NAACL 2007, pp. 212–219 (2007)

Domain Specific Opinion Retrieval

Guang Qiu, Feng Zhang, Jiajun Bu, and Chun Chen

Zhejiang Key Laboratory of Service Robot, College of Computer Science
Zhejiang University, Hangzhou 310027, China
{qiuguang,zhangfeng,bjj,chenc}@zju.edu.cn

Abstract. Opinion retrieval is a novel information retrieval task and has attracted a great deal of attention with the rapid increase of online opinionated information. Most previous work adopts the classical two stage framework, i.e., first retrieving topic relevant documents and then re-ranking them according to opinion relevance. However, none has considered the problem of domain coherence between queries and topic relevant documents. In this work, we propose to address this problem based on the similarity measure of the usage of opinion words (which users employ to express opinions). Our work is based on the observation that the opinion words are domain dependent. We reformulate this problem as measuring the opinion similarity between domain opinion models of queries and document opinion models. Opinion model is constructed to capture the distribution of opinion words. The basic idea is that if a document has high opinion similarity with a domain opinion model, it indicates that it is not only opinionated but also in the same domain with the query (i.e., domain coherence). Experimental results show that our approach performs comparatively with the state-of-the-art work.

Keywords: Opinion Retrieval, Domain Coherence, Opinion Model, Opinion Similarity.

1 Introduction

With the rapid development of novel Web applications (e.g. blogs) and users' accessibility to the Internet, opinionated information is turning to be an increasing important type of online content. Compared with traditional content (e.g. news reports), people can benefit more from such information. Take a major type of opinionated information, i.e., product reviews, for example, consumers can make decision on whether to purchase a product by reading the reviews, while product manufacturers could keep track of their products in order to improve the quality.

However, the volume of opinionated information has recently become extremely large, which consequently prevents people from finding opinions on some entity (i.e. products, events, organizations, etc.) manually. Current search engines, like Google, Yahoo!, Live Search, have provided excellent performance in traditional search tasks. Unfortunately, they perform poorly in this novel *opinion retrieval* task, i.e., retrieving documents which are both relevant to the queries and containing opinions on them. Obviously, the keyword-based matching strategy in those search engines does not

G.G. Lee et al. (Eds.): AIRS 2009, LNCS 5839, pp. 318–329, 2009.
© Springer-Verlag Berlin Heidelberg 2009

take into account the constraint of the occurrence of opinions. Consequently, lots of researchers have recently conducted intensive work to address this opinion retrieval task [1, 2, 3, 4, 5, 6, 7, 8, 9, 10, 11]. Most previous work proposes solutions following the classical two stage framework, i.e., retrieving topic relevant documents using traditional retrieval models and then re-ranking the candidates according to opinion relevance [3, 4, 5, 6, 7, 8, 9, 10, 11]. There is also another line of recent work to solve this problem using a unified framework [1, 2].

One obvious drawback of some work is that as they were designed for the Blog track in TREC (http://trec.nist.gov) [12, 13] (especially those perform well in this track), they have heavy dependence on the unique properties of blogspace. However, these evidences can not be utilized in ordinary Web pages. Another problem of all previous work is that they neglected the problem of domain coherence between topic relevant candidate documents and queries. Problems would rise if we consider any candidate documents which contain opinion words as the opinionated results for a query. There are cases that the documents may contain opinions on topics of other domains rather than the queries themselves. Another example is that for the classical ambiguous query "*Apple*", it could be a kind of food or the well-known company, so the expressive opinions of the documents may be any one of these two kinds. Without imposing the domain coherence constraint, documents of different domains will be mixed up. Such results would not satisfy users' information need consistently.

In this paper, we propose to take use of only textual characteristics and address the second issue based on the idea of query classification. Specifically, in our work, we do not focus on how to classify queries into different categories but on addressing the mentioned issues using query category information. We make the assumption that the category information is known in advance. In practice, users could specify the category manually with little additional efforts when issuing the query. In detail, our work falls into the traditional two stage framework and the major difference with previous work is that we propose to re-rank the topic relevant candidates using the *opinion similarity* measure between the candidate *opinion models* and corresponding domain *opinion models*. An opinion model captures the distribution of opinion words in document(s) and the opinion similarity measures the similarity between two opinion models. We will give detailed definitions later. The basic idea is that if a document has high opinion similarity with a domain opinion model, it is regarded as containing content that is not only opinionated but also coherence with the issued query in domain category. This idea is based on the observation that the opinion words, which users employ to express their opinions, are domain dependent [14]. For example, in domain of "*Food*", opinion words like "*delicious*", "*yummy*" would be used at large while in domain of "*Sports*", frequent opinion words would be "*winning*", "*athletic*". So far as we know, we are the first to concern the domain coherence problem in opinion retrieval. Note that the category information can also be utilized in the first step (i.e., topic relevant retrieval), which turns the issue to a traditional information retrieval problem. Experiments on the Blog track testing data show that our framework performs comparatively with the state-of-the-art approaches.

The remainder of this paper is organized as follows. In Section 2, we give an overview on the related work. In Section 3, we describe the domain taxonomy used in our work. In Section 4 we elaborate our framework. The experimental results and analysis are demonstrated in Section 5, and conclusions are drawn in Section 6.

2 Related Work

Lots of research has been done on opinion retrieval in last few years. The major part of previous work is proposed for the Blog track [3, 4, 5, 6, 7, 8, 9, 10, 11] initiated by TREC. Additionally, there is also another line of work to solve the problem using a general opinion retrieval framework [1, 2].

Since 2006, TREC has launched a special track on opinion retrieval in blogspace (i.e., Blog track), which aims to retrieve opinionated information for given queries from Blog data [12, 13]. Lots of organizations have participated in this task, and the prevailing systems adopt the two stage framework. In the first step, topic relevant documents are retrieved using the classical information retrieval models (e.g., TFIDF, Okapi, Language Modeling, etc. [15]) and each document is assigned with a topic relevance score. In the second step, opinion relevant ranking is performed and opinion relevance score is calculated for each document. There are two main strategies for the second step, namely lexicon-based and machine learning approaches. Lexicon-based methods take use of an existing opinion lexicon [3, 4, 5] or the word distribution over the data set [3, 4], and the opinion relevance score is calculated based on the occurrences of the opinion words in the documents, with constraint on the distance of topic words and the opinion words. Machine learning approaches treat the opinion relevance identification problem as a classification problem by employing a bunch of features learned from the training data [6, 10, 11]. The final re-ranking of documents is performed through some combination (mostly linear [3, 4, 5]) of the scores of topic relevance and opinion relevance. Our work falls into this category. Due to limited space, we only show some representative work following such framework.

In [3], Mishne adopted the language modeling based retrieval model to select the topic relevant documents as the candidates, and then employed lexicon-based way to compute the opinion relevance values of those candidates. He also considered the post quality as a component in determining the final ranking scores. In his following work [7], he refined the system and improved the performance by considering the temporal and comment information and enhancing the quality measure according to the query dependent context. Zhang et al. [10] proposed to use concept-based topic retrieval as the topic retrieval method and built the support vector machine classifier to identify the opinionated documents. The final ranking component utilized the NEAR operator to find the query relevant opinions. In [11], they proposed to improve the system effectiveness by employing techniques like spam filtering.

In addition to the two stage framework, Zhang et al. proposed to solve the problem under a unified framework [1]. They started from the classical probabilistic based retrieval model and considered the information need of opinion searchers as a subset of opinionated ones in the original topic relevant documents. They characterized this information need as the implied opinionated expressions towards the query and re-formulated the posterior probability of fitting document to a particular query to the query and the opinionated expressions. In their framework, the final ranking is a quadratic combination of topic and opinion scores. In the work of [2], Eguchi and Lavrenko also proposed a generative model for this task but did not get good results.

However, all previous work neglected the domain coherence between queries and candidate documents. In cases of ambiguous queries or documents containing opinions on more than one topic (partly solved by the distance constraint of topic words and opinion words), the mismatch of domains would result in inconsistent performance. Our work is the first to address such issues in opinion retrieval through the idea of *opinion similarity*. Besides, similar to [1, 2], we only exploit the textual information in ranking, thus our approach can be applied to the general Web opinion retrieval task.

3 Domain Taxonomy

As described above, the foundation of our work is the classification of queries into different domains. This classification task is known as the query classification in traditional information retrieval area [16], which can be regarded as a type of short text classification task. However, it is still a challenging problem considering few words in queries. In our work, we focus on the utilization of the domain information rather than the classification task itself. Additionally, the domain information can be easily provided by users in practical applications. We argue that users would be willing to submit such information in addition to queries in order to refine the search results. Therefore, in following sections we make the reasonable assumption that the domain information of a query is known in advance. The automatic classification task can be investigated in future work.

Another critical issue with domain classification is the design of a proper domain taxonomy. Some efforts have already been made in the area of query classification. In [16], they use a taxonomy containing 67 hierarchal categories, including "*Entertainment\Movies*", "*Entertainment\TV*", "*Computers\Software*", etc. Unfortunately, this fine-grained taxonomy is not suitable for our task. For example, in opinionated documents of "*Entertainment\Movies*" and "*Entertainment\TV*", the opinion words may have high overlap. Our taxonomy is aimed to discriminate the usage of opinion words of different categories. Consequently, in this paper, we define a new coarse-grained taxonomy for our task by investigating the information need of opinion searchers. Table 1 shows the details.

Table 1. Domain taxonomy designed for the opinion retrieval task

Domains	Explanations
Society	Topics on policies, politics, cultures, histories, etc.
Entertainment	Topics on movies, music, books, TV programs, etc.
Sports	Topics on football, basketball, etc.
Food and Health	Topics on food, health.
Technology	Topics on digital products, high technology, etc.
Other	Other topics

As can be seen, we only define totally six categories in our taxonomy. One important reason of using such small size taxonomy is to minimize users' efforts when providing category information for the query. This kind of taxonomy also makes the future automatic query classification easier and thus practical. It is known that classification is extremely hard using fine-grained taxonomies. We will show the wide coverage of this taxonomy in a publicly available collection of queries for opinion retrieval.

4 Domain Specific Opinion Retrieval

Given the taxonomy defined above, user queries are assigned with corresponding categories (which are provided by users in current work). In other words, our opinion retrieval system (**D**omain **S**pecific **Op**inion **R**etrieval, **DS-OpR**) has queries and their category information as input. In following sections, we first give an overview of our retrieval system, and then elaborate the critical components in detail.

4.1 Overview of the DS-OpR System

As stated before, our retrieval system falls into the classical two stage opinion retrieval framework, which typically contains a classical task of retrieving topic relevant documents and a re-rank module to sort the document candidates considering the opinion relevance. In this way, the resulted documents are both topic and opinion relevant to the queries. Figure 1 illustrates the framework of our retrieval system **DS-OpR** in detail.

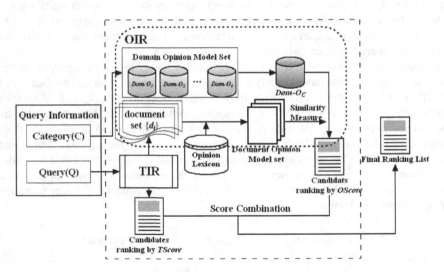

Fig. 1. The framework of our **DS-OpR** opinion retrieval system

In the framework, given the query Q and its category C, we first employ a classical retrieval model to get documents relevant to the query Q in the first stage **TIR**. The returned documents $\{d_i\}$ are those matching the keywords in Q and each document d_i is assigned with a topic relevance score $TScore_i$. Several retrieval models are available for this stage, such as TFIDF, Okapi, Language Modeling, etc. One main advantage of the two stage framework is that there is no constraint on the selection of retrieval model as long as the returned documents are assigned with $TScores$.

In our second stage **OIR**, we first construct the opinion model O_i of each candidate document d_i given the opinion lexicon (in practice, the modeling can be done in the indexing phase for **TIR**.), and then compute the opinion relevance score $OScore_i$ for d_i by measuring the opinion similarity OS_i of O_i and the domain opinion model of category C $Dom\text{-}O_C$. Note that the domain models are computed offline in advance using a collection of documents of the corresponding domain and the opinion lexicon. Specifically, an opinion model is represented as a probability distribution of opinion words and thus the opinion similarity is measured between two distributions.

To generate the final ranking list, $TScore_i$ and $OScore_i$ of the document d_i are combined in some way. Most previous work adopts the linear strategy, while Zhang et al. take a quadratic combination [1]. In our work, we follow the first combination strategy as shown in equation 1. Alpha balances the importance of $OScore$ and $TScore$. Documents are then ranked by the final combination $Score$.

$$Score = (1-\alpha) \times TScore + \alpha \times OScore, \quad 0 \le \alpha \le 1 \tag{1}$$

From the above description of our system, we can see that the most critical components in the framework are (1) Opinion Modeling for domains and documents; (2) Opinion Similarity measure between opinion models. In next section, we describe the details of these two components.

4.2 Opinion Modeling and Opinion Similarity Measure

So far, no previous work in opinion retrieval has considered the opinion word distribution in computing the opinion relevance scores. We propose that the distribution is helpful in both measuring the opinion relevance scores and discriminating the domain ambiguity of queries (recall the example queries "*Apple*"). In this work, we call a distribution of opinion words as an *Opinion Model* (O). The generation of O is thus named as opinion modeling. The main intuition of this idea is that opinionated documents can be regarded as a mixture of opinion models and topic models (i.e., distribution of topic relevant words other than opinion ones), which is enlightened by the work of topic modeling [17, 18]. Formally, we define opinion model as follows.

Definition (Opinion Model): An *opinion model O* for a collection of documents $\{d_m\}$ or a single document d_m is a probability distribution $\{P(w_i|O)\}$ of opinion words $\{w_i\}$ representing either positive or negative polarities.

These opinion words $\{w_i\}$ are taken from an existing opinion lexicon. The construction of opinion lexicon has already received lots of attention in the research area of opinion mining and there are several publicly available lexica, such as General Inquirer [19], SentiWordNet [20], etc. When separately applying opinion modeling to $\{d_m\}$ and d_m, we have two specific kinds of opinion models correspondingly.

Definition (Domain Opinion Model): A *domain opinion model $Dom\text{-}O_k$* for domain D_k is a general opinion model for documents of that domain. In other words, it captures the probability distribution $\{P(w_i|D_k)\}$ of opinion words $\{w_i\}$ in those documents.

In this paper, we totally construct six domain opinion models according to the categories defined in our taxonomy. Clearly, the construction of a domain opinion model demands a collection of documents belonging to the corresponding domain.

Definition (Document Opinion Model): Similar to domain opinion model, a *document opinion model Doc-O_m* is a specific opinion model for a document d_m, represented as a probability distribution $\{P(w_i|d_m)\}$ of opinion words $\{w_i\}$.

Specifically, we compute $P(w_i|O)$ (O can be D_k or d_m) using following formula:

$$P(w_i \mid O) = \frac{TF(w_i, O)}{\sum_w TF(w, O)} \tag{2}$$

where $TF(w_i, O)$ is the frequency of w_i in O and $\sum_w TF(w, O)$ is the frequency of all words (including both opinion and non-opinion ones) in O.

According to our definitions, for different domains, we have different domain opinion models, which means that the distributions of opinion words are different. This is quite natural in opinion expressions of different domains, i.e., people choose different words to express their attitudes towards topics of different domains. For example, considering domains of "*Food and Health*" and "*Entertainment*", people would choose "*yummy*" to express their satisfaction with some food, while using "*moving*" for some movie. Therefore, if the document opinion model of an opinionated document is similar to a domain opinion model, it indicates that there is high possibility that the opinionated document belongs to that domain. In this way, we can guarantee the domain coherence of the returned opinionated documents with the issued query with designated domain. From this perspective, we argue that previous work can be regarded as using a universal opinion model to cover opinions of all domains, which is not reasonable according to our analysis.

As opinion models are represented as probability distributions of opinion words, we compute the similarity between two opinion models based on the well known probability distribution similarity measure Kullback-Leibler (KL) divergence [21]. Formally, given two opinion models O_i $\{P(w_k|O_i)\}$ and O_j $\{P(w_k|O_j)\}$, their opinion similarity is measured using following formula:

$$OSim(O_i, O_j) = \frac{1}{2}(KL(O_i \parallel O_j) + KL(O_j \parallel O_i)) \tag{3}$$

$$KL(O_i \parallel O_j) = \sum_k P(w_k \mid O_i) \log \frac{P(w_k \mid O_i)}{P(w_k \mid O_j)} \tag{4}$$

$KL(O_j \parallel O_i)$ is defined similarly as equation 4. KL divergence measures the similarity asymmetrically; we take the average value to make it symmetric.

5 Experiments and Results

In our experiments, we use the Blog track data set in TREC as the testing data as previous work. Details of the data set can be found in [12]. The 100 topics in the 2006 Blog track (Blog06) and the 2007 Blog track (Blog07) (topics 851-900 in Blog06 and topics 901-950 in Blog07) are evaluated. We only take the titles of topics as queries as in [1]. This strategy also suits the scenery of traditional search engines in which the issued queries are always composed of several keywords. The opinion lexicon used in our experiments is provided by Hu and Liu [22], which comprises of 1752 opinion words including 1098 negative ones and 654 positive ones. For the first stage of our framework (i.e., topic relevance retrieval), we employ Lemur (www.lemurproject.org) as the retrieval tool and use the simple retrieval model TFIDF. Top 1000 documents are kept as the candidates for re-ranking as in previous work.

To construct the domain opinion model for each category in our taxonomy, we use the TREC labeled results of the queries other than the testing ones as the documents for different domains, i.e., if we are testing queries of Blog06, we use the results of Blog07 and Blog08 for model construction. Each of the 150 queries is assigned with one of the six categories and thus the labeled document results of a query can be categorized into corresponding domain. Note that the 50 topics in the 2008 Blog track (Blog08) are only utilized for the construction of domain opinion models.

The system performance is evaluated using the traditional evaluation metrics, i.e., precision in top 10 results (P@10), mean average precision (MAP) and R-precision (R-Prec). All these results are computed using the program in Lemur. In following sections, we first show the results of the category distribution in the 150 queries of Blog06, Blog07 and Blog08, and then demonstrate the detailed evaluation results of our system.

5.1 Taxonomy Examination

The domain taxonomy is fundamental in our work. To validate the coverage of the taxonomy, we conduct an examination on the category distribution of Blog06, Blog07 and Blog08 queries (i.e., titles of the topics). The 150 queries are analyzed by annotators and the final category of a query is determined by voting. Note that the examination should have been carried out in a large scale query collection for justified results. However, it is almost impossible for us to collect large number of such queries that users issue for opinionated information. Additionally, till now, there is no such query log publicly available. We believe that the results of the Blog track data from TREC would be capable of reflecting the ground truth to some extent.

Figure 2 shows the detailed distribution of the queries among different domains. From the results we can see that majority queries (90%) can be categorized into one of the five categories, i.e., "*Technology*", "*Food and Health*", "*Society*", "*Entertainment*" and "*Sports*". The results indicate that our definition of the domain taxonomy is reasonable. Among these five ones, queries of "*Society*" make up the large percentage (29.3%) and then those of "*Technology*" and "*Entertainment*". This observation shows that when users are seeking opinions, they are mostly concerned with popular topics like politics, digital products and movies. This conclusion is quite obvious in real life. When checking the queries of the "*Others*" category, we find extremely ambiguous queries such as "*brrreeeport*" of Topic No. 907 in Blog07, which are hard to be assigned to any one of the five categories.

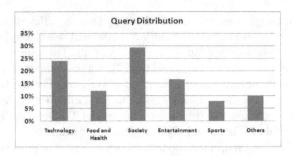

Fig. 2. Query distribution over the six domains

5.2 Performance Evaluation

In our final combination of topic relevance and opinion relevance scores, we adopt the linear way. The optimal value of the weight parameter alpha is tuned empirically. Hence, we give the results of using different values for alpha first.

5.2.1 Parameter Selection

We conduct the parameter selection experiments on queries of Blog06. Totally, we conduct 11 comparative experiments using different values of alpha ranging from 0 to 1 with a step length of 0.1. Furthermore, we argue that alpha may take different values for queries of different domains, which means that the opinion relevance plays different roles in different domains for the final ranking. Hence, we perform the same experiments for queries of each of the six domains individually. Figures 3 and 4 show the results of MAP and P@10 respectively.

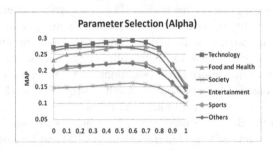

Fig. 3. MAP results of our approach using different values of alpha

From the results we can see that queries belonging to "*Society*" reach the best P@10 and MAP when alpha is set to 0.2; while queries of the other five categories get their best performance when alpha takes the value of 0.6. The results show that we should impose less weight on the opinion relevance score for queries of "*Society*". One possible reason might be that "*Society*" includes topics about politics, policies, etc., whose topic relevant documents are mostly already opinionated in blogspace. Therefore, it makes sense to lower the weight of opinion relevance scores, because once the documents are retrieved as topic relevant, they are more likely to be opinionated already. However, documents of the other five categories are relatively objective in blogspace; therefore the weight of opinion relevance scores should be enhanced.

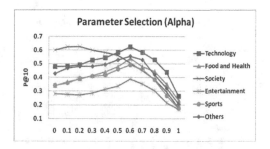

Fig. 4. P@10 results of our approach using different values of alpha

5.2.2 Evaluation Results and Discussions

We take the optimal values of alpha from above parameter tuning for the performance comparison of our system with others. Table 2 demonstrates the results of the best title-run at Blog tracks, Zhang et al. (**Zhang**) [1] and ours (**DS-OpR**). From the results, we can see that our approach outperforms the other approaches in P@10 in the Blog06 topics and has close result to the best one in Blog07. In the MAP measure, we outperform the best title-run and have close result to **Zhang** in Blog06. Unfortunately, we have relatively poor MAP in Blog07 compared to others, although not far from **Zhang**. However, as the retrieval performance of the opinion retrieval task is strongly dominated by the performance of the underlying topic relevance task [12], we argue that our approach still has large room for improvement considering the naïve retrieval model employed in current experiments. Another reason is that we currently only take use of one simple opinion lexicon. The experiments of Zhang et al. [1] show that different opinion lexica would affect the final performance. In all, the current results show that our approach of considering the domain coherence of queries and documents is promising and comparative to others in some measures.

Table 2. P@10, MAP, R-Prec Results of different approaches in topics of Blog06 and Blog07

Data set	Method	P@10	MAP	R-Prec
Blog06	Best title-run at Blog06	0.512	0.1885	0.2771
	Zhang	0.507	0.2257	0.3038
	DS-OpR	0.514	0.2064	0.2689
Blog07	Best title-run at Blog07	0.690	0.4341	0.4529
	Zhang	0.606	0.3371	0.3896
	DS-OpR	0.614	0.2915	0.3363

Figure 5 illustrates the effect of opinion modeling re-ranking per topic in P@10. The results show that 22 out of 100 queries get improvement by over 50%, including 11 queries with improvement over 200%. Only 7 out of 100 get adverse effects. It indicates that the re-ranking method is helpful. Specific examples are also given in Table 3. These two queries demonstrate the effectiveness of our approach in re-ranking. On one hand, it proves the correctness of our re-ranking method; on the other hand, it indicates the potential of getting a better topic relevance retrieval model as well.

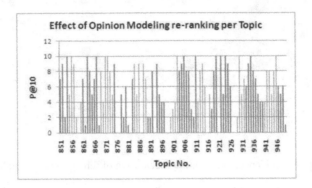

Fig. 5. Illustrations of effects of opinion modeling re-ranking per topic

Table 3. Results of example queries of Blog06 and Blog07

Topic	Title	Category	Description		
TREC06_ID886	West Wing	Entertainment	Provide opinion concerning the television series West Wing.		
	MAP	**P@10**	**P@30**	**P@100**	**P@1000**
Before re-ranking	0.0869	0.1000	0.2000	0.1600	0.1670
After re-ranking	0.1241	0.9000	0.5667	0.2400	0.1670

Topic	Title	Category	Description		
TREC07_ID936	grammy	Entertainment	Find opinions regarding the Grammy awards.		
	MAP	**P@10**	**P@30**	**P@100**	**P@1000**
Before re-ranking	0.0661	0.0000	0.0000	0.0000	0.1660
After re-ranking	0.1144	0.8000	0.5000	0.2100	0.1660

6   Conclusions

Opinion retrieval has received a great deal of attention from both areas of information retrieval and natural language processing. Lots of work has been proposed to solve this problem; however none of them considers the domain coherence between topic relevant documents and queries. In this paper, we follow the classical two stage framework in opinion retrieval and propose to compute the opinion relevance scores according to the opinion similarity between the opinion models of candidate documents and corresponding domain opinion model of the query. High opinion similarity means that the documents are both opinionated and coherent with the query from the perspective of domain specific opinion word distributions. Experiments are carried out on the Blog track data set, and the results show that our framework performs comparatively with state-of-the-art methods.

Acknowledgments. This work is supported by National Key Technology R&D Program(2008BAH26B00 & 2007BAH11B06).

References

1. Zhang, M., Ye, X.: A generation model to unify topic relevance and lexicon-based sentiment for opinion retrieval. In: Proceedings of SIGIR 2008, pp. 411–418 (2008)
2. Eguchi, K., Lavrenko, V.: Sentiment Retrieval using Generative Models. In: Proceedings of EMNLP 2006, pp. 345–354 (2006)
3. Mishne, G.: Multiple Ranking Strategies for Opinion Retrieval in Blogs. In: Online Proceedings of TREC (2006)
4. Yang, K., Yu, N., Valerio, A., Zhang, H.: WIDIT in TREC 2006 Blog track. In: Online Proceedings of TREC (2006)
5. Liao, X., Cao, D., et al.: Combing Language Model with Sentiment Analysis for Opinoin Retreival of Blog-Post. In: Online Proceedings of TREC (2006)
6. Zhang, W., Yu, C.: UIC at TREC 2006 Blog Track. In: Online Proceedings of TREC (2006)
7. Mishne, G.: Using blog properties to improve retrieval. In: Proceedings of the International Conference on Weblogs and. Social Media, ICSWM (2007)
8. He, B., Macdonald, C., He, J., Ounis, I.: An effective statistical approach to blog post opinion retrieval. In: Proceedings of CIKM 2008 (2008)
9. He, B., Macdonald, C., Ounis, I.: Ranking opinionated blog posts using OpinionFinder. In: Proceedings of SIGIR 2008 (2008)
10. Zhang, W., Yu, C., Meng, W.: Opinion retrieval from blogs. In: Proceedings of CIKM 2007 (2007)
11. Zhang, W., Jia, L., Yu, C., Meng, W.: Improve the effectiveness of the opinion retrieval and opinion polarity classification. In: Proceeding of CIKM 2008 (2008)
12. Ounis, I., de Rijke, M., Macdonald, C., Mishne, G., Soboroff, I.: Overview of the TREC 2006 Blog Track. In: Online Proceedings of TREC (2006)
13. Macdonald, C., Ounis, I.: Overview of the TREC 2007 Blog Track. In: Online Proceedings of TREC (2007)
14. Turney, P.D.: Thumbs Up or Thumbs Down? Semantic Orientation Applied to Unsupervised Classification of Reviews. In: Proceedings of ACL 2002, pp. 417–424 (2002)
15. Baeza-Yates, R., Ribeiro-Neto, B.: Modern Information Retrieval. ACM Press, New York (1999)
16. Shen, D., Pan, R., et al.: Q2C@UST: our winning solution to query classification in KDDCUP 2005. SIGKDD Explorations 7(2), 100–110 (2005)
17. Blei, D.M., Ng, A.Y., Jordan, M.I.: Latent dirichlet allocation. The Journal of Machine Learning Research 3, 993–1022 (2003)
18. Hofmann, T.: Probabilistic Latent Semantic Analysis. In: Proceedings of UAI 1999 (1999)
19. Stone, P., Dunphy, D., Smith, M., Ogilvie, D.: The General Inquirer: A Computer Approach to Content Anaysis. MIT Press, Cambridge (1966)
20. Esuli, A., Sebastiani, F.: Determining the semantic orientation of terms through gloss classification. In: Proceedings of CIKM 2005, pp. 617–624 (2005)
21. Kullback, S., Leibler, R.A.: On Information and Sufficiency. The Annals of Mathematical Statistics 22(1), 79–86 (1951)
22. Hu, M., Liu, B.: Mining and Summarizing Customer Reviews. In: Proceedings of SIGKDD 2004 (2004)

A Boosting Approach for Learning to Rank Using SVD with Partially Labeled Data

Yuan Lin, Hongfei Lin, Zhihao Yang, and Sui Su

Department of Computer Science and Engineering,
Dalian University of Technology, Dalian 116023, China
yuanlin@student.dlut.edu.cn, hflin@dlut.edu.cn,
yangzh@dlut.edu.cn, beast5117@yahoo.com.cn

Abstract. Learning to rank has become a hot issue in the community of information retrieval. It combines the relevance judgment information with the approaches of both in information retrieval and machine learning, so as to learn a more accurate ranking function for retrieval. Most previous approaches only rely on the labeled relevance information provided, thus suffering from the limited training data size available. In this paper, we try to use Singular Value Decomposition (SVD) to utilize the unlabeled data set to extract new feature vectors, which are then embedded in a RankBoost leaning framework. We experimentally compare the performance of our approach against that without incorporating new features generated by SVD. The experimental results show that our approach can consistently improve retrieval performance across several LETOR data sets, thus indicating effectiveness of new SVD generated features for learning ranking function.

Keywords: Information Retrieval; Learning to rank; Machine learning; SVD.

1 Introduction

Learning to rank is a cross field of machine learning and information retrieval. It aims to learn the rank function with the relevance judgments from training set, while the rank function is used to sort by the document relevance. The core issue of the learning to rank is how to construct a model or function to predict the relevance of the documents. Ranking task is defined as follows. The training data (referred as D), which consists of a set of records of the form $< q, d, r >$, where q is a query, d is a document (represented as a list of features $\{f_1, f_2, \ldots, f_m\}$), and r is the relevance judgment of d to q. The relevance draws its values from a discrete set of possibilities (e.g., 0, 1). The test set (referred as T) consists of records $< q, d, ? >$ in which only the query q and the document d are known, while the relevance judgment of d to q is unknown. The training data is used to construct a model which relates features of the documents to their corresponding relevance in order to predict the relevance of the document of test data for getting ranking list.

Although many algorithms and models based on labeled data set were introduced to learning to rank, the research on unlabeled data is still insufficient. In view of this,

G.G. Lee et al. (Eds.): AIRS 2009, LNCS 5839, pp. 330–338, 2009.
© Springer-Verlag Berlin Heidelberg 2009

this paper deals with the data set composed of labeled and unlabeled data by SVD, introducing the information of unlabeled data into the training set, in order to improve the performance of the RankBoost and the ranking result.

The paper is divided as follows: Section 2 introduces the related work about learning to rank. Section 3 proposes the idea that to use SVD introduce the unlabeled data information. Then section 4 presents the experimental results. Finally, we conclude this work and point out some directions for future research.

2 Related Work

Many models and theories have been introduced to the field of learning to rank for obtaining the ranking function. Freund et al. use AdaBoost to the web meta-search task, and propose the RankBoost algorithm [1] for learning to rank. Herbrich et al. propose a learning algorithm for ranking on the basis of Support Vector Machines, called RankSVM [2]. Zhe Cao et al. propose the Listnet [3] based on neural network and gradient descent. Guiver et al. get satisfactory ranking result for introduce Gaussian processes [4] to the task. Adriano Veloso et al. extract the interval association rules from features to improve the ranking performance [5]. Although they are rewarded by a pretty success, none of them utilize the unlabeled data to improve the results.

Information retrieval and machine learning algorithms used to ranking task achieve a great success, most of the experts and scholars on the basis of using labeled data sets for researching. It appears inadequate on the study of unlabeled data. However, some scholars try to improve the performance of ranking on unlabeled data set. They propose some methods to introduce unlabeled corpus to training model. Massih-Reza Amini et al, [6] assume that an unlabeled instance that is similar to a labeled instance should have similar label, and introduce the instance to the training set. The experimental results show it can improve the ranking result. Kevin Duh et al. apply kernel PCA (kernel-based principal component analysis) [7] on the test set mode principal component pattern extraction, and use this pattern to extract new features from the training set. And thus it is adding the information of test set to training set implicitly, and gets good ranking result. This method is only adding test set of information to the training set, and it does not take introducing the training set information to the test set to improve the performance of ranking model prediction. In addition, it does not take the impact of the feature normalization in to account for principal component pattern extraction. These two issues are the main research contents of this paper.

In this paper, drawing on previous research methods, we try to make more efficient apply unlabeled corpus to improve the predicting model for the relevance using a new idea originated from transfer learning. SVD is more suitable for mining information and relationships in the document features. In this paper, there are two phases to ranking task. In phase 1, we use SVD to look for the new features as a preprocessing step. The training set is combined with test set; the combined collection is processed by singular value decomposition. In this way not only it can introduce the unlabeled information but also the two feature sets can be projected to a same dimension space for extracting new set of feature vectors. Further more we deal with the feature vector by normalizing at query level, then select the appropriate feature subset to SVD processing; finally extract new feature vectors from decomposed feature set. In phase 2, we use the RankBoot to learn a new ranking model from the feature set added new features, so as to improve the ranking performance.

3 A SVD and RankBoost-Based Approach

Feature-representation-transfer is one of transfer learning methods. Its aim is to look for "good" features in order to reduce the difference of source domain and target domain and lower error rate of model predicting. These features are generated by the pivot features that are contained in both source domain and target domain. In this paper we propose an approach similar to the SCL (Structural Correspondence Learning) [8] that is often used to structure new features for transfer learning, we use SVD to do that to introduce the unlabeled data information. The training set and test set are seemed as source domain and target domain; all the features of them are used as pivot features. it is different that we simply use SVD to process the set(has been normalized) merged by training set and test set in order to construct association relation with the labeled data(training set) and unlabeled data(test set),and introduce the information to each other at the same time. Unlabeled data information can be used in Rank-Boost model training process, while it is useful for predicting the relevance of the document in the test set. Meanwhile, SVD is a method for extracting principal component features; this method can implicitly obtain the more effective feature vectors. By adding such identical distribution feature vectors to the training set and test set it is meaningful to the iterative training and the relevance predicting. After SVD, the eigenvectors with the largest eigenvalues can form a projection matrix. RankBoost is based on training the features to obtain the ranking model for predicting the document relevance, so it is helpful to expend the feature set, especially by adding the features which contains unlabeled data information for improving the performance of ranking model. Therefore, this paper apply RankBoost for training ranking model with new feature set to study whether it is helpful to introduce the SVD to rank.

3.1 SVD for Extracting Features

Singular value decomposition (SVD) is a feature extraction methods based on matrix transformation. After SVD, we can get tree matrixes: one is the document- potential feature matrix U, the second is the eigenvalue matrix S, and the third is the potential feature-original feature matrix P. The eigenvalue shows how much information of original matrix, which its corresponding eigenvector contains. The bigger the eigenvalue is, the more information the eigenvector contains. So we choose the SVD-feature subset F based on the eigenvalues. The method is as table 1 showing.

Algorithm 1 shows the pseudo code for this SVD-feature extracting. After that, we obtain the new training set which is introduced the test set information, and the new test set also contains the information of training set. Naturally, we project the new features to the same dimensional space for next process of RankBoost training. Step 2 is used to normalize the feature value into scaling of (0, 1) at query level. The documents associated with the same query may have the same value with respect to one feature, after Step 2 we can see that all of the values for that feature become zero. It will not be used for SVD, so we used Step 3 to filter the features like that. In the next part of our experiment we can use the U to choose the most appropriate new feature subset to learn the rank model, taking the result and time-cost into account.

Table 1. SVD for Extracting Features - algorithm

Algorithm 1. SVD for Extracting Features -algorithm
Input: Training set D ; Test set T;
Output: New training set D′ ;New test set T′;
Start: 1: Combine(D,T) => DT;
2: Normalize_Querylevel(DT)=>(DT)*;
3: Filter((DT)*)=> (DT)**
4: SVD((DT)**) => U, S, P;
5: Extract_newFeature(U)=>F; //F denotes SVD-feature subset;
6: Merge((DT)*,F)=>(DT)′;
7: Decomposition((DT)′, D.length, T.length)=> D′, T′;
//|D|=| D′|,|T|=| T′|;
End

3.2 RankBoost Algorithm

RankBoost is a kind of pairwise approach which can reduce ranking to classification on document pairs with respect to the same query, no longer assume absolute relevance. Its primary task is to make document pairs based on relevance judgments with respect to the same query. $<x_0, x_1>$ denotes a document pair, and x_0 is unrelevance document, x_1 is relevance document. Meanwhile, RankBoost is one of the improved AdaBoost algorithms for ranking task. A high weight assigned to a pair of documents indicates a great importance that the weak learner orders that pair correctly. We also assign to the weak learner to show its performance of predicting the document relevance. Once iterative it generates a weak learning. Reserving weak learning of each iterative and accumulating them multiplied by their weights; we can obtain the final ranking model. RankBoost Algorithm [1] we use is as table 2 showing.

Weak rankings have the form $h_t(x)$. We think of these as providing ranking information in the same manner as ranking features and the final ranking. The weak learners we used in our experiments are based on the original given ranking features and new features extracted by SVD. We focus in this section and in our experiments on {0, 1}-valued weak rankings that use the ordering information provided by the ranking features, but ignore specific scoring information. In particular, we will use weak rankings h of the form: if $f_i > \theta$, $h(x) = 1$; else $h(x) = 0$.where $\theta \in R$ .That is a weak ranking derived from a ranking feature f_i by comparing the score of f_i on a given instance to a threshold θ. α_t is the weight of the weak learning which can be generated at each iteration and computed by RankBoost. It shows the importance of the corresponding weak learning, with respect to predicting the relevance of document. Z_t is a normalization factor (chosen so that D_{t+1} will be a distribution).Let T equal to D′ , and we can get the ranking model by introducing the new features extracted by SVD.

Table 2. RankBoost for learning ranking model

Algorithm 2. RankBoost for learning ranking model

Input: Training set T; Relevance judgments R;
Output: Ranking Function H(x);
Start:
    1: Pair(T,R) => Document pair set D;
    2: Initialize(D)=> D_t : $\forall\ D_1(x_0,x_1)=1/|D|$;
    3: For t=1,…,T:
        1) Train weak learner using distribution D_t;
        2) Get weak ranking h_t;
        3) Choose $\alpha_t \in R$;
        4) Update: $D_{t+1}(x_0,x_1) = \dfrac{D_t(x_0,x_1)\exp(\alpha_t(h_t(x_0)-h_t(x_1)))}{Z_t}$;
 	
    4: Output the final ranking: $H(x) = \sum\limits_{t=1}^{T}\alpha_t h_t(x)$。

End

4 Experiments

4.1 Data Description

We used the Letor2.0 [9] data set released by Microsoft Research Asia. This data set contains three data sets: the OHSUMED data set, the TREC2003 data set (TD2003) and the TREC2004 data set (TD2004). The OHSUMED data set derived from medicine retrieval task, while TD2003 and TD2004 come from TREC task.

Letor2.0 is based on many query-document features. For each query-document pair of OHSUMED, there is a 25-dimensional feature vector that contains the most frequently used features in information retrieval, for example tf-idf, BM25 [10] score etc., while that in TD2003 and TD2004 is represented by a 44-dimensional feature vector, the feature in the vector such as HITS[11]、Page Rank[12] and LMIR[13] etc .

In this paper, we deal the matrix constructed by 44 features in TREC data set with SVD, to choose 10 new features based on its quantity of information with respect to the original matrix [14], and for the OHSUMED data set we choose 5 new features adding to training set.

4.2 Experimental Results

In order to evaluate the performance of the proposed approach, we adopt MAP [15] as evaluation method. The average precision of a query is the average of the precision scores after each relevant document retrieved. Average precision (AP) takes the positions of relevance documents on ranking list into account to give scores of the list with respect to one query. Finally, MAP is obtained by the mean of the average precision over a set of queries. There are three data subsets in the Letor2.0, and 5 groups of training set and test set in each subset. Our Experiments give the result of SVD-RankBoost (using SVD features and original features) compared with a baseline released by letor2.0 (using original features only).There are tree tables: table 3, table 4 and table 5 used for showing the results of OHSUMED, TD2003 and TD2004.

Table 3. MAP of the test sets in OHSUMED

OHSUMED	Baseline	SVD
Fold1	0.3391	0.3618
Fold2	0.4470	0.4678
Fold3	0.4456	0.4565
Fold4	0.5059	0.5147
Fold5	0.4637	0.4668
Average	0.4403	0.4535

Table 3 shows the results of baseline and adding five SVD features with the top 5 largest eigenvalues, according to the ranking model used to OHSUMED.

Table 4. MAP of the test sets in TD2003

TD2003	Baseline	SVD
Fold1	0.1444	0.1452
Fold2	0.2572	0.3001
Fold3	0.2174	0.2109
Fold4	0.2523	0.2275
Fold5	0.1912	0.1938
Average	0.2125	0.2155

Table 4 shows the results of baseline and adding ten SVD features with the top 10 largest eigenvalues, according to the ranking model used to TD2003.

Table 5. MAP of the test sets in TD2004

TD2004	Baseline	SVD
Fold1	0.4133	0.4914
Fold2	0.3402	0.3885
Fold3	0.4397	0.4293
Fold4	0.3467	0.2992
Fold5	0.3777	0.3666
Average	0.3835	0.3950

Table 5 shows the results of baseline and adding ten SVD features with the top 10 largest eigenvalues, according to the ranking model used to TD2004.

We compare the performance of other boosting method for ranking with our approach using SVD. For comparisons of the different models, we report the performance measured by both the MAP and NDCG at position 1, 5 and 10.We average these performance measures over 5 folds for each data set. The results are shown in Table 6.

Table 6. Performance comparisons over Letor2.0 data set

		MAP	NDCG@1	NDCG@5	NDCG@10
	RankBoost	0.4403	0.4977	0.4502	0.4356
OHSUMED	AdaRank	0.4419	0.5420	0.4554	0.4385
	SVD	0.4535	0.5131	0.4529	0.4435
	RankBoost	0.2125	0.2600	0.2789	0.2851
TD2003	AdaRank	0.1373	0.4200	0.2424	0.1940
	SVD	0.2155	0.3800	0.2761	0.2798
	RankBoost	0.3835	0.4800	0.4368	0.4716
TD2004	AdaRank	0.3308	0.4133	0.3932	0.4063
	SVD	0.3950	0.4933	0.4368	0.4392

4.3 Experimental Analysis

The experiment results show that it is helpful to improve the documents ranking list according to the relevance to the query in most of test set. We use the both the SVD and original features for ranking model training and relevance predicting. These SVD features reduced the difference of the training set and test set, so they can improve the ranking model trained by training set and used to predict the relevance of the documents in test set. Therefore we can get better ranking list in this way. There are still some results of test sets that aren't improved, it is may be caused by the features that aren't chosen furthermore, and the experiment is based on the average the number of new features extracted by SVD. Especially for some test set the average number are not perfectly fit for them, we focus on that average result is best based on the same number of new features added to training set.

In order to get most appropriate features and the number of them, our work is as follows. Firstly, we merge the training set and test set for SVD in order to introduce the information to each other, for the training set it is introduced the unlabeled information for RankBoost to training ranking model to improve its relevance prediction effect. Meanwhile, the training set information is added to the test set, which project the new features of the training set and test on the same dimensional vector space, which make the ranking model more meaningful for using the new features. Secondly, on the detail of experiment, we normalize every feature to (0, 1) scaling. After that we can see the values of a few features are equal to zero with respect to the same query. We think that these features are no use to training ranking model by Rank-Boost, so in the process of the SVD, we introduce the original features without them, and the table 4 show the differences of results between the features after normalizing and filtering and not ,that used for SVD. Finally, we choose 10 SVD features according to top 10 eigenvalues for TD2003 and TD2004, 5 features according to top 5 eigenvalues for OHSUMED, after taking the performance of final ranking model and the cost of training time into account.

For RankBoost, there are two factors that should be considered to implement algorithm: 1) the number of iterations; 2) the initial weight of document pairs. We use cross validation to select the number of iterations. In this paper, the initial weight distribution of document pairs is set to uniform distribution as $D_1(x_0, x_1)=1/|D|$. Table 7 shows the results of all the methods we used to rank. The approach SVD-1 denotes we only use the original features to SVD process compared with SVD-2 that the features are processed with normalizing, while we use SVD-3 to show the results after normalizing and filtering. Finally we list the result of RankSVM and the result of the PCA approach [7] as compare. We can see that SVD-1, SVD-2, SVD-3 are all helpful to improve the performance of ranking, and for the different test set, they have different effects. However, in our work SVD-3 get the best result. It is our further work to study how to merge the different ranking model to improve the result of ranking.

Table 7. Average MAP of the test sets in OHSUMED☐TD2003 and TD2004

	Baseline	RankSVM	PCA	SVD-1	SVD-2	SVD-3
OHSUMED	0.4403	0.4469	0.4455	0.4530	0.4519	0.4535
TD2003	0.2125	0.2564	0.3226	0.2066	0.2075	0.2155
TD2004	0.3835	0.3505	0.3703	0.3729	0.3903	0.3950

5 Conclusions

In this paper, we propose an approach whose idea is derived from transfer learning to use the unlabeled data for ranking. We apply SVD to extract new features from training set and test set, and add these features to each other. The difference between training set and test set can be decreased, so as to improve the ranking model. The experiment shows that it is meaningful for the learning to rank to introduce the SVD features and unlabeled data information to training sets, but it can't improve the results of all the test set. It may be caused by the detail of the experiments. For all that, this method improves the ranking performance in most of test sets. We will continue to research on the application of unlabeled data set in order to perfect the approach proposed in this paper. In addition, the other methods will be also our research contents. Possible extensions and future work include: using ties [16] (the documents pairs with the same relevance judgment) to learning to rank, which can expand the sample space, whose key is how to combine this ranking model with the model trained by document pairs with the different relevance judgment. Furthermore, it is an important research direction to apply the other model and algorithm used in the field of information retrieval and machine learning.

Acknowledgement

This work is supported by grant from the Natural Science Foundation of China (No.60373095 and 60673039) and the National High Tech Research and Development Plan of China (2006AA01Z151).

References

1. Freund, Y., Iyer, R., Schapire, R., Singer, Y.: An efficient boosting algorithm for combining preferences. Journal of Machine Learning Research 4, 933–969 (2003)
2. Herbrich, R., Graepel, T., Obermayer, K.: Support vector learning for ordinal regression. In: International Conference on Artificial Neural Networks, Edinburgh, UK, vol. 1, pp. 97–102 (1999)
3. Cao, Z., Qin, T., Liu, T.-Y., Tsai, M.-F., Li, H.: Learning to Rank: From Pairwise Approach to Listwise Approach. In: International conference on Machine learning, Corvalis, Oregon, USA, pp. 129–136 (2007)
4. Guiver, J., Snelson, E.: Learning to Rank with SoftRank and Gaussian Processes. In: ACM Special Interest Group on Information Retrieval, Singapore, pp. 259–266 (2008)
5. Veloso, A., Almeida, H.: Learning to Rank at Query-Time using Association Rules. In: ACM Special Interest Group on Information Retrieval, Singapore, pp. 267–274 (2008)
6. Amini, M.-R., Truong, T.-V., Goutte, C.: A Boosting Algorithm for Learning Bipartite Ranking Functions with Partially Labeled Data. In: ACM Special Interest Group on Information Retrieval, Singapore, pp. 99–106 (2008)
7. Duh, K., Kirchhoff, K.: Learning to Rank with Partially-Labeled Data. In: ACM Special Interest Group on Information Retrieval, Singapore, pp. 251–258 (2008)
8. Blitzer, J., McDonald, R., Pereira, F.: Domain Adaptation with Structural Correspondence Learning. In: Proceedings of the 2006 Conference on Empirical Methods in Natural Language Processing, Sydney, Australia, pp. 120–128 (2006)
9. Liu, T.-Y., Qin, T., Xu, J., Xiong, W., Li, H.: LETOR: Benchmark dataset for research on learning to rank for information retrieval. In: SIGIR 2007 Workshop on Learning to Rank for IR, ACM SIGIR Forum, vol. 41(2), pp. 58–62 (2007)
10. Robertson, S.E.: Overview of the okapi projects. Journal of Documentation 53(1), 3–7 (1997)
11. Kleinberg, J.: Authoritative sources in a hyperlinked environment. Journal of the ACM 46(5), 604–622 (1999)
12. Page, L., Brin, S., Motwani, R., Winograd, T.: The PageRank citation ranking: bringing order to the Web, Technical report. Stanford University (1998)
13. Zhai, C., Lafferty, J.: A study of smoothing methods for language models applied to Ad Hoc information retrieval. In: Proceedings of SIGIR 2001, pp. 334–342 (2001)
14. Lin, H.-F., YAO, T.-S.: Text Browsing Based on Latent Semantic Indexing. Journal of Chinese Information Processing 14(5), 49–56 (2000)
15. Järvelin, K., Kekäläinen, J.: IR evaluation methods for retrieving highly relevant documents. In: The 23rd Annual International ACM SIGIR Conference on Research and Development in Information Retrieval, New York, USA, pp. 41–48 (2000)
16. Zhou, K., Xue, G.-R., Zha, H.-Y., Yu, Y.: Learning to Rank with Ties. In: ACM Special Interest Group on Information Retrieval, Singapore, pp. 275–282 (2008)

Opinion Target Network and Bootstrapping Method for Chinese Opinion Target Extraction

Yunqing Xia[1], Boyi Hao[1,2], and Kam-Fai Wong[3]

[1] Tsinghua National Laboratory for Information Science and Technology,
Tsinghua University, Beijing 100084, China
yqxia@tsinghua.edu.cn
[2] Department of Computer Science and Technology, Tsinghua University
Beijing 100084, China
haoby@cslt.riit.tsinghua.edu.cn
[3] Department of System Engineering and Engineering Management,
The Chinese University of Hong Kong, Shatin, Hong Kong
kfwong@se.cuhk.edu.hk

Abstract. Opinion mining systems suffer a great loss when unknown opinion targets constantly appear in newly composed reviews. Previous opinion target extraction methods typically consider human-compiled opinion targets as seeds and adopt syntactic/statistic patterns to extract opinion targets. Three problems are worth noting. First, the manually defined opinion targets are too large to be good seeds. Second, the list that maintains seeds is not powerful to represent relationship between the seeds. Third, one cycle of opinion target extraction is barely able to give satisfactory performance. As a result, coverage of the existing methods is rather low. In this paper, the opinion target network (OTN) is proposed to organize atom opinion targets of component and attribute in a two-layer graph. Based on OTN, a bootstrapping method is designed for opinion target extraction via generalization and propagation in multiple cycles. Experiments on Chinese opinion target extraction show that the proposed method is effective.

Keywords: Opinion target extraction, opinion mining, information extraction.

1 Introduction

Opinion mining is a text understanding technology that assists users to automatically locate relevant opinions from within a large volume of review collection. A typical opinion mining system integrates two modules, i.e. opinion extraction and sentiment analysis. This paper addresses the opinion target extraction task, which seeks to find out the subject of the opinions. The opinion targets discussed in this paper are equivalent to attribute and features discussed in [1] and [2].

Many research works on target extraction have been reported. Popular approaches attempt to combine lexicon and corpus statistics in various manners to achieve good results. Lexicon is a handcrafted opinion dictionary or a review corpus that covers opinion targets, sentiment keywords, modifiers and negations. To improve coverage,

G.G. Lee et al. (Eds.): AIRS 2009, LNCS 5839, pp. 339–350, 2009.

the lexicon is further expanded with synsets. The corpus statistics are usually frequencies of co-occurrences. Ghani et al. (2006) proposes a direct solution [1], where opinion targets in the dictionary are considered to be seeds to detect attribute-value pairs within the review texts using *co-EM* algorithm. It is due to these works that a fundamental framework for opinion target system has been setup. However, there is a lot of room for improvement to achieve better performance. Study on the previous works on opinion target extraction leads to following observations.

Human-compiled opinion targets are used as seeds in the previous researches. For instance, *brightness of image* is a human-compiled opinion target involving both *brightness* and *image*. However, they are too large to be considered effective seeds so far as granularity is concerned. One undesirable consequence of this is low coverage. It is suggested that smaller and more general seeds should be defined. With better seeds, more powerful patterns for opinion target extraction can be made. Meanwhile, the seeds should work well with synsets so that the synsets can be fully used to improve coverage.

Opinion targets i.e. seeds and opinion target candidates have been organized with independent lists in previous works. However, lists provide limited ability to model relations. There are always exceptions and new occurrences of opinions despite a try to build a complete dictionary. Hence, one cycle of opinion target extraction cannot give a satisfactory performance.

In our work, manually compiled opinion targets are viewed as compound opinion targets and an atomization algorithm is designed to extract atom opinion targets from the compound opinion targets. We define atom opinion target and compound opinion target as follows.

Definition 1: Atom opinion target (AOT)

An atom opinion target is an opinion target that is:

(1)   internally cohesive, i.e., words within the opinion target strongly adhere to each other; and
(2)   externally flexible, i.e., the opinion target can combine with many different opinion targets to form opinion targets.

Definition 2: Compound opinion target (COT)

A compound opinion target is an opinion target that combines atom opinion targets in particular patterns.

For example, *brightness of image* is a human-compiled opinion target. We define *brightness of image* a compound opinion target and define *brightness* and *sensitivity* atom opinion targets. Intention to define atom opinion target lies in generalization demand. As they are smaller and more general, the atom opinion targets are deemed better seeds for opinion target extraction.

With atom opinion targets, the opinion target patterns (OTP) can be generated accurately. The atom opinion targets work effectively with synsets, so synsets can be used fully to improve coverage. Furthermore, atom opinion targets are classified into components (COM) and attributes (ATT) from the perspective of ontology to further improve power of opinion target patterns. A novel framework to organize atom opinion targets, compound opinion targets, synsets and opinion target patterns, the opinion target network (OTN) is proposed in this work, where component atom opinion targets and attribute targets are located in different layers, synsets for the atom opinion targets are considered as nodes, and patterns are represented by paths that connect the nodes. To construct OTN, the generalization algorithm is designed to 1) extract atom opinion targets from compound opinion targets, 2) assign the atom opinion targets appropriate synsets, 3) classify the atom opinion targets, and finally 4) generate the opinion target patterns. To make the OTN more generalized, propagation algorithm is designed to utilize of atom opinion targets as seeds to synthesize new compound opinion targets with opinion target patterns. In order to improve performance, a bootstrapping algorithm is designed to call the generalization algorithm and the propagation algorithm iteratively for a few years so that the OTN can be updated and verified. Experiments on Chinese opinion target extraction show that the method outperforms the baseline by 0.078 on f-1 score in the first cycle and by 0.112 in the last cycle.

The rest of this paper is organized as follows. In Section 2, related works are addressed. In Section 3, the opinion target network is presented. In Section 4, opinion target extraction method is presented. We present evaluation in Section 5 and conclude the paper in Section 6.

2 Related Works

Opinion target extraction research is either viewed as a subtask of opinion mining or taken as an independent information extraction task. We summarize the related works as follows.

Hu and Liu (2004) proposed to find frequent opinion targets with association miner and infrequent opinion targets with syntactic patterns based on opinion words [2]. To improve coverage of opinion target extraction, Hu and Liu (2006) further adopted *WordNet* to find synonyms for the known opinion targets [3]. Popescu and Etzioni (2005) considered opinion targets as concepts forming certain relationships with the product and proposed to identify the opinion targets connected with the product name through corresponding meronymy discriminators [4]. Ghani et al. (2006) developed a system that is capable of inferring implicit and explicit opinion targets using *co-EM* algorithm [1]. Generic and domain-specific opinion targets are considered as seeds and the seeds were applied to locate real product opinion targets with attribute-value pairs. Kobayashi et al. (2007) adopted machine learning techniques to extract *aspect-of* relation from a blog corpus, which is viewed as statistical patterns for opinion target extraction [5]. Xia et al. (2007) proposed to make use of collocations of opinion targets and sentiment keywords to find unknown opinion targets [6].

Our work is closely related to but different from the aforementioned spellings.

The idea of seed-based learning is followed. However, unlike previous works, we do not consider human-compiled opinion targets as seeds directly. In this work, we consider these opinion targets as compound opinion targets and attempt to generate atom opinion targets automatically. The atom opinion targets, being smaller and more general than the compound opinion targets, are used as seeds to find new opinion target candidates. Synsets are also adopted in this work to organize synonyms for the atom opinion targets. Also, in addition, it is used to classify the atom opinion targets into components and attributes. This work is deemed necessary because, in real reviews components and attributes make different contributions in forming opinion targets.

Opinion targets including seeds and opinion target candidates are organized in lists in previous works. We design a graph-based opinion target network to maintain atom opinion targets, compound opinion targets, synsents and opinion target patterns. The opinion target network shows obvious advantages over list in representing seeds, candidates and patterns.

An iterative mechanism of the *co-EM* algorithm is mentioned in [1]. But their opinion target extraction method is still one-cycle. In this work, a bootstrapping method is designed to optimize the opinion target extraction task in multiple cycles and this method has proven worthwhile which will be shown later in the paper.

3 Opinion Target Network

3.1 Formalism

The opinion target network is a two-layer graph G^{OTN} defined as follows,

$$G^{OTN} =< V^{COM}, E^{COM}; V^{ATT}, E^{ATT}; E^{\Theta} >$$

where V^{COM} and V^{ATT} represents component node set and attribute node set, respectively; E^{COM} and E^{ATT} denotes component edge set and attribute edge set, respectively; E^{Θ} denotes the set of edges that connect component nodes and attribute nodes. Note that the edges are all directed, which point from the subsequent atom opinion target to the antecedent atom opinion targets. Opinion target network follows the two-layer architecture and the edges bridging components and attributes are of peculiar interest. Meanwhile, paths in the opinion target network forms compound opinion targets and often discloses opinion target patterns.

Note that a node within the opinion target network is a synset, which covers all alternative names of the component or attribute. In this work, synsets for attributes are extracted from HowNet [7] while those for components are extracted from the Opinmine corpus [8].

3.2 An Illustrative Case

An illustrative opinion target network is given in Figure 1.

In this case, the nodes on the component layer, i.e. *DC*, *lens*, *button* and *shuttle*, are component atom opinion target synsets, and the paths, e.g. *lens* of *DC*, *shuttle* of *lens*, *button* of *lens* and *button* of *panel*, are compound opinion targets.

On the attribute layer, the node *scale* is the root attribute, which is used for categorization purpose. The nodes, i.e. *price*, *voice* and *speed* are attribute synsets, and the edges merely reflect the categories they belong to. Normally, there is no edge between sibling nodes, e.g., *voice of speed* is not a valid attribute. Thus, the attribute layer forms a tree.

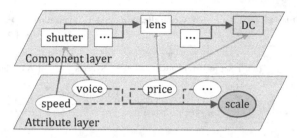

Fig. 1. An illustrative opinion target network for digital camera (DC) contains a component layer and an attribute layer

Note that it is beyond the scope of this work to investigate the semantic relation that each edge in the component layer represents. The edges are just naturally created based on opinion target patterns that some atom opinion target follows to form compound opinion target.

3.3 The Advantages

Three advantages of opinion target network have been found. (i) The inference ability comes into being as seeds become cohesive and patterns effective. With atom opinion target synsets and opinion target patterns, hundreds of opinion target candidates can be inferred. This greatly helps to improve coverage of the opinion target extraction method. (ii) The opinion target network created by our opinion target extraction method reveals some interesting information. The opinion target network shows that, in real reviews, attributes and components must work with each other in order to form opinion targets. Another finding is that the core part of opinion is the attribute, which contacts opinion keyword explicitly or implicitly. These findings help understand how opinion is formed. (iii) Synsets in opinion target network help find concepts, and patterns within the component layer disclose *part-of* semantic relations. We assume that the opinion target network may help to construct domain-specific ontology automatically.

4 Opinion Target Extraction

4.1 The General Workflow

Overall, our opinion target extraction method comprises two major parts, i.e. generalization and propagation, and executes in a bootstrapping manner (see Fig.2).

In the workflow, the annotation review corpus provides the initial compound opinion target set for generalization and propagation. After the first cycle, more compound opinion targets are extracted from the raw review corpus. Then after a few cycles, size of the compound opinion target set starts to stabilize and the opinion target extraction task is deemed accomplished.

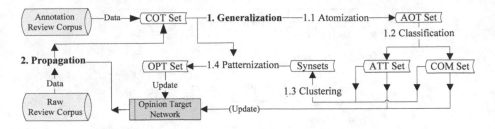

Fig. 2. Workflow for construction of the opinion target network

In the generalization algorithm, compound opinion targets are first resolved into atom opinion targets, referred to as *atomization* process. Then atom opinion targets are classified into components and attributes, referred to as *classification* process, and are further assigned proper synset labels, referred to as *clustering* process. Finally, opinion target patterns are generalized from the compound opinion targets, referred to as *patternization* process. In the propagation algorithm, synsets and patterns are applied on raw review corpus to find compound opinion targets.

Note that in every cycle, parsers should be updated with newly obtained atom opinion targets so that atom opinion targets would be considered normal words in lexical analysis and dependency parsing.

4.2 Generalization

The generalization algorithm seeks to extract atom opinion targets from compound opinion targets, then to classify atom opinion targets to components and attributes, and then to find the synset label that every new atom opinion target carries, and finally to generate opinion target patterns from the compound opinion targets.

Atomization
Atom opinion targets are extracted based on degree of cohesion and degree flexibility degree. The cohesion degree is obtained by calculating the point-wise mutual information [4].
The flexibility degree of word W is calculated as follows,

$$F(W) = \frac{1}{2}\left(\frac{\sum_{W_i \in N^L} \frac{1}{N^R(W_i)}}{N^L(W)} + \frac{\sum_{W_i \in N^R} \frac{1}{N^L(W_i)}}{N^R(W)} \right) \qquad (1)$$

where N^L denotes the set of neighboring words to the left, N^R is the set of neighboring words to the right. The function $N^L(x)$ returns number of unique left-neighboring words, and $N^R(x)$ returns those of right-neighboring words. It can be seen from Equation (2) that flexibility degree reflects flexibility from both sides. We select the words as atom opinion targets if cohesion degree and flexibility degree both satisfy empirical thresholds, which are obtained in our experiments.

Classification

A probabilistic classifier is designed to recognize components and attributes considering the following two features.

(1) Average Edit Distance (d^{AVG})

Average edit distance measures string similarity which is calculated as follows,

$$d^{AVG}(t \mid X) = \frac{1}{|X|} \sum_{x_i \in X} d(t, x_i) \qquad (2)$$

where t denotes the to-be-classified atom opinion targets, $X = \{x_i\}$ represents atom opinion targets set of component or attribute, $|X|$ represents size of set X and $d(t, x_i)$ is the function to measure edit distance between t and x_i. With Equation (2), we are able to calculate how likely it is for an atom opinion target to be from component set C or attribute set A.

(2) Overall Position Tendency (t^{OVA}).

Overall position tendency measures how likely an atom opinion target is to be a component or an attribute according to position heuristics. In certain language, the attributes tend to appear at the end of compound opinion targets. So, the overall position tendency is calculated as follows,

$$t^{OVA}(t) = \frac{count(t, A)}{count(C, t)} \qquad (3)$$

where $count(t, A)$ returns number of compound opinion targets in which t appears before the attributes, and $count(C, t)$ returns number of compound opinion targets in which t appears after the components.

Note that the initial component and attribute sets are extracted from annotation review corpus. To improve coverage, we extract human-compiled attribute words from WordNet and HowNet. Finally, atom opinion target can be classified as component or attribute by simply comparing d^{AVG} and t^{OVA}.

Clustering

To assign synset labels to every new atom opinion target, we first apply k-means clustering algorithm to group the atom opinion target set into a few clusters. Then we adjust parameters to find a cluster that satisfies the following two conditions:

(i)   The cluster contains more than three atom opinion targets carrying same synset label;

(ii)   The cluster contains at least one new atom opinion targets.

Once such a cluster is found, the atom opinion targets with unknown synset label are considered synonyms of the other atom opinion targets. This updates the atom opinion target set. We repeat the clustering process until no unknown atom opinion targets can be assigned any synset label. The following two features are considered in atom opinion target clustering.

(1)   Opinion words neighboring the atom opinion targets in the raw review sentences.
(2)   Edit Distance between unknown atom opinion targets and known ones.

There must be some unknown atom opinion targets that cannot be assigned any synset label. We run the clustering algorithm merely on these unknown atom opinion targets and attempt to find new synsets. A new synset is created if one cluster is found satisfying the following two conditions.

(i)   The cluster contains more than three atom opinion targets.
(ii)   Number of atom opinion target occurrences is over three on average for this cluster.

Now a synset label should be selected for the new synset. We choose the atom opinion target with most occurrences.

After the above procedure, there might still be some atom opinion targets with unknown synset label. These atoms could be catered for in the next iteration.

Patternization
The opinion target patterns follow regular expressions of the form

$$\{A_c\}^*\{string\{B_c\}^*\}^* \ ,$$

where A_c and B_c represent constant synset labels for the atom opinion target; *string* is constant in the pattern. For example, we have a pattern *color* of *image*. In this pattern, *image* is the synset label named after image, *color* the synset label named after color, and *of* the pattern string.

4.3 OTN Update

Once new atom opinion targets and new opinion target patterns are found, we create nodes and/or edges in the OTN. Note that no node will be created if a new atom opinion target carries known synset label. New edges should be drawn when new opinion target patterns are found.

4.4 Propagation

The propagation algorithm aims to extract unknown opinion targets with OTN. To do so, the algorithm first makes use of OTN as well as dependency parsing tool to infer the opinion target candidates. Then the raw review corpus is used to filter out false candidates.

Inference with Opinion Target Network
OTN is capable of inferring new opinion targets with atom opinion targets and opinion target patterns. In other words, if an edge exists between atom opinion target synsets A and B, combinations between atom opinion targets in synset A and in synset B probably exist. Based on this assumption, a large number of opinion target candidates can be inferred.

This assumption might lead to false candidates. To filter them, we apply sequence confidence measure on the raw review corpus to estimate how likely atom opinion target A appears before atom opinion target B. Given that a candidate X contains N atom opinion targets, i.e. $\{A_i\}_{i=1,...,N}$, the sequence confidence (SC) is calculated as follows,

$$SC(X) = \sum_{i<j} \frac{count(A_i, A_j)}{C_N^2} \qquad (4)$$

where $count(A_i, A_j)$ denotes number of occurrences of A_i, before A_j, and C_N^2 is the number of binary combinations. An empirical threshold for sequence confidence is set in our experiments.

Inference with Dependency Relations

OTN helps to find new atom opinion targets but new synset can hardly be detected. To boost propagation ability, we further adopt dependency parsing tool to find unknown atom opinion targets that the patterns fail to cover. In real reviews, known atom opinion targets may syntactically depend on some words or vice versa. We follow the dependency relations to find atom opinion target candidates. To make the dependency-based propagation reliable, we setup the following constraints:

(1)   No more than four dependency relations, i.e. ATT (modifying), COO (coordinating), QUN (numbering) and DE (DE structure[1]).

(2)   Atom opinion target candidates should adjoin the known atom opinion targets accept when a conjunction or 的(de0) appears in the middle.

(3)   An atom opinion target candidate is not adjective or pronoun.

Experiments show that inference with dependency relations does help to find high quality atom opinion targets that carry unknown synset labels.

4.5 Bootstrapping

The bootstrapping algorithm dispatches the generalization and propagation algorithm and updates the opinion target network in multiple cycles. The incremental learning process terminates when the opinion target network is stabilized. In other words, when no new atom opinion target or pattern is extracted any longer. This goal can be achieved in a few cycles.

5 Evaluation

5.1 Setup

Two corpora are used in our work. Opinmine corpus [8] contains 8,990 human-judged opinions on digital camera. The raw review corpus contains 6,000 reviews in the same domain. We divide Opinmine corpus into training set and test set evenly and adopt precision (p), recall (r) and f-1 score (f) in this evaluation.

---

[1] In Chinese DE structure refers to the clause that contains 的(de0, of).

We apply HIT Language Technology Platform [9] for Chinese word segmentation and dependency paring.

5.2 Experiments

Baseline Method

The baseline method is defined as the one that uses human-compiled opinion targets as seeds. To find more opinion targets, similar dependency relations and rules are also considered. The intention is to prove necessity and effectiveness of OTN for the opinion target extraction task. The fundamental difference is that OTN views atom opinion targets as seeds.

Our Method

Our method is described in Section 4. The thresholds in our method are determined empirically, e.g., threshold for cohesion degree is set as 0.001, for flexibility degree as 0.333, and that for sequence confidence as 0.8. The thresholds are configured based on parameter optimization.

Results

Providing no generalization ability, the baseline method finishes in one cycle. Our method concludes in eight cycles. Experimental results are shown in Fig.3.

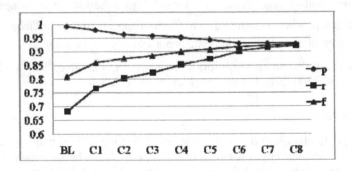

Fig. 3. Experimental results on precision (p), recall (r) and f-1 measure (f) of the baseline method (BL) and our method in eight cycles

5.3 Discussions

Two comparisons are done based on the experimental results. Firstly, we compare our method in the cycle $C1$ against the baseline method. Our method outperforms the baseline by 0.051 on f-1 score, in which recall is improved by 0.085 while the loss of precision was an insignificant 0.014. This proves that OTN is necessary and suitable for the generalization and propagation algorithms to improve recall without significant loss of precision.

Secondly, we compare our method against the baseline in terms of final performance. Our method outperforms the baseline by 0.117 on f-1 score, in which recall is improved by a huge 0.239 with loss of precision of 0.063. This reveals enormous significance of the bootstrapping process.

To understand how the bootstrapping process contributes, we present statistics on component set, attribute set, compound opinion target set and opinion target pattern set for every cycle in Table 1.

Table 1. Size of component (COM) set, attribute (ATT) set, compound opinion target (COT) set and opinion target pattern (OTP) set for the cycles. Note that cycle C0 refers to the human annotation phase

Cycle	COM	ATT	COT	OPT
C0	–	–	978	–
C1	213	84	13898	294
C2	324	84	22706	5008
C3	476	100	28434	6902
C4	636	132	34519	7601
C5	803	148	41306	8159
C6	942	180	46161	8817
C7	1133	223	48917	9027
C8	1291	254	51742	9077

It is shown in Table 1 that component set, attribute, compound opinion target and opinion target pattern sets are all extended in every cycle, in which the compound opinion target set is extended to the largest extent. Another observation is that the extension tends to converge in five cycles. Then we review Table 1 and Fig.3 at the same time. We find that with extension of the above sets, performance of our method tends to converge. The above observations reveal that the cycles are limited and the bootstrapping process may finally produce an optimal performance of our method. This proves that OTN is reasonable and effective.

6 Conclusions

Two issues are addressed in this paper. First, the opinion target network is proposed to manage synsets of atom opinion targets as nodes in two layers and the atom opinion targets interact with each other via paths that represent opinion target patterns. Second, an OTN-based opinion target extraction method is presented to extract opinion targets from raw reviews via generalization and propagation in multiple cycles. The experimental results show the OTN is necessary and suitable for the generalization and propagation algorithms to improve overall performance in limited cycles.

This work is still preliminary and future work is planned as follows. Firstly, we intend to refine the OTN-based opinion target extraction method further in pattern generation. Secondly, we intend to conduct more experiments to evaluate how our method fits other domains such as mobile phone, movie and hotel scenarios.

Acknowledgement

Research work in this paper is partially supported by NSFC (No. 60703051), MOST (2009DFA12970) and Tsinghua University under the Basic Research Foundation (No. JC2007049). We thank the reviewers for the valuable comments.

References

1. Ghani, R., Probst, K., Liu, Y., Krema, M., Fano, A.: Text mining for product attribute extraction. SIGKDD Explorations Newsletter 8(1), 41–48 (2006)
2. Hu, M., Liu, B.: Mining opinion features in customer reviews. In: Proc. of AAAI 2004, pp. 755–760 (2004)
3. Hu, M., Liu, B.: Opinion Extraction and Summarization on the Web. In: Proc. of AAAI 2006 (2006)
4. Popescu, A., Etzioni, O.: Extracting product features and opinions from reviews. In: Proc. of HLT-EMNLP 2005, pp. 339–346 (2005)
5. Kobayashi, N., Inui, K., Matsumoto, Y.: Extracting Aspect-Evaluation and Aspect-Of Relations in Opinion Mining. In: Proc. of EMNLP-CoNLL 2007, pp. 1065–1074 (2007)
6. Xia, Y., Xu, R., Wong, K.-F., Zheng, F.: The Unified Collocation Framework for Opinion Mining. In: Proc. of ICMLC 2007, vol. 2, pp. 844–850 (2007)
7. Dong, Z., Dong, Q.: HowNet and the Computation of Meaning. World Scientific Publishing, Singapore (2006)
8. Xu, R., Xia, Y., Wong, K.-F.: Opinion Annotation in Online Chinese Product Reviews. In: Proc. of LREC 2008 (2008)
9. Ma, J., Zhang, Y., Liu, T., Li, S.: A statistical dependency parser of Chinese under small training data. In: Proc. of Workshop: Beyond shallow analyses-formalisms and statistical modeling for deep analyses, IJCNLP 2004, pp. 1–5 (2004)

Automatic Search Engine Performance Evaluation with the Wisdom of Crowds*

Rongwei Cen, Yiqun Liu, Min Zhang, Liyun Ru, and Shaoping Ma

State Key Laboratory of Intelligent Technology and Systems,
Tsinghua National Laboratory for Information Science and Technology,
Department of Computer Science and Technology, Tsinghua University, Beijing, China
crw@mails.tsinghua.edu.cn

Abstract. Relevance evaluation is an important topic in Web search engine research. Traditional evaluation methods resort to huge amount of human efforts which lead to an extremely time-consuming process in practice. With analysis on large scale user query logs and click-through data, we propose a performance evaluation method that fully automatically generates large scale Web search topics and answer sets under Cranfield framework. These query-to-answer pairs are directly utilized in relevance evaluation with several widely-adopted precision/recall-related retrieval performance metrics. Besides single search engine log analysis, we propose user behavior models on multiple search engines' click-through logs to reduce potential bias among different search engines. Experimental results show that the evaluation results are similar to those gained by traditional human annotation, and our method avoids the propensity and subjectivity of manual judgments by experts in traditional ways.

Keywords: Performance evaluation, click-through data analysis, Web search engine, the wisdom of crowds.

1 Introduction

How to evaluate the performance of search engines promptly, accurately and objectively, is of vital importance for Web search users, online advertisers and search engine system engineers. The evaluation of information retrieval (IR) systems, which is the procedure of assessing how well a system satisfies user information requirements [19], is one of the most important research topics in IR research field. As noted by Saracevic [15], "Evaluation became central to R&D in IR to such an extent that new designs and proposals and their evaluation became one."

Kent et al. [10] was the first to propose the criterion of relevance and the measures of precision and relevance (later renamed recall) for evaluating IR systems. While most current IR evaluation researches, including the famous workshop TREC (Text Retrieval Conference), are based on the Cranfield methodology [4]. A Cranfield-like

---

* Supported by the Chinese National Key Foundation Research & Development Plan (2004CB318108), Natural Science Foundation (60621062, 60503064, 60736044) and National 863 High Technology Project (2006AA01Z141).

G.G. Lee et al. (Eds.): AIRS 2009, LNCS 5839, pp. 351–362, 2009.

approach is based on a query set and corresponding answers (called qrels). Queries in the query set are processed by an IR system. Then results are compared with qrels using evaluation metrics. Finally, the performance of the IR system is represented in the value of the relevance metrics.

The annotation of qrels for a query set is usually the most difficult part in IR system evaluations. Manual assessment is such a time-consuming task that Voorhees [19] estimated that about nine person-months is required to judge one topic for a collection of 8 million documents. For the task of search engine performance evaluation, the content of Web is significantly larger than TREC-based corpuses, and real time character is of vital importance for engineering concerns. Therefore, the evaluation method should not be one in which lots of human efforts are involved.

Recently, the *wisdom of crowds* is paid much attention to in the area of Web research (eg. [6][9]). This paper focuses on inferring implicit preference of Web search users by observing their querying and clicking behavior and the *wisdom of crowds* is used to evaluate the performance of Web search engines reliably and timely.

The contributions of the paper are:

1. A framework of automatic search engine evaluation is proposed, which is based on click-through data analysis and estimate the performance reliably and timely.

2. An automatic qrel annotation algorithm is designed to annotate Web search queries with user behavior analysis.

3. A user behavior model (called *Multiple Click-through Rate model, MCTR*) which combines user behavior information from multiple search engines is proposed to reduce potential bias in information collected from a single search engine.

2 Related Work

Several attempts have been made towards automatic IR system evaluation to tackle difficulties related to manual assessment. Some of methods avoid manual assessments by adopting pseudo-relevance feedback information. Soboroff et al. [17] randomly selected a subset of documents from the results of each topic as relevant documents. Nuray and Can [12] made similar attempts by selecting documents with the highest RSV values from the result pool as relevant documents. While these methods bypass manual assessment, they have to compromise the loss of the accuracy and reliability because pseudo-relevance feedback cannot provide reliable relevance judgment.

Oard and Kim [13] were among the first to model Web users' information acquisition process by behavior analysis. They presented a framework for characterizing observable user behaviors in order to understand the underlying behavior goals and the capability scope provided by information systems. Later, through the eye-tracking equipments, Joachims et al. [8] pointed out that clicks reflect relative relevance of queries and results. In 2006, Agichtein et al. [1] proposed an idea of aggregating information from many unreliable user search session traces instead of treating each user as a reliable "expert" and pointed out that user behaviors were only probabilistically related to relevance preferences. Dou et al. [9] studied the problem of using aggregate click-through logs, and found that the aggregation of a large number of user clicks provided a valuable indicator of relevance preference.

Different from the previous work, our approach assesses the relevant documents by employing the click-through data of statistically significant users in real Web search settings. Hence, rather than manually judged by few experts, our assessment is determined by thousands even millions concerned users, robust to the noise inherently in individual interactions. A Cranfield-like evaluation framework is constructed to evaluate the performance of search engines automatically with the *wisdom of crowds*.

3 Automatic Web Search Engine Evaluation Framework

As outlined in the introduction, our evaluation method measures an engine's quality by examining users' querying and clicking behavior extracted from real world search logs. The general process is similar to the Cranfield-like approach in the adoption of a query topic set and corresponding answers (qrels). But both the topic set and the answer set are generated automatically according to user click-through logs.

An advantage of adopting the Cranfield-like framework is that there is no need to design new evaluation metrics. Existing metrics such as MAP, MRR, P@N and Bpref have been used in the evaluation of IR systems for a long time and lots of research work is based on them. Once the topic and qrel sets are generated automatically, Cranfield approach is a better choice than adopting new architectures and metrics.

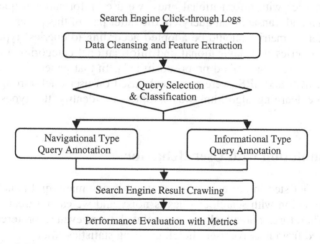

Fig. 1. The framework of the automatic search engine evaluation method

The flow chart of our automatic evaluation framework is shown in Fig. 1. First, click-through logs are pre-processed and several user-behavior features are extracted. These features are then adopted in query selection, classification and annotation steps. Queries are sampled and grouped into two categories, navigational type and informational type, because their qrels are annotated with different strategies and their performance is evaluated using different metrics. After annotating relevant documents respectively, search engines' result of these queries are crawled and extracted. Finally, traditional Web IR evaluation metrics are used to assess the performance of search engines by comparing their result lists and the automatic annotated answers.

4 Query Set Construction

How to construct a suitable query set is challenging for Cranfield-like evaluation methods. For traditional methods such as those adopted in TREC, query topic sets are developed by assessors or participants. Sometimes, queries are selected from search engine query logs by human efforts but the query set usually contains no more than several hundred query topics.

For the evaluation purpose, we construct a query set representing the information needs of Web users. Since our qrel annotate method needs to aggregate a reasonable number of user clicks to provide a valuable indicator of relevance and we cannot extract the reliable relevance preference from an individual click, we select queries with a number of user requests to construct the query set from click-through data.

After the query set is constructed, the information need of each query should be annotated automatically. According to Broder [2] and Rose et al. [14], there are three major types of Web search queries: navigational, informational and transactional. For evaluation purposes, these three query types should be treated respectively. MRR (Mean Reciprocal Rank) is adopted in the evaluation of navigational queries. For both informational and transactional queries, there is usually more than one correct answer. Mean average precision (MAP) and precision at top N documents (P@N) are therefore used in evaluation of these two types of queries. Because transactional queries use the same metrics with informational ones, we use "informational query" instead of "informational and transactional query" in the latter part of this paper.

Because different metrics should be adopted according to queries' types, we have to classify user queries into navigational and informational categories. In this paper, we adopt the classification method proposed in [11]. In that paper, two new user behavior features, nCS and nRS, were proposed based on user click-through logs, and the decision tree learning algorithm was adopted to identify the types of queries automatically.

5 Qrels Annotation with User Behavior

The most important step in our evaluation method is automatic qrel annotation. Web search user interaction with search engines is noisy and we cannot treat each user as an "expert". Therefore, not all clicks are related to relevance preferences which should be derived from massive user click logs using statistics strategy.

5.1 The Feature of Click Distribution

Click-through logs provide detailed and valuable information about users' interaction with search engines. Based on previous studies [8][9], we have the intuitive idea that one result with a large number of clicks might be more relevant to the query than the one with less clicks probabilistically.

We defined the number of clicks on result d for query q, $N(d|q)$, and the probability that d is relevant to q, $P(d\ is\ relevant\ to\ q)$, then we have the following assumption:

Assumption 1. *If $N(d_1|q) > N(d_2|q)$, then $P(d_1\ is\ relevant\ to\ q) > P(d_2\ is\ relevant\ to\ q)$ for query q.*

Assuming that we have a search query q, assumption 1 says that a single click on a document d does not mean d is relevant to q absolutely. However, if d_1 is clicked M times and d_2 is clicked N (M>N) times by users who propose q, d_1 may be more relevant to q than d_2 probabilistically.

Due to different queries with different frequency of user visits, we normalize $N(d|q)$ as $CTR(d|q)$ (Click-through Rate) for query q:

$$CTR(d \mid q) = \frac{N(d \mid q)}{\sum_{d_i} N(d_i \mid q)} .$$

(1)

According to Joachim's work [8], the assumption 1 does not hold well, and the rank bias shows that the results at top positions have more chance to be viewed and clicked. Fig. 2 shows the relevance precision of query-doc pairs with different CTR values. From the figure, we can know that the relevance precision of pairs with large CTR values higher than the ones with lower CTR values, and it indicates that the pairs with large CTR values is more likely to be relevant in a probabilistic notion.

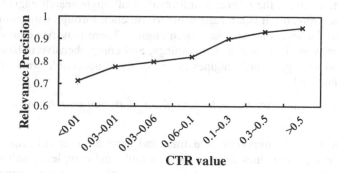

Fig. 2. The relevance precision of query-doc pairs with different CTR values

5.2 Multiple Click-Through Rate (*MCTR*) Model

Due to differences in crawling, indexing and ranking strategies, search engines may return completely different results for a same query. In addition, the quality of the summarizations or snippets of results is a factor to affect user's click behavior. Hence, users may click different result pages using different search engines.

For example, Fig. 3 shows the click distributions of one sample query in four different Chinese search engines (referred to as SE1, SE2, SE3 and SE4). We collected ten of the most-frequently clicked result from each of these four engines and constructed a result pool which contained 27 unique pages. The "Click Distribution" axis is the CTR value of result pages and the curves show the different click distributions. Meanwhile, each point on the "Clicked Pages" axis represents one click result. According to Fig. 3, we note the discrepancy of user clicks in different engines and find that no search engine covers all these 27 pages. For instance, most of the clicks in SE1 focus on the 1st clicked page, while SE2 focuses on the No.3 and No.10.

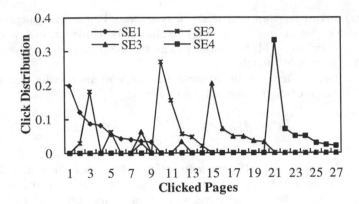

Fig. 3. Click distributions of a sampled query in different search engines

Since click distributions in different search engines are discrepant and biased to the search engine of itself, the relevance information of single search engine logs is not able to be used directly. It is not reasonable to use click-through data from one single search engine for assessing another search engine. There may be some strategies to decrease or prevent the bias and shortcomings, and comprehensive utilization of user behavior from multiple search engines is one of these accessible ways. We give the second assumption here:

Assumption 2. *The click-through data from multiple search engines are more informational.*

Now, we see that it is important to extract and combine user behavior information from multiple search engines to get a more reliable and complete result set for each query. However, there remains a problem, how to exploit user interaction information from multiple engines. Different search engines have different market shares and user visits. If click-through logs are aggregated simply, there would be biased to the search engines with relatively larger user visits. Therefore, we propose a model of integrating user behavior information based on click distributions of individual search engines to generate multiple click distribution. To our best knowledge, there are no studies about click-though data of multiple search engines.

In section 5.1, we mentioned the *CTR* feature normalized in probability, which is seen as a probability function to approximate the probability that a click locates randomly on a result document. Supposing that there is a virtual meta-search engine which integrates all search engines' results, a similar probability function is adopted to characterize the click-through information from multiple search engines. This virtual engine indexes more Web data than any subsistent search engines, has lower bias to single search engine, and gathers more accurate user behavior information.

The user click-through information is characterized as a triple $<q, d, MCTR>$ expression, where q is a submitted query, d is a click result, and $MCTR$ is a multiple click-through rate with integrating click distributions from single search engines. Since $MCTR(d|q)$ is a probability distribution function, it follows some probability principles and is rewritten as $Pi(d|q)$, which means the probability that a click locates randomly on a result d giving query q. According to the full probability distribution, the equation is constructed as follows:

$$Pi(d \mid q) = \sum_{\forall se_j \in Set\ of\ SEs} Pi(d \mid se_j, q) \square Pi(se_j \mid q). \qquad (2)$$

where $Pi(d|se_j,q)$ is the probability of clicking d given q on search engine se_j, equaling to the CTR value of result d for query q on engine se_j. $Pi(se_j|q)$ is the probability weight of search engine se_j for query q and we use Bayes theorem and maximum likelihood estimation method to rewrite it as:

$$Pi(se_j \mid q) = \frac{Pi(q \mid se_j) Pi(se_j)}{Pi(q)} \propto Pi(q \mid se_j). \qquad (3)$$

where, $Pi(q)$ is the proportion of queries submitted by users in whole Web search, which is regarded as a constant value for query q. $Pi(se_j)$ is considered as the probability weight of search engine se_j, which is difficult to be estimated. $Pi(q|se_j)$ is described using the proportion of user sessions of query q to ones of all queries for engine se_j. Combining expressions (2) and (3), we obtain:

$$Pi(d \mid q) \propto \sum_{\forall se_j \in Set\ of\ SEs} Pi(d \mid se_j, q) Pi(q \mid se_j). \qquad (4)$$

After normalized by all $Pi(d|q)$ for all documents, we have the following formula:

$$MCTR(d \mid q) = Pi(d \mid q) = \frac{\sum\limits_{\forall se_j \in Set\ of\ SEs} Pi(d \mid se_j, q) Pi(q \mid se_j)}{\sum\limits_{r_k \in AllClickedResults} \sum\limits_{\forall se_j \in Set\ of\ SEs} Pi(r_k \mid se_j, q) Pi(q \mid se_j)}. \qquad (5)$$

This means that $MCTR$ of each result d for query q is able to be achieved from CTR of single search engines with some additional user session information.

5.3 Automatic Qrel Annotation for Query Set

According to [16], users only view or click a few top-ranked pages of the result list. Therefore, it is difficult for an evaluation method to find all search target pages for submitted queries. In order to solve this problem, pooling technology [3][18] supposes that only the documents in the result pool are relevant. Similar to this assumption, we assume that relevant Web pages should appear in search engines' result lists and be clicked by users. In our experiments, we only care whether the results annotated by our method are really qrels.

For informational queries, there is a set of frequently-clicked results, due to the fact that informational queries don't have a fixed search target. A search engine may return several related results, and users click some of them according to their information needs. In Section 5.1 and 5.2, the CTR and $MCTR$ features are described to reveal the relative relevance of clicks statistically based on two assumptions. According to the relative relevance, queries are able to be annotated and the most relative relevant pages are regarded as the results of corresponding queries. We give the third assumption here:

Assumption 3. *Search target pages for a corresponding query are several results with the largest CTR or MCTR values.*

According to this assumption, the annotation process is described as follows:

Algorithm 1. Query annotation with *CTR* or *MCTR*

1:For a given query q in the Query Set and its clicked result list $r_1, r_2, ..., r_M$
2:$R_1, ..., R_M$ = Sort $r_1, r_2, ..., r_M$ according to *CTR or MCTR*;
3:FOR $(i=1; i<=N; i++)$
4:    IF $R_i < T$ or $i=N$
5:        Annotate q with $R_1, ..., R_{i-1}$
6:        EXIT
7:    END IF
8:ENDFOR
9:q cannot be annotated.

In the annotation process, there are two parameters: T and N. T is used as a lower limit for that each target page has sufficient *CTR* or *MCTR* value to guarantee the annotating accuracy. N is used as an upper limit for the qrels numbers of queries. The choice of T and N selects the tradeoff between recall and relevant reliability.

For navigational queries, there is a fixed search target and the purpose of the query is to reach a particular Web page. For most navigational request, the target is unique. Therefore, the result R_1, which is the most frequently clicked in search engines' result set, is likely to be the correct answer probabilistically. Thus, navigational queries are considered as special cases of informational queries and we use the process described in Algorithm 1 to annotate navigational queries by setting N to 1.

When the search engines failed to return correct answers in result lists, it is almost impossible for users to click those answers and as such the annotation process return "q cannot be annotated".

6 Experiments and Results

6.1 Experiment Settings

With the help of a popular commercial Chinese search engine, click-through logs of four most frequently-used Chinese search engines were collected from October 1st to November 20th, 2008. These click-through logs (recording altogether 53,367,427 querying and clicking events, about 33.0M, 9.4M, 9.0M and 1.9M respectively) were applied in our evaluation experiment. The four search engines are referred to as SE1, SE2, SE3 and SE4.

In order to verify the reliability of the proposed qrel annotation method and user behavior model, we first performed experiment to examine the correctness of the automatically-annotated qrels. After that, effectiveness of our evaluation method was examined by comparison with manual-based evaluation results.

As for the manual annotation process, we had three product engineers to work as assessors. All of them are familiar with search engine products and is able to estimate

user information needs according to context of queries. The correctness of their annotation was also examined by co-checking each other's judgment results and the final
labeling results were determined by the majority opinion of assessors.

6.2   Qrel Annotation Experiments

With the query set construction strategy proposed in Section 4, we randomly selected
7000 queries from query logs to evaluate search engines' performance. After automatic classification and annotation, our methods successfully annotated 5996 queries
with qrels (the others were regarded as "cannot be annotated" by algorithm 1).
Among these queries, 1225 are navigational and the rest are informational.

Annotation of Navigational Queries
One qrel was automatically annotated to each of these navigational queries using the
algorithm 1 with setting N to 1, and the results were compared with manual annotation. Table 1 shows the consistent and inconsistent results number.

Table 1. The number of consistent and inconsistent annotation results

#(navigational query)	#(consistent results)	#(inconsistent results)
1225	1093 (89.2%)	132 (10.8%)

The inconsistent results are mainly caused by two reasons, and most of these inconsistent queries meet users' information need in real world scenario:

[1] When people refer a certain Web site with its name, it is not necessarily that
they want to visit the homepage. For example, the query "163" is annotated with
http://mail.163.com/ instead of http://www.163.com/ with our method. The reason is
that http://mail.163.com/ is the most famous free e-mail service provider in China.
People are more likely to visit this site to get free mail service rather than navigating
the main Web site. Therefore, most users who query "163" visit the mail sub-site and
the wisdom of crowds annotates the frequently visited URL as the qrel. For this kind
of queries, we believe that the automatically-annotated answer is more suitable than
the manual ones because it meets users' information need more closely.

[2] There are also several cases in which the automatically-annotated qrels are not
so suitable. Some of these wrongly-annotated cases are due to query-type misclassifi-
cation. These queries are then reclassified as informational ones. Other problems are
caused by mirror sites, which mean our algorithm annotates a certain URL but the
assessors considers a different URL with the same content.

Annotation of Informational Queries
In our experiments, altogether 4771 informational queries were sent to be annotated
and assessors were asked to check the correctness of annotation results. 1000 (about
21%) queries were selected by random sampling and labeled by assessors.

Fig. 4 compared the precision of qrels annotated by *MCTR* and *CTR* of four search
engines. We find that the proposed *MCTR* strategy from multiple search engines has
better performance than the *CTR* strategies from individual search engines.

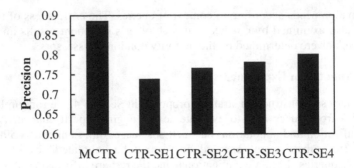

Fig. 4. Precision of Qrels annotated by *MCTR* and *CTR* strategies of four search engines

In order to verify the reliability of our automatic method, we compared our method's annotation results with those of human assessors. The annotated results of our method and all three assessors were checked with a cross validation method. In Table 2, the result annotated by a certain assessor was used to evaluate another assessor's or our automatic method's performance. The results in Table 2 show that human assessors do not agree with each other totally. The results of our automatic method are similar to that of assessors. With assessor 2's annotation results as the correct answer, the precision of assessor 3's annotation is about 92.2%, while our method gains a precision of 88.6%, which means the automatic annotated results are close to the one of assessor 3's using assessor 2's results as the correct answer.

Table 2. Comparison between automatic annotation results and manually annotated ones

	Assessor 1's results as the correct answer	Assessor 2's results as the correct answer	Assessor 3's results as the correct answer
Assessor 1	1	0.953	0.956
Assessor 2	0.966	1	0.925
Assessor 3	0.989	0.922	1
Our method	0.880	0.886	0.858

6.3 Performance Evaluation Results

With the query set constructed according to the descriptions in Section 4 and the qrel set annotated in Section 5, search engine performance was evaluated with traditional IR metrics. MRR (Mean Reciprocal Rank) and P@10 (precision of the top 10 documents) were used for evaluating navigational and informational type queries, separately. Besides that, MAP (Mean Average Precision) was adopted in both query types' evaluation processes.

In Fig. 5(a), MAP values equal to MRR because queries are all navigational and annotated with one correct answer only. The differences in MAP between Fig. 5(a) and (b) reveal that search engines have better performance while processing navigational type queries than informational ones. This conclusion accords with previous studies in [5][7].

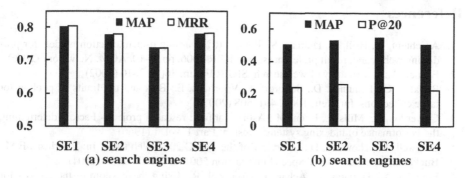

Fig. 5. Automatically Performance Evaluation of four commercial search engines (a: performance of navigational type queries; b: performance of informational type queries)

6.4 The SearchE System

A demo called SearchE System (http://searchE.thuir.cn/) is constructed to evaluate online commercial search engines based on the framework and *MCTR* model proposed in this paper. It collects click-through information of multiple search engines, generates query set, annotates them with *MCTR*, collects results from online search engines, and show evaluation results on the Web. The performances of these search engines and top queries with corresponding results are provided every day.

7 Conclusions and Future Work

In this paper, we introduced an automatic method to evaluate the Web search engines' performance with the *wisdom of crowds*. It involves a statistically significant number of users for unbiased assessment, is robust to the noise in individual interactions and evaluates the performance reliably and timely.

First, we proposed a framework of automatic performance evaluation for Web search engines. It is a Cranfield-like approach based on users' click-through behavior information. Second, the method of large scale query set construction was introduced and queries were classified into informational and navigational categories automatically. Third, qrel annotation algorithm was described to annotate queries. The *MCTR* strategy combined user behavior information from multiple search engines.

Experimental results show that most of the qrels are annotated correctly and automatic evaluation results are highly correlated with manual-based evaluation results. By this approach, different with traditional, automatic search engine evaluation based on large scale queries for real world is easy to conduct.

Future study will focus on the following aspects: How much click-through data is needed for a reliable and efficiency evaluation method? How to improve the evaluation reliability using more user behavior except click-through data?

References

1. Agichtein, E., Brill, E., Dumais, S., Ragno, R.: Learning user interaction models for predicting web search result preferences. In: SIGIR 2006, pp. 3–10. ACM, New York (2006)
2. Broder, A.: A taxonomy of web search. SIGIR Forum 36(2), 3–10 (2002)
3. Buckley, C., Dimmick, D., Soboroff, I., Voorhees, E.: Bias and the limits of pooling for large collections. Inf. Retr. 10(6), 491–508 (2007)
4. Cleverdon, C., Mills, J., Keen, M.: Aslib Cranfield research project - Factors determining the performance of indexing systems; Design; Part 1, vol. 1 (1966)
5. Craswell, M., Hawking, D.: Overview of the TREC 2003 Web track. In: Voorhees, E.M., Buckland, L.P. (eds.) NIST Special Publication 500-261: TREC 2004 (2004)
6. Fuxman, A., Tsaparas, P., Achan, K., Agrawal, R.: Using the wisdom of the crowds for keyword generation. In: Proc. of WWW 2008, pp. 61–70. ACM, New York (2008)
7. Hawking, D., Craswell, N.: Overview of the TREC 2003 Web track. In: Voorhees, E.M., Buckland, L.P. (eds.) NIST Special Publication 500-255: TREC 2003 (2003)
8. Joachims, T., Granka, L., Pan, B., Hembrooke, H., Gay, G.: Accurately interpreting click-through data as implicit feedback. In: SIGIR 2005, pp. 154–161. ACM, New York (2005)
9. Dou, Z., Song, R., Yuan, X., Wen, J.R.: Are click-through data adequate for learning web search rankings? In: CIKM 2008, New York, NY, pp. 73–82 (2008)
10. Kent, A., Berry, M., Leuhrs, F.U., Perry, J.W.: Machine literature searching VIII. Operational criteria for designing information retrieval systems. American Documentation 6(2), 93–101 (1955)
11. Liu, Y., Zhang, M., Ru, L., Ma, S.: Automatic Query Type Identification Based on Click Through Information. In: Ng, H.T., Leong, M.-K., Kan, M.-Y., Ji, D. (eds.) AIRS 2006. LNCS, vol. 4182, pp. 593–600. Springer, Heidelberg (2006)
12. Nuray, R., Can, F.: Automatic ranking of retrieval systems in imperfect environments. In: Proc. of SIGIR 2003, pp. 379–380. ACM, New York (2003)
13. Oard, D.W., Kim, J.: Modeling information content using observable behavior. In: Proc. of ASIST 2001, Washington, D.C., USA, pp. 38–45 (2001)
14. Rose, D.E., Levinson, D.: Understanding user goals in web search. In: Proc. of WWW 2004, pp. 13–19. ACM, New York (2004)
15. Saracevic, T.: Evaluation of evaluation in information retrieval. In: Proc. of SIGIR 1995, pp. 138–146. ACM, New York (1995)
16. Silverstein, C., Marais, H., Henzinger, M., Moricz, M.: Analysis of a very large web search engine query log. SIGIR Forum 33(1), 6–12 (1999)
17. Soboroff, I., Nicholas, C., Cahan, P.: Ranking retrieval systems without relevance judgments. In: Proc. of SIGIR 2001, pp. 66–73. ACM, New York (2001)
18. Soboroff, I., Voorhees, E., Craswell, N.: Summary of the SIGIR 2003 workshop on defining evaluation methodologies for terabyte-scale test collections. SIGIR Forum 37(2), 55–58 (2003)
19. Voorhees, E.M.: The Philosophy of Information Retrieval Evaluation. In: Peters, C., Braschler, M., Gonzalo, J., Kluck, M. (eds.) CLEF 2001. LNCS, vol. 2406, pp. 355–370. Springer, Heidelberg (2002)

A Clustering Framework Based on Adaptive Space Mapping and Rescaling

Yiling Zeng[1], Hongbo Xu[1], Jiafeng Guo[1], Yu Wang[1], and Shuo Bai[1,2]

[1] Institute of Computing Technology, Chinese Academy of Sciences, Beijing 100080, China
[2] Shanghai Stock Exchange, Shanghai, 200120, China
{zengyiling,hbxu,guojiafeng,wangyu2005}@software.ict.ac.cn,
sbai@sse.com.cn

Abstract. Traditional clustering algorithms often suffer from model misfit problem when the distribution of real data does not fit the model assumptions. To address this problem, we propose a novel clustering framework based on adaptive space mapping and rescaling, referred as M-R framework. The basic idea of our approach is to adjust the data representation to make the data distribution fit the model assumptions better. Specifically, documents are first mapped into a low dimensional space with respect to the cluster centers so that the distribution statistics of each cluster could be analyzed on the corresponding dimension. With the statistics obtained in hand, a rescaling operation is then applied to regularize the data distribution based on the model assumptions. These two steps are conducted iteratively along with the clustering algorithm to constantly improve the clustering performance. In our work, we apply the M-R framework on the most widely used clustering algorithm, i.e. k-means, as an example. Experiments on well known datasets show that our M-R framework can obtain comparable performance with state-of-the-art methods.

Keywords: Document Clustering; Space Mapping; Data Representation.

1 Introduction

With the explosion of documents on the Web, there has been increasing need for efficient and effective analysis methods to manage massive text collections. Document clustering, as one of the primary analysis techniques in text mining area, has then been applied in different kinds of IR tasks [2], e.g. speeding up the information retrieval procedure [1], improving the precision or recall in information retrieval systems [3], browsing a collection of documents [4], and organizing the search results for a given query [5].

However, the performance of clustering algorithms often suffers from the model misfit problem [7]. Most clustering algorithms are based on some underlying model assumptions. When real data fits the assumptions well, the performance of the clustering algorithm would be reasonably good, otherwise not. Typically, there are two kinds of approaches to addressing the model misfit problem: adjusting algorithms' strategies to deal with the real data distributions [7, 8], or applying space transformation (e.g., kernels) to alter the data representation (or distributions) [9, 10]. The first kind

G.G. Lee et al. (Eds.): AIRS 2009, LNCS 5839, pp. 363–374, 2009.

of approaches usually makes refinement on local regions where training errors occur, while the global characteristics of data distribution are often ignored. However, such global characteristics should be reasonably considered because they may be directly related to the model misfit problem. The second kind of approaches tries to apply space transformation to alleviate the problem. However, since space transformation is usually proposed based on some assumptions without the consideration of algorithm's model, the performance improvement may be limited. In this paper, we present a framework in the way of space transformation. However, unlike the previous approaches, our transformation is proposed with respect to the clustering algorithm's model assumptions.

The framework we propose is referred as M-R framework. It is a clustering framework based on adaptive space mapping and rescaling. Considering that real data distribution usually may not fit the model assumptions of clustering algorithms very well, our solution is to regularize the data distribution such that it is more consistent with the model assumptions. Specifically, the M-R framework consists of two important steps as follows.

Step 1: Space Mapping. Since the distribution features of data in high dimensional space are usually complicated and hard to analyze, we choose to map all documents into a low dimensional coordinate which is constructed with respect to the cluster centers. In this way, the distribution statistics of each cluster could be analyzed on the corresponding dimension.

Step 2: Rescaling. With those distribution statistics obtained in hand, we apply a rescaling operation to regularize the data distribution based on the model assumptions. In this way, we are able to make the data distribution more consistent with the model assumptions to help make better clustering decisions.

By conducting these two steps iteratively along with the clustering algorithm, we are able to constantly improve the clustering performance. In our paper, we apply our M-R framework on the most popular clustering algorithm, i.e. k-means, to verify the effectiveness of our framework.

The rest of this paper is organized as follows. Section 2 discusses related work and Section 3 proposes the M-R framework in detail. We apply our framework on k-means in Section 4 and present the experimental results in Section 5, which is followed by some concluding remarks in the last section.

2 Related Work

Different clustering algorithms [6] take different point of views of data spaces. Hierarchical clustering algorithms hold the assumption that any cluster is composed of smaller sub clusters that semantically related to each other, while partitioning algorithms (e.g., k-means) believe that clusters obey isotropic Gaussian distributions that are distributed in spherical regions with the same radius. Density-based algorithms and grid-based algorithms, however, focus on local attributes that restricted in an ε-neighborhood or in a small unit named grid, and make clustering decisions with these attributes. Actually, such ideal models more or less misfit real data distribution. Therefore, model misfit becomes a common problem in text mining. To address this problem, researchers make their efforts to figure out solutions in both algorithm layer and data presentation layer.

In algorithm layer, most research approaches require training errors to refine the model, and thus are mainly proposed in supervised learning area. Wu et al. [7] retrain a sub-classifier using the training errors of each predicted class with the same learning method to refine the algorithm models. Tan et al. [8] propose an effective and yet efficient refinement strategy to enhance the performance of text classifiers by means of on-line modification of the base classifier models. However, as in the clustering field, little work has been done for solving the model misfit problem due to the lack of labels. Generally, this kind of approaches improves performance by achieving local refinements, but the global feature of real data distribution is usually ignored.

As for approaches in the data representation layer, the basic idea is to apply space transformation such that certain problems in the old feature space could be properly solved. Dumais [1] proposes LSI decomposition to rectify the deficiency in VSM model that takes correlated terms as independent dimensions. Kernel method [9] aims to find a proper implicit mapping φ via kernel functions such that in the new space, problem solving is much easier. Another novel and important approach in feature space transformation for unsupervised learning is spectral clustering [10]. The basic idea of spectral clustering is to model the whole dataset as a weighted graph, and aims to optimize some cut value (e.g. Normalized Cut [11], Ratio Cut [12], Min-Max Cut [13]). Since those criterions could be guaranteed global optimums via certain eigen-decompositions, spectral clustering algorithms often achieve better results than traditional clustering algorithms. On the whole, this kind of approaches concentrates on solving existing problems in the current feature space. Most of them seldom account in the algorithm's model assumptions during the mapping procedure. And the computational and memory requirements are usually high.

3   M-R Framework

3.1   Model Misfit in Clustering

A simple but direct example about model misfit problem of clustering algorithm (e.g., k-means) is shown in Figure 1. There are two intrinsic clusters in the dataset that are marked out with dashed lines (labeled with "+" and "○" respectively). Points with label "+" distribute in a narrow elliptic region while points with label "○" distribute in a circular region. Obviously, the distribution of the dataset violates the underlying assumption of k-means. Therefore, without knowledge of the distribution characteristics, k-means probably organizes all the points into two clusters which are marked with the two solid circles, i.e. Cluster1 and Cluster2. As a result, part of the points with label "+" is assigned to the wrong cluster.

How can we avoid such kind of mistakes caused by model misfit? According to the decision criterion of k-means, a point should be assigned to the nearest cluster. Take a point $\mathbf{x}$ in Figure 1 as an example. Assume the distance from $\mathbf{x}$ to the centroid of Cluster1 is $d1$, and the distance to the centroid of Cluster2 is $d2$. Since $d2 < d1$, document x will be assigned to Cluster2 improperly. To avoid the mistake, we can make a transformation according to the distribution characteristics, such that in the new scenario $d1 < d2$, and thus point x will be assigned to Cluster1 more reasonably.

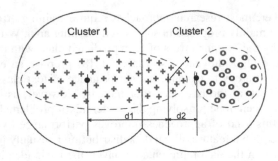

Fig. 1. An example of model misfit in K-means. Point x which should belong to Cluster1 is assigned to Cluster2 as a result of improper distance comparison.

An intuitive solution is to use Gaussian Mixture Model together with Expectation-Maximization [14] which would estimate the parameters of different distributions. However, the disadvantage is that GMM could not be directly applied in a sparse and high dimensional feature space and therefore high computational operation (i.e., LSI) is introduced to reduce the dimensionality. While in this paper, the M-R framework, as our proposed solution to address the model misfit problem, would not bring in any time consuming operations.

3.2 Rationale

In our solution of M-R framework, we analyze the data distribution and transform it to better fit the model assumptions. However, the analysis work in a high dimensional space is complicated and the sparsity feature of data distribution makes the analysis impractical. To solve this problem, we first map all documents into a lower dimensional space which is suitable for distribution analysis.

Suppose we have a dataset of n documents which contains k classes/clusters marked as $C_1,...,C_i...,C_k$ ($1 \leq i \leq k$). The corresponding document numbers of these clusters are $n_1,...,n_i...,n_k$. Let $\mathbf{m}_i = \frac{1}{n_i}\sum_{\mathbf{x} \in C_i}\mathbf{x}$ be the centroid of C_i, and $\mathbf{m} = \frac{1}{n}\sum_{\mathbf{x}}\mathbf{x}$ be the mass center of the dataset. According to the Fisher Linear Discriminant [15] for multiple classes, a matrix $\mathbf{W}$ constructed with the best set of discriminant vectors can be obtained by maximizing the *between-class scatter* $|\mathbf{W}^T\mathbf{S}_B\mathbf{W}|$ and minimizing the *within-class scatter* $|\mathbf{W}^T\mathbf{S}_W\mathbf{W}|$. This is equivalent to solve the criterion function defined as $J(\mathbf{W}) = |\mathbf{W}^T\mathbf{S}_B\mathbf{W}|/|\mathbf{W}^T\mathbf{S}_W\mathbf{W}|$, where $\mathbf{S}_B = \sum_{i=1}^{k}n_i(\mathbf{m}_i - \mathbf{m})(\mathbf{m}_i - \mathbf{m})^T$ is the *between-class scatter matrix* and $\mathbf{S}_W = \sum_{i=1}^{k}\sum_{\mathbf{x} \in C_i}(\mathbf{x} - \mathbf{m}_i)(\mathbf{x} - \mathbf{m}_i)^T$ is the *within-class scatter matrix*. In our framework, we assume that the rescaling operation would constrain the within-class scatter properly. Therefore, to simplify the criterion function, only the maximization of the between-class scatter will be considered. Thus, our criterion function is defined as

$$J_{M-R}(\mathbf{W}) = \left| \mathbf{W}^T \mathbf{S}_B \mathbf{W} \right| = \left| \mathbf{W}^T \sum_{i=1}^{k} n_i (\mathbf{m}_i - \mathbf{m})(\mathbf{m}_i - \mathbf{m})^T \mathbf{W} \right| . \tag{1}$$

The columns of $\mathbf{W}$ could be obtained by solving the eigenvectors of $\mathbf{S}_B$. Specifically, the space spanned by the eigenvectors is the same spanned by $\mathbf{m}_1 - \mathbf{m}, ... \mathbf{m}_i - \mathbf{m}, ..., \mathbf{m}_k - \mathbf{m}$ *(1≤i≤k)*. Therefore, we may directly use this set of directions to construct a new coordinate as well.

M-R Coordinate: For a dataset containing k clusters, a coordinate system could be constructed by taking the mass center $\mathbf{m}$ of the dataset as the origin and $\mathbf{m}_1 - \mathbf{m}, ... \mathbf{m}_i - \mathbf{m}, ..., \mathbf{m}_k - \mathbf{m}$ *(1≤i≤k)* as the directions of coordinate axes, This coordinate system is called M-R coordinate. The coordinate value of point x_j in M-R coordinate is marked as $(\mathbf{x}_{j,1}^c, ..., \mathbf{x}_{j,i}^c, ..., \mathbf{x}_{j,k}^c)$, where

$$\mathbf{x}_{j,i}^c = (\mathbf{x}_j - \mathbf{m})^T \frac{\mathbf{m}_i - \mathbf{m}}{\left\| \mathbf{m}_i - \mathbf{m} \right\|_2} . \tag{2}$$

Although the choice of axes in M-R coordinate may not be optimal (better axes could be obtained by applying orthogonalization to those directions), there are proper features of these directions that may facilitate the rescaling operation. While the whole set of directions $\mathbf{m}_1 - \mathbf{m}, ... \mathbf{m}_i - \mathbf{m}, ..., \mathbf{m}_k - \mathbf{m}$ *(1≤i≤k)* optimizes the criterion function (1), any single direction $\mathbf{m}_i - \mathbf{m}$ *(1≤i≤k)* in this set is also a discriminative direction to distinguish corresponding cluster C_i and other clusters. Suppose a partition containing two clusters, one of the clusters consists of documents in C_i and the other one consists of documents out of C_i, referred as $C_{i'}$, and the centroid of $C_{i'}$ is marked as $\mathbf{m}_{i'}$. Significantly, the best direction to maximize the between-class scatter of C_i, and $C_{i'}$ is $\widehat{\mathbf{d}}_i = \mathbf{m}_i - \mathbf{m}_{i'}$. And this direction is identical with $\mathbf{m}_i - \mathbf{m}$ because

$$\widehat{\mathbf{d}}_i = \mathbf{m}_i - \mathbf{m}_{i'} = \mathbf{m}_i - \frac{1}{n - n_i} \sum_{x \notin C_i} \mathbf{x} = \mathbf{m}_i - \frac{n\mathbf{m} - n_i\mathbf{m}_i}{n - n_i} = \frac{n}{n - n_i}(\mathbf{m}_i - \mathbf{m}).$$

Therefore, for every cluster in the dataset, there is a corresponding axis which gives a discriminative direction to distinguish the current cluster and other clusters by maximizing the between-class scatter. The benefit is that the rescaling operation on this axis could be applied according to the distribution statistics of the current cluster.

We apply the rescaling operation by designing rescaling functions on all dimensions according to the distribution statistics of data. Assume the rescaling functions of different axes in M-R coordinate are $R_1(\bullet), ..., R_i(\bullet), .., R_k(\bullet)$ *(1≤i≤k)*, where $R_i(\bullet)$ could be either linear functions or nonlinear functions. Especially, if the adopted functions are linear functions, i.e., $R_i(x_{.,i}) = \xi_i x_{.,i} + \ell_i (1 \leq i \leq k)$, and we directly introduce them into traditional distance measure, more simplified form of the rescaling operation could be obtained.

M-R Distance: If the adopted rescaling functions on all dimensions are linear functions i.e., $R_i(x_{.,i}) = \xi_i x_{.,i} + \ell_i$ *(1≤i≤k)*, we may directly introduce them into traditional Euclidian distance to form a new distance measure, referred as M-R distance:

$$d_{M-R}(\mathbf{x}_i, \mathbf{x}_j) = \sqrt{\sum_{t=1}^{k} \left(R_t(\mathbf{x}_{i,t}^c) - R_t(\mathbf{x}_{j,t}^c) \right)^2} = \sqrt{\sum_{t=1}^{k} \left(\xi_t(\mathbf{x}_{i,t}^c - \mathbf{x}_{j,t}^c) \right)^2} . \tag{3}$$

where ξ_i $(1 \le i \le k)$ could be regarded as the rescaling coefficients of different axes. The scales of axes are expanded or shrunken according to the rescaling coefficients such that the coordinate values on different axes are more comparable.

Considering that the rescaling operations are applied to regularize the data distribution, an intuitive but effective solution is to choose a statistic that reflects the distribution characteristics on the corresponding directions as the rescaling coefficient. Since the standard deviation is a parameter that reflects how a distribution spreads out, it is a reasonable choice for the rescaling coefficient. Thus, we can calculate the standard deviation of a cluster's projection on the corresponding axis, and take the reciprocal value of the standard deviation, $\xi_i = 1/\sigma_i$, as the axis' rescaling coefficient. Therefore, Formula (3) could be represented as:

$$d_{M-R}(\mathbf{x}_i, \mathbf{x}_j) = \sqrt{\sum_{t=1}^{k} (\mathbf{x}_{i,t}^c - \mathbf{x}_{j,t}^c)^2 / \sigma_t^2} . \tag{4}$$

where

$$\sigma_t = \sqrt{\frac{1}{n_t} \sum_{\mathbf{x} \in C_t} \left(\mathbf{x}_{j,i}^c - \|\mathbf{m}_t - \mathbf{m}\|_2 \right)^2} . \tag{5}$$

By using $\xi_i = 1/\sigma_i$ as the rescaling coefficient, the scales of clusters on the corresponding directions are regularized according to their standard deviations. As a result, the scales of different axes are more comparable and the distance measure is more reasonable.

From another perspective, the rescaling operation could be regarded as a kind of space transformation. The feature space is transformed according to the rescaling coefficient, such that under the new scale the distribution of the dataset fits the model assumptions of the algorithm better. It is worth noting that the choice of $\xi_i = 1/\sigma_i$ is only a special case of all possible rescaling functions. Any kinds of dedicated and reasonable rescaling function could be constructed according to the real distribution of datasets to bring better clustering results.

3.3 The Framework

The remaining question is that how we can obtain information of different clusters. Our solution is to execute the given clustering algorithm to generate a rough partition that reflects the intrinsic clusters to some extent. With this rough partition obtained, M-R coordinate could be constructed and rescaling coefficients could be calculated via statistical analysis. With more reasonable distance measure in the M-R coordinate, all documents could be re-clustered under the given clustering algorithm. The new result will be more appropriate than the initial rough partition. As a result, we may conduct the M-R framework iteratively along with the clustering algorithm to constantly improve the clustering performance. And the optimization procedure is quite similar to that of the co-clustering algorithm [16]. By taking iteration strategy, our framework and the clustering partition will be improved simultaneously until the final result is obtained.

For any clustering algorithm, the M-R framework works as bellow:

Step 1. *Start up the given clustering algorithm to generate a rough partition;*
Step 2. *Construct/reconstruct the M-R coordinate according to the current partition. Calculate/recalculate the rescaling coefficients of different axes in the M-R coordinate;*
Step 3. *Execute the given clustering algorithm in the M-R coordinate with M-R distance to generate a new partition;*
Step 4. *If the stop criterion is achieved, then stop the iteration. Otherwise, go to Step 2.*

Fig. 2. Applying M-R framework on a given clustering algorithm

4 M-R K-means

To give a direct example, we apply our M-R framework on the traditional k-means to form the M-R k-means algorithm. For a dataset with n documents, the M-R k-means clustering algorithm works as follows.

---

1 *Run r iterations of k-means to generate a rough partition of k rough clusters;*
2 *Re-calculate the centroids of all clusters and Re-construct the M-R coordinate;*
3 *For i=1 to n do:*
   3.1 *Calculate coordinate value of* x_i *in the new M-R coordinate;*
   3.2 *Find the nearest cluster for* x_i *in the M-R coordinate and assign* x_i *to it.*
4. *Repeat steps 2 and 3 until no documents shift to other clusters or maximum iteration time is achieved.*

Fig. 3. M-R K-means algorithm

The parameter r that controls the initial k-means iteration time is a small integer to guarantee fast generation of the rough partition. Usually, $r=2$ or 3.

Obviously, the time complexity of generating the rough partition of k-means is $O(krn)$, where k is the cluster number, r is the iteration time and n is the total document account. In each iteration of M-R k-means, there are mainly three kinds of operations: adjusting the M-R coordinate, calculating the coordinate values of every document, and calculating the document-to-cluster distances while finding the nearest cluster. The time complexity of adjusting the M-R coordinate is $O(n)$ which is mainly induced by updating centroids of all clusters. When calculating the coordinate values of all documents, k dot products per document is required according to Formula (2), so the time complexity is $O(kn)$. With new coordinate values calculated, the dimensionality is reduced to k, i.e. the number of clusters. Comparing with the operations in the original space with tens of thousands of dimensions, the complexity of the document-to-cluster distance computation in the new coordinate could be omitted. Therefore, the time complexity of each iteration is $O(n)+O(kn)=O(kn)$. Assume that the total iteration time of M-R k-means is T (including the r iterations while generating the rough partition), the time complexity of M-R k-means is $O(knT)$, which remains the same as k-means.

5 Experiments

In this section, we describe the datasets and evaluation measures used in our experiments, analyze the model misfit problem on the real datasets and eventually conduct experiments to prove the effectiveness of our M-R framework.

5.1 The Datasets

In our experiments, we use two corpora: RCV1-v2 [17] and 20Newsgroup [18].

RCV1-v2. The RCV1 dataset contains a corpus of more than 800,000 newswire stories in 103 classes from Reuters. From RCV1 we randomly select documents to construct a series of datasets (R1, R2, R3, R4 and R5) with cluster numbers vary from 7 to 15. And the document numbers vary from 1932 to 3330.

20NewsGroup. The 20Newsgroup (20NG) contains approximately 20,000 articles evenly divided into 20 Usenet newsgroups. From 20NG, we construct another series of datasets (N1, N2, N3, N4 and N5) by randomly select classes. The cluster numbers of these datasets vary from 6 to 15, and document numbers vary from 2112 to 3406.

The overview of the 10 datasets is illustrated in Table 1. Both RCV1 series and 20NG series are designed to keep a large range on cluster numbers and document numbers. The reason is that we want to give a global view of performance comparison on datasets of different sizes. Stop words are removed, and simple feature selection is applied, e.g., words appear in less than three documents or more than 80% of the documents are automatically removed. Finally, the normalized VSM vector of every document is calculated.

Table 1. Overview of the datasets (corpus type: R- RCV1 series, N-20NG series)

Dataset	R1	R2	R3	R4	R5	N1	N2	N3	N4	N5
#Classes	7	9	11	13	15	6	9	11	13	15
#Documents	1932	2482	2932	3220	3330	2112	3168	3248	3398	3406
Avg. class size	276	257.8	266.5	247.7	222	352	352	464	377.6	227.1

5.2 Evaluation Measures

For clustering, there are many different quality measures, among which the most commonly used ones are F-measure and entropy.

F-measure combines the precision and recall metrics from information retrieval. For a given class in the dataset, the F-measure is determined by the most similar cluster in the result. F-measures of all classes, weighted by the size of each class, are finally averaged to form the total F-measure.

Entropy is a measure that analyzes the homogeneity of all clusters (with the caveat that the best entropy is obtained when each cluster contains exactly one data point).

The total entropy for a set of clusters is calculated as the sum of the entropy of each cluster weighted by the size of each cluster.

Both F-measure and entropy are used to evaluate our experimental results.

5.3 Model Misfit Problem Verification

As presented before, the M-R framework is proposed since the data distribution often misfit the model assumptions of the clustering algorithms. Here we conduct experiments to show this problem on the real datasets. Specifically, standard deviation is used to illustrate the irregular data distribution of clusters on the directions of corresponding axes defined in the M-R coordinate. Therefore, standard deviation is also used to illustrate the model misfit problem in our experiment.

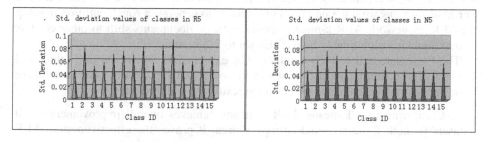

Fig. 4. Std. deviation values of classes in R5 and N5

As an example, we choose the datasets with most clusters and documents (R5 and N5) to verify the model misfit problem. For both datasets, we first construct the M-R coordinate according to the document labels and then calculate the standard deviations of clusters' projections on corresponding axes. Those standard deviations are plotted in the form of column sections in Figure 4. As shown in this figure, for dataset R5, the standard deviation values vary from 0.045 (class ID=1) to 0.092 (class ID=11). And for N5, the values vary from 0.038 (class ID=8) to 0.077 (class ID=3). The results clearly show the differences of standard deviation values, which reflect the misfit between data distribution and algorithm model. Therefore, it is improper to run clustering algorithms with original model assumptions while ignoring the irregular nature of real data distributions. It is better to combine clustering algorithms with M-R framework which is capable of normalizing the irregular data.

5.4 Performance Improvement

In this section, we conduct experiments to verify the performance of M-R framework. Three clustering algorithms are included in our experiments for performance comparison. They are k-means, M-R k-means and spectral clustering with normalized cut (Ncut) [11]. K-means and M-R k-means are compared to verify the performance improvement of M-R framework. Ncut is selected since it is one of the most successful clustering algorithms proposed by researchers in recent years. We can demonstrate the effectiveness of our M-R framework through the comparison between our approach and the state-of-the-art clustering method.

The k-means and M-R k-means in our experiments are implemented with C++ language while the Ncut code is obtained from Spectral Clustering Toolbox provided by University of Washington[1]. For all algorithms in our experiments, i.e. k-means, Ncut and M-R k-means, the algorithm results are affected by the selection of initial points. To avoid the influence of initial points, we run 10 times of all algorithms. For every run, k-means, Ncut and M-R k-means are launched with the same set of randomly selected initial points. At last, we average the F-measure and entropy scores of 10 runs for comparison. For experiments of every dataset, the cluster number of all algorithms is set the same with the category numbers in the dataset. We use Euclidean distance in k-means for comparison with our M-R distance. The k-means algorithm stops when the difference value of the sum-of-square criterion function in the recent two iterations is less than 0.001, or maximum iteration time (which is set to 20) is achieved. For Ncut, the convergence criterion is set to the same value with K-means. For M-R k-means we run 3 iterations of k-means to generate a rough partition to setup the M-R framework. M-R k-means iterates until no documents shift to other clusters or maximum iteration time (which is also set to 20) is achieved.

The F-measure and entropy results of the experiments are shown in Table 2 and Table 3. Those results are also drawn in Figure 5 and Figure 6 to give an intuitive view of performance comparison. From the results we observe that:

1. Comparing with k-means, M-R k-means achieves overall improvement in all datasets on both F-measure and entropy scores. It proves the effectiveness of M-R framework.

2. As for comparison with Ncut, the result is quite interesting. When cluster number is small, the performance of Ncut is superior to M-R k-means. However, with the growth of cluster numbers, M-R k-means is capable of getting comparable results with Ncut. In detail, the results of M-R k-means are comparable to or slightly weaker than Ncut on datasets of RCV1 series, and are comparable to or even better than Ncut on datasets of 20NG series when the cluster number is large. Here we give some brief explanation on the result. On one hand, Ncut constructs the new feature space with the best k eigenvectors. Therefore, when k is small, the quality of new feature space would be very good. However, when k increases, the quality of new feature space would decrease. On the other hand, for our M-R k-means, the improvement comparing with k-means is quite significant and stable. As a result, for large datasets, M-R k-means would generate comparable results with Ncut.

We omitted the execution time comparison in this section because we have proved that the time complexity of M-R k-means remains the same as k-means in Section 4. That is to say, M-R k-means is capable of executing as fast as k-means but generating better result.

Table 2. Average F-measure of the clustering result for all datasets

Dataset	R1	R2	R3	R4	R5	N1	N2	N3	N4	N5
k-means	0.568	0.534	0.529	0.483	0.468	0.702	0.672	0.647	0.605	0.632
Ncut	**0.698**	0.567	**0.578**	**0.550**	0.521	**0.811**	0.691	0.662	0.652	0.701
M-R k-means	0.630	**0.577**	0.562	0.528	**0.535**	0.792	**0.737**	**0.699**	**0.673**	**0.713**

---

[1] http://www.cs.washington.edu/homes/sagarwal/code.html

Table 3. Average entropy of the clustering result for all datasets

Dataset	R1	R2	R3	R4	R5	N1	N2	N3	N4	N5
k-means	0.545	0.548	0.513	0.538	0.514	0.424	0.399	0.388	0.408	0.362
Ncut	**0.403**	0.472	**0.446**	0.474	0.488	0.324	0.407	0.393	0.382	0.317
M-R k-means	0.468	**0.463**	0.474	**0.470**	**0.476**	**0.313**	**0.319**	**0.318**	**0.311**	**0.269**

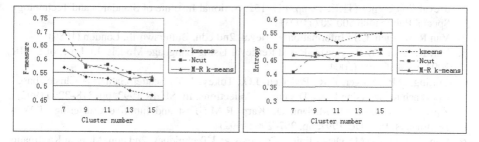

Fig. 5. F-measure and entropy scores of experiments on datasets of RCV1 series

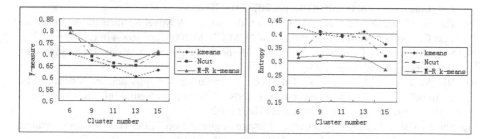

Fig. 6. F-measure and entropy scores of experiments on datasets of 20NG series

6   Conclusions and Future Work

In this paper, we propose a novel clustering framework based on adaptive space mapping and rescaling, referred as the M-R framework. The M-R framework maps all documents into a low dimensional coordinate which is constructed with respect to the cluster centers. In this way, the distribution statistics of each cluster could be analyzed on the corresponding dimension. A rescaling operation is then conducted with respect to such statistics to regularize the data distribution based on the model assumptions. By conducting the framework iteratively along with the clustering algorithm, we are able to constantly improve the clustering performance. It is worth noting that M-R framework does not introduce any time consuming operation to the original algorithm. Experiments on well known dataset show that by combining the M-R framework, traditional algorithm like k-means is capable of achieving comparable clustering performance with respect to the state-of-the-art methods.

The distribution regularization idea introduced by M-R framework is novel and interesting, and is applicable to more text mining areas. In future work, we will focus on theoretical study of our M-R framework and try to apply the framework to supervised learning area like text classification.

References

1. Dumais, S.T.: LSI Meets TREC: A Status Report. In: Harman, D. (ed.) The First Text REtrieval Conference (TREC1), pp. 137–152. National Institute of Standards and Technology Special Publication 500-207 (1993)
2. Van Rijsbergen, C.J.: Information Retrieval, 2nd edn. Buttersworth, London (1989)
3. Liu, X., Croft, W.B.: Cluster-Based Retrieval Using Language Models. In: Proc. of SIGIR 2004, pp. 186–193 (2004)
4. Cutting, D.R., Karger, D.R., Pedersen, J.O., Tukey, J.W.: Scatter/Gather: A Cluster-based Approach to Browsing Large Document Collections. In: SIGIR 1992, pp. 318–329 (1992)
5. Zamir, O., Etzioni, O., Madani, O., Karp, R.M.: Fast and Intuitive Clustering of Web Documents. In: KDD 1997, pp. 287–290 (1997)
6. Han, J., Kamber, M.: Data Mining: Concepts and Techniques, 2nd edn. Morgan Kaufmann Publishes, San Francisco (2006)
7. Wu, H., Phang, T.H., Liu, B., Li, X.: A Refinement Approach to Handling Model Misfit in Text Categorization. In: SIGKDD, pp. 207–216 (2002)
8. Tan, S., Cheng, X., Ghanem, M.M., Wang, B., Xu, H.: A Novel Refinement Approach for Text Categorization. In: Proc. of the 14th ACM CIKM 2005, pp. 469–476 (2005)
9. Shawe-Taylor, J., Cristianini, N.: Kernel Methods for Pattern Analysis. Cambridge University Press, Cambridge (2004)
10. Ng, A., Jordan, M., Weiss, Y.: On Spectral Clustering: Analysis and an Algorithm. In: Dietterich, T., Becker, S., Ghahramani, Z. (eds.) Advances in Neural Information Processing Systems, vol. 14. MIT Press, Cambridge (2002)
11. Shi, J., Malik, J.: Normalized Cuts and Image Segmentation. IEEE Transactions on Pattern Analysis and Machine Intelligence 22(8), 888–905 (2000)
12. Chan, P.K., Schlag, D.F., Zien, J.Y.: Spectral K-way Ratio-Cut Partitioning and Clustering. IEEE Trans. Computer-Aided Design 13, 1088–1096 (1994)
13. Ding, C., He, X., Zha, H., Gu, M., Simon, H.D.: A Min-Max Cut Algorithm for Graph Partitioning and Data Clustering. In: Proc. of ICDM 2001, pp. 107–114 (2001)
14. Liu, X., Gong, Y.: Document Clustering with Cluster Refinement and Model Selection Capabilities. In: Proc. of SIGIR 2002, pp. 191–198 (2002)
15. Duda, R.O., Hart, P.E., Stork, D.G.: Pattern Classification, 2nd edn. Wiley-Interscience Publishes, Hoboken (2000)
16. Dhillon, I.: Co-clustering Documents and Words using Bipartite Spectral Graph Partitioning (Technical Report). Department of Computer Science, University of Texas at Austin (2001)
17. Lewis, D.D., Yang, Y., Rose, T., Li, F.: RCV1: A New Benchmark Collection for Text Categorization Research. Journal of Machine Learning Research (2004)
18. 20 Newsgroups Data Set,
   http://www.ai.mit.edu/people/jrennie/20Newsgroups/

Research on Lesk-C-Based WSD and Its Application in English-Chinese Bi-directional CLIR*

Yuejie Zhang[1] and Tao Zhang[2]

[1] School of Computer Science,
Shanghai Key Laboratory of Intelligent Information Processing,
Fudan University, Shanghai 200433, P.R. China
[2] School of Information Management and Engineering,
Shanghai University of Finance and Economics, Shanghai 200433, P.R. China
yjzhang@fudan.edu.cn, taozhang@mail.shufe.edu.cn

Abstract. Cross-Language Information Retrieval (CLIR) combines the traditional Information Retrieval technique and Machine Translation technique. There are many aspects related to the problem of polysemy, which are good cut-in points for the application of WSD in CLIR. Therefore, an attempt in this paper is to apply WSD in English-Chinese Bi-Directional CLIR. The query expansion and the proposed Lesk-C WSD strategy are explored. Although limited improvement on WSD can be obtained, query expansion and disambiguation based on the related strategies of WSD are beneficial to CLIR, and can improve the whole retrieval performance. Specially, by considering the "Coordinate Terms", the Lesk-C algorithm shows the better performance and has more extensive applicability on CLIR.

Keywords: Cross-Language Information Retrieval (CLIR); Word Sense Disambiguation (WSD); WordNet; Lesk-C algorithm; query expansion.

1 Introduction

Cross-Language Information Retrieval (CLIR) provides a convenient way that can solve the problem of crossing the language boundary, and users can submit queries which are written in their familiar language and retrieve documents in another language. Among current CLIR systems, most of them select Query Translation as the main strategy [1]. Query Translation refers to translate users' queries written in a single language into queries in different languages used by retrieval documents from a multilingual pool of documents [2]. This pattern has become a method with the lowest cost and the smallest difficulty. In order to improve the performance of CLIR, it is necessary to solve the problem of ambiguities that exists in the translation of query [3]. For the same query term in source language, although there may be several translations, in

---

* This paper is supported by National Natural Science Foundation of China (No. 60773124), National Science and Technology Pillar Program of China (No. 2007BAH09B03) and Shanghai Municipal R&D Foundation (No. 08dz1500109). Tao Zhang is the corresponding author.

G.G. Lee et al. (Eds.): AIRS 2009, LNCS 5839, pp. 375–386, 2009.

which only part of them are suitable. The disambiguation for query translation is to decide the specific sense in the particular context of query term [4]. With the development and growth of large-scale corpus and machine-readable dictionary, there is a good platform for the research on Word Sense Disambiguation (WSD) and its application in CLIR [5].

CLIR combines the traditional IR technique and Machine Translation (MT) technique. There are many aspects related to the problem of polysemy, which are good cut-in points for the application of WSD in CLIR. Therefore, this paper attempts to apply WSD in English-Chinese Bi-Directional CLIR and make in-depth discussion from two points of view. On the one hand, WSD is applied in query expansion, which aims to make the expanded content having the higher relativity with the original query. On the other hand, WSD is also applied in retrieval process, which aims to improve the precision of retrieval. Although limited improvement on WSD can be obtained, query expansion and disambiguation based on the related strategies of WSD are beneficial to CLIR, and can improve the whole retrieval performance. Specially, by considering the "Coordinate Terms", the proposed Lesk-C WSD algorithm shows better performance and has more extensive applicability on CLIR.

2   Related Research Work

In recent decades, the performance of WSD has been increasing steadily. Corpus-based unsupervised methods are usually a clustering task, and can efficiently avoid the reality of inadequate training corpus [6]. Besides corpus, dictionary is another kind of knowledge source. Its development trends to contain more and more semantic information. Hence, dictionary-based methods have become the hotspot research. The classical thesauruses include WordNet, HowNet, etc. Although there are many different kinds of WSD methods, the difference between their performance and the upper bound acquired by human (>90%) is still very large. Therefore, it is difficult for these methods to be put into real application. In the 3rd Sense Evaluation Conference (SEN-SEVAL-3), the precision of the best system based on the English samples is 72.9% (Fine-grained). There is still a large space to improve the whole performance.

For a long time, researchers in the field of Natural Language Processing (NLP) have been considering ambiguity of word sense as one of the main factors which affect the performance of IR. However, in the earlier research work, there are a few attempts to accurately evaluate the performance of WSD alone. In the same way, there are a few reports about the combination of WSD and IR. A few works about integration pattern show that the application of WSD in IR cannot bring the improvement in performance, but has some negative effects. This kind of case attributes to the lack of useful resource that is suitable for evaluation. Generally the evaluation process can only be established based on small-scale training and testing corpus which lacks of representativeness. In recent years, with the stepwise development of large-scale training corpus which has been disambiguated manually and the emergence of many famous international evaluation conferences related to the field of IR, such as Text Retrieval Conference (TREC), not only large-scale IR testing sets become efficient and available, but also a series of strict and reasonable evaluation criteria have been proposed. Hence, the resolution of available resource can provide a feasible route for

WSD evaluation and reliability validation that WSD can improve IR performance. Recently, more and more researchers concern the ambiguity problem in IR corpus and try to probe into the benefits of performing WSD in IR. However, for CLIR that integrates MT and monolingual IR, there is still little related research work. As an important technique in MT and bottom processing of NLP, WSD can benefit from CLIR, which is a good application platform [7].

3 Lesk-C-Based Unsupervised WSD Pattern

3.1 Lesk Algorithm and Its Variation

The classical Lesk algorithm views a sense definition as an unordered bag of words based on the assumption that related word senses are often defined in a dictionary using similar words. The score of a sense S_i is computed by counting the number of common words in the definition of S_i and those of its neighboring words. The main shortcoming of Lesk algorithm is that it may cause data sparseness problem due to short definitions. To solve this problem, some extended Lesk algorithms are proposed.

Existing extended Lesk algorithms overcome the data sparseness problem to some extent by expanding the definition of a word sense. EKEDAHL and GOLUB modified the Lesk algorithm by adding two nearest hypernyms' definitions to the original synset, which obviously enhanced the chances of overlapping [8]. However, their experimental result (45%) was still below the baseline (60%). Pedersen adopted another extended method that utilized the definitions of all the synsets directly connected to the target synset, including hypernyms, hyponyms and more [9]. In addition, they assigned more weights to phrases. The authors declared that their method was much more efficient than the Lesk algorithm.

The information commonly used in extended Lesk algorithms is hypernym information, that is, parent node in WordNet. This information is the further abstraction of word sense. The hypernyms used in the experiment of this paper include direct hypernym and two nearest hypernyms.

3.2 Extended Lesk Algorithm Based on Coordinate Terms (Lesk-C)

Existing extended Lesk algorithms mainly concentrate on direct information connected to a synset, especially hypernyms, but neglect some useful indirect information. The Lesk Algorithm based on Coordinate Terms (Lesk-C) expands further the sense definition of a synset by using coordinate terms, that is, sibling nodes in Word-Net hierarchies. The coordinate terms of English noun "*basketball*", for example, are "*football*" and "*volleyball*". It is obvious that a synset and its coordinate terms have a common parent.

We suppose that any feature has the consistent effect with its coordinate terms on determining the sense of an ambiguous word. Based on this assumption, the Lesk-C algorithm expands the sense definitions of a synset by adding all (or part of) the definitions of its coordinate terms. Hence, the ambiguous word "*play*" as in "*play basketball*", "*play football*" or "*play volleyball*" should be of the same sense, i.e., "*participate in games or sport*". For the Lesk algorithm, if there is no overlap between the sense *"participate in games or sport"* and the sense definition of *"volleyball"*, then the correct resulting sense can hardly be acquired due to the data sparseness problem. In comparison with the Lesk algorithm, the Lesk-C algorithm may work better because the coordinate terms can expand the definition of *"volleyball"*. The formalization description of the Lesk-C algorithm is shown in Algorithm 1.

FUNCTION Lesk-C (c: context, w: the ambiguous word in the context c)

DEFINE

Def_W[i]: the *i-th* definition of word sense for ambiguous word w;
Def_F[i]: the definition of the *i-th* feature word and its coordinate term;
GetDef(w): search the definition of word sense w in WordNet;
Overlap(String *str*1, String *str*2): compute the number of common words between *str*1 and *str*2.

BEGIN

Stemming(c); //Through stopword filtering, the rest words in c except w are considered as Feature
                Words, represented as the set F
QTag(c); //Part-of-Speech Tagging (because it is necessary to know the word and its part-of-speech
                before searching the related words)
FOR each sense $w[i]$ of w
{   //Search the definitions of ambiguous words in the database of WordNet
    Def_W[i] = *GetDef*($w[i]$);
}
FOR each word $F[i]$ in F
{   //Search the definitions of the feature words and their coordinate terms in the database of WordNet
    FOR each sense $f[j]$ of $F[i]$
    {
        Def_F[i] = *GetDef*($f[j]$);
        The Coordinate Term Set C is found;   //A coordinate term is just a synset
        FOR each element $C[k]$ in C
            Def_F[i] = *Def_F*[i] + *GetDef*($C[k]$);   //Concatenate two strings
    }
}
FOR each sense $w[i]$ of w
{   //Compute the score for every word sense, that is, the similarity value between the word sense and the
    corresponding feature word
    FOR each *Def_F*[j]
        score[i] += *Overlap*(*Def_W*[i], *Def_F*[j]);
}
RETURN $w[i]$;   //$i = \arg_j\max(score(j))$
END

Algorithm 1. Formalization description of Lesk-C algorithm

Coordinate Terms are sibling nodes in WordNet. They are not the abstraction of the original word sense, even there are not direct relations between both. However, under the assumption that the effect of any word sense and its coordinate terms on certain context is equal, they are all helpful for WSD. The experiment in this paper proves that coordinate terms and hypernyms are both significant in the same way.

4   The Application of WSD in CLIR

4.1   WSD for Query Expansion

At first, based on query disambiguation, word senses in query are determined. Secondly, the expansion proceeds according to the related resources, such as thesaurus and corpus. Because restricting the specific word senses, the result with expansion is better than the result without expansion. However, there are some drawbacks. One is that errors in disambiguation will influence the performance seriously. The other is that the run speed of system will be reduced. Take *"net income"* as an example, in order to compare the effect of WSD on query expansion, the cases based on WordNet and corpus are considered respectively, as shown in Table 1. Being a polysemous word, *"net"* can not only denote the meaning of *"net income"*, but also *"network"*, *"fishing net"*, etc. The column of **"Without Disambiguation"** describes the result after direct expansion. It can be observed that senses and their associated words can be expanded based on WordNet, but the expansion result based on corpus is more extensive and noisy. Through the disambiguation process for *"net income"*, the word sense for *"net"* can be confirmed and then expanded further. The final result is improved to a great extent.

Table 1. Example of comparable analysis for the application of WSD in Query Expansion

Query *"net income"*	Without Disambiguation	With Disambiguation
Expansion based on WordNet	*Meshwork, web, clear, profit, lucre, Internet, net, profit, cyberspace, income, earning, ...*	*Net income, profit, lucre, earnings, ...*
Expansion based on Corpus	*Net income, earnings, reported quarter serono, fertility, gain, trust, sale, drugs, rose, ...*	*Net income, earnings, gain, ...*

4.2 WSD for Query and Document

IR models commonly used, such as Boolean Model and Space Vector Model, all ignore word senses of feature terms in documents. These models often emphasize particularly on whether feature terms exist, and do not care whether word senses match. There are some deficiencies in this manner. Take *"net income"* as an example, many documents related to *"network"* will be returned by these models above, rather than documents related to *"net income"*. Therefore, WSD is introduced into CLIR model, and used to disambiguate the translation of query and retrieval results. The retrieval precision can be improved through the irrelevant document filtering. Only when feature terms in a document are matching with the corresponding word senses of query terms, this document will be merged into the final retrieval document list. Otherwise, such a document will be filtered out.

5 Experiment and Analysis

5.1 Experiment on WSD Algorithms

Data Set and Evaluation Metrics. The corpus of the English Lexical Sample Task in SENSEVAL-3 is utilized as the data set, which contains 57 ambiguous words (composed of 20 nouns, 32 verbs and 5 adjectives). At the same time, two evaluation parameters are introduced to evaluate the performance of WSD algorithms. Precision (P) is computed by summing the scores over all the instances handled through WSD processing, and divided by the number of the handled instances. Recall (R) is computed by summing the system's scores over all the instances including the unhandled instances, and divided by the total number of the instances in the evaluation data set. If all the instances are attempted or handled, the precision is identical to the recall. Therefore, the recall is omitted in our experimental results, because all the instances are handled.

Experimental Results. Based on the selected mainstream WSD algorithms and the established Lesk-C algorithm, ambiguous words with different POSs are tested under different preprocessing strategies. The experimental results are shown in Table 2, in which "Fine" represents "fine-grained", "Coarse" means "coarse-grained", and "Stopwords/Stemming (Y/N)" represents whether the contexts are processed through stopword filtering and stemming. "Lesk1" is the extended Lesk algorithm that uses its direct hypernyms, and "Lesk2" uses two nearest hypernyms. "WP" and "JC" represent the other two kinds of classic unsupervised WSD algorithms, in which WP is Wu and Palmer's approach based on path information [10] and JC is Jiang and Conrath's measurement based on information content [11].

Table 2. Experimental results for various WSD algorithms

WSD Pattern	Stopwords/ Stemming	Noun&Verb		Noun		Verb	
		P (Fine)	P(Coarse)	P (Fine)	P(Coarse)	P (Fine)	P(Coarse)
Lesk	Y / Y	0.312	0.401	0.357	0.490	0.271	0.319
	Y / N	0.318	0.407	0.356	0.489	0.284	0.334
	N / Y	0.310	0.398	0.354	0.490	0.269	0.315
	N / N	0.312	0.402	0.348	0.486	0.279	0.326
Lesk1	Y / N	0.348	0.437	0.396	0.536	0.303	0.366
Lesk2	Y / N	0.325	0.419	0.379	0.515	0.298	0.358
Lesk-C	Y / N	0.377	0.455	0.426	0.543	0.331	0.374
WP	Y / N	0.358	0.441	0.404	0.538	0.311	0.367
JC	Y / ~	/	/	0.504	0.572	/	/

Analysis and Discussion. It can be observed from Table 2 that the performance of the Lesk's variations is better than that of the Lesk algorithm. This is because the Lesk algorithm often leads to the data sparseness problem, and then the most frequently used word sense will be chosen automatically. At the same time, it also shows that the extended Lesk algorithms are more accurate than the Lesk algorithm itself. According to the experimental results, whether using the stemming process or not, the stopword filtering will obviously improve the precision. This indicates that the stopword filtering can eliminate some useless information. The performance of Lesk-C is better than those of Lesk1 and Lesk2, which proves the role coordinate term plays is equivalent to that hypernyms do. Through introducing the coordinate terms, the Lesk-C algorithm can not only get the better performance, but also be suitable for the disambiguation for both noun and verb. The performance of the WP algorithm is similar to Lesk1, better than Lesk2, but still worse than Lesk-C. The JC algorithm outperforms the other unsupervised algorithms on nouns, and its precision reaches 50.4%. However, it is unable to handle words with other POSs except noun, which is the main restriction and shortage of this algorithm.

5.2 Experiment on the Application of WSD in CLIR

Data Set. Based on the English Corpus of *Financial Times* (1991-1994) from the Ad hoc task in TREC-7, the performance of English Monolingual IR and Chinese-English CLIR is tested. This corpus includes 210,158 documents (about 560M), and the average length of a document is 2,338 bytes. Based on the Chinese Corpus from the CLIR task in TREC-9, the performance of Chinese Monolingual IR and English-Chinese CLIR is tested. This corpus is made up of three news document sets from Hongkong Commercial Daily, Hongkong Daily and Ta Kun Pao respectively, which contains 127,938 documents (about 360M) in all. The English Query Set is established based on the query set from the Ad hoc task in TREC-7, and the Chinese Query Set based on the query set from the CLIR task in TREC-9.

For English-Chinese CLIR, the corresponding Chinese translation of English query is provided in TREC-9. Therefore, it is not necessary to translate query manually. For Chinese-English CLIR, English corpus and query set are reused. The title of English query is manually translated into Chinese, and then the acquired Chinese query is used in Chinese-English query translation, retrieval and the final evaluation.

Experimental Results. For English Monolingual IR, four runs are tested based on the related strategies of WSD. Every run utilizes the content in the title field of English query as the original query, and takes the English corpus as the retrieval object.

(1) *E-E_Base* – without disambiguation and query expansion;
(2) *E-E_WnAll* – using WordNet to expand all the word senses of query terms in original query;
(3) *E-E_WnFirst* – using WordNet to expand the most frequently used word sense of each query term;
(4) *E-E_Lesk-C* – firstly using the Lesk-C WSD algorithm based on WordNet to disambiguate the original query, and then making expansion.

The results about all the runs above are shown in Figure 1.

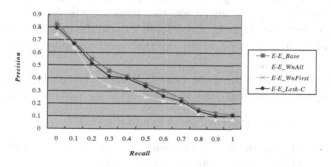

Fig. 1. *P/R* curve about the performance of English Monolingual IR based on WSD strategy

For Chinese-English CLIR, four runs are implemented based on the related strategies of WSD. These runs are tentative application for WSD in CLIR. Every run utilizes the content in the title field of Chinese query as the original query in source language, which will be converted into the query in target language. The English corpus is taken as the retrieval object.

(1) *C-E_Base* – without disambiguation and query expansion;
(2) *C-E_WnAll*– using WordNet to expand all the word senses of query terms in the query in target language;
(3) *C-E_WnFirst* – using WordNet to expand the most frequently used word sense of each query term in the query in target language;
(4) *C-E_Lesk-C* – firstly using the Lesk-C WSD algorithm based on WordNet to disambiguate the query in target language, and then making expansion.

The results about all the runs above are shown in Figure 2.

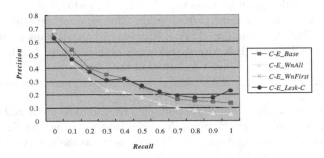

Fig. 2. *P/R* curve about the performance of Chinese-English CLIR based on WSD strategy

For English-Chinese CLIR, four runs are implemented based on the related strategies of WSD. Every run utilizes the content in the title field of English query as the original query in source language, which will be converted into query in target language. The Chinese corpus is taken as the retrieval object.

(1) *E-C_Base* – without disambiguation and query expansion;
(2) *E-C_WnAll* – firstly using WordNet to expand all the word senses of query terms in the query in source language, and then translating them;
(3) *E-C_WnFirst* – firstly using WordNet to expand the most frequently used word sense of each query term in the original query in source language, and then translating them;
(4) *E-C_Lesk-C* – firstly using the Lesk-C WSD algorithm based on WordNet to disambiguate the original query in source language, and then making expansion.

The results about all the runs above are shown in Figure 3.

Analysis and Discussion. It can be observed from Figure 1 that there are the following cases in the runs on English Monolingual IR.

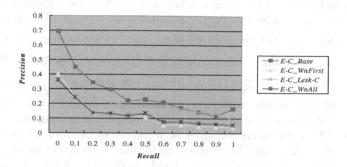

Fig. 3. *P/R* curve about the performance of English-Chinese CLIR based on WSD strategy

✧ *E-E_Base* has the best performance, and its average precision is up to 38.2%.

✧ *E-E_WnAll* has the worst performance, and its average precision is only 30.7%.
Maybe this phenomenon is caused by introducing too much noise. For example,
given an initial English query *"Nobel prize winners"*, the word set after expansion is {*"pry"*, *"Alfred Nobel"*, *"loot"*, *"Alfred Bernhard Nobel"*, *"plunder"*,
"award", *"success"*, *"lever"*, *"pillage"*, *"trophy"*, *"succeeder"*, *"victor"*,
"achiever", *"esteem"*, *"value"*, *"respect"*, *"booty"*, *"appreciate"*, *"jimmy"*,
"treasure", *"dirty"*, *"money"*, *"swag"*}. Hence, the most important three words
{*"Nobel"*, *"prize"*, *"winner"*} are submerged by these 22 words. In addition, because the proper name identification is not adopted, the initial English query is
split into single words. For example, the initial English query *"El Nino"* is split
into two single words {*"El"*, *"Nino"*}. However, the synset of *"El"* is {*"elevation"*, *"altitude"*, *"ALT"*, *"elevated railway"*, *"elevated railroad"*, *"overhead
railway"*}. The words and phrases acquired after expansion have no direct relation with the original word senses.

✧ *E-E_WnFirst* only expands the most frequently used word senses. The word set
after expansion is a subset of the word set acquired by the *E-E_WnAll* pattern.
However, the result is obviously better than that by the *E-E_WnAll* pattern. This
fully shows that too much useless information is introduced in the *E-EWnAll*
pattern.

✧ *E-E_Lesk-C* is similar to *E-E_WnFirst*. Firstly, the specific word senses of query
words are determined based on WSD, and then expanded by using synset information in WordNet. For example, for the initial English query *"Nobel prize winners"*, *"Nobel"* has only one word sense in WordNet. In the *E-E_Lesk-C* pattern,
this single sense is chosen. The synset of this sense is {*"Nobel"*, *"Alfred Nobel"*,
"Alfred Bernhard Nobel"}, and then two words are expanded. The noun *"prize"*
has three word senses, and the first one is chosen in the *E-E_Lesk-C* pattern, that
is, the sense used the most frequently. This case is equivalent to *E-E_WnFirst*.
The corresponding synset is {*"prize"*, *"award"*}. The noun *"winner"* also has
three senses, and the third one is chosen in the *E-E_Lesk-C* pattern. Its synset is
{*"achiever"*, *"winner"*, *"success"*, *"succeeder"*}. Therefore, for the query above,
the query word set is expanded to ten words in the *E-E_Lesk-C* pattern. It can be
seen that *E-E_Lesk-C* can use WordNet to make expansion by selection.

For Chinese-English CLIR, because Chinese-English query translation module is not perfect, the effect of Chinese query translation is not very satisfied. At present, the translation module can not correctly translate many proper names, such as "领海 (*marginal sea*)", "氰化物(*cyanide*)", "厄尔尼诺(*El Nino*)", "病态建筑综合症(*sick building syndrome*)", "亚马逊河(*Amazon*)" and "热带雨林(*rain forest*)", etc. This disadvantage directly results in the whole performance of CLIR falling rapidly. In order to reduce the influence of the quality of Chinese-English translation on the performance of CLIR and measure the whole performance of CLIR based on WSD strategy, some manual adjustments have been done for the translation of the original query in source language beforehand. On the one hand, the correct or basically correct translations for the original query are reserved as much as possible. For example, for the translation result of the query "诺 贝 尔 奖 获 得 者 (*Nobel Laureate*)", "*win Nobel prize*", any modification will not be made. On the other hand, some words with more ambiguity are removed. For example, "厄尔尼诺(*El Nino*)" is translated into "*be in distress nun that promise*", then the translation should be adjusted. It can be observed from Figure 2 that in comparison with Figure 1, *C-E_Base* no longer has the best performance and the average precision is lower than those of and *C-E_Lesk-C*. At the same time, this case fully shows that query expansion and disambiguation can indeed improve the whole performance of Chinese-English CLIR. The comparison of the average precision between English Monolingual IR and Chinese-English CLIR based on WSD strategy is shown in Table 3.

Table 3. The performance comparison between English Monolingual IR and Chinese-English CLIR

Testing Pattern	English Monolingual IR Average Precision	Chinese-English CLIR Average Precision	Comparison between Both
Base	38.2%	30.2%	79.1%
WnAll	30.7%	22.7%	73.9%
WnFirst	36.7%	27.7%	75.5%
Lesk-C	36.0%	30.5%	84.7%

It can be seen from Table 3 that although *E-E_Base* is the best one in English Monolingual IR, *C-E_Base* falls to the third one in Chinese-English CLIR. The main reason is that the lost or noisy information has the direct influence on the performance of *C-E_Base*. However, the other patterns except *C-E_Base* can make up this problem. Thus it can be seen that the effect of query expansion is very obvious. Among the methods based on WordNet, *C-E_Lesk-C* based on WSD is the best one, and its performance is evidently better than those of *C-E_WnAll* and *C-E_WnFirst*. This shows that WSD is useful to some extent in Chinese-English CLIR, but this effect is not obvious in English Monolingual IR.

Because Chinese query is not expanded by Chinese dictionary or other resources, Chinese Monolingual IR is only tested in the *C-C_Base* manner based on Chinese query set and corpus. The average precision of this manner is 27.7%, which is less 10.5% in comparison with the average precision 38.2% of English Monolingual IR. Certainly, owing to the corpus used differently, these two manners have no direct comparableness. Similarly, in English-Chinese CLIR, the direct Chinese query expansion is still not used. Firstly, English query is expanded by using WordNet, and then converted into Chinese query through query translation. Due to the faultiness of the query translation module used, too much noisy information is introduced. Therefore, as shown obviously in Figure 3, even if some noisy information is cut down manually, the precision reduces sharply. The comparison of the average precision between Chinese Monolingual IR and English-Chinese CLIR based on WSD strategy is shown in Table 4.

Table 4. The performance comparison between Chinese Monolingual IR and English-Chinese CLIR

Testing Pattern	Chinese Monolingual IR Average Precision	English-Chinese CLIR Average Precision
Base	27.7%	27.7%
WnAll	/	13.4%
WnFirst	/	12.7%
Lesk-C	/	21.6%

6 Conclusions and Future Work

Through specially introducing and considering the "Coordinate Terms", the Lesk-C algorithm exhibits the better performance among various popular unsupervised WSD algorithms, and has more extensive applicability. This paper attempts to apply WSD in CLIR and evaluate the whole retrieval performance in the round. Although being restricted by the performance of WSD methods, query expansion and disambiguation based on the related strategies of WSD are beneficial to CLIR, and can improve the whole performance of CLIR to some extent.

In addition, the relatively objective evaluation metrics are also proposed. In order to measure the performance of various WSD methods, the corpus from SENSEVAL-3 are used in training and testing. In order to measure the performance after applying WSD in CLIR, the query set, corpus and result set provided by TREC are utilized to make general evaluation. This pattern makes our experimental results relatively fair and objective, and comparable.

WordNet is a thesaurus annotated manually. The existing algorithms do not fully utilize semantic information in WordNet, and more useful semantic information should be mining from it. At the same time, current research work focuses on English query expansion and WSD, Chinese query expansion and WSD based on Chinese thesaurus (e.g., HowNet) are not concerned. Though there is some commonness between English and Chinese WSD, they have their own characteristics and Chinese WSD is likewise important. These aspects above will be the focus problems in our future research work.

References

1. Peters, C.: Cross-Language Information Retrieval and Evaluation. LNCS, vol. 2069, pp. 261–272. Springer, Germany (2001)
2. Oard, W., Ertunc, F.: Translation-Based Indexing for Cross-Language Retrieval. In: Crestani, F., Girolami, M., van Rijsbergen, C.J.K. (eds.) ECIR 2002. LNCS, vol. 2291, pp. 324–333. Springer, Heidelberg (2002)
3. Gao, J., Nie, J.-Y., He, H., Chen, W., Zhou, M.: Resolving Query Translation Ambiguity Using a Decaying Co-Occurrence Model and Syntactic Dependence Relations. In: Proc. of the 25th Annual International ACM SIGIR Conference on Research and Development in Information Retrieval (SIGIR 2002), pp. 183–190. ACM Press, New York (2002)
4. Stokoe, C., Oakes, M.P., Tait, J.: Word Sense Disambiguation in Information Retrieval Revisited. In: Proc. of the 26th Annual International ACM SIGIR Conference on Research and Development in Information Retrieval (SIGIR 2003), pp. 159–165. ACM Press, New York (2003)
5. Monz, C., Dorr, B.J.: Iterative Translation Disambiguation for Cross-Language Information Retrieval. In: Proc. of the 28th Annual International ACM SIGIR Conference on Research and Development in Information Retrieval (SIGIR 2005), pp. 520–527. ACM Press, New York (2005)
6. Witten, I.H., Frank, E.: Data Mining: Practical Machine Learning Tools and Techniques, 2nd edn. Morgan Kaufmann, San Francisco (2005)
7. Liu, Y., Jin, R., Chai, J.Y.: A Statistical Framework for Query Translation Disambiguation. ACM Transaction on Asian Language Information Processing (TALIP) 5(4), 360–387 (2006)
8. EKEDAHL, J., GOLUB, K.: Word Sense Disambiguation using WordNet and the Lesk Algorithm. Projektarbeten 2004. Institutionen för Datavetenskap, Lunds University (2004)
9. Pedersen, T., Banerjee, S., Patwardhan, S.: Maximizing Semantic Relatedness to Perform Word Sense Disambiguation. University of Minnesota Supercomputing Institute Research Report UMSI 2005/25 (2005)
10. Wu, Z., Palmer, M.: Verb Semantics and Lexical Selection. In: Proc. of the 32nd Annual Meeting of the Association for Computational Linguistics (ACL 1994), Las Cruces, New Mexico, pp. 133–138 (1994)
11. Jiang, J.J., Conrath, D.W.: Semantic Similarity based on Corpus Statistics and Lexical Taxonomy. In: Proc. of International Conference on Research in Computational Linguistics (ROCLING X), pp. 19–33. Scandinavian University Press, Taiwan (1997)

Searching Polyphonic Indonesian Folksongs Based on N-gram Indexing Technique

Aurora Marsye and Mirna Adriani

Faculty of Computer Science, University of Indonesia, Depok Campus,
Depok 16424, Indonesia
aurora.marsye@gmail.com, mirna@cs.ui.ac.id

Abstract. Availability of enormous number of digital music presents challenge to organize and retrieve it in an effective way. We explore polyphonic Indonesian folksongs retrieval based on pattern matching such as n-gram in searching the songs. We compare the pattern matching results to regular text-based information retrieval system. The folksongs are either fully or partially indexed. The results of the experiments show that using text-based IR system or n-gram matching technique, both are effective in retrieving the polyphonic songs, regardless of the query length or position where the query fragment is taken. However, to achieve a better performance, fully indexed songs is preferable than partially indexed songs.

Keywords: Music Information Retrieval.

1 Introduction

Nowadays the availability of digital information is very enormous, especially on the Internet that easily accessible to users around the world. The information comes in various formats such as text, audio, image, and video. Research interest in multimedia data such as music has increased in recent years as there is a need to better organize these music files. Music Information Retrieval (MIR) focuses on categorizing, searching, and retrieving music files in various formats [1].

Before MIR systems exist, one used to utilize a text-based search engine to find a particular musical data by entering keywords. The search engine will identify the relevant music document by its text properties such as the title or the lyric. This searching process has a limitation in representing musical data. An example of the limitation is when a user would like to find a particular song by expressing melody as query. Therefore, a content-based MIR system is preferable as a music search engine.

Musical data can be categorized into two types: monophonic and polyphonic. Monophonic musical data has only one melody, usually played by a single musical instrument with one note at a time. Polyphonic musical data has a number of notes played at one time. In the beginning, MIR research only focuses on monophonic music, however one deals with polyphonic music most of the time.

In this work we explore polyphonic Indonesian folksongs. Research and development on MIR system grows rapidly. However, almost all of the systems developed are

G.G. Lee et al. (Eds.): AIRS 2009, LNCS 5839, pp. 387–396, 2009.

based on Western music [2]. Several MIR studies have started to work on folksongs but only a few researches done on Indonesian folksongs [3]. Indonesian folksongs are of great variety in characteristics as the country has many different ethnicities spread across more than 15,000 islands.

2 Related Works

One of the first published works on MIR is Michael Kassler's in the mid-1960. For many years since then, very little work on MIR is done. Then from late 1990's the interest in MIR explodes.

In his research, Downie [4] evaluates a simple approach to MIR by conceiving melodic n-grams as text. He considers the interval representation of monophonic melodies as words followed by applying traditional text information retrieval techniques using SMART system. The experimental results show that the n-gram approach performs well. The n-gram length affect performance, long queries perform better than short queries, the location of the query does not affect retrieval effectiveness, and minor query errors affect performance.

The use of n-gram representation is also used in polyphonic MIR studies. Suyoto and Uitdenbogerd [5] implements n-grams indexing for polyphonic melody retrieval based using a symbolic melody matching by Uitdenbogerd and Zobel [6]. The approach reduces a polyphonic file into several monophonic files for n-gram construction. Subsequently, the searching process is done by string matching approach. This simple approach works effectively in retrieving symbolic melody such as MIDI.

Another approach to polyphonic MIR is done by Doraisamy [7]. She compares indexing techniques for polyphonic music data with n-grams without reducing the polyphonic files and by extending the n-grams to include rhythmic information in addition to intervallic information. The results show that the n-gram approach to polyphonic MIR is a promising and robust approach for indexing large collections of music.

MIR works on folksongs are investigated by several studies such as: Thai music classification and retrieval [8], MIR system for Malaysian music [2], Korean traditional music notation representation [9], Greek and African traditional music retrieval [10], and Indonesian folksongs classification [3].

3 Music Information Retrieval (MIR)

In this section we discuss MIR in detail, covering digital music, MIR system, and MIR techniques used in our experiments.

3.1 Digital Music

The discovery of recording technology, digital information storage, and internet, all have transformed music documentation form from hand written music scores into uncountable digital music.

Music data can be encoded digitally in one of several digital music formats. These digital formats are categorized as structured format such as Hundrum, semi-structured format such as MIDI, and unstructured format such as MP3. In structured format, the musical information encoded is detail. However in a semi-structured format such as MIDI not every musical information is encoded. It encodes event information of digital sound, for instance when the notes start and stop. In unstructured format, the raw audio encodes sample of sound energy level [7].

MIDI (Musical Instrument Digital Interface) format is a standard protocol of musical information that connects electronic music instrument to computer. A MIDI document may consist of several channels which each represents a play of certain musical instrument. For that reason, MIDI format is also known as a symbolic music format.

Most MIR research dealing with MIDI files because the processing of an unstructured format, also known as audio format, such as MP3 requires complex computation. Although currently audio MIR research is not as many as the symbolic's, a few approaches such as symbolic query [11] and energy distribution [12] are explored.

3.2 MIR System

An IR system is designed to retrieve document or information that is needed by user. This system gives the right information to the right user along with its role, bridging the information creator and information user.

An MIR system is an IR system which deals with musical data and typically the system is content-based MIR system. Typke et al. [13] divides content-based MIR system into two general categories: audio music search engine (e.g. to search MP3 music) and symbolic music search engine (e.g. to search MIDI music). There are also hybrid systems which convert the query from audio signal to symbolic description then do the search process on symbolic music collection [14, 15] and vice versa which use symbolic query to do search process on audio music collection that already converted to symbolic music collection [11].

The development of a content-based MIR system is aimed to many goals and users. Usually someone who wants to buy a copy of certain piece of music only remember the melody instead of the title, artist, or composer. A system that is able to identify melody hummed by the user will be really helpful in this case. Content-based MIR system also help musicologist to know how the composers affect each other or how their current works related to their past works or even to other composer's works. Copyright problem is possible to be solved since the composers may utilize the system to detect whether someone has plagiarized their work or whether their new work is potentially considered as plagiarism [13].

3.3 MIR Techniques

Melody Extraction. Polyphonic music might have several notes played at one time, but only particular notes are the main melody that representing the content of music. Therefore, a melody extraction is required to choose which notes are important in a polyphonic music. The melody extraction process is somewhat similar to stopwords removal in text-based IR, and to do it one need an effective and efficient technique [16].

Uitdenbogerd and Zobel [17] evaluates some polyphonic melody extraction techniques on MIDI format and concludes that certain technique performs well on certain case but not well on another case. The most consistent melody extraction technique is *all-mono*. *All-mono* technique combines all musical parts and includes the highest pitch note starting at any instance as part of the melody.

All-mono technique does not give satisfactory result when the main melody is located on lowest notes and also when the main melody is distributed on several channels. One approach to solve this problem is by eliminating channels those do not contain melodic information before applying the *all-mono* technique [16].

Text Representation. Music computing, especially symbolic music such as MIDI format, generally transform musical information into text before being processed to the next step. This is a common technique for both monophonic music [4] and polyphonic music [7].

Musical information such as pitch and duration are able to be represented as text. This representation might be absolute measures or relative measures. However, most MIR researchers favor relative measures because it is more flexible when dealing with change in tempo or transposition across keys [18]. Common techniques in representing pitch are contour and interval; both are relative measures.

Contour representation can be constructed using U for up (if the next note is higher), D for down (if the next note is lower), and S for same pitch (if the next note has the same pitch). For example, the contour for "1 2 3 4 5 5 4 3 2 1" and "C4 D4 E4 F4 G4 G4 F4 E4 D4 C4" is "U U U U S D D D D".

Note interval is a distance between notes. As illustrated in Fig. 1, the distance between C to the next C, often called an octave, is 12 semitones.

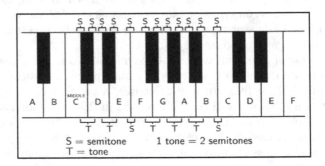

Fig. 1. Intervals

In Fig. 2, interval between F4 and F4 is 0, interval between F4 and A5 is 16 semitones up (+16), interval between A5 and F5 is 4 semitones down (-4), and so forth. Using *exact intervals* technique, the melody in Fig. 2 is represented as "0 0 16 -4 -3 -5 -4 -1 -2". *Directed modulo-12 intervals* technique attempts to modify *exact intervals* technique by mapping the melody to one octave [5]. For example, "F4 F4 F4 A5 F5 D5 A4 F4 E4 D4" is mapped into "F F F A F D A F E" so that the *directed modulo-12 intervals* representation is "0 0 4 -4 -3 -5 -4 -1 -2".

In content based MIR, although duration information is possible to be represented as text, using only pitch representation is preferable. A study shows that note duration information does not increase the retrieval effectiveness [19].

Fig. 2. Example of Melody Fragment

Indexing and Similarity Measurement. After being represented as text, MIR process continues with indexing and similarity measurement. The techniques are vary one to another depends on the representation used. A commonly used indexing technique in MIR is n-gram indexing which use n-grams of text representation as glossary in inverted index. N-grams are substrings of a string with length n. For example, 5-grams of "RETRIEVAL" are "RETRI", "ETRIE", "TRIEV", "RIEVA", and "IEVAL".

N-gram is effectively used for comparing text documents because every string in text is decomposed into small parts so that if error exists, it tends to affect only a limited number of those parts. Besides that, n-gram approach is language independent so that it is more universal to compare text documents [20].

To simplify the text representation of musical information, we map the value of *directed modulo-12 intervals* sequence "-12, -11, ..., -2, -1, 0, 1, 2, ..., 11, 12" into "a, b , ..., k, l, m, n, o, ..., x, y". For example, "F F F A F D A F E D" whose *directed modulo-12 intervals* sequence is "0 0 4 -4 -3 -5 -4 -1 -2", will be mapped into "mmqijhink" and the n-grams (5-grams for example) are "mmqij", "mqijh", "qijhi", "ijhin", and "jhink". Next, these n-grams are used as glossary in index document.

In content-based MIR, generally the query is a music fragment. To continue to the searching process, this music fragment should also be transformed into text.

Two similarity measurement techniques that commonly used to process the text representation of musical information are language model, as implemented in text-based IR system, and pattern matching such as n-gram matching. In text-based IR system, the text representation of songs is indexed and retrieved just like text documents. In n-gram matching, the similarity measure between a document and a query is obtained by counting the distinct n-grams that they have in common.

In this study we apply *all-mono* melody extraction technique for each channel of polyphonic MIDI files. We use *directed-modulo-12* sequence as the text representation and then apply 5-gram indexing to this sequence. The use of n-gram with length five is based on a study that concludes 5-gram as the best approach for ranking melody similarity [6]. We compare the performance of language model, as implemented in text-based IR system, and pattern matching such as n-gram matching in searching the polyphonic Indonesian folksongs. We use both distinct and duplicate n-gram in document and query while evaluate the use of text-based IR system. We also evaluate the performance of a full and partial indexing techniques.

In full-indexing technique, we index all part of the song, from the beginning until the end. In partial-indexing technique, we index only partial part of the song which taken from the beginning of each channel until a half part (in a half indexing technique) and one third part (in one third indexing technique). The half and one third parts refer to part of the text representation of the song (it is not part of the songs according to its timeline). Fig. 3 shows the examples of indexed part of a song that consist of three channels in full and partial indexing technique.

Full-indexing
Channel 1: `nomjnojnmojkrqmnmlmkmimlmuqnomjnojnmojkrqm`
Channel 2: `nomjnojnmojkrqmnmlmkmimlmu`
Channel 3: `qcwcwcwiqcwcwiqfulkqxjepulnbxjepulcyxjepulcyxj`

A half indexing
Channel 1: `nomjnojnmojkrqmnmlmkm`
Channel 2: `nomjnojnmojkr`
Channel 3: `qcwcwcwiqcwcwiqfulkqxje`

One third indexing
Channel 1: `nomjnojnmojkrq`
Channel 2: `nomjnojn`
Channel 3: `qcwcwcwiqcwcwiq`

Fig. 3. Examples of Indexed Part of A Song

4 Experiments

Until this research is done, no standard music collection exists yet to evaluate MIR methods. We construct a song collection, 2158 polyphonic MIDI, including 191 Indonesian folksongs. This collection gathered from many sources such as websites of MIDI repository (MIDI Database[1], Midi Directory[2], and Mutopia Project[3]), Balai Budaya Minomartani Yogyakarta collection, websites of Indonesian folksongs, and personal collection.

We randomly select 30 Indonesian folksongs from our collection as queries to be used in the experiments. Each of these songs has at least one different version of the same title in the collection. We construct five query sets which each consists of 30 monophonic queries described as follows: Query Set 1 consists of long queries from 30 selected songs, i.e., the whole main melody track from each of the 30 songs; Query Set 2 consists of short queries taken from the beginning part of each song, i.e., beginning part of the main melody track; Query Set 3 consists of short queries taken from the middle part of each song; Query Set 4 consists of short queries taken from the end part of each song; Query Set 5 consists of short queries played and recorded by virtual keyboard representing the beginning part of each song that usually perceived by human. MIR flow in the experiments is illustrated in Fig. 4. We use RMIT MIRT Fanimae MIREX 2005 (FM05) [5] as MIDI parser that transcribes the collection and queries into *directed modulo-12 intervals* sequences. We then apply 5-gram indexing to these sequences. We implement a program based on a pattern matching technique

---

[1] http://www.mididb.com
[2] http://www.freemidi.org
[3] http://www.mutopiaproject.org

to measure the similarity between 5-gram indexed texts. We use Indri search engine [21], a text-based IR system, to index and retrieve the folksongs. Indri's retrieval model is a combination of language model and inference network.

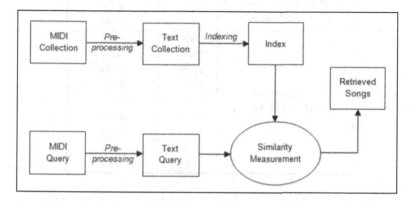

Fig. 4. Music Information Retrieval Flow

Evaluation. In this work, we use Mean Reciprocal Rank (MRR) to evaluate the retrieval performance. MRR is initially used to evaluate a question-answering system. But now MRR is also used to evaluate IR system [22]. A reciprocal rank can be defined mathematically as $1/r$ where r is the highest rank of relevant documents retrieved. MRR is the average of reciprocal rank in a query set. Higher value of MRR indicates more effective retrieval.

5 Results

Table 1 presents the results of the experiments. The pattern matching technique is represented by 5G. Text-based IR system represented by T followed by NN for allowing 5-gram duplication in both document and query, NU for allowing 5-gram duplication in document and using distinct 5-gram in query, UU for using distinct 5-gram in both document and query. The last digit of each initial represents the indexing technique; 1 for full-indexing technique, 2 for a half indexing technique, 3 for one third indexing technique.

The experiments show that 5G, TNN, TNU, and TUU work very well in retrieving the musical data either in full-indexing or partial-indexing techniques. However, fully indexed songs give better performance than partially indexed songs. Using full-indexing techniques, all aproaches give MRR score 0.99 (i.e. only one query fail). MRR score of a half indexing technique is 0.935 and one third indexing technique is 0.895. The retrieval performance of the text-based IR system is not significantly affected even though a document or query contains 5-gram text duplication or not. Overall, both pattern matching technique and text-based IR system are able to retrieve the polyphonic MIDI with MRR average score 0.94.

Table 1. MRR of Similarity Measurement Techniques

Similarity Measurement	Initial	MRR
Pattern Matching (5-gram Matching)	5G-1	0.99
	5G-2	0.93
	5G-3	0.89
Language Model (Text-Based IR System)	TNN-1	0.99
	TNN-2	0.94
	TNN-3	0.90
	TNU-1	0.99
	TNU-2	0.94
	TNU-3	0.90
	TUU-1	0.99
	TUU-2	0.93
	TUU-3	0.89

In regard to the Query Sets we use, Query Set 1 and 2 both give MRR score 1 in full-indexing and partial-indexing techniques. MRR score for Query Set 3 is 0.92 with MRR score 1 in full-indexing technique, 0.91 in a half indexing technique, and 0.86 in one third indexing technique. MRR score for Query Set 4 is 0.82 with MRR score 1 in full-indexing technique, 0.79 in a half indexing technique, and 0.66 in one third indexing technique. Query Set 5 give MRR score 0.97 in full-indexing and partial-indexing techniques. These scores imply that partial-indexing technique works as well as full-indexing technique does if the query contains the beginning part of the song. Otherwise, although folksongs tend to be short and repetitive, full-indexing technique performs better because it indexes all part of the song. In general, long query gives better result than short query.

In the experiments we find an interesting case where some relevant songs are not found because of different musical arrangements. For example, when a percussive instrument such as *Kolintang* (North Celebes traditional music instrument) is used, a long note is played by repetitive hits of the same note. Such patterns result in musical sequences that do not match the query.

6   Conclusions

This paper has presented a work on content-based polyphonic MIR for Indonesian folksongs. The results of the experiments show that the pitch interval can be represented as text. In retrieving the songs, using text-based IR system or n-gram matching technique, both are effective in retrieving the polyphonic songs, regardless of the query length or position where the query fragment is taken. However, a better performance is shown using fully indexed songs rather than partially indexed songs.

In the future we will use a bigger music collection in our study. We have a difficulty in creating Indonesian folksongs in MIDI format because most Indonesian music is currently available in MP3 format. Audio transcription is difficult to apply to Indonesian folksongs, especially the *Javanese* and *Balinese* music due to the microtonality (musical scale) problem. We are in the process of tweaking our MIR system, which is based on the diatonic scale, to process Indonesian folk and traditional music which typically uses pentatonic scales.

References

1. Shakra, I., Frederico, G., El Saddik, A.: Music Indexing and Retrieval. In: IEEE International Conference on Virtual Environments, Human-Computer Interfaces, And Measurement Systems, Boston (2004)
2. Doraisamy, S., Adnan, H., Norowi, N.M.: Towards A MIR System for Malaysian Music. In: International Symposium on Music Information Retrieval 2006, pp. 342–343. University of Victoria, Victoria (2006)
3. Indah, N., Adriani, M.: Klasifikasi Otomatis Lagu-Lagu Daerah Indonesia dalam Kerangka Music Retrieval. In: Seminar Nasional Sistem dan Teknologi Informasi (SNASTI). Surabaya (2006)
4. Downie, J.S.: Evaluating A Simple Approach To Music Information Retrieval: Conceiving Melodic N-Grams As Text. The University of Western Ontario, London (1999)
5. Suyoto, I.S., Uitdenbogerd, A.L.: Simple Efficient N-gram Indexing for Effective Melody Retrieval. In: Proceedings of the Annual Music Information Retrieval Evaluation exchange (2005)
6. Uitdenbogerd, A.L., Zobel, J.: Music Ranking Techniques Evaluated. In: Proceedings of Australasian Computer Science Conference, Melbourne, pp. 275–283 (2002)
7. Doraisamy, S.: Polyphonic Music Retrieval: The N-gram Approach. University of London, London (2004)
8. Nopthaisong, C., Hasan, M.M.: Automatic Music Classification and Retrieval: Experiments with Thai Music Collection. In: International Conference on Information and Communication Technology ICICT 2007, Dhaka (2007)
9. Lee, J.H., Downie, J.S., Renear, A.: Representing Korean Traditional Musical Notation in XML. In: Proceedings of the 3rd International Conference on Music Information Retrieval. IRCAM, Paris (2002)
10. Antonopoulos, I., Pikrakis, A., Theodoridis, S., Cornelis, O., Moelants, D., Leman, M.: Music Retrieval by Rhythmic Similarity Applied on Greek and African Traditional Music. In: The 8th International Symposium on Music Information Retrieval, ISMIR 2007. Austrian Computer Society, Vienna (2007)
11. Suyoto, I.S., Uitdenbogerd, A., Scholer, F.: Searching Musical Audio Using Symbolic Queries. IEEE Transactions on Audio, Speech and Language Processing, 372–381 (2008)
12. Liu, C.-C., Tsai, P.-J.: Content-Based Retrieval of MP3 Music Objects. In: Conference on Information and Knowledge Management Proceedings of The Tenth International Conference on Information And Knowledge Management, pp. 506–511. ACM, Atlanta (2001)
13. Typke, R., Wiering, F., Veltkamp, C.: A Survey of Music Information Retrieval Systems. In: Proceedings of The International Symposium on Music Information Retrieval, ISMIR 2005, pp. 153–160 (2005)
14. Pickens, J., Bello, J.P., Monti, G., Crawford, T., Dovey, M., Sandler, M., Byrd, D.: Polyphonic Score Retrieval Using Polyphonic Audio Queries: A Harmonic Modelling Approach. In: Proceedings of the 3rd International Conference on Music Information Retrieval. IRCAM, Paris (2002)

15. Hu, N., Dannenberg, R.B., Tzanetakis, G.: Polyphonic Audio Matching and Alignment for Music Retrieval. In: Proceedings of The 2003 IEEE Workshop on Application of Signal Processing to Audio and Acoustics, pp. 185–188. IEEE, New Paltz (2003)
16. Ozcan, G., Isikhan, C., Alpkocak, A.: Melody Extraction on MIDI Music Files. In: Proceedings of The Seventh IEEE International Symposium on Multimedia (ISM 2005), pp. 414–422. IEEE, Washington (2005)
17. Uitdenbogerd, A.L., Zobel, J.: Manipulation of Music for Melody Matching. In: Proceeding ACM International Multimedia Conferences, pp. 235–240. ACM Press, Bristol (1998)
18. Pickens, J.: A Survey of Feature Selection Techniques for Music Information Retrieval. CIIR Technical Report, Amherst (2001)
19. Suyoto, I.S., Uitdenbogerd, A.L.: Effectiveness of Note Duration Information for Music Retrieval. In: Proceeding Tenth International Conference on Database Systems for Advanced Application, pp. 265–275. Springer, Beijing (2005)
20. Cavnar, W.B., Trenkle, J.M.: N-Gram-Based Text Categorization. In: Proceedings of SDAIR 1994, 3rd Annual Symposium on Document Analysis and Information Retrieval, Las Vegas, pp. 161–175 (1994)
21. Strohman, T., Metzler, D., Turtle, H., Croft, W.B.: Indri: A Language-Model Based Search Engine for Complex Queries (Extended Version). CIIR Technical Report, Amherst (2005)
22. Shah, C., Croft, W.B.: Evaluating High Accuracy Retrieval Techniques. In: Proceedings of the 27th Annual International ACM SIGIR Conference on Research and Development in Information Retrieval (SIGIR 2004), pp. 2–9. ACM, Sheffield (2004)

Study on the Click Context of Web Search Users for Reliability Analysis[*]

Rongwei Cen, Yiqun Liu, Min Zhang, Liyun Ru, and Shaoping Ma

State Key Laboratory of Intelligent Technology and Systems,
Tsinghua National Laboratory for Information Science and Technology,
Department of Computer Science and Technology, Tsinghua University, Beijing, China
crw@mails.tsinghua.edu.cn

Abstract. User behavior information analysis has been shown important for optimization and evaluation of Web search and has become one of the major areas in both information retrieval and knowledge management researches. This paper focuses on users' searching behavior reliability study based on large scale query and click-through logs collected from commercial search engines. The concept of reliability is defined in a probabilistic notion. The context of user click behavior on search results is analyzed in terms of relevance. Five features, namely query number, click entropy, first click ratio, last click ratio, and rank position, are proposed and studied to separate reliable user clicks from the others. Experimental results show that the proposed method evaluates the reliability of user behavior effectively. The AUC value of the ROC curve is 0.792, and the algorithm maintains 92.8% relevant clicks when filtering out 40% low quality clicks.

Keywords: User behavior analysis; click reliability; search user; search engine.

1 Introduction

User feedback provides useful information for analyzing, estimating and optimizing the performance of Web retrieval systems. It is an important topic for both IR researchers and search engine system engineers. Previous studies indicate that users are unwilling to provide explicit feedback for search engines [3]. Therefore, more studies (e.g. [4][5][19]) looked into implicit feedback information extracted from click-through data. This kind of feedback information has developed into an important research topic in the area of information retrieval and knowledge management, and has also been emphasized by commercial search engine community.

Unfortunately, practical Web data sources as well as click through logs contain lots of noise. Individual users may behave irrationally or maliciously, or may not even be real users, and we cannot treat each user as an individual "expert" [4]. By performing eye-tracking study, Joachims et al. [5] showed that individual user clicks include bias

---

[*] Supported by the Chinese National Key Foundation Research & Development Plan (2004CB318108), Natural Science Foundation (60621062, 60503064, 60736044) and National 863 High Technology Project (2006AA01Z141).

G.G. Lee et al. (Eds.): AIRS 2009, LNCS 5839, pp. 397–408, 2009.

and was not able to be used as judgments of absolute relevance directly. Therefore, state-of-the-art approaches require a large volume of click-through data to extract credible user feedback information based on the wisdom of the crowd (eg. [1][2][4][5] [12][13][18][19]). The consequent problem is that these methods only deal with hot queries with large number of access users and are not applicable for long-tail queries with rare user accesses.

To solve the problem, this paper defines the click reliability as whether a click is treated as one of an individual "expert" for relevance labeling probabilistically. Based on large scale log analysis, we study the contexts of individual user clicks and look into user decision process. Five context features of user clicks are proposed and employed to estimate click reliability.

The remaining part of the paper is as follow. Section 2 introduces the related work. Section 3 gives definition on users' click reliability in a probabilistic notion. In section 4, context of search behavior is been studied empirically and several features are proposed and verified for reliability analysis. Conclusions and future work are drawn in the section 5.

2 Related Work

In recent years, using implicit feedback information has been receiving much attention in the information retrieval area. Several approaches are proposed to mine relevant information from click-through data and some applications are implemented based on user behavior information, such as re-ranking search results [18], learning ranking [19], evaluating performance automatically [12], et al.

Tan et al. [6][7] detected robot behaviors for increasing the robustness of data and mining techniques applied to Web logs. Baeza-Yates et al. [8] and Kammenhuber et al. [9] modeled user clicks, query formulations and pages visited and revealed several interesting aspects of user behavior: users tend to formulate short queries, click on few pages and majority of users refine their initial queries in order to retrieve relevant documents. Sadagopan et al. [17] identified typical and atypical user sessions in click streams based on detecting outliers using Mahalanobis distance. These studies interpreted and modeled general user behaviors, detected robot behaviors or atypical ones, without analyzing and comparing different behaviors of one user.

In 2005, Joachims et al. [5] studied a work called "Eye-tracking", and mined implicit feedback information through analyzing user decision process. Based on this work, Agichitein et al. [4] depicted the rank bias of clicks, proposed background model and several user behavior models to mine relevant query-doc pairs. Recently, Craswell et al. [10] and Guo et al. [11] drew a cascade model, in which users view results from top to bottom and leave as soon as they see a worthwhile document, for explaining position bias of user behavior. However, some of these studies were performed over controlled lab conditions, which is not clear whether these techniques and rules will work for general real-world search. Some studied user behavior using statistics methods without analyzing individual user behaviors, which needed a large volume of user clicks for each query and cannot be adapted to long tail queries.

This paper analyzes user behaviors based on the contexts of user click process, looking into user decision process, studying user click preference, and estimating click reliability.

3 Search Process and Click Reliability

Before analyzing user behaviors, we assume the interactive process between user and search engine and look into user decision process. The concept of click reliability is defined, which has applications in performance evaluation of context features.

3.1 User Search Process

There is an interactive process between user and search engine. Generally, a user submits a query to a search engine. Then search engine presents results. By comparing the information of search result list, such as title, snippets, URL, other results, the searcher clicks results. When satisfied, he might be left. Otherwise, he returns to the result list, clicks more pages or refines his query and keeps on searching.

Table 1. A case of an interaction between user and search engine for the same topic

No.	Time	Query	Rank	Page Clicked
1	20:58:58	丰田(*Toyata*)	6	www.autohome.com.cn/526/
2	21:02:34	丰田(*Toyata*)	5	www.autohome.com.cn/110/
3	21:03:23	丰田(*Toyata*)	6	www.autohome.com.cn/526/
4	21:04:11	上海大众 (*Shanghai Volkswagen*)	5	www.che168.com/che168/cardb/brand/brand_58.html
5	21:06:14	广州本田 (*Guangzhou Honda*)	3	car.autohome.com.cn/brand/32/
6	21:09:23	丰田(*Toyata*)	2	car.autohome.com.cn/brand/63/
7	21:10:20	丰田(*Toyata*)	4	price.pcauto.com.cn/brand.jsp?bid=31
8	21:11:20	丰田(*Toyata*)	10	www.che168.com/che168/cardb/brand/brand_24.html
9	21:12:43	丰田卡罗拉 (*Toyota Corolla*)	1	www.autohome.com.cn/526/
10	21:19:12	丰田卡罗拉 (*Toyota Corolla*)	11	www.autohome.com.cn/526/options.html

Table 1 shows a case of user click process, from which we conjecture that a user want to find information for purchasing a car, and it is likely that the goal is *Toyota Corolla*(丰田卡罗拉). In the search processing, the user refers to other cars, *Shanghai VolksWagen*(上海大众) and *Guangzhou Honda*(广州本田). He finally clicks a page of *Toyota Corolla*'s configuration, and we can guess the user's need is the information about car configuration. Based on user search process analysis, we have the idea that each click happens in the context of interactive process between user and search engine system, which is able to derive user decision process, and user click reliability can be estimated and judged to mine the preference information.

3.2 User Click Reliability

By performing eye-tracking studies and analyzing users' decision process, Joachims et al. [5] show that clicks are informative but biased, and it is difficult to make the interpretation of clicks as absolute relevance and relative preferences derived from clicks are reasonably accurate on average. This paper estimates the click relevance using click reliability.

Definition: *User Click Reliability* $\Re$ is an estimated probability of the relevance between query q and document d, given the context F of click c, and is formalized as:

$$\Re(c(q,d)) = P(R(c(q,d)) = 1 \mid F).$$ (1)

where $c(q,d)$ is a click c of query q and document d, $R(c(q,d))$ presents the relevance of q and d, when $R(c(q,d)=1$ means relevant and $R(c(q,d))=0$ means irrelevant, and F is the click context, such as other clicks and queries in the current user session.

The concept defined here is different from the traditional studies which select relevant clicks by counting large scale logs statistically. This definition evaluates the relevance of query-document pairs from individual user clicks.

Based on Bayesian theorem, we have:

$$P(R(c) = 1 \mid F) = \frac{P(F \mid R(c) = 1)}{P(F)} P(R(c) = 1).$$ (2)

Here $P(R(c)=1)$ is the likelihood of relevant clicks in whole click set. If we just compare the values of click reliability in a given click corpus, $P(R(c)=1)$ can be regarded as a constant value and wouldn't affect the comparative results. The equation is rewritten as:

$$P(R(c) = 1 \mid F) \propto \frac{P(F \mid R(c) = 1)}{P(F)}.$$ (3)

Now consider the terms in equation (3), $P(F)$ is the probability of context feature F which can be estimated using the proportion of F in a given click corpus. $P(F|R(c)=1)$ is the probability of feature F in relevant click set and equals to the proportion of clicks with feature F in relevant click set. According to equation (3), the reliability of click c with context feature F is proportional to $P(F|R(c)=1)$ and inversely proportional to $P(F)$. Therefore, the expression $P(F|R(c)=1)/P(F)$ is able to estimate the performance of feature F and we define the concept of *Click Reliability Value (CRV)* as follows:

$$ClickReliabilityValue(CRV) = \frac{P(F \mid R(c) = 1)}{P(F)}.$$ (4)

According to equation (2), when *CRV* is larger than 1, namely $\dfrac{P(F \mid R(c)=1)}{P(F)} > 1$,
then we have $P(R(c)=1|F)>P(R(c)=1)$, and it means that the clicks with feature F is more reliable than the clicks in whole corpus in a probabilistic notion.

4 Empirical Study on the Context of User Behavior

Traditionally, user clicks are considered as a proof of relevance between queries and documents, and state-of-the-art approaches requires extensive user interaction data to guarantee statistical reliability. However, for long-tail queries, the challenge is that there is insufficient click data for statistical analysis. To assure our approach working for long-tail queries, we extracted features based on individual user clicks instead of relying on global statistics oriented features. Hence, we look into user decision process, analyze user click behaviors, then observe and propose several features from individual user at click level.

4.1 Data

The former study [4] showed that the Web search is not controlled and the techniques from the controlled lab may not work for general real world. To study user behavior in the real world, we collected search engine access logs from Sep. 10, 2008 to Oct. 24, 2008, with the help of a commercial search company. These access logs contain more than 194 million user clicks, 91 million unique queries, and 58 million user search sessions. Information extracted from the access logs is shown in Table 2.

Table 2. Information sources in the click-through logs collected

Item	Record Content
Query	The user query submitted
URL	URL of the result clicked by the user
Rank	The rank of the result clicked by the user
Order	The order of the result in the click sequence
User ID	Automatically assigned user's identification code
Time	Data and time of the clicking or querying event

For evaluating different context of relevant clicks and irrelevant ones, we randomly sampled 3000 queries from query logs. For each query, the top 20 results returned by 5 search engines in China were manually annotated as relevant or not by three assessors from a search engine company. Each assessor annotated about one third of the whole pooling set. The correctness of their annotation was also examined by co-checking each other's judgment results over a small subset containing 1000 query-doc pairs, and the kappa coefficient [20] measures agreement among these three assessors. Table 3 shows that the kappa value between any two assessors is large than 0.8, which means that our annotation is good reliability.

Table 3. The kappa statistic value of the manual annotation

	Assessor 1	Assessor 2	Assessor 3
Assessor 1	1.00	0.84	0.86
Assessor 2	0.84	1.00	0.87
Assessor 3	0.86	0.87	1.00

After annotation process, we have 89 thousands relevant query-doc pairs. With these pairs, 1.295 million click logs are picked out which have the same query-doc pairs. These clicks are treated as relevant ones reliably and annotated as Rel-Set and the whole click logs in our click-through data are annotated as Whole-Set.

4.2 Click Context Features

In the interaction between Web user and search engine systems, there are uncertainties in the process of user search and click. Firstly, we study the uncertainties in user query and click process and it is a possibility that there are two types of uncertainties, query uncertainty and click uncertainty. When a user clicks an irrelevant result, he/she is not satisfied and still need more information. Then the query tends to be refined and resubmitted to search engine, or more results will be clicked. We summarize these two types of uncertainties of click context as query number feature and click entropy feature.

(I) *QueryNum*: the unique query number submitted in current search session.

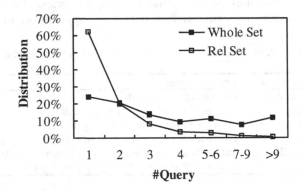

Fig. 1. QueryNum distributions of Rel-Set and Whole-Set. The category axis represents unique query number in current search session.

Fig. 1 shows that the context of 62% relevant clicks only contain one submitted query, which is larger than Whole-Set. The *QueryNum* of most clicks (82.1%) in Rel-Set is less or equal than 2 (only 44.9% for Whole-Set). Based on Fig. 1, we are able to derive that clicks with more submitted queries in their session context are less reliable. The *Click Reliability Value (CRV)* is 2.55, when *QueryNum* equals to 1, according to equation (4), which means that it works well for identifying click reliability. This context feature illustrates the existence of query uncertainty in process of query summiting.

(II) *ClickEntropy***:** the information entropy (proposed by Shannon in 1948 [15]) of user click distribution in current search session, which is calculated as follows:

$$ClickEntropy = -\sum_{p_i} p_i \log(p_i) .\tag{5}$$

Here, p_i is the click distribution, estimated using the proportion of click on result i in all clicks in a user session and calculated as:

$$p_i = \frac{\#(click\ on\ result\ i)}{\#(total\ clicks\ in\ a\ user\ session)} .\tag{6}$$

p_i is different from the traditional Click-through Rate (CTR) metric [4][19]. The CTR is a statistic metric based on all user clicks, while p_i is based on current session.

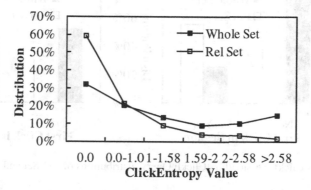

Fig. 2. ClickEntropy distributions of Rel-Set and Whole-Set. The category axis represents ClickEntropy value.

Fig. 2 shows that the *ClickEntropy* of Rel-Set is lower than Whole-Set's. When *ClickEntropy* equals to 0, namely user only click one page (one time or click the same page several times), it is 60% for Rel-Set, while it is only 32.2% for Whole-Set. Similar to *QueryNum* feature, the *ClickEntropy* of most clicks (81.1%) in Rel-Set is less than or equal to 1 (only 52.4%% for Whole-Set). According to *ClickEntropy* feature, it is able to derive that the click with more pages clicked in its session context is less reliable and *CRV*(*ClickEntropy*=1) is 1.85. This context feature illustrates the existence of click uncertainty in process of clicking result pages.

In a search process, each click is able to be sorted according to click time. Secondly, we conclude that the reliability is different for different order of click sequence. In [5], Jochims et al. proposed a rule, "Click > Earlier Click", which means that the later click is more relevant than earlier click, though the result of [13] shows that the rules of contradicting the existing search order perform worse compared to the rules that fully or partially reinforce the existing order of search result.

(III) *FirstClickRatio***:**
which is deined as *FirstClickInSession/FirstClickInQuery*: a click is the first one of a click sequence of session or query, or not. Here, we observe user click in two different scales of click context, session scale for the same user session and query scale for

the same query submitted by user. Since a user may submit several queries in current session, the scale of query is smaller than the one of session.

Fig. 3 shows that the 62% clicks in Rel-Set is the first ones of click sequence in user session, while it is 25.8% for clicks in Whole-Set. Similarly, 55.4% clicks in Rel-set are the first ones in query click sequence (25.8% for clicks in Whole-Set). The $CRV(FirstClickInSession=yes)$ is 2.15 and the $CRV(FirstClickInQuery=yes)$ is 1.42, which means that clicks with first position of click sequence in user session or query are more reliable than other ones. These two context features may be interpreted by the phenomenon that users may prefer to pay more attention to result lists and compare more information of each result before clicking any results and after first clicks, users' clicks tend to be less informative.

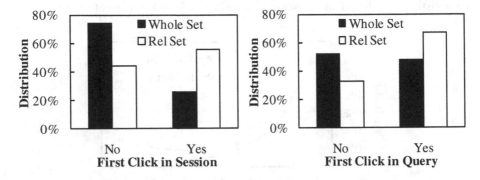

Fig. 3. FirstClickInSession/FirstClickInQuery distributions of Rel-Set and Whole-Set

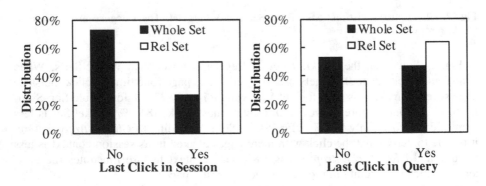

Fig. 4. LastClickInSession/LastClickInQuery distributions of Rel-Set and Whole-Set

(IV) *LastClickRatio*:
which is defined as *LastClickInSession/LastClickInQuery*: a click is the last one of a click sequence of user session or query, or not, analogous to the features *FirstClickInSession* and *FirstClickInQuery*.

Fig. 4 displays the distributions of these two context features and shows that 50% clicks in Rel-Set are the last ones in user session (63.6% in user query), while it is 26.7% in user session for Whole-Set (36.4% in user query). These two features can be interpreted by that users tend to stop interaction with search engines when they finally take the satisfying documents. These two features is special cases of the rule "Click > Earlier Click" in [5].

These four sequence context features look into different orders of click sequences based on session and query scale respectively, and the performances of session scale are better than the ones of query scale. By log analyzing, 71% users submit query only once in user session (Yu et al. have similar findings in [14]). For sessions with one query, the features, *FirstClickInSession* and *LastClickInSession*, is consistent with the other two, *FirstClickInQuery* and *LastClickInQuery*. Therefore, there are high correlations between these two groups of features with different scale. By studying the logs, the correlation between *FirstClickInSession* and *FirstClickInQuery* is 0.654 and it is 0.654 between other two features, which shows that these two groups of features are dependent.

Due to the high correlations, the first/last clicks of session sequences are filtered out from the sets of first/last clicks of query sequences. Fig. 5 shows the distributions of modified features of first/last clicks in query sequences, which are almost the same for both sets. According to the distributions in Fig. 5, we can derive that the first/last clicks of session sequences perform well, while the ones of query sequences fail to filter out reliable clicks. The user decisions under first/last clicks of query sequence, not session sequence, have the similar properties to middle clicks in sequence.

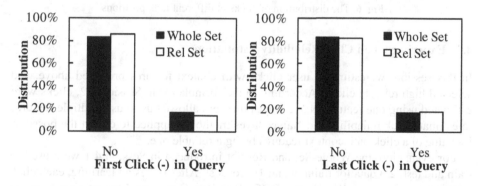

Fig. 5. The distributions of first clicks and last clicks in query sequences, not in sessions

Besides the features above, there are other context features, such as rank position of results, which is considered as rank bias [4][5][10][11] and causes the difficulties of log mining. To rank position factors, there are two different viewpoints. One is that the results at top positions have more possibilities to be viewed and clicked, the other is that search engines are experts in ranking the relevant documents at top positions. Here, we study user clicks with different result positions to look into the rank feature.

(IV) *RankPosition*: the rank position of click result.

Fig. 6 presents the rank distribution of Rel-Set and Whole-Set. For both of the sets, the results at top positions have more chance to be viewed and clicked than lower positions, and this phenomenon is defined as rank bias [4][5][10][11]. 30.3% users click the result at first position for Whole-Set, while it is 47.8% for Rel-Set, which means that the search engine may supply more intelligence for the first position than general rank bias. The reason may be that search engineers pay more attention to first rank position, and employ more rules or strategies.

According to the above analysis of five context features, we find out that processes of user clicks are influenced by result lists returned, and user decisions are applicable for interactions between users and search engines. By studying the context features in user click logs, we can look into user decision process and mine performing information to estimate click reliabilities.

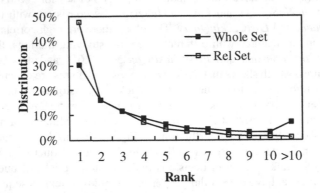

Fig. 6. The distribution of clicks at different rank positions

4.3 Experiments on Click Reliability Estimation

In this session, we estimated user clicks with context features proposed above, and selected high reliable clicks. According to the formula (3) in Session 3.2, clicks were estimated using one feature. Naïve Bayes theorem allowed us to use multiple features to estimate click reliability and Naïve Bayes method is applied to export the possibility value of a click with context features being a reliable one.

For evaluation, the Whole-Set and Rel-Set introduced in Session 4.1 were used to train and test, 2/3 data for training and 1/3 for test. After Bayesian learning, each click was assigned a score. We choose ROC (Receiver Operating Characteristic) curves and corresponding AUC (Area under the ROC Curve) values to evaluate the performance of our method. ROC is a useful technique for organizing classifiers and visualizing their performance and it is also adopted by quality estimation [16].

After learning and testing, ROC curves of our algorithm are shown in Fig. 7. From this figure, we see that our method is able to select reliable clicks probabilistically, which is better than random selecting method. The AUC value for the algorithm's ROC curve is 0.792, which means our estimation algorithm has 79.2% chances to rank a reliable click higher than a non-reliable click, while the AUC value for the random curve is 0.5.

The ROC curve shows that the high reliable clicks are able to be selected using our algorithm probabilistically. Table 4 lists that when we filter out 80% low reliable clicks, the algorithm can maintain 60% relevant click and we filter out 40% low reliable clicks, the algorithm can maintain 92.8% relevant click, which shows the effectiveness performance of our algorithm.

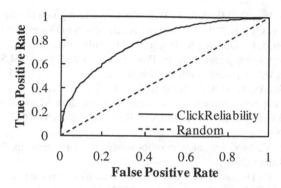

Fig. 7. ROC curves to evaluate the performance of click selecting method, compared with random method

Table 4. Different cleansed data size and corresponding relevant click recalls (the proportion of retained relevant clicks)

Cleansed data size	20.0%	40.0%	60.0%	80.0%
Relevant click recall	60.0%	81.4%	92.8%	98.4%

5 Conclusion and Future Work

In this paper, we analyze the contexts of user click behavior in interaction processes between users and search engine systems, and look into user decisions. The definition of click reliability is defined in a probabilistic notion. Five user context features are proposed and analyzed. The main conclusions are listed as follows:

[1] There are uncertainties in user query and click process, and clicks with more certainties are more reliable;

[2] The first and last clicks in click sequences have higher click reliability than the others;

[3] User decision process and search results influence user click behaviors and context features are effective for finding reliable clicks.

In the future, more work will be done on the application of reliability click estimation, such as improving Web search ranking, evaluating search engine performance, detecting click spam, finding bad search cases, etc.

References

1. Yates, R., Tiberi, A.: Extracting semantic relations from query logs. In: Proceedings of the 13th ACM SIGKDD international Conference on Knowledge Discovery and Data Mining, pp. 76–85. ACM, New York (2007)
2. Fuxman, A., Tsaparas, P., Achan, K., Agrawal, R.: Using the wisdom of the crowds for keyword generation. In: Proceeding of the 17th international Conference on World Wide Web, pp. 61–70. ACM, New York (2008)

3. Joachims, T., Freitag, D., Mitchell, T.: WebWatcher: a tour guide for the world wide Web. In: IJCAI 1997, vol. 1, pp. 770–777. Morgan Kaufmann, San Francisco (1997)
4. Agichtein, E., Brill, E., Dumais, S., Ragno, R.: Learning user interaction models for predicting web search result preferences. In: Proceedings of the 29th ACM SIGIR Conference on Research and Development in information Retrieval, pp. 3–10. ACM, New York (2006)
5. Joachims, T., Granka, L., Pan, B., Hembrooke, H., Gay, G.: Accurately interpreting click-through data as implicit feedback. In: Proceedings of the 28th ACM SIGIR Conference on Research and Development in information Retrieval, pp. 154–161. ACM, New York (2005)
6. Tan, P., Kumar, V.: Modeling of web robot navigational patterns. In: Proceedings ACM WebKDD Workshop (2000)
7. Tan, P., Kumar, V.: Discovery of web robot sessions based on their navigational patterns. Data Mining and Knowledge Discovery 6, 9–35 (2002)
8. Yates, R., Hurtado, C., Mendoza, M., Dupret, G.: Modeling user search behavior. In: Proceedings of the 3th Latin American Web Congress. LA-WEB, p. 242. IEEE Computer Society, Los Alamitos (2005)
9. Kammenhuber, N., Luxenburger, J., Feldmann, A., Weikum, G.: Web search clickstreams. In: Proceedings of the 6th ACM SIGCOMM Conference on internet Measurement, pp. 245–250. ACM, New York (2006)
10. Craswell, N., Zoeter, O., Taylor, M., Ramsey, B.: An experimental comparison of click position-bias models. In: Proceedings of the international Conference on Web Search and Web Data Mining, pp. 87–94. ACM, New York (2008)
11. Guo, F., Liu, C., Wang, Y.M.: Efficient multiple-click models in web search. In: Proceedings of the 2nd ACM international Conference on Web Search and Data Mining, pp. 124–131. ACM, New York (2009)
12. Liu, Y., Cen, R., Zhang, M., Ru, L., Ma, S.: Automatic Search Engine Evaluation Based On User Behavior Analysis. Journal of Software 19(11), 3023–3032 (2008)
13. Agrawal, R., Halverson, A., Kenthapadi, K., Mishra, N., Tsaparas, P.: Generating labels from clicks. In: Baeza-Yates, R., Boldi, P., Ribeiro-Neto, B., Cambazoglu, B.B. (eds.) Proceedings of the 2nd ACM international Conference on Web Search and Data Mining (2009)
14. Yu, H., Liu, Y., Zhang, M., Ru, L., Ma, S.: Research in Search Engine User Behavior Based on Log Analysis. Journal of Chinese Information Processing 21(1), 109–114 (2007)
15. Shannon, C.E.: A Mathematical Theory of Communication. Bell System Technical Journal 27, 379–423, 623–656 (1948)
16. Svore, K., Wu, Q., Burges, C., Raman, A.: Improving Web Spam Classification using Rank-time Features. In: Proceedings of AIRWeb 2007, pp. 9–16. ACM, New York (2007)
17. Sadagopan, N., Li, J.: Characterizing typical and atypical user sessions in clickstreams. In: Proceedings of the 17th international Conference on World Wide Web, pp. 885–894. ACM, New York (2008)
18. Agichtein, E., Brill, E., Dumais, S.: Improving web search ranking by incorporating user behavior information. In: Proceedings of the 29th Annual international ACM SIGIR Conference on Research and Development in information Retrieval, pp. 19–26. ACM, New York (2006)
19. Dou, Z., Song, R., Yuan, X., Wen, J.: Are click-through data adequate for learning web search rankings? In: Proceeding of the 17th ACM Conference on information and Knowledge Management, pp. 73–78. ACM, New York (2008)
20. Carletta, J.: Assessing Agreement on Classification Tasks: The Kappa Statistic. Computational Linguistics 22(2), 249–254 (1996)

Utilizing Social Relationships for Blog Popularity Mining

Chih-Lu Lin, Hao-Lun Tang, and Hung-Yu Kao

Department of Computer Science and Information Engineering,
National Cheng Kung University, Tainan, Taiwan, R.O.C.
hykao@mail.ncku.edu.tw

Abstract. Due to the ease of use in blogs, this new form of web content has become a popular online media. Detecting the popularity of blogs in the massive blogosphere is a critical issue. General search engines that ignore the social interconnection between bloggers have less discrimination of blogs. This study extracts real-world blog data and analyzes the interconnection in these blog communities for blog popularity mining. The interconnections reveal the consciousness of bloggers and the popularity of blogs which may refer to blog qualities. In this paper, we propose a blog network model based on the interconnection structure between blogs and a popularity ranking method, called BRank, on the constructed model. Several experiments are conducted to analyze the various explicit and implicit interconnection structures and discover variances of the impact of interactions in different communities. Experiments on several real blog communities show that the proposed method could detect blogs with great popularity in the blogosphere.

Keywords: blog, link analysis.

1 Introduction

In recent years, blogging has become a trend for people to publish content on the internet. With the ease of use in blogs, users can rapidly share their daily dairies, discuss the latest news, and express their opinions. As an emerging online medium, many blog hosting sites now provide free services. Given this convenient platform, the number of blogs is increasing in a dramatically fast manner.

A blog consists of the title, subscription information, multiple posts which are displayed in a descending order by the publish date. A general blog post is combined with the post date, text, hyperlinks, images and other media. The user writing a blog is a blogger. There may be some comments or trackbacks from other bloggers below a post, and that means those users are interested in the topics of that blog. In addition, bloggers can add favorite blogs to their blogrolls which indicate the subscriptions or links listed in the front page of a blog. These are the most common interactions between bloggers. Also, hyperlinks contained within a blog post give additional information for readers who would like to read some more related news, or blog posts.

Bloggers may form virtual communities through several kinds of interactive behaviors among bloggers and their audience. In comparison with general web contents, blogs show the particular characteristics themselves. Many researchers consider blogs

G.G. Lee et al. (Eds.): AIRS 2009, LNCS 5839, pp. 409–419, 2009.
© Springer-Verlag Berlin Heidelberg 2009

to be valuable resources. As a result of the massive growth of blogs, blog readers may find out a great deal of hot blogs on the blog hosting sites, yet do not know which ones contain the most informative contents. Some Blog service providers (BSPs) list hot blogs based on the number of visitors, but this indicator is weak for determining the quality of blogs. Many web sites have developed some blog-related technologies.

Google Blog Search[1] and Technorati[2] are now available for users to search blogs by query words. BlogLook[3] provides a ranking service for blogs. It allows bloggers to enter the URL of his blog, and multiple features along with the corresponding rankings are computed. The features are obtained from the search engines. However, the detailed information such as content, comments, links, and citations within the blogs are not considered comprehensively in these services. Though the number of blog interactions is used to calculate the authority score for Technorati, the comments are ignored and the various interactions are all seemed as the same.

The PageRank [16] is a popular ranking algorithm for web pages. For a hyper-linked set of pages, PageRank assigns a score to each page representing its relative importance among pages. The linking structure in the blogosphere is similar to that in web sites but with additional characteristics. Thus, general ranking method for web pages is not appropriate for the structure of the blogosphere. Some ranking algorithms for blogs are presented in recent studies. EigenRumor [7] is based on eigenvector calculation of the adjacency matrix of links, which is similar to HITS [10] and PageRank. The algorithm focus on the behaviors of bloggers on blog posts. We find some interactions between bloggers should be taken into account. BlogRank [4] is a generalized form of PageRank which use similarity features to make the link graph denser. The study does not consider the comments. Comments are conversational information for blog post. We regard this kind of interactions as an important factor for analysis. This work presents a comprehensive study on the interactive behaviors in the blog community. The behaviors could be seemed as connections between bloggers. We think that the interconnections reveal the consciousness of bloggers, and the impact of blogs can be derived by analyzing this information.

A blog with the high popularity produce more interactive behaviors to other bloggers. For a popular blog, the contents are normally informative for some readers. The posts could be insightful discussions or innovative opinions depending on subjective judgments. However, analyzing the contents is a time-consuming work. We therefore aim to use other obvious factors which reveal the consciousness of readers. In a popular blog, there may be more support relationships made by other different bloggers than in general ones. In addition, bloggers play different roles in the blogosphere. These criteria should be taken into account to measure the blog popularity. Users may cite a blog or a blog post in a web page including personal web sites, forums, or any other web contents. When a blog is frequently referenced on the internet, we can say that the blog receives a high reputation. According to these observations, it is claimed that the popularity of blogs could be measured by their impact on induced social interconnections.

---

[1] http://blogsearch.google.com.tw
[2] http://technorati.com
[3] http://look.urs.tw

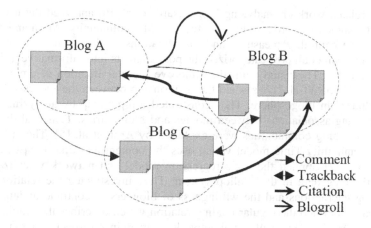

Fig. 1. The linking structure in blogosphere

PageRank is used to analyze the page reputation or popularity in the linking graph. With the linking structure between web pages, the ranking score is generated in a random walking model for PageRank. The probabilities for a random surfer moving from a web page A to another web page B is a critical concept in the algorithm. We use blog interactive behaviors to construct a blog network model which describes the blog-specialized linking structure (as illustrated in Fig. 1). For ranking bloggers in a blogosphere, a weighted and interaction-aware PageRank algorithm called "BRank" is proposed in this paper. In addition to the explicit relationships for blog interactions, common links in the posts which are regarded as the similarity relationship between blogs is also employed in the blog network model.

The rest of this paper is organized as follows. We introduced some related work in Section 2. The detail of the method is described in Section3. In Section 4, we presented some experimental results and discussions and this work is concluded in Section 5.

2 Related work

PageRank is a link analysis algorithm that calculates the importance score for each element of a group of hyperlinked documents. We think the linking structure in the blogosphere is different with that in web sites. Thus, our BRank would be designed based on the characteristics of blogs.

Kleinberg [10] proposed the Hyperlink-Induced Topic Search (HITS) algorithm. This is another prominent algorithm for ranking web pages. The critical concepts are the hubs and authorities. The hub score presents the quality of links to other pages about that topic, while the authority score indicates the quality of the page content. Besides web pages, Yupeng et al. [19] presented an expertise propagation algorithm to find the potential experts for a specific topic by using the co-occurrences of people from web pages and communication patterns from emails.

In the related work on analyzing blogs, Nardi et al. [6] analyzed the text of blog posts and comments in order to identify feelings of a community. Alvin and Mark [3] employed the centrality measures and visualizations to detect the communities in blogs. Xiaochuan et al. [18] categorized the post contents into informative articles and affective articles. Their study aimed to improve the classification of emotions from blog posts. Tirapat et al. [17] studied the relationship between the success of a movie and the "buzz" in blogosphere. They constituted a topic map by analyzing the blog posts capturing associations between movies and blog entries. Temporal discussions and stories in blog communities were exploited by Arun et al. [5]. They proposed a Content-Community-Time model which uses the content and timestamps of entries. Lento et al. [12] applied the logic regression model and network visualizations to analyze the blog data in the Wallop system. They investigated the relationship between blog interactions and the willingness of bloggers to continue updating blogs. Furukawa [8] defined the regular reading relation which describes the reading behavior for bloggers. Four kinds of social network are examined respectively to predict the regular reading relation and analyze the information diffusion. Ali-Hason and Adamic [2] studied the bloggers' online and real life relationships in three blog communities. They analyzed the different kinds of links including blogrolls, citations, and comments and discovered that few blog interactions reflect the real life relationships.

Gill et al. [9] discussed the effect of blogs on mass media and politics. It also compares some websites such which provide blog ranking services. Nakajima et al. [14] aimed to capture critical conversation topics from blogs. They proposed a blog thread model and define the roles of bloggers as agitators and summarizers. The agitators are considered to have a great impact on the discussions in the blogosphere. The timestamps of posts are examined to discriminate these two kinds of bloggers.

Some algorithms were proposed in the following studies. Adar et al. [1] introduced the implicit links representing similarity between blog posts. With the implicit links, a blog ranking algorithm called iRank was presented. Apostolos et al. [4] add implicit links to increase the density of the blogs' graph based on the similarity in topics and users. They modified the PageRank into an algorithm for ranking blogs called BlogRank. Comments are not considered in their work. As the most frequent interactions, comments show their importance in our work. The B2Rank [13] presented by Mohammad is also modified from PageRank. Nitin et al. [15] investigated the behaviors of influential bloggers and presented a preliminary model to quantify them. Several experiments were conducted to identify influential bloggers from many distinct points. The work focuses on the community blogs, and our study targets the individual blogs. Fujimura et al [7] proposed the EigenRumor algorithm for ranking blogs. The algorithm is based on HITS and can be applied on communities with observable membership identities. They define the information providers as agents. The agent property consists of the authority score and the hub score.

Some Web sites also provide the blog ranking service. Technorati is real-time blog search engine which watches over 100 million blogs. It allows users to know what is being discussed now in the blogosphere. Its authority ranking uses the blog authority which counts the number of blogs linking to a specific blog. BlogLook is a prominent blog ranking service in Taiwan. Its data contains more than 60,000 blogs. Instead of analyzing the real blog contents, blog information is retrieved from several blog search engines including Technorati authority, and the number of links and subscriptions. These services provide a general view of the blog reputation. Our method gives insight into various interactions inside a blog community. In addition, the number of links obtained from search engines will be adopted in this work for global blog ranking.

3 Method

Our method is based on the characteristics of blogs, which includes the linking structure and blog similarity features. We first introduce the idea on measuring the blog popularity and present how the proposed blog network model is constructed. The proposed ranking algorithm will be detailed in the latter subsection.

3.1 Blog Social Relationships and Network

The linking structure of a blogosphere is different from that of general web pages. When a blogger performs blog interactions like comments, trackbacks, the implicit link information is generated in the blog pages. The blog-specific linking structure is then constructed in this way. For example, when a blogger A leaves comments to a blog post in blog B, a link to blog A is presented in the pages of blog B. For PageRank, blog A receive a vote from blog B. However, the original linking structure is not reasonable to represent the relationship. Indeed, the sender of the comment, blog A, makes a vote to B in this case. This describes how the linking structure of blogosphere differs from that of general web pages. Blog readers can determine topics of blogs while browsing the blog posts. If readers are interested in a certain topic, they are likely to read blogs containing similar contents. We hence assume that if Blog A and Blog B are similar, there will be some probability for readers of A to read B. Links mentioned frequently may refer to some hot topics. In our work, the co-occurrence of hyperlinks between blogs is regarded as a similarity relationship in our model.

Two genres of blog social relationships between blogs are then defined in the proposed blog network model. Interactions or links from Blog A to Blog B indicate that A is a reader of B, and we take these relationships as *support relationships*. Four kinds of interactive behaviors, i.e., comments, trackback, blogrolls and citations, are all regarded as support relationships in our model. Common links in Blog A and Blog B make a *similarity relationship* between A and B. If Blog A and Blog B are similar, there will be some probability for readers of A to read B.

Other than the blog relationships, several characteristics, e.g., the number of posts, comments and trackbacks and the average length of posts, for blogs should be taken into account. For each blog in the data set, the features are combined as a *blog quality score*. By using the score in the BRank algorithm, the impact score granted by a blog will be revised.

3.2 BRank

The proposed social-based link analysis algorithm, BRank, computes the popularity scores in the blog network. BRank is a modified PageRank algorithm. Based on the original PageRank, we adjust the probability of a blog reader to follow a link in blog

A to another blog B. The probability is given by a new formula as shown in Table 1. The probabilities for the bloggers from blog A to blog B ($P_{A \rightarrow B}$) in PageRank is decided by the out degree of blog A. For BRank, it is determined by the relationship scores ($R_{A \rightarrow B}$). In formula 2, set S represents the set of blogs linking to blog A.

Table 1. The formula calculating the surfing probability

Algorithm	Formula	
PageRank	$$P_{A \rightarrow B} = \frac{1}{Outdegree\ of\ blogA}$$	(1)
BRank	$$P_{A \rightarrow B} = \frac{R_{A \rightarrow B}}{\sum_{X \in S(A)} (R_{A \rightarrow X})}$$	(2)

The relationship score for the bloggers from Blog A to Blog B is decided by three factors. The first is the type of blog relationships. Different blog relationships are given different weights (W_{Rtype}) since they have distinct meanings for a blogger. This value will be set experimentally. The second factor is the number of the corresponding relationship. Here we simply aim to use the degree of the number (RN_{Rtype}) to express the strength of the relationship. Instead of the actual numbers, we adopt the natural log values. The last one is the blog quality score (BQ_k) which is combination of the normalized blog features. The blog quality score show the general activity of a blog. We assumed that the probability of moving to a blog with a higher quality score is more than others. This quality score is also converted to the natural log value for calculation. The relationship score combines all kinds of relationship between two blogs, and it can be calculate as follows: The list of relationship types between bloggers A and K:

$$R_{A \rightarrow K} = \sum_{Rtype} W_{Rtype} * RN_{Rtype} * BQ_k \tag{3}$$

We compute a relationship score for each edge in the blog network. A blog edge is a composition of the blog relationships. To recall, the blog relationships includes comments, trackbacks, blogrolls, and common links. Among them, common links create a virtual relationship, and the comment cause a support relationship which is opposite in direction with the actual link in the blog page. Each directional relationship has a corresponding weight, and the adjustment of weights will be discussed in the experiments. According to the above modifications, the formula of BRank can be defined as:

$$BRank(A) = \frac{1-d}{n} + d * \sum_{X \in S(A)} BRank(X) * P_{X \rightarrow A}, \tag{4}$$

where d is the damping factor defined as the original PageRank.

4 Experiments

In our experiments, we focus on 5 well-known BSPs in Taiwan: Wretch Blogs[4], Yahoo Blogs[5], Yam Blogs[6], Xuite Blogs[7] and Pixnet Blogs[8]. The crawling process was started from the top blogs since September, 2007 to May, 2008. These crawled blogs are chosen from the top-bloggers list in several authoritative blog sharing and ranking sites. Thus, we can guarantee that our dataset contains the popular blogs and blogs which interact with them. After the blog pages have been retrieved, we then extract blog posts, comments, trackbacks, citations, and blogrolls in pages. The extraction program is customized for each type of blog pages since they are laid out in a variety of HTML templates. The statistics of our date set is presented in Table 2.

Table 2. Information of Blogs and important interactions

	#Blog	#Post	Comment	Trackback	Citation	Blogroll
Wretch	592,123	6,880,087	16,527,101	316,263	154,190	236,168
Yahoo	294,352	727,335	1,589,940	137,232	253,928	110,837
Yam	84,536	1,895,319	2,318,052	104,594	65,125	15,583
Xuite	27,320	1,270,830	822,398	21,053	325,854	12,791
Pixnet	41,507	2,511,188	4,356,075	14,336	57,504	27,602

In the conducted experiments, we retrieve the information from BlogLook for the purpose of experimental evaluation. The ranking data will be represented as $Rank_{BL}$ in the following sections. Besides the existed ranking service, we prepare three sets of manually ranking data. The first set consists of 400 blogs selected randomly from BlogLook for four BSPs including Wretch, Yahoo, Yam, and Pixnet. Several manual features are derived by four researchers and the summation scores will be our human ranking result for the 400 blogs. The manual features are: Detailed Discussion, Abundant Information, Clear and Suitable Self-defined Categories, Update Frequency, The quality of subscriptions, Suitable Advertisements and Expert Popularity Score. These features are established based on our criteria for determining the quality and popularity of blogs. All the manually rankings are named as $Rank_H$ for convenience.

To quantify the comparisons between two kinds of rankings, we use the correlation coefficient and kappa coefficient in our experiments. The correlation coefficient of 1 indicates a perfect positive linear relationship. On the contrary, −1 refers to a perfect negative linear relationship. A value 0 represents that there is no correlation. We denote this coefficient as CC in the following sections.

Cohen's kappa coefficient is a statistical measure for comparing the degree of consensus between raters. Cohen's kappa measures the agreement between two raters who each classify N items into C mutually exclusive categories. To use kappa coefficient

---

[4] http://www.wretch.cc/blog/
[5] http://tw.blog.yahoo.com/
[6] http://blog.yam.com/
[7] http://blog.xuite.net/
[8] http://www.pixnet.net/blg/

for our ranking results (N=100), we simply divide the blogs into two disjoint categories (C=2). Thus, the kappa coefficient can be employed on the two categories between different ranking results. We adopt three ways of division, i.e., two-fold, third-fold (1:2) and third-fold (2:1) in our experiments and denoted as K1, K2 and K3 respectively.

4.1 Effects of Blog Relationships

In order to test the effects of different blog relationships, several ranking processes with various settings of weights in the algorithm are done. In Fig. 2, we show the results for different weighting schemes. The comparison with the original PageRank is shown in Fig. 3 and Fig. 4. For the average kappa value, our algorithm outperforms the PageRank in Wretch blogs. Although the kappa values are similar for Yam and Pixnet, the correlation coefficients show the difference. Kappa value including K1, K2, and K3 show a general view for detecting a group of better blogs.

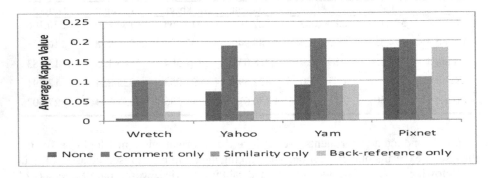

Fig. 2. The average kappa value for 4 series of weights

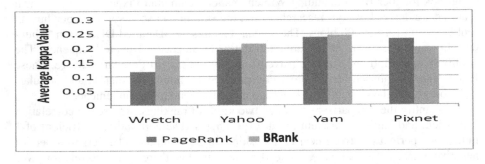

Fig. 3. Kappa value comparison for PageRank and BRank

4.1 The Analysis of Ranking Results

In the following experiments, we will compare results of BRank ($Rank_L$) and ranks in BlogLook ($Rank_{BL}$) for each BSP. $Rank_H$ is considered as the ground truth in our evaluation. Table 3 represents the agreement analysis for Wretch, Yahoo, Yam and Xuite. According to the AvgK values, the proposed BRank outperforms BlogLook in these four BSPs.

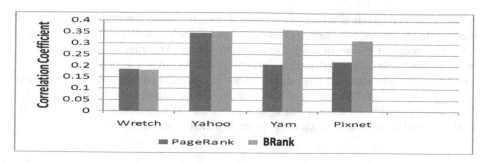

Fig. 4. Correlation coefficient comparison for PageRank and BRank

BRank attains the lower agreements and weaker correlation in Wretch. It is caused by the large amount of comment relationships. In Wretch, some diaries may be commented by a great number of friends and the comments are short and non-informative in general. This kind of noise comments will increase the ranking score for blogs in a small but highly connective community. Thus, our results may be biased.

Table 3. Agreement Evaluations

Wretch	CC	K1	K2	K3	Avg K
Rank$_H$ & Rank$_{BL}$	0.342	0.187	0.005	0.194	0.129
Rank$_H$ & **Rank$_L$**	0.181	0.103	**0.229**	0.192	0.174
Rank$_L$ & Rank$_{BL}$	0.071	0.077	-0.038	0.000	0.013
Yahoo					
Rank$_H$ & Rank$_{BL}$	0.103	-0.012	-0.063	-0.047	-0.040
Rank$_H$ & **Rank$_L$**	**0.350**	**0.177**	**0.308**	**0.163**	**0.216**
Rank$_L$ & Rank$_{BL}$	0.196	0.106	0.095	0.000	0.067
Yam					
Rank$_H$ & Rank$_{BL}$	0.124	0.077	-0.029	-0.007	0.014
Rank$_H$ & **Rank$_L$**	**0.359**	**0.297**	**0.322**	0.115	**0.245**
Rank$_L$ & Rank$_{BL}$	0.145	0.243	0.322	0.000	0.188
Xuite					
Rank$_H$ & Rank$_{BL}$	0.373	0.029	0.009	0.436	0.158
Rank$_H$ & **Rank$_L$**	0.297	**0.272**	**0.182**	0.318	**0.257**
Rank$_L$ & Rank$_{BL}$	0.705	0.393	0.455	0.000	0.283

5 Conclusion

The proposed blog popularity ranking algorithm BRank detects the popularity of blogs in a BSP community by the social relationships. Using the proposed blog network model, we comprehensively examine the interactive behaviors in some BSPs.

As the content analysis is a time-consuming work, the co-occurrence of hyperlinks is used to determine the similarity relationships between blogs. BRank is applied on a BSP community by investigating the interconnection. The ranking process can be accomplished efficiently and provide effective results. In the experiments, we discover that the importance of interactions varies in different communities. Information of blogs is spread in different ways based on the characteristics of a blog community. The comparison results show a fair agreement with the judgment of human and our method could detect blogs which are popular in a BSP community.

Reference

1. Adar, E., Zhang, L., Adamic, L.A., Lukose, R.M.: Implicit structure and the dynamics of blogspace. In: Workshop on the Weblogging Ecosystem, 13th International World Wide Web Conference (2004)
2. Ali-Hasan, N.F., Adamic, L.A.: Expressing Social Relationships on the Blog through links and comments. In: The International Conference for Weblogs and Social Media 2007, Boulder, CO (2007)
3. Alvin, C., Mark, C.: A social hypertext model for finding community in blogs. In: Proceedings of the seventeenth conference on Hypertext and hypermedia, Odense, Denmark. ACM, New York (2006)
4. Apostolos, K., Martha, S., Iraklis, V.: BlogRank: ranking weblogs based on connectivity and similarity features. In: Proceedings of the 2nd international workshop on Advanced architectures and algorithms for internet delivery and applications, ppisa, Italy. ACM, New York (2006)
5. Arun, Q., Belle, T., Edward, Y.C.: Mining blog stories using community-based and temporal clustering. In: Proceedings of the 15th ACM international conference on Information and knowledge management, Arlington, Virginia, USA. ACM, New York (2006)
6. Bonnie, A.N., Diane, J.S., Michelle, G., Luke, S.: Why we blog. Commun. ACM 47(12), 41–46 (2004)
7. Fujimura, K., Inoue, T., Sugisaki, M.: The EigenRumor Algorithm for Ranking Blogs. In: WWW 2005 Workshop on the Weblogging Ecosystem (2005)
8. Furukawa, T., Matsuo, Y., Ohmukai, I., Uchiyama, K., Ishizuka, M.: Social Networks and Reading Behavior in Blogosphere. In: The International Conference for Weblogs and Social Media (2007)
9. Gill, K.: How can we measure the influence of the blogosphere? In: The Workshop on the Weblogging Ecosystem at the 13th International World Wide Web Conference, New York (2004)
10. Jon, M.K.: Authoritative sources in a hyperlinked environment. J. ACM 46(5), 604–632 (1999)
11. Kumar, R., Novak, J., Raghavan, P., Tomkins, A.: On the Bursty Evolution of Blogspace. World Wide Web 8(2), 159–178 (2005)
12. Lento, T., Welser, H., Gu, L., Smith, M.: The Ties that Blog: examining the Relationship Between Social Ties and Continued Participation in the Wallop Weblogging System. In: 3rd annual workshop on the Weblogging Ecosystem at the World Wide Web Conference, Edimburgh (2006)
13. Mohammad, A.T., Hashemi, S.M., Ali, M.: B2Rank: An Algorithm for Ranking Blogs Based on Behavioral Features. In: Proceedings of the IEEE/WIC/ACM International Conference on Web Intelligence. IEEE Computer Society, Los Alamitos (2007)

14. Nakajima, S., Tatemura, J., Hara, Y., Tanaka, K., Uemura, S.: Identifying agitators as important blogger based on analyzing blog threads. In: Zhou, X., Li, J., Shen, H.T., Kitsuregawa, M., Zhang, Y. (eds.) APWeb 2006. LNCS, vol. 3841, pp. 285–296. Springer, Heidelberg (2006)
15. Nitin, A., Huan, L., Lei, T., Philip, S.Y.: Identifying the influential bloggers in a community. In: Proceedings of the international conference on Web search and web data mining, ppalo Alto, California, USA. ACM, New York (2008)
16. Page, L., Brin, S., Motwani, R., Winograd, T.: The pagerank citation ranking: Bringing order to the web. 1998, Technical report, Stanford Digital Library Technologies Project (1998)
17. Tapanee, T., Cleo, E., Eleni, S.: Taking the community's pulse: one blog at a time. In: Proceedings of the 6th international conference on Web engineering, ppalo Alto, California, USA. ACM, New York (2006)
18. Xiaochuan, N., Gui-Rong, X., Xiao, L., Yong, Y., Qiang, Y.: Exploring in the weblog space by detecting informative and affective articles. In: Proceedings of the 16th international conference on World Wide Web, Banff, Alberta, Canada. ACM, New York (2007)
19. Yupeng, F., Rongjing, X., Yiqun, L., Min, Z., Shaoping, M.: Finding Experts Using Social Network Analysis. In: Proceedings of the IEEE/WIC/ACM International Conference on Web Intelligence. IEEE Computer Society, Los Alamitos (2007)

S-node: A Small-World Navigation System for Exploratory Search

Satoshi Shimada[1], Tomohiro Fukuhara[2], and Tetsuji Satoh[1]

[1] Graduate School of Library, Information and Media Studies, University of Tsukuba,
Kasuga 1-2, Tsukuba, Ibaraki, 305-8550, Japan
[2] Research into Artifacts, Center for Engineering, University of Tokyo,
Kashiwanoha 5-1-5, Kashiwa, Chiba, 277-8568, Japan

Abstract. In the retrieval of newspapers or weblogs in which particular terms and expressions are used frequently, it is not easy to remind the user of appropriate query terms. For this case, it is necessary to present typical feature terms or documents in the document set without depending on the user's input. In this paper, we propose the navigation system 'S-node' for documents. The system extracts two kinds of words that show exhaustivity or specificity from documents written in Japanese based on repetition index, and constructs hyperlinks between documents that can reach as short as possible to various documents based on co-occurrence of terms. We describe the implementation of the system, and the results of evaluation.

1 Introduction

In information retrieval, it is not easy for users to recollect appropriate query terms to obtain retrieval results that users expected. Though a lot of information recommendation system have been proposed, the recommendation for users who want to find out novel information or unexpected relativity between documents is quite difficult. Additionally in recent years, there are users who retrieve documents on the Web without having a clear search request. It might be not irrelevant to daily use of the search engine and the expansion of use from a cell phones that has only poor input method.

In a community that composed by users who reside in certain regions or who interests to something special, e.g. weblogs, BBS, and mailing lists, the word that runs only among companions tends to be multiused, and the content is not understood easily for an outside user.

To deal with those problem, a new approach like exploratory search is needed 1. Navigation is one of the methods of bringing users a new retrieval experience without demanding a specified input.

We propose a web navigation system that aims to access various documents by the small distance. In the previous research, the existence of the network that can be able to reach various nodes in the small distance, called small-world network, is known 2. The small-world character is defined as the case that average path length(L) is almost equal, and that average clustering coefficient(C) is very

G.G. Lee et al. (Eds.): AIRS 2009, LNCS 5839, pp. 420–431, 2009.

large, comparing with random graph that has equal number of nodes and links. In recent study, it is clarified that the co-occurrence word graph in the document collection shows a small-world character 3.

The proposal system applies this character. To make the user reachable to various documents in just a small distance, our method generates small-world network that consists of hyperlinks between each document and related documents, each keyword and related keywords, and each document and each keyword. And we show the effectiveness of the method in exploratory search by the experiment with subjects.

In this paperCwe describe our aproach in section 2, experiment and evaluation in section 3, and conclude.

2 Approach to Web Navigation

There is a problem that is difficult to specify an appropriate retrieval word by users themselves. To deal with this problem, a number of query expantion systems was proposed. However, it is still difficult for users who didn't choose the apt word to select the apt word from among words that the system presents. In addition, if user's query is vaguer, the recommendation becomes harder. On the other hand, being classified as expected by user is rare in topic clustering systems because overlaps of each topic expressed in documents are large, and it extends over two or more classifications usually. So, we propose a navigation system to present relation of those overlapped topics as a small-world network based on co-occurence words graph of documents.

The proposed system has following advantages:

1. Because only the amount of characteristic observed within document set is used, no dictionaries are needed, and it is easy to apply to any document sets. (The document need not have the hyperlink.)
2. By enabling to transit between related documents tied by co-occurence words locally, it becomes easy to reach the document that users wouldn't expect. (The difference in usage frequency between documents can be levelled.)
3. The system presents only some hyperlinks for the item that user focused. The transition of users will converge with constant routes, and an excessive personalizing is avoided. (The difference of experience through the retrieval between users that be derived from difference of retrieval skill can be levelled.)

The system is intelligent in that it can automatically mine the network of co-occurring words and construct the navigation network based on only features that the document set has. Then we describe about each processing on our system along the flow shown in Fig.1.

2.1 Extraction and Classification of Keywords

In the proposed system, character strings that might be keyword are divided in every turn of the kind of characters (e.g. Kanji, Hiragana, and Katakana), and they are filtered with their repetition level and document frequency.

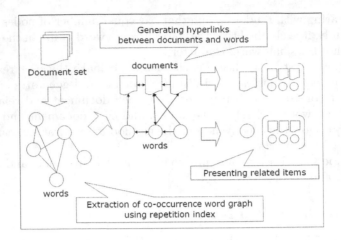

Fig. 1. The flow of proposed system

Table 1. Threshold for keyword division and score for related level calculation

Division	Adaptation	Document Frequency	Score
I Specific Words	≥ 0.6	> 2	10
II Comprehensive Words	$\geq 0.35, < 0.6$	$> 9C< 19$	1
III Border Words	≥ 0.1	> 3	0.1
IV Other Words	(Other)	(Other)	0.01

The word that has $df < 3$ is excluded in all division.

Index terms are extracted by morphological analysis generally, but it is needed to deal with unknown words or complex words. We gave priority to being able to extract keywords stably rather than precisely because we are trying to apply the proposal system to documents like weblogs in which unknown words and complex words occur frequently.

In Japanese text, the words expressing concept, operation, or relation are often written in Kanji, Katakana, and alphanumeric characters. This system extracts the character string that consists of Kanji and Katakana characters or of alphanumeric characters and assumes them as keyword. But, the string that can be interpreted as a URL, or that consists only of numeric characters is excluded. The string that consists of only one Kanji or Katakana character or only two or less alphanumeric characters are excluded, too.

Although the pointed measures (like *tfidf*) that sift out a few 'top' terms has been used to extract index terms in traditional information retrieval, in the web navigation that intends to exploratory comprehension, comprehensive words (the words that leads exhaustive search results) have to be able to extract independently from specific words (the words that identifies topic or concept). For that reason, it is appropriate in weighting terms to use the index that shows the amount of characteristic of each word as directly as possible.

In this system, we use document frequency (df) and repetition index (adaptation) of terms often used in extraction of technical terms 4 or automatic summarization 5 that is not intended to narrow the retrieval result. Especially, it is known that the repetition index has strong relation with whether it is content word 6. In this paper, we use the repetition index that is shown as the formula (1). $df(t)$ means the number of documents where word t appears once or more. $df_2(t)$ means the number of documents where word t appears twice or more.

$$Adaptation(t) = \frac{df_2(t)}{df(t)} \tag{1}$$

The proposal system classifies the extracted keywords into four divisions by the range of each value indicated in Table 1. 'Specific words' is useful to tie local related documents. 'Comprehensive words' is useful to bridge between topics that have unexpected relation. 'Border words' is vaguer than previous two, but it have the level of repetetion index that shouldn't be disregarded. 'Other words' is all of the rest. Though it is thought that these roles are originally distributed continuous and have overlap, but to simplify implementation, the overlap of division is not considered in this paper.

Then, a phased score is given to each word. The experimental system uses the threshold and the score that is shown in the table.

2.2 Generation of Hyperlinks

The proposal navigation system generates hyperlinks between each keyword, each document, and between keywords and documents by using the co-occurrence and the score given to each word.

First, the hyperlink between documents is generated by following methods. All documents including any words that co-occurs between the base document that becomes a starting point of navigation are assumed as the candidate of related document. The rank of each related document is decided according to sum of scores co-occurred words have. The relation level of each document is calculated by the formula (2).

$$r(d_i, d_j) = \sum_{k=1}^{n} w_k \tag{2}$$

d_i is a base document. d_j is a related document. $r(d_i, d_j)$ is the level of relation between d_i and d_j. $T_{ij} = \{t_1, ..., t_n\}$ is co-occurred words between d_i and d_j. w_k is the score of t_k. Related level r is calculated about all the related documents in which one or more words co-occurred with the base document d_i. The system generates constant number of hyperlinks from d_i to d_j that r is high. In the experimental system, 8 or less hyperlinks are generated within 20% of all related documents per base document. This number was decided in consideration of the visibility for users.

Next, the hyperlink between keywords is generated by following methods. The relation level of each word is calculated by the formula (3).

$$r(t_i, t_k) = mw_k \tag{3}$$

t_i is a base word. t_k is a related word. $r(t_i, t_k)$ is the level of relation between t_i and t_k. D_i is documents in which t_i appears. $T_i = \{t_1, ..., t_n\}$ is words occurred in D_i. $D_{ik} = \{d_1, ..., d_m\}$ is documents in which t_i and t_k co-occurred. w_k is the score of t_k. Related level r is calculated about all the related words that appears in one or more documents in D_i. The system generates constant number of hyperlinks from t_i to t_k that r is high. In the experimental system, 8 or less hyperlinks are generated within 20% of all related words per base word.

Finally, the hyperlink between documents and keywords is generated by following methods. In generating the hyperlink from document to keyword, all words that occurs in the base document are assumed as the candidate of keyword for transit to another document. The importance level of each keyword is decided according to sum of the score of the word and its document frequency.

d_i is a base document. $T_i = \{t_1, ..., t_n\}$ is words occurred in d_i. w_k is the score of t_k in T_i. $df(t_k)$ is document frequency of t_k. t_k is a related word. $r(t_i, t_k)$ is the level of relation between t_i and t_k. D_i is documents in which t_i appears. The keyword is acquired from T_i in order of division I, II, III, and IV. The keywords of division I are sorted in the descending order of df, and the keywords of other division are sorted in the ascending order of df. The system generates constant number of hyperlinks from t_i to t_k that is early in the order. In the experimental system, 8 or less hyperlinks are generated within 20% of all keywords per base document.

Then, in generating the hyperlink from keyword to document, the importance level of each document is calculated by the formula (4).

$$S_j = \sum_{k=1}^{n} w_k \tag{4}$$

t_i is a base word. $D_j = \{d_1, ..., d_m\}$ is documents in which t_i appears. $T_j = \{t_1, ..., t_n\}$ is words occurred in d_j that is included in D_j. S_j is the importance score of d_j. w_k is the score of t_k. Importance level S is calculated about all documents in D_j. The system generates constant number of hyperlinks from t_i to d_j that S is high. In the experimental system, 8 or less hyperlinks are generated within 20% of all documents per base word.

As a result, the network that has the nodes consists from documents and keywords and has hyperlinks between each document and each keyword is generated.

2.3   Presenting Related Items

The proposal system presents related documents and keywords according to the context of the retrieval by users.

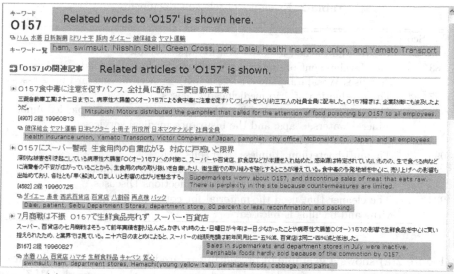

(a) Related words and articles to the keyword 'O157' are presented

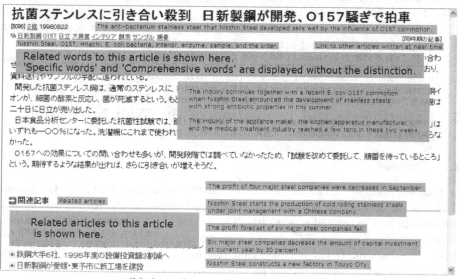

(b) An individual article is displayed

Fig. 2. Screen shots of experimental system

Related words are presented for a document or a word that user focused, and related documents are presented for a document or a word that user focused. Related words are also presented for a document that is related to a word that user focused.

Table 2. Examples of relevant documents

No.	Score	Subject
6846	4	Yogurt and pickled ume sell well in supermarkets.
5783	4	The development of material for keeping freshness of foods that uses Wasabi(Japanese horseradish) or oriental mustard progresses.
5657	4	The production of drip medicines increases.
5642	2	The anti-bacterium coated lunchbox, vegetables, yogurt, pickled ume, mineral waters, soaps, and the insurance medical services (inspection of the bacterium) sell well.
5096	2	The anti-bacterium stainless steel sells well.
5255	1	The Ministry of Health and Welfare demanded the budget that related to National Institute of Infections Diseases.
4934	1	The cleaner that used the material for controlling proliferation of virus and bacillus is released.

Figure 2(a) shows a example of presenting related items for a word that user focused. In this screen, related words and articles to the keyword 'O157' are presented. Some keywords that relate to presented articles are also presented at the same time. Users reach this screen by a click of green hyperlink that is related words from current word or article.

Figure 2(b) shows a example of presenting related items for a document that user focused. An individual article is displayed. Some keywords that relate to this article are presented under the subject. Some related articles are presented under the text. In this system, 8 keywords or articles are presented in maximums.

3   Experiments and Evaluation of the Web Navigation System

3.1   Dataset

In this experiment, 7,770 news articles on the economic pages of 'the Asahi Shimbun' in 1996 were used as a document set. The number of keywords that the system had extracted was 20,103. The system had generated 61,987 of arcs between documents, 158,251 arcs between words, and 366,126 arcs between documents and words. It is confirmed that the network that consists of these arcs is a small-world structure by our previous work 7.

3.2   Problems and Evaluation Method

It is the purpose of this study to verify effectiveness of the navigation system in exploratory search.

We got four cooperators for this experiment, had them impose a retrieval task, and do the search that used the proposal system. The task is 'discover articles about commodities or services that sales increased because of influence

Table 3. Outline of system used for user experiment

Function	Equipped by System	
	Proposal	Baseline
'Article' (screen for viewing individual article)	x	x
Related word presentation to article	x	
Embedding of link to related word in article body	x	
'Retrieve' (simple full-text search)	x	x
'Archive' (sequential view of titles of articles)	x	x
'Navi' (related items presentation to selected word)	x	
'Keywords' (abstract of keywords on all documents)	x	

Table 4. Display frequency of each function

User	Proposal System					Baseline System		
	Article	Retrieve	Archive	Navi	Keywords	Article	Retrieve	Archive
A	24	21	0	7	0	22	31	0
B	30	5	10	32	0	28	55	0
C	14	2	13	4	0	14	21	64
D	17	7	0	15	0	14	24	0
Average	21.3	8.8	5.6	14.5	0	19.5	32.8	16.0

of O157'. This means that the testees have to find out the unexpected articles while articles that sales fell occupied the great number. Additionally, it was not explained that 'O157' is a Escherichia coli.

We prepared the relevant document set for this experiment by following method. We judged relevance of 68 articles that include the word of 'O157', 'food poisoning', 'anti-bacterium', or 'kin'(one Japanese Kanji character that means bacilli). Table 2 is list of relevant documents. Higher score is given to documents that are unexpected or harder to find. Whether to find the article easily is judged from whether the keyword appear in title or first sentence.

3.3 Experimental System

The proposal system and the baseline system were used for the experiment. Table 3 shows functions that each system has.

The usage of each function is shown in table 4. On the proposal system, 'retrieve' is frequently requested by user A, and 'archive' is by user C. All users never requested 'keywords'. On the baseline system, 'retrieve' has the majority. The most of 'retrieve' is requested by user B.

3.4 Results

Table 5 is shown statistics of user's behavior. The number is the one of distinct documents. d(rel)/trial shows the efficiency of the retrieval per trial frequency.

Table 5. Statistics of user's searching behavior

User	System	#Trials	#Documents		Sum of Score	d(rel)/Trial	s(rel)/Trial
			Accessed	Relevant			
A	1	12	5	2	6	0.167	**0.500**
	2	21	7	4	9	**0.190**	0.429
B	1	24	8	3	8	**0.125**	**0.333**
	2	77	23	9	25	0.117	0.325
C	1	33	7	6	11	**0.182**	0.333
	2	16	4	2	6	0.125	**0.375**
D	1	30	8	5	10	**0.167**	**0.333**
	2	14	1	1	1	0.071	0.071
Average	1	24.8	7.0	4.0	8.8	**0.160**	**0.375**
	2	32.0	8.8	4.0	10.3	0.126	0.300

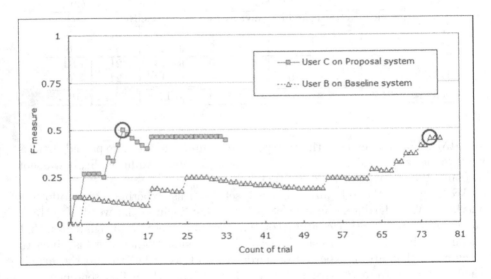

Fig. 3. Comparison of F-measure in each trial point

s(rel)/trial shows it of considering the relevance score. The efficiency in proposal system roughly exceeds in baseline system. Especially, difference of user D is large.

User A and B used in order of proposal system and baseline system, while user C and D used in order of baseline system and proposal system. It entrusted to the user's judgment how long time is spent on each task. The proposal system has the navigation, and the baseline system doesn't have it. Four testees (from A to D) use both proposal system and baseline system alternately.

Figure 3 shows F-measure in each trial point of two remarkable users. Although F-measure is used as an index that shows efficiency of search engines usually, we are using it as an index that shows the efficiency of user's search

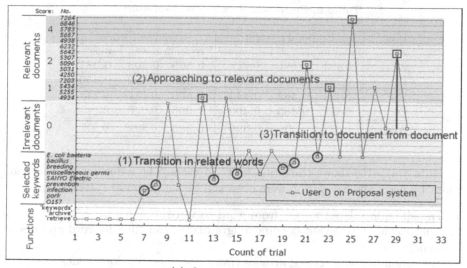

(a) On proposal system

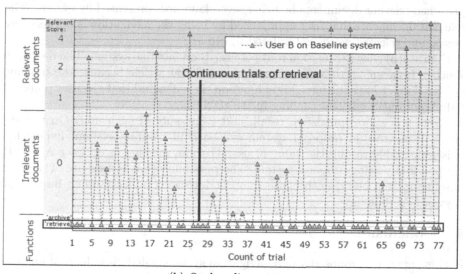

(b) On baseline system

Fig. 4. Transition route of remarkable users observed in the experiment

behavior in this paper. Invariable points mean display of inrelevant document or retrieval result, or return to article that has already been browsed. Decreasing points mean that user accessed distinct but inrelevant document. In proposal system, standing up is early. In baseline system, to exceed proposal system, time and a lot of long trials were required. User B (on baseline system) had needed 6.25 times of trials to reach same level as user C (on proposal system).

Table 6. Precisions, Recalls, and F-measures

User	Proposal System			Baseline System		
	Precision	Recall	F-measure	Precision	Recall	F-measure
A	0.400	0.154	0.222	**0.500**	**0.308**	**0.381**
B	0.273	0.231	0.250	**0.333**	**0.692**	**0.450**
C	0.429	**0.462**	**0.444**	**0.500**	0.154	0.235
D	0.417	**0.385**	**0.400**	**1.000**	0.077	0.143
Average	0.379	0.308	**0.329**	**0.583**	0.308	0.302

Figure 4(a), 4(b) shows transition route of users. The following special transitions by using the navigation were observed in the proposal system.

1. Transition in ralated words (from word to word)
2. Approaching to relevant documents gradually (shuttling in word and document)
3. Transition between related articles (from document to document)

Figure 4(a) shows transition route of user D on proposal system. The user approached relevant documents by using seven keywords presented by system. The relevant documents that user approached by using the navigation is bordered by boxes, and selected words is bordered by circles.

Figure 4(b) shows transition route of user B on baseline system. The user only shuttles between 'retrieve' function and individual article because the baseline system doesn't have the navigation function. It is demand that the user prepares a different query whenever failing in the retrieval.

Table 6 is shown precision, recall, and F-measure of each user. On the average, F-measure in proposal system exceeded in baseline system. Especially, it is remarkable raising the value of user D to an average level. User A and B are previously using proposal system. The possibility that F-measure of them in baseline system has risen because they retrieve with understanding important keywords is thought.

4   Conclusion

In this paper, we proposed a web navigation system called 'S-node' that enables users to transit related words or documents effciently. The system offers users an average search experience using small-world navigation network that consists of constant number of hyperlinks. We showed in the experiment that users viewed more documents in less time by proposal system than by baseline system.

References

White, R., Kales, B., Ducker, S., Schraefel, M.: Supporting exploratory search. Communications of the ACM 49(4), 36–39 (2006)

Watts, D., Strogatz, S.: Collective dynamics of small-world networks. Nature 393(6684), 440–442 (1998)

Matsuo, Y., Ohsawa, Y., Ishizuka, M.: Keyworld: Extracting keywords from document s small world. In: Jantke, K.P., Shinohara, A. (eds.) DS 2001. LNCS (LNAI), vol. 2226, pp. 271–281. Springer, Heidelberg (2001)

Shirai, S., Torii, O., Kanai, T.: Keyword extraction from documents based on the string repetition acyclic graph. DBSJ Letters 4(1), 77–80 (2005)

Yamada, K., Komine, H., Kinukawa, H., Nakagawa, H.: Abstract of abstract: A new summarizing method based on document frequency and clause length. In: SCI 2004. The 8th World Multi-Conference on Systemics, Cybernetics and Informatics, vol. XIV, pp. 56–61 (2004)

Takeda, Y., Umemura, K.: Selecting indexing strings using adaptation. In: SIGIR 2002: Proceedings of the 25th annual international ACM SIGIR conference on Research and development in information retrieval, pp. 427–428. ACM, New York (2002)

Shimada, S., Fukuhara, T., Satoh, T.: Evaluation of comprehensive web navigation method using social network analysis. In: WebDB Forum 2008, Tokyo, Japan, 5A–2 (2008) (in Japanese)

Efficient Probabilistic Latent Semantic Analysis through Parallelization

Raymond Wan[1,3], Vo Ngoc Anh[2], and Hiroshi Mamitsuka[1]

[1] Bioinformatics Center, Institute for Chemical Research, Kyoto University,
Gokasho, Uji, 611-0011, Japan
[2] Department of Computer Science and Software Engineering, University of Melbourne,
Victoria, 3010, Australia
[3] Computational Biology Research Center, AIST, 2-42, Aomi,
Koto-ku, Tokyo, 135-0064, Japan
r.wan@aist.go.jp, vo@csse.unimelb.edu.au,
mami@kuicr.kyoto-u.ac.jp

Abstract. Probabilistic latent semantic analysis (PLSA) is considered an effective technique for information retrieval, but has one notable drawback: its dramatic consumption of computing resources, in terms of both execution time and internal memory. This drawback limits the practical application of the technique only to document collections of modest size.

In this paper, we look into the practice of implementing PLSA with the aim of improving its efficiency *without* changing its output. Recently, Hong et al. [2008] has shown how the execution time of PLSA can be improved by employing OpenMP for shared memory parallelization. We extend their work by also studying the effects from using it in combination with the Message Passing Interface (MPI) for distributed memory parallelization. We show how a more careful implementation of PLSA reduces execution time and memory costs by applying our method on several text collections commonly used in the literature.

1 Introduction

Probabilistic latent semantic analysis (PLSA) has been shown to be a competitive technique for document retrieval compared to other methods such as the vector space model (VSM). While the VSM calculates the cosine similarity between query and document vectors, PLSA operates on the co-occurrence information between the set of words W and the set of documents D in the collection. By mapping this information on to a set of latent states Z (where $|Z|$ is user-specified and is smaller than both $|W|$ and $|D|$), relationships between the words and documents can be discovered. These latent states can be seen as clusters, or "concepts".

One of the limiting factors of PLSA is the amount of time and memory required for its execution, making it difficult for practical use. This limitation is related to the sizes of W, D, and Z. Even with the growth in main memory size, only moderately-sized document collections of 1 to 2 MB can be processed [Deerwester et al., 1990]. Other researchers have proposed vocabulary sampling as a means of processing larger collections [Kim et al., 2003].

G.G. Lee et al. (Eds.): AIRS 2009, LNCS 5839, pp. 432–443, 2009.

In this paper, we investigate alternative techniques for addressing the memory and running time constraints of PLSA *without* changing its output. We consider three different methods which we use in concert: (1) augmenting the data structures for faster access; (2) modifying how the EM algorithm for parameter estimation is implemented; and (3) parallelizing using a shared and distributed memory framework. We assume that our base line system is a naïve implementation derived from directly translating the description of PLSA.

The remainder of this paper is structured as follows. In the next section, we describe some background related to our work. In Section 3, we report on work related to the efficiency of PLSA. In Section 4, we outline our three suggested improvements. Results from experiments using both real and synthetic data are reported in Section 5 which demonstrate the level of improvement in terms of running time and memory. Section 6 summarizes our findings and provides some directions for future work.

2   Background

Probabilistic latent semantic analysis (PLSA) is an extension of latent semantic analysis (LSA) [Deerwester et al., 1990]. In the context of information retrieval (IR), the latter is also dubbed latent semantic indexing (LSI). The purpose of LSA is to associate two types of data through a set of latent (hidden) states. If the sizes of the two types of objects are m and n, respectively, the starting point of LSA is a co-occurrence table X with m rows and n columns. The cell X_{ij} (where $0 \leq i < m$ and $0 \leq j < n$) is a non-negative integer indicating the number of times the two objects occur together.

In IR, the two types of data are the words W and the documents D, such that X_{ij} represents the frequency of word i is in document j. Moreover, the matrix X is generally sparse, especially in the IR domain, since a document only contains a small subset of words. LSA then decomposes the matrix by performing a singular value decomposition (SVD) on X. If the number of latent states is $|Z|$, then the $|Z|$ largest singular values and their corresponding vectors from the decomposition are chosen to form the next reincarnation X' of X.

PLSA [Hofmann, 2001, 1999] re-interprets LSA within a probabilistic framework by defining the joint probability between a word $w \in W$ and a document $d \in D$ across Z latent states as:

$$p(w, d) = \sum_{z \in Z} p(d|z)p(w|z)p(z). \tag{1}$$

These probabilities are obtained by maximizing the log-likelihood:

$$\mathcal{L} = \sum_{w \in W} \sum_{d \in D} n(w, d) \log p(w, d), \tag{2}$$

where $n(w, d)$ is the co-occurrence matrix X. The parameters of Equation (1) are determined through applying the Expectation-Maximization (EM) algorithm [Dempster et al., 1977]. The EM algorithm is an iterative procedure for finding solutions to the maximum likelihood of models with latent variables, such as PLSA. Each

Table 1. Comparison of parallel programming using shared and distributed memory across various categories/tools

	Shared	Distributed
Machines	Single CPU	Multiple CPU
Network	N/A	Possible
Granularity	Fine (loop-level)	Coarse (function-level)
Data structures	Shared	Explicitly transmitted
Disk access	Competing	Separate
Standard	OpenMP	MPI
API	OpenMP	Open MPI

iteration consists of two steps: the Expectation step (E-step) and the Maximization step (M-step). In the E-step, the expected value of the log-likelihood is estimated based on the values of the parameters. In the M-step, these parameters are re-estimated in order to maximize $\mathcal{L}$. Thus, the E and M-steps are:

E-step:

$$p(z|w,d) \propto p(d|z)p(w|z)p(z) \tag{3}$$

M-step:

$$p(w|z) \propto \sum_{d \in D} n(w,d)p(z|w,d) \tag{4}$$

$$p(d|z) \propto \sum_{w \in W} n(w,d)p(z|w,d) \tag{5}$$

$$p(z) \propto \sum_{d \in D} \sum_{w \in W} n(w,d)p(z|w,d) \tag{6}$$

Since these values are all probabilities, they have to be normalized accordingly. The algorithm stops when a condition such as a small change in the log-likelihood is obtained or a pre-defined number of iterations has been completed.

Generally, the EM algorithm is computationally intensive and one way of addressing this problem is through parallelization. Parallelization refers to the distribution of independent sections of an algorithm across multiple processors. The term generally refers to one of two flavors: shared and distributed memory. A comparison of these two paradigms is summarized in Table 1. Shared memory processing operates on a single computer and offers a more fine-grained distribution of work. A common example of it is in a loop structure which perform n iterations, where each iteration is independent of all others. The advantage of this method is that data structures do not need to be distributed between each thread of execution since they execute on the same computer using the same memory space. As the name suggests, distributed memory processing refers to multiple CPUs each operating more independently at a coarser-grain level, such as functions instead of loops. These CPUs can be all within the same computer or spread across a network. In this case, data structures need to be sent and received explicitly. However, if the CPUs are spread across a network, each CPU will have access to their own disk storage device.

Two standards for these paradigms are OpenMP for shared memory processing [OpenMP Architecture Review Board, 2008] and the Message Passing Interface (MPI) for distributed memory processing [Message Passing Interface Forum, 2008]. The MPI standard has been interpreted by many – the application programming interface (API) that we will make use of is Open MPI [Gabriel et al., 2004]. Both paradigms can be used in conjunction in a single implementation. Further details about this and these two methods in general can be found in other sources [Quinn, 2003].

3   Related Work

Implementing PLSA is straightforward given Equations (2) to (6). However, a more careful implementation of PLSA could improve its efficiency. Some related work in this area is summarized next.

Hong et al. [2008] recently considered parallelizing PLSA. Their main idea is to partition the co-occurrence matrix into blocks using one of several blocking strategies. Each partitioned block is queued and processed one-by-one by each available CPU in a round-robin fashion. This strategy keeps each CPU busy, and minimizes the overall CPU idle time. Our method of partitioning is different, as we explain below, and we also investigated both shared and distributed memory parallelization, while they only considered the former.

Another way of improving the efficiency of PLSA is to reduce one dimension of X by removing words from W which do not co-occur with many documents. This process of vocabulary reduction is similar to applying a stop word list in a document retrieval system and, consequently, has the advantage of also improving the quality of the output of PLSA [Kim et al., 2003]. The focus of this work is primarily on improving the efficiency of PLSA computation without altering the output, and hence orthogonal to vocabulary reduction. In fact, both vocabulary reduction and the ideas outlined in this paper can be combined together to handle larger document collections.

After PLSA has completed, the output is a probabilistic latent semantic *index* (PLSI) X' whose dimensions is equal to that of X. However, the dimensions are the only common characteristics of the two structures. Since X is a table of co-occurrence counts, its values are integers, and, in practice, with a high proportion of zeroes. In contrast, since X' is a table of probabilities, storing these floating point values for efficient retrieval is a more important concern – a problem which has been addressed recently by Park and Ramamohanarao [2009]. Thus our work and their's refer to two different aspects of PLSA efficiency: calculation of X' and subsequent retrieval with it.

4   Methods

We investigated three methods for improving the efficiency of PLSA compared to a straightforward implementation. We focus our attention on the data structures used, the EM algorithm, and parallelization of PLSA.

Table 2. Sizes of data structures employed by the original and new implementations of the EM algorithm. The set of words, documents, and latent states are represented as W, D, and Z.

Category	Original size	New size
$n(w, d)$	$|W| \times |D|$	$|W| \times |D|$
$p(d|z)$	$|D| \times |Z|$	$2 \times |D| \times |Z|$
$p(w|z)$	$|W| \times |Z|$	$2 \times |W| \times |Z|$
$p(z)$	$|Z|$	$2 \times |Z|$
$p(z|w, d)$	$|D| \times |W| \times |Z|$	–
$p(w, d)$	–	$|W| \times |D|$

4.1 Data Structures

The data structures required by PLSA are summarized in the first column of Table 2. The space requirements depend on the number of words W, documents D, and latent states Z, with this latter value chosen by the user.

In Equations (2) and (4) to (6), note that only the non-zero co-occurrence counts contribute to the total sum. The solution adopted by Hong et al. [2008] was to balance each block by first permuting the rows and columns of X as a pre-processing step.

In our case, due to a different partitioning scheme, we keep only the entries in $n(w, d)$ which are non-zero in memory (not shown in Table 2). In the next section, we examine the EM algorithm and consider the sizes of the other data structures.

4.2 The EM Algorithm

Our second improvement was initiated by Hong et al. [2008], but comparatively little attention was given to its importance. The EM algorithm iterates between two steps which are separated for clarity. However, in the interest of space efficiency, the two steps can be combined. Old and new versions of $p(d|z)$, $p(w|z)$, and $p(z)$ are now kept, with the latter ones indicated by the superscript "new". This has the advantage of eliminating the largest data structure ($p(z|w, d)$) entirely, as illustrated by comparing the two columns of Table 2.

Replacing $p(z|w, d)$ in Equations (4) to (6) with Equation (3) yields the following:

E-step + M-step:

$$p^{(new)}(w|z) \propto \sum_{d \in D} n(w, d) \frac{p(d|z)p(w|z)p(z)}{p(w, d)} \tag{7}$$

$$p^{(new)}(d|z) \propto \sum_{w \in W} n(w, d) \frac{p(d|z)p(w|z)p(z)}{p(w, d)} \tag{8}$$

$$p^{(new)}(z) \propto \sum_{d \in D} \sum_{w \in W} n(w, d) \frac{p(d|z)p(w|z)p(z)}{p(w, d)} \tag{9}$$

$$\text{where } p(w, d) = \sum_{z \in Z} p(d|z)p(w|z)p(z) . \tag{10}$$

Algorithm 1. The steps that the master CPU (CPU 0) and the k slave CPUs (CPU k) have to perform under a distributed memory environment. Let $P()$ represent $(p(w|z)$, $p(d|z)$, $p(z))$ and $P^{(new)}()$ stand for $(p^{(new)}(w|z), p^{(new)}(d|z), p^{(new)}(z))$. The functions prefixed by an asterisk (*) have had OpenMP (shared memory parallelization) enabled for their main loops.

	CPU 0	CPU k
1	Initialize $P()$	
2	Broadcast $P()$	Receive $P()$
3		* Calculate $p(w, d)$
4	* Compute $\mathcal{L}$	
5	Broadcast exit condition	Evaluate exit condition
5	Broadcast $p(w, d)$	Receive $p(w, d)$
6		* Compute $P^{(new)}()$
7	Receive $P^{(new)}()$	Send $P^{(new)}()$
8	* Normalize probabilities	
9	Send $P^{(new)}()$	Receive $P^{(new)}()$
10	Return to step 3.	

While an explicit data structure for $p(w, d)$ is now needed, it is still a considerable savings compared to $p(z|w, d)$. Again, since these values are probabilities, normalization is required.

4.3 Parallelization

Our last improvement is the parallelization of PLSA, where we consider both distributed and shared memory models. Of the two, distributed memory is potentially more complex since it requires communication between processors to be explicitly taken care of. Among the possible forms of communication between CPUs, two extremes are possible. In the first one, all CPUs are equal in role and communicate directly with all of their peers. In the second model, one of the CPUs is designated as the master and all other CPUs as slaves which communicate directly only with the master and not to each other. While the first variant has the potential advantage of equally dividing the workload, some tasks such as normalization to obtain probabilities is best done by one CPU.

Adopting the first model for distributed memory, the steps that the master CPU (CPU 0) and the k slave CPUs (CPU k) have to perform are given in Algorithm 1. For clarity, we represent $(p(w|z), p(d|z), p(z))$ and $(p^{(new)}(w|z), p^{(new)}(d|z), p^{(new)}(z))$ as $P()$ and $P^{(new)}()$, respectively.

The two main tasks performed by CPU 0 are initializing $P()$ with random values and normalization. During these times, the other CPUs are idle. In our implementation, we prevent CPU 0 from being idle at steps 3 and 6 by giving it an equal unit of work to perform. All transmissions are blocked so that CPUs do not proceed until what they need to send or receive has succeeded. This approach reduces concurrency, but facilitates easier management.

An important question is how are the data structures partitioned. We contrast our approach to that of Hong et al. [2008] using Figure 1 by envisioning the three variables

Table 3. The systems under consideration, with the first being the baseline, naïve approach. The last system has the number of threads fixed at 4.

Program	Only store non-zero $n(w, d)$?	Combined EM?	OpenMP?	MPI?
Baseline	Yes	No	No	No
None	Yes	Yes	No	No
OpenMP	Yes	Yes	Yes	No
MPI	Yes	Yes	No	Yes
All	Yes	Yes	Yes (4)	Yes

(W, D, and Z) as the sides of a cube. Dotted lines indicate the manner in which work is partitioned into work units. Hong et al. elected to divide $n(w, d)$ according to W and D (Figure 1(a)). As noted earlier, a consequence of this decision is a pre-processing step to distribute the non-zero co-occurrence counts among blocks. In contrast, we chose to divide the work according to latent states (Figure 1(b)). This gives the same amount of work to each CPU without any pre-processing, provided the number of latent states is evenly divisible by the number of processors.

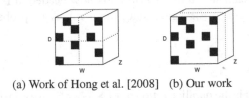

(a) Work of Hong et al. [2008]   (b) Our work

Fig. 1. Difference in partitioning the workload between Hong et al. [2008] and our's, as viewed as a cube with the set of words W, documents D, and latent states Z as the three sides. Regions in black indicate non-zero co-occurrence counts.

In addition to distributed memory, we also consider shared memory parallelization. This is relatively easier than distributed parallelization since only the necessary compiler instructions are inserted before the loops to be parallelized. The functions which have their main loops parallelized with OpenMP are indicated by asterisks in Algorithm 1. We emphasize that how we make use of OpenMP is different from Hong et al. and, because of this, our methods are not comparable. In fact, given their blocking strategy minimizes CPU idle time, their technique may be more efficient.

5   Experiments

We implemented two versions of PLSA – both with and without the modifications to the EM algorithm. Taking into account parallelization, we obtain five systems, as shown in Table 3. All versions only store the non-zero co-occurrence counts. We have made all attempts to keep the differences between the two versions at a minimum. Both programs are written in C and compiled using GNU gcc v4.1.2 under Linux with

Table 4. The dimensions and percentage of non-zeros for the data sets for our experiments

Data set	Size (MB)	Dimensions		Percentage of non-zeros
		Words	Documents	
CRAN	1.57	7,479	1,398	1.23 %
CISI	2.31	9,814	1,460	1.36 %
MED	1.04	10,673	1,033	0.81 %
CACM	2.08	12,195	3,204	0.52 %

the -O3 optimization flag. OpenMP is included with the compiler while v1.2.4 of Open MPI was used. Due to potential underflow problems caused from the large number of multiplications of probabilities, all values are stored in log-space.

The hardware used was a cluster of eight 3.0 GHz Dual-Core AMD Opteron Processor 2222 SE with 18 GB RAM connected by a 16 Port 10M/100M Switching Hub. All computers were generally idle at the time of the experiments.

Times are reported as elapsed time in seconds and averaged across four trials. For every system variation, the same initial values to the EM algorithm were used. The times reported represent just over a *single* iteration of the EM algorithm from step 3 to step 10 of Algorithm 1 (i.e., steps 3 to 5 are executed twice before exiting). We perform only a single iteration because different initial values may change the number of iterations performed, resulting in greater variations in total running time. The number of latent states is fixed at $|Z| = 32$.

We considered two sets of experiments. First, we evaluated both versions of PLSA on real data. These experiments are conducted first on a single computer and then again on the network of 8 computers. We then used Baseline and the fastest system and applied them to one of the data sets with varying levels of co-occurrence counts added. The purpose of this experiment is to assess to what extent more non-zero co-occurrence counts would affect the overall execution time.

5.1 Real Data

We selected four medium-sized data sets that have been used in the PLSA literature [Deerwester et al., 1990, Hofmann, 1999]. They were downloaded from http://www.cs.utk.edu/~lsi/corpa.html. In processing the text, we applied case-folding and a simple stemming scheme, but not stopping. Some statistics of the parsed collections are summarized in Table 4. We expect alternative pre-processing steps could change these statistics and affect the size of $n(w, d)$. The table also shows the percentage of the matrix which is non-zero. As this table shows, the matrix is very sparse with well below 1.5% of non-zero elements.

We begin by estimating the amount of memory used by the two versions of PLSA (Baseline and None) by using the information from Table 2 and Table 4, including the percentage of non-zeros, with all four data sets shown together. Figure 2 shows that Baseline uses a significant amount of memory, due to the size of $p(z|w, d)$.

Figure 3 presents a set of graphs, one for each of the data files. On the vertical axes, the average elapsed time in seconds is shown. Along the horizontal axes is the number

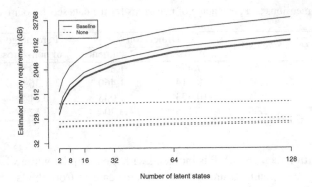

Fig. 2. Amount of memory used by the original and new implementations of PLSA, according to Table 2, for the four data sets. All values are stored as 8-byte doubles except for $n(w, d)$, which is stored as 4-byte integers.

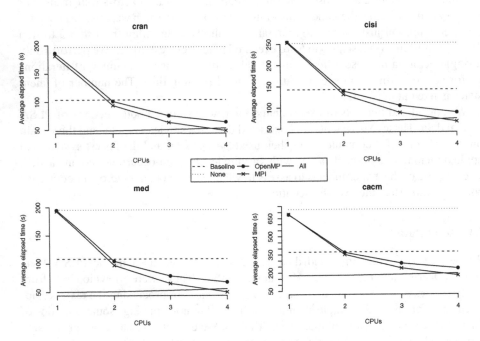

Fig. 3. Executing PLSA on a single computer through parallelization for the four test collections. Elapsed time in seconds are averaged across four trials and plotted against the number of CPUs.

of cores in our computers, ranging from 1 to 4. The changes we have introduced, without parallelization, increases the execution time (None). Profiling this system indicates that this is due to the calculation of $p(w, d)$ (step 3 of Algorithm 1). Fortunately, this additional computational cost is alleviated through parallel programming. When only one of OpenMP or MPI is used, the execution times improves with the increase in the

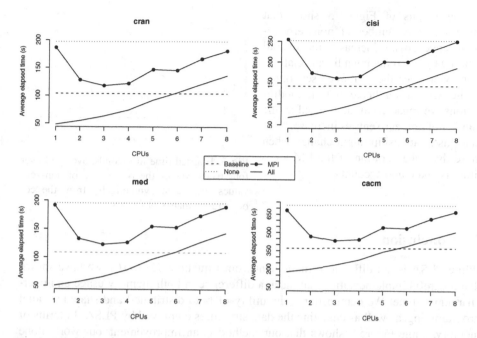

Fig. 4. Executing PLSA on a network of 8 computers through parallelization for the four test collections. Elapsed time in seconds are averaged across four trials and plotted against the number of CPUs.

number of processors, with MPI performing slightly better in all cases. The best method is All, when both are used together. With All, the number of OpenMP threads is fixed at 4, so when the number of CPUs (MPI processors) increase, the system resources are strained. Hence, this line actually increases and at 4 CPUs, MPI improves on it slightly.

In Figure 4, we extend the experiments to our network of 8 computers, removing OpenMP from consideration and focussing on MPI and All. The results show an increase in running time due to the overhead of communicating over a network. This is in contrast to Figure 3, where all MPI processes communicated to each other within the same computer. A faster network switch is expected to reduce this overhead.

If we compare MPI with All, MPI performs worse. The difference between the two lines is due to the use of OpenMP. The best performance of MPI occurs when the number of computers is 3 or 4.

5.2   Synthetic Data

As a final experiment, we selected CRAN and increased the percentage of non-zero values from 1.23% (see Table 4). We changed zero values to 1 by selecting cells at random, until a fixed percentage of non-zero values was reached. We then applied Baseline and OpenMP to this data for a single iteration of the EM algorithm.

The results of Figure 5 show that increasing the number of non-zero co-occurrence counts increases the overall running time, producing a linear relationship. Moreover, the continued separation of the two lines indicate that the modifications we made continue to hold as the number of non-zero entries increase. Note that this relationship might change when more than one iteration of the EM algorithm is taken into account.

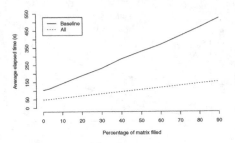

Fig. 5. Elapsed time in seconds, averaged over four trials, versus the percentage of non-zero values, generated synthetically from the co-occurrence matrix of CRAN

6 Conclusion

While PLSA is not difficult to implement from Equations (2) to (6), we have shown that a careful implementation can make a difference in both memory usage and overall running time. We demonstrated the utility of both distributed and shared parallel programming, as well as changing the data structures employed by PLSA. In terms of memory, while Figure 2 shows that our method is an improvement, our work alone seems unable to process data sets larger than CACM.

In order to tackle this problem and to further reduce running time, the techniques reported by others can be coupled with our work. This includes vocabulary reduction, as described earlier, and careful consideration of the number of latent states and iterations of the EM algorithm used.

A possible avenue of future research is extending our work to GPGPU (general-purpose computing on Graphics Processing Units) (see Owens et al. [2008] for details). Though GPGPUs are a recent trend in parallel computing, our work represents a first step in parallelizing PLSA without any specialized hardware.

While PLSA was originally proposed for IR, our work is equally important to other areas which operate on co-occurrence data. Some applications which have found a use for PLSA include machine translation [Kim et al., 2003], analysis of proteins [Chang et al., 2008, Mamitsuka, 2003], and image analysis [Hanselmann et al., 2008]. It is likely that these data sets possess different properties from those of text data (i.e., number of non-zero values, etc.) and it would be interesting to evaluate our techniques to those problems.

Availability. The source code of all 5 systems in Table 3 are distributed under the GNU Public License and available for download from http://www.cbrc.jp/~rwan/

Acknowledgements. Financial support for RW was provided by INTEC Systems Institute, Inc. and a post-doctoral fellowship from the Japan Society for the Promotion of Science (JSPS). VNA was supported by a short-term fellowship, also from JSPS, and a grant from the Australian Research Council. This work has been funded in part by BIRD of the Japan Science and Technology Agency (JST).

References

Chang, J.-M., Su, E.C.-Y., Lo, A., Chiu, H.-S., Sung, T.-Y., Hsu, W.-L.: PSLDoc: Protein subcellular localization prediction based on gapped-dipeptides and probabilistic latent semantic analysis. Proteins 72(2), 693–710 (2008)

Deerwester, S., Dumais, S.T., Furnas, G.W., Landauer, T.K., Harshman, R.: Indexing by latent semantic analysis. Journal of the American Society of Information Science 41(6), 391–407 (1990)

Dempster, A., Laird, N., Rubin, D.: Maximum likelihood from incomplete data via the EM algorithm. Journal of the Royal Statistical Society 39(1), 1–38 (1977)

Gabriel, E., Fagg, G.E., Bosilca, G., Angskun, T., Dongarra, J.J., Squyres, J.M., Sahay, V., Kambadur, P., Barrett, B., Lumsdaine, A., Castain, R.H., Daniel, D.J., Graham, R.L., Woodall, T.S.: Open MPI: Goals, concept, and design of a next generation MPI implementation. In: Kranzlmüller, D., Kacsuk, P., Dongarra, J. (eds.) EuroPVM/MPI 2004. LNCS, vol. 3241, pp. 97–104. Springer, Heidelberg (2004), http://www.open-mpi.org/

Hanselmann, M., Kirchner, M., Renard, B.Y., Amstalden, E.R., Glunde, K., Heeren, R.M.A., Hamprecht, F.A.: Concise representation of mass spectrometry images by probabilistic latent semantic analysis. Analytical Chemistry (November 2008), ISSN 1520-6882

Hofmann, T.: Unsupervised learning by probabilistic latent semantic analysis. Machine Learning 42(1–2), 177–196 (2001)

Hofmann, T.: Probabilistic latent semantic indexing. In: Proc. 22nd ACM International Conference on Research and Development in Information Retrieval (SIGIR), pp. 50–57. ACM Press, New York (1999)

Hong, C., Chen, W., Zheng, W., Shan, J., Chen, Y., Zhang, Y.: Parellelization and characterization of probabilistic latent semantic analysis. In: Proc. 37th International Conference on Parallel Processing, pp. 628–635 (2008)

Kim, Y.-S., Chang, J.-H., Zhang, B.-T.: An empirical study on dimensionality optimization in text mining for linguistic knowledge acquisition. In: Whang, K.-Y., Jeon, J., Shim, K., Srivastava, J. (eds.) PAKDD 2003. LNCS (LNAI), vol. 2637, pp. 111–116. Springer, Heidelberg (2003)

Mamitsuka, H.: Hierarchical latent knowledge analysis for co-occurrence data. In: Proc. 20th International Conference on Machine Learning, pp. 504–511 (2003)

Message Passing Interface Forum. MPI: A message-passing interface standard, version 2.1(June 2008), http://www.mpi-forum.org/docs/docs.html

OpenMP Architecture Review Board. OpenMP application programming interface, version 3.0 (May 2008), http://openmp.org/wp/openmp-specifications/

Owens, J.D., Houston, M., Luebke, D., Stone, J.E., Philips, J.C.: GPU computing. Proc. IEEE 96(5), 879–899 (2008), http://gpgpu.org/

Park, L.A.F., Ramamohanarao, K.: Efficient storage and retrieval of probabilistic latent semantic information for information retrieval. The VLDB Journal 18(1), 141–155 (2009)

Quinn, M.J.: Parallel Programming in C with MPI and OpenMP. McGraw-Hill, New York (2003)

Author Index

—